Computational Intelligence and Decision Making

International Series on
INTELLIGENT SYSTEMS, CONTROL, AND AUTOMATION:
SCIENCE AND ENGINEERING

VOLUME 61

For further volumes:
http://www.springer.com/series/6259

Ana Madureira • Cecilia Reis • Viriato Marques
Editors

Computational Intelligence and Decision Making

Trends and Applications

Springer

Editors
Ana Madureira
GECAD Research Group
School of Engineering
Polytechnic Institute of Porto (ISEP/IPP)
Porto, Portugal

Cecilia Reis
GECAD Research Group
School of Engineering
Polytechnic Institute of Porto (ISEP/IPP)
Porto, Portugal

Viriato Marques
School of Engineering
Polytechnic Institute of Coimbra
(ISEC/IPC)
Coimbra, Portugal

ISBN 978-94-007-4721-0 ISBN 978-94-007-4722-7 (eBook)
DOI 10.1007/978-94-007-4722-7
Springer Dordrecht Heidelberg New York London

Library of Congress Control Number: 2012951848

Printed on acid-free paper

Springer is part of Springer Science+Business Media (www.springer.com)

Preface

Advances in Computational Intelligence and Decision Making: Trends and Applications provides an overview and original analysis of new developments and applications in several areas of Computational Intelligence and Information Systems, in general.

Computational Intelligence became the roadmap for engineers to develop and analyze novel techniques to solve problems in basic sciences such as physics, chemistry, biology, engineering, environment and social sciences. The material contained in this book addresses the foundations and applications of Artificial Intelligence and Decision Support Systems, Complex and Biological Inspired Systems, Simulation and Evolution of real and artificial life forms, Intelligent Models and Control Systems, Knowledge and Learning Technologies, Web Semantics and Ontologies, Intelligent Tutoring Systems, Intelligent Power Systems, Self-Organized and Distributed Systems, Intelligent Manufacturing Systems and Affective Computing. The contributions are written by international experts, who provide up-to-date aspects of the topics discussed and present recent, original insights of their own experience in these fields.

Its aim is the presentation of state-of-the-art technologies in the field of Computational Intelligence as well as the discussion of new research findings in this field. The book is suitable for scientists, engineers, educators and students, as it addresses a large diversity of subjects presented in a broad band of complexity, ranging from simple natural language, found in some state-of-the-art articles, to some more complex mathematical issues, found in control applications, robotics and power systems, such as fractional calculus, fuzzy systems and rough-sets theory. Data Mining techniques, such as Support-Vector Machines and Neural Networks, are presented in application of the Biomedical and Bioinformatics fields, among others.

Briefly, we believe that this book provides a good window on most of the subjects that directly or indirectly make use of Computational Intelligence, also showing the contributions that it is already giving, or can be given in a near future, for solving some of the more pressing problems of today's world, such as energy and environment, society and economy.

We would like to thank all those that made possible the edition of this book, especially Ivo Pereira for the editing process. Finally, the editors would like to acknowledge FCT (Portuguese Science and Technology Foundation) and IEEE Computational Intelligence Society Portuguese Chapter for its support and sponsorship.

Ana Madureira
Informatics Department
School of Engineering – Polytechnic Institute of Porto

Cecilia Reis
Electronics Engineering Department
School of Engineering – Polytechnic Institute of Porto

Viriato Marques
Computer Science and Systems Engineering
School of Engineering – Polytechnic Institute of Coimbra

Contents

Chapter 1
The Process of Industrial Bioethanol Production Explained by Self-Organised Maps

Miguel A. Sanz-Bobi, Pablo Ruiz, and Julio Montes

Abstract Bioethanol is produced on an industrial scale by means of fermentation of a sugar substrate by Saccharomyces cerevisiae. Models for the detection of anomalies and their possible evolution are difficult to elaborate due to the biological nature of the fermentation process. This paper describes a method able to characterize patterns for explaining industrial bioethanol production using self-organised maps. Also, this method allows for an estimation of the probabilities of evolution to any pattern that the process may have from its last recognized state, therefore helping to take measures to correct a possible problem as soon as possible.

1.1 Introduction

Bioethanol is produced on an industrial scale by means of fermentation of a sugar substrate by Saccharomyces cerevisiae [1]. This is a biological process which is difficult to model with accuracy in order to detect some anomaly in the process as soon as possible. Industrial bioethanol production from sugar based products is a biochemical process that can be affected by different events that may result in a decrease of its performance and contamination which has to be prevented. All these possible problems can cause important costs and undesired shortages. New methods for a better monitoring of the process are being proposed [2].

This paper describes a method based on the analysis and investigation of real data taken from the process of bioethanol production in order to detect a deviation as soon as possible with respect to the normal evolution expected for the process. The method proposed is based on the identification of typical patterns using self-organised maps from historical information observed during the normal process of ethanol production.

M.A. Sanz-Bobi (✉) • P. Ruiz • J. Montes
IIT- Institute of Technological Research, Engineering School,
Comillas Pontifical University, Madrid, Spain
e-mail: masanz@upcomillas.es

A. Madureira et al. (eds.), *Computational Intelligence and Decision Making: Trends and Applications*, Intelligent Systems, Control and Automation: Science and Engineering 61,
DOI 10.1007/978-94-007-4722-7_1, © Springer Science+Business Media Dordrecht 2013

The paper is organised in the following sections. First, a brief description of bioethanol production is included. This is followed by a description of the main source of information used for the identification of performance patterns. The next sections describe the methods followed to identify and use the patterns found in the propagation and fermentation processes. Finally, a set of conclusions has been included in the last section.

1.2 Brief Description of Bioethanol Production

The process for the industrial production of bioethanol is similar to that of any industrial production of an alcoholic beverage, but in order to give some idea to the reader, a brief description follows.

Essentially, industrial bioethanol is obtained from cereals or other sugar based products by yeast (Saccharomyces cerevisiae) fermentation of a substrate, which consists of mainly sugar (glucose) obtained through an enzymatic action on the starch contained in the cereal (usually wheat or barley). In this way, bioethanol and carbon dioxide are produced via the following reaction:

$$C_6H_{12}O_6 \rightarrow 2C_2H_5OH + 2CO_2 \tag{1.1}$$

The data used in this paper have been collected at a type batch plant where the production process consists of the following steps:

1. First the sugar based product or prime matter input is ground in the dry-milling line. When it reaches a suitable grain size distribution, the resulting flour is fed to the mixing tank.
2. The product resulting from the previous step is mixed with a condensed fraction recycled from distillation columns downstream from the fermentation stage.
3. The mash obtained is pumped to the liquefaction tank where the temperature is raised to allow for a deeper fragmentation of the cereal grains, and therefore higher enzymatic efficiency in the following processes.
4. At the liquefaction tank some enzymatic pre-dose is also added in order to start the saccaharification process and to avoid possible viscosity problems in the pipe system due to gelatinization of the starch.
5. The majority of the sugar-enriched liquid produced is pumped to fill one of the four fermentation tanks. The remainder is used to fill the propagation tank, where the inoculum of yeasts is prepared.
6. While the fermentation tank is being filled, the yeast inoculum and the saccaharification enzymes are added. There are four fermentation tanks that are filled in sequence.

Then, after about 50 h, the beer produced with about 10% content of ethanol is sent to the distillation system, where alcohol is distilled with a purity of up to 95%.

It is necessary to take into consideration some important remarks about the physical process:

- The process of industrial production of bioethanol is not deterministic. The reason for this is that a biological process is the core of the bioethanol production and it is not possible to make an accurate measurement of factors that can affect the dynamics of the whole process.
- It is only possible to obtain a few measurements on-line and the most part are indirect measurements made by chemical analysis that require some time to be performed.
- In some situations the process could degrade till the point of wasting all the production because it is not useful and in the best cases the removal process of all bad product takes hours or even days. This causes an important loss of production and time. Sometimes this could be worse because some pipes and devices can result damaged and then they have to be replaced.

All these factors justify the investigation of a procedure able to detect some deviations observed as soon as possible by comparison with other previous situations in which the process ended successfully. This is the objective of the research described in this paper. In order to reach this objective, a method based on the use of machine learning techniques, such as neural networks, has been tested and described in the next sections.

1.3 Data Used

The data used for the study presented in this paper came from different types of chemical analyses performed by the Chemical Analysis Department of the production plant in order to monitor and characterize the evolution of the process. They mainly correspond to chromatograms that are based on chemical analyses. They include variables that are used in order to know the evolution of the propagation and fermentation processes. Some of the variables measured by chemical analyses are pH and concentrations in weight of dextrines, maltotrioses, maltoses, glucose, sum of residual sugars, lactic acid, glycerol and ethanol.

The propagation process takes about between 3 and 10 h after its beginning and usually two samples per batch are taken, one at the beginning and another near the end. Furthermore, during this process some of these samples are analysed under the microscope in order to characterize the health of the yeast population and also the inoculum feasibility which is a key factor for the success of the next process corresponding to fermentation.

Also, during the fermentation process several chemical analyses are done. These are similar to those performed during propagation, but more samples are taken to analyze. Usually five samples are taken per batch approximately every 12 h. An estimation of the process performance is carried out using the values of the variables measured.

The training set used consists of data from approximately 600 batches processed at the plant (almost 150 per fermentation tank) for both propagation and fermentation cases.

1.4 Patterns Extracted from the Propagation and Fermentation Processes

As was mentioned, the biochemical nature of the bioethanol production makes it difficult to define in terms of behaviour models based on clear and simple equations from which a possible evolution could be estimated. If these models were available, anomalies could be detected when a deviation with respect to the normal behaviour is observed [3] and the operator of the bioethanol production plant could activate some corrective actions or stop the process before major damage progresses. These models are not possible to create due to the fact that not all the key variables of the process can be measured. Only chemical analyses are available and those require some time to obtain results.

In this context the solution adopted has been to learn patterns of behaviour based on the observed values from the chemical analyses performed in the propagation and fermentation processes. It is convenient to remember at this point that the objective of this investigation is to reach the detection of deviations with respect to behaviour patterns of successful cases of bioethanol production.

According to the information available and the goal to be reached, the learning method used has been based on neural networks, and in particular self-organised maps [4, 5]. Other studies about the bioethanol production based on the use of other types of neural networks are in the literature [6–8]. The reasons for selecting self-organised maps were its ability to automatically create clusters of homogeneous behaviour by selecting few parameters, no sophisticated computing methods are required, and there is an easy and intuitive interpretation of the patterns obtained, using the observation of the cases clustered in each neuron and the opinion of process operators experts with a deep knowledge of the bioethanol plant.

The method of automatic learning is based on data of which special circumstances have to be taken into account for better understanding. They are the following:

1. There are two chemical analyses per batch during the propagation process and five per batch during the fermentation phase. According to this, two types of patterns have been considered.
2. The time duration of each batch is long: approximately 10 h for propagation and 12 for fermentation. The samples taken from them to be analyzed correspond to approximately the same times with respect to the beginning of the processes of propagation and fermentation.
3. The chemical analyses take times but not as much as is needed to take the next sample to analyse.

The most part of the available batches correspond to a successful bioethanol production process, but some of them do not. The idea is to learn behavior patterns for the bioethanol production through self-organised maps created using the values of the chemical analysis.

Several alternative options could be adopted for learning the desired patterns, but the final decision was to elaborate two types of patterns: one for the propagation and another one for the fermentation process. For both types of patterns all the samples collected from the same batch were used because there are several advantages expected from the patterns obtained. They are:

- In the case of the fermentation process there are five samples per batch taken at different moments of the process progress, and for this reason at least five different patterns are expected in the neural network once it is trained. Each one will correspond to a different moment within the fermentation process. The self-organized map would include at least five neurons, but also, it is necessary to take into account that the biological process is not always progressing in the same way and some unsuccessful cases are also included. For this reason more neurons would be included in the architecture of the neural network to be trained. Similar approach was done for the propagation process.
- The use of unsuccessful cases in the training set is considered to be an advantage in the learning process because they will generate patterns in the neural network but representing few cases. They will be used as a reference that is undesirable to reach.
- According to the main chemical Eq. (1.1) of the process, the bioethanol production is essentially an autoregressive process [9] and for this reason the self-organising nature of the neural network chosen fits perfectly to the problem handled. According to this principle, the five analyses taken in the fermentation process (two in the case of propagation) correspond to five different moments within the industrial process progression and they would correspond to different profiles. Also, the autoregressive nature fits well to the self-organised character of the neural network. Therefore it would be expected that the possible patterns of the first chemical analysis will be followed by the patterns of the second one and so on till the fifth analysis. According to this, once the neural network has been trained, the reading of the weight values of each neuron will correspond to the typical values of analyses, and they will be ordered or they will be neighbours from the first sample to the fifth one.
- The values of the variables observed resulting from the chemical analyses are a consequence of what is happening inside the biochemical processes of propagation and fermentation and for this reason they can be used for the creation of behaviour patterns and also for the detection of possible anomalies.

The application of the method proposed will cover two important aspects for the diagnosis of the bioethanol production:

- Automatic discovery of behaviour patterns at different moments during the evolution of the propagation and fermentation phases.

- Automatic discovery of possible transitions between patterns. The self-organised nature of the neural network used is a key point in this aspect because the pattern defined by a neuron can only transit to the neighbour patterns if the process is going according to the typical progression of the bioethanol production, (auto-regressive condition). An unexpected transition from a neuron to another which is not in the neighbouring proximity will indicate that an anomaly could be present.

This information helps to determine if the current situation corresponds to a desirable evolution or not in comparison with the patterns discovered, and also, if the next step of the propagation or fermentation process has the probability to reach an undesirable pattern or not. An undesirable pattern corresponds to cases where the production of bioethanol was lower than normal or the whole production batch had to be rejected.

1.5 Patterns Learned from the Propagation Process

The objective of the propagation process is to obtain an adequate inoculum of yeast. This is important because an inadequate inoculum can produce unsuccessful fermentation. The propagation process requires around 10 h to obtain an inoculum of the desired quality. Once obtained, it is sent to the tank being filled among the four downstream fermentors. The whole propagation process takes around 10 h and one inoculum is sent to a fermentor every 15 h.

Behaviour patterns have been investigated and automatically obtained for the propagation process based on certain regularities observed in the values of the chemical variables analysed along the propagation process of the bioethanol production for all the batches which are passed through it. They can be used for the classification of a new batch of product in a pattern previously labelled, and for the detection of possible anomalies. The generation of such patterns was based on the use of unsupervised self-organised maps. The set of inputs used to train the neural network were the following sequence of variables: time from the beginning of the propagation process, pH and concentration in weight of dextrines, maltotrioses, maltoses, glucose, residual sugars, succinic acid, lactic acid, acetic acid, glycerol and ethanol.

The training of the self-organised map was based on real data from the propagation process in order to establish the typical behaviour patterns, whether they are normal or not. Before to the training process, a pre-screening of available records of information was done excluding unreliable measurements. After that, 200 records of information were used to train the neural network. A normalization process was done in all the inputs of the neural network.

Three different sizes of lattices were tested for the architecture of the self-organised maps: 3×2, 3×3 and 3×4 neurons respectively. In any case the results were similar. In the obtained maps there are two clear behaviour patterns, one corresponding to the usual profile of the first sample collected approximately

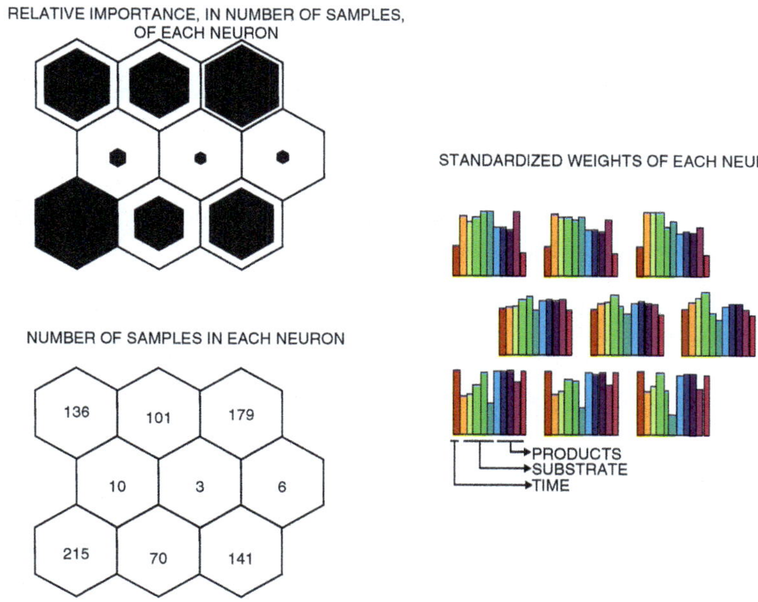

Fig. 1.1 Patterns of the propagation phase

3 h from the beginning of the propagation and the other one corresponding to the second sample taken approximately 7 h later. Figure 1.1 shows the results from the 3 × 3 map obtained using the SOM toolbox [10].

The two captions farthest on the left in Fig. 1.1 show the occupancy of the neurons or number of analysis covered by each of them. The caption to the right shows several bar diagrams that correspond to the weights of each neuron. As is shown in the 3rd (the bottom left-hand side) neuron, the bar on the left in every neuron represents the time from the beginning of the propagation phase and longer values of this bar are observed in the three neurons on the bottom rather than in the three neurons on top. The five next bars are related to the characterization of the substrate and the rest deal with the possible products of the reactions that may take place. The three neurons in the middle are border neurons and contain very few non-representative cases. The map obtained was tested with new samples of batches not used during the training phase and they were classified in neurons according to the profile expected for the new cases and for this reason the map obtained was considered to be well-fitted.

The model obtained can be used to verify if the new taken samples correspond to the expected time sequence or not. If the profile of values obtained in a sample does not correspond to the typical pattern according to the amount of time passed, an anomaly could be present and this must be taken into account before continuing on to the next phase of fermentation.

1.6 Patterns Learned from the Fermentation Process

The fermentation is the main process in the bioethanol production process. In the case of the plant studied, this process takes place in four tanks filled in batch mode. Each one is operated in a design sequence of 60 h divided into 15 h for filling with liquefied mash as a substrate, 45 h for fermentation and 15 h for cleaning, as the tank emptying takes place very quickly if there is no anomaly in the process. Enzymes, some additional nutrients and antibiotics are added at the beginning of the fermentation process. The propagation inoculums are also added after around 1 h of the tank filling. When a batch sequence is completed, the so-called resulting beer is sent to the distillation system. Distillation yields ethanol with 98% purity from the beer fed. A great part of the condensed fraction obtained is recycled to feed the beginning of the process.

In a similar way as the propagation process was studied, an investigation was done to discover if behaviour patterns could be inferred from the fermentation data. During normal plant operation, the evolution of each batch is characterized through the results of analyses of five or six samples which are taken around every 12 h for each ongoing process of fermentation. Sample frequency has great variability because it is a manual procedure, which is greatly influenced by working-shifts at the plant. Over 100 batches were analysed per fermentation tank.

Self-organised maps were also trained using the fermentation data previously normalized. A map was constructed for each of the four tanks because some differences were observed in the fermentation of each tank. Different sizes of lattices and sets of variables were tried until satisfactory results were obtained. These were reached when the maps offered a robust enough representation of the process and plenty of significance from a technical point of view and according to the expert criteria of the plant operators. This turned out to be a 4×4 neurons map structure with the input variables: time from the beginning of the fermentation, dextrines, maltotrioses, maltose, glucose, succinic, lactic and acetic acids and ethanol.

The maps obtained showed several interesting properties and behaviours. Although showing peculiarities and different layouts for each fermentation tank, they turned out to obtain very similar inlay process descriptions of the phenomena taking place in the four existing vessels. Also, according to their in-process meaning, similar in-map gradient structures can be observed for different sets among the selected variables. Moreover, the maps showed great physical process-sense.

The maps obtained with the behaviour patterns were reviewed and a cross-validation process was developed for each one. Patterns with similar qualitative profiles were obtained in the four fermentation tanks. However their quantitative values are not the same.

The most important result from the maps obtained is the knowledge included. The analysis performed shows that the batch fermentation process involves two simultaneous steps [1]. In the first step, the enzymes added broke the dextrines chains into sugar, and in the second one, this sugar is fermented to ethanol by yeast. These two steps can be observed in the change of the neuron weights (patterns) along the neurons of the map. This change is progressive in a gradient mode along the neighbour neurons of the map. Notice how dextrines, maltotrioses, maltose and glucose (which are

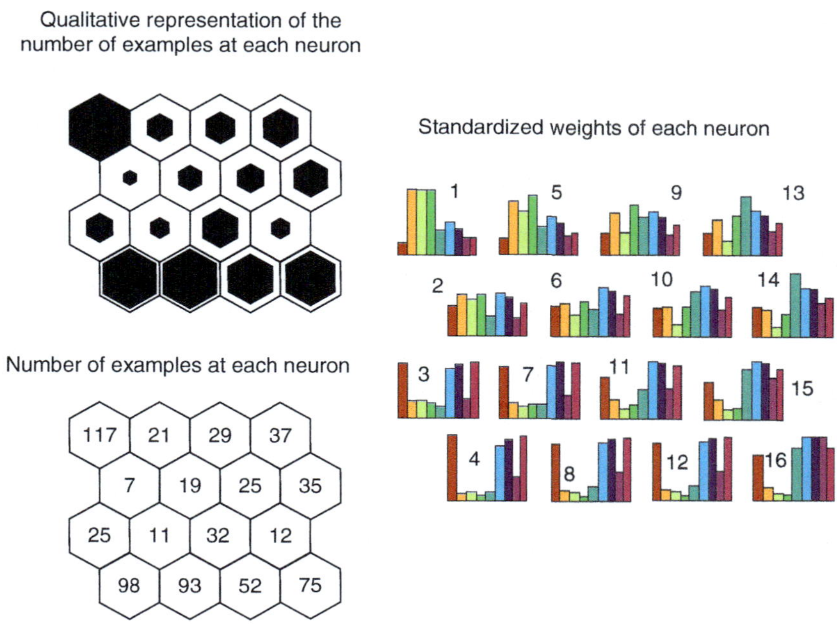

Fig. 1.2 Fermentation patterns for the tank A

variables related with the performance of the enzymatic action) have a very similar profile, while the same happens with ethanol and succinic acid (both are products from yeast digestion) and with acetic and lactic acid, which are products related with another kind of bacterial action which may interfere with yeast activity [1].

As a result of this in-map variable gradient distribution, when the map is used to classify, all the samples obtained from a whole batch fermentation process pass through a sequence of activated neurons (pattern identified) which goes from the upper zones of the map to the lower ones. The self-organised nature of the neural network used has been a key point for the automatic generation of this knowledge. If the process goes well, the activation sequence tends to end in the left hand bottom neurons. They represent a successful fermentation process with lower lactic and acetic acids and, therefore, higher concentrations of ethanol. On the other hand, if the sequence tends to end in the right lower zone of the maps, this involves higher levels of acids being produced and, therefore, lower levels of ethanol can be expected.

The behaviour patterns obtained are very useful because according to the number of examples classified in each neuron, it is possible to obtain an estimation of the probability of transition from one neuron (pattern) to the next potential ones. Then, if a new sample from a new batch is obtained, it is possible to use the probabilities formerly calculated to obtain an estimation of the future type of evolution of the process that can be expected, which is very helpful information to decide if it is necessary to take corrective measures in order to obtain the best possible process of bioethanol production. In order to explain this, two batches with different evolutions were selected and evaluated. Both of them started with a sample classified such as "neuron 1" on the map shown in Fig. 1.2. For the first batch, the subsequent evaluation of samples resulted in the activation of neurons 5-2-3-3-4, resulting in a good process with a high concentration of ethanol obtained quickly. The second batch analyzed resulted in an activation sequence of 1-5-16-16-16 which turned out to be a batch contaminated with bacteria.

All the characteristics and conclusions obtained from analysing the self-organised maps are very important instruments for a better and efficient intelligent control of the plant and for the detection of anomalies.

1.7 Conclusions

This paper has presented a method for automatic extraction of behavior patterns of an industrial process for bioethanol production using self-organised maps. The patterns obtained can be used for quick anomaly detection in the progress of the propagation and fermentation processes. Real data taken from a reasonable number of chemical analyses of the bioethanol production have been the main basis for the identification of regularities or patterns in the process. The maps are able to diagnose the current process of bioethanol production and its most probable evolution. This will make it possible to take actions as soon as possible to correct possible anomalies, if they appear. Also, the maps obtained are an excellent instrument for a better understanding of the biochemical process in general.

Further work has to be done in order to consolidate the reliability of the patterns obtained. Also the analysis of other variables taken in real-time has to be done in order to combine the knowledge extracted from all the different sources of information. Finally, the evaluation of the influence that the propagation has over the subsequent fermentation is being studied by analysing other peculiarities of the models developed.

Acknowledgments The authors are very grateful to the entire staff of the Bioetanol Galicia plant and ABENGOA for their valuable collaboration during all this work. This project has been partially supported by the research program PROFIT promoted by the Spanish Government.

References

1. Ingledew WM (2004) Alcohol production by Saccharomyces cerevisiae: a yeast primer. In: Jacques K, Lyons TP, Kelsall DR (eds) The alcohol text book, 3rd edn. Nottingham University Press, Nottingham, pp 49–87
2. Ccopa Rivera E, Farias F Jr, Pires Atala DI, Ramos de Andrade R, Carvalho da Costa A, Maciel R (2009) Development and implementation of an automated monitoring systems for improved bioethanol production. Chem Eng Trans 18:445–450
3. García MC, Sanz-Bobi MA, del Pico J (2006) SIMAP: intelligent system for predictive maintenance: application to the health condition monitoring of a windturbine gearbox. Comput Ind 57(6):552–568
4. Kohonen T (1990) The self-organizing map. Proc IEEE 78(9):1464–1480

5. Kohonen T, Oja E, Simula O, Visa A, Kangas J (1996) Engineering applications of the self-organizing map. Proc IEEE 84(10):1358–1384
6. Ccopa Rivera E, Candida Rabelo S, dos Reis GD, Maciel Filho R, Carvalho da Costa A (2010) Enzymatic hydrolysis of sugarcane bagasse for bioethanol production: determining optimal enzyme loading using neural networks. J Chem Technol Biotechnol 85:883–992
7. Emmanuel AN, David A, Benjamin YK, Yannick ET (2009) A hybrid neural network approach for batch fermentation simulation. Aust J Basic Appl Sci 3(4):3930–3936
8. Eyng E, da Silva FV, Palú F, Fileti AMF (2009) Neural network based control of an absorption column in the process of bioethanol production. Int J Braz Arch Biol Technol 52(4):961–972
9. Box GP, Jenkins GM, Reinsel GC (2008) Time series analysis. Forecasting and control, 4th edn. Wiley series in probability and statistics. Wiley, Chichester
10. SOM Toolbox (2011) http://www.cis.hut.fi/projects/somtoolbox/

Chapter 2
Towards a Further Understanding of the Robotic Darwinian PSO

Micael S. Couceiro, Fernando M.L. Martins, Filipe Clemente, Rui P. Rocha, and Nuno M.F. Ferreira

Abstract This paper presents a statistical significance analysis of a modified version of the Particle Swarm Optimization (*PSO*) on groups of simulated robots performing a distributed exploration task, denoted as *RDPSO* (Robotic *DPSO*). This work aims to evaluate this novel exploration strategy studying the performance of the algorithm under communication constraints while increasing the population of robots. Experimental results show that there is no linear relationship between the number of robots and the maximum communication range. In general, the decreased performance by the developed algorithm under

M.S. Couceiro (✉)
Institute of Systems and Robotics, University of Coimbra, Pinhal de Marrocos - Polo II, 3030-290 Coimbra, Portugal

RoboCorp, Department of Electrical Engineering, Engineering Institute of Coimbra, Rua Pedro Nunes, Quinta da Nora, 3030-199 Coimbra, Portugal
e-mail: micaelcouceiro@isr.uc.pt; micael@isec.pt

F.M.L. Martins
Instituto de Telecomunicações (Covilhã), Coimbra College of Education, Rua Dom João III – Solum, 3030-329 Coimbra, Portugal
e-mail: fmlmartins@esec.pt

F. Clemente
RoboCorp, Faculty of Sport Sciences and Physical Education, University of Coimbra, Estádio Universitário – Pavilhão III Stª Clara, 3030-156 Coimbra, Portugal
e-mail: fclemente@esec.pt

R.P. Rocha
Institute of Systems and Robotics, University of Coimbra, Pinhal de Marrocos - Polo II, 3030-290 Coimbra, Portugal
e-mail: rprocha@isr.uc.pt

N.M.F. Ferreira
RoboCorp, Department of Electrical Engineering, Engineering Institute of Coimbra, Rua Pedro Nunes, Quinta da Nora, 3030-199 Coimbra, Portugal
e-mail: nunomig@isec.pt

A. Madureira et al. (eds.), *Computational Intelligence and Decision Making: Trends and Applications*, Intelligent Systems, Control and Automation: Science and Engineering 61, DOI 10.1007/978-94-007-4722-7_2,

communication constraints can be overcome by slightly increasing the number of robots as the maximum communication range is decreased.

2.1 Introduction

The biological world abounds in collective phenomena that have important adaptive functions, ranging from coordinated movement to nest building and all the way to communication [1]. The principles of self-organization are appealing for explaining biological collective phenomena where the resulting structures and functionalities greatly exceed in complexity the perceptual, physical, and cognitive abilities of the participating organisms. Examples of biological self-organization include the construction of beehives, the foraging strategies of ants, and the regulation of colony life in social insects. In all these cases, the resulting structure emerges from the collective work of individual organisms that execute simple behaviors based on local information and do not possess a global plan of the end result or a central coordinator.

The examples of behavior-based collective approaches described above inspired the design of novel machine-learning techniques and swarm robotics [2, 3]. This area of research, also known as *swarm intelligence* [4, 5], studies large collections of relatively simple agents that can collectively solve problems that are too complex for a single robot or that can display the robustness and adaptability to environmental variation displayed by biological agents.

One of the most well-known swarm algorithms is the Particle Swarm Optimization (*PSO*) developed by Kennedy and Eberhart [6]. This optimization technique models a set of potential problem solutions as a swarm of particles moving around in a virtual search space. However, a general problem with the *PSO* and other optimization algorithms is that of becoming trapped in a local optimum, such that it may work in some problems but may fail on others. In search of a better model of natural selection using the *PSO* algorithm, the Darwinian Particle Swarm Optimization (*DPSO*) was formulated by Tillet et al. [7] enhancing the ability to escape from local optima.

Just like in Multi-Robot Systems (*MRS*), where groups of robots interact to accomplish their goals [8], both *PSO* and *DPSO* use groups of interacting virtual agents (*aka*, particles) in order to achieve their optimization. However, contrarily to virtual agents, robots are designed to act in the real world where communication constraints and obstacles need to be taken into account.

In our previous work [9], an extension of the *DPSO* to *MRS* was proposed. Each robot is then responsible for each virtual agent, which it needs to evaluate at each iteration. After each set of evaluations, the robots communicate to share the objective information (cost or fitness) needed to progress to the next iteration of the algorithm.

However, the design of new swarm robotics systems cannot provide us with quantitative prediction of the collective performance. Real robot experiments and simulations are the most direct way to observe the behavior of the system under different conditions (*i.e.*, population size and communication constraints).

However, trials with real or simulated robots do not scale well as the size of the system grows. Therefore, it is hard to predict the ideal number of robots necessary for a given task.

Bearing these ideas in mind, this paper carries out a statistical analysis of the previously proposed algorithm in order to evaluate the relationship between the population of robots in the *RDPSO* and the maximum the communication range between robots. To that end a Multivariate Analysis of Variance technique (*MANOVA*) is used to evaluate the performance of the algorithm based on the number of robots and the communication distance.

The paper is organized as it follows. Section 2.2 presents an overview of the *R-DPSO* algorithm. A brief description of the *MANOVA* is given in Sect. 2.3. Experimental results analyzing the performance of the algorithm are demonstrated Sect. 2.3 while Sect. 2.4 outlines the main conclusions.

2.2 RDPSO Overview

This section briefly presents the *RDPSO* algorithm proposed in [9]. The *DPSO* [7] is an evolutionary algorithm that extends the well-known *PSO* [6] using natural selection, or survival-of-the-fittest, to enhance the ability to escape from local optima. Since the *RDPSO* approach is an adaptation of the *DPSO* to real mobile robots, four general features are proposed: (i) a novel "punish"-"reward" mechanism to emulate the deletion and creation of robots; (ii) an obstacle avoidance algorithm to avoid collisions; (iii) an enforcing multi-hop network connectivity algorithm to ensure that the mobile ad hoc network (*MANET*) remains connected throughout the mission; and (iv) a novel methodology to establish the initial planar deployment of robots preserving the connectivity of the *MANET* while spreading out the robots as most as possible (*cf.*, [9] and [10] for a more detailed description).

In the common *DPSO*, "punish" means the deleting of particles and swarms, while "reward" means the spawning of new particles and swarms. In order to adapt *DPSO* to mobile robotics, the deleting and spawning of a robot are modelled by the mechanisms of *social exclusion* and *social inclusion*, respectively.

The *RDPSO* is then represented by multiple swarms, *i.e.*, multiple groups of robots that altogether form a population. Each swarm individually performs just like an ordinary *PSO* in search for the solution and some rules governs the whole population of robots. If there was no improvement in a swarm's objective over a period of time, the swarm is punished by excluding the worst performing robot, which is added to a socially excluded group. The worst performing robot is evaluated by the value of its objective function compared to other members in the same swarm. In other words, if the objective is to maximize the fitness function, the robot to be excluded will be the one with the lower fitness value. Those socially excluded robots, instead of searching for the objective function's global optimum like the other robots in the active swarms do, they basically randomly wander in the scenario. This approach improves the algorithm making it less susceptible of becoming trapped in a local optimum. Note, however, that they are always aware of their individual solution and the global solution of the socially excluded group.

The *RDPSO* also considers obstacle avoidance. To that end, a new cost or fitness function is defined in such a way that it would guide the robot to perform the main mission while avoiding obstacles. For this purpose it is assumed that each robot is equipped with sensors capable of sensing the environment for obstacle detection within a finite sensing radius r_s. A monotonic and positive *sensing function* $g(x_n)$ that depends on the sensing information (*i.e.*, distance from the robot to obstacle) is defined. In most situations the *sensing function* $g(x_n)$ can be represented as the relation between the analog output voltage of distance sensors and the distance to the detected object.

Besides obstacle avoidance, robots' position also needs to be controlled in order to maintain the communication based on constraints such as maximum distance or minimum signal quality. The way network will be forced to preserve connectivity depends on the characteristics of the communication (*e.g.*, multi-hop, biconnectivity). Assuming that the network supports multi-hop connectivity, the communication between two end nodes (*i.e.*, robots) is carried out through a number of intermediate nodes whose function is to relay information from one point to another (note that any robot may be used as a relay node independently of their swarm). Considering that nodes are mobile, it is necessary to guarantee the communication between all nodes. The nodes' position (*i.e.*, robots' position) are initialized and updated by means of the enforcing *MANET* connectivity algorithm further described in [10].

2.3 Experimental Results

In this section, it is explored the effectiveness of the *RDPSO*, while performing distributed unsupervised learning with local and global information, under communication constraints while increasing the population of robots.

The number of robots is set equal to the number of particles and the number of nodes in the network. Robots are deployed in the search space in a spiral manner (as previously described in [10]) where the radius depends on the maximum communication distance and the number of robots in the population. Since the *R-DPSO* is a stochastic algorithm, every time it is executed it may lead to different trajectory convergence. Therefore, multiple test groups of 100 trials, of 300 iterations each were considered. It will be used a minimum, initial and maximum number of 1, 3 and 6 swarms (represented by different colors in Fig. 1.2), respectively, independently of the population of robots. The search space is represented by a Gaussian distribution consisting on a function of two variables of the search space, x and y-axis, which represents the position of the robot in meters. The particles will then move in an outdoor scenario of 300×300 m where the z-axis represents the value of the objective function. In order to improve the interpretation of the algorithm performance, results were normalized in a way that the objective of robots team is to find the optimal value of 1 while avoiding obstacles and enforcing the *MANET*

Table 2.1 Typical maximum communication distances of the WiFi, ZigBee and Bluetooth protocols

	No limit	WiFi	ZibBee	Bluetooth
d_{max} [m]	∞	100	55	10

connectivity (*cf.*, Sect. 2.2). Trying to maintain the network connectivity by only taking into account the communication range does not match reality since the propagation model is more complex – the signal depends not only on the distance but also on the multiple paths from walls and other obstacles. However, in simulation, the communication distance is a good approach and it is easier to implement.

The maximum communication distance d_{max} will then vary depending on the chosen wireless protocol. Four conditions were described: (1) Existence of a communication infrastructure (*i.e.*, without communication constraints $\equiv d_{max} \rightarrow \infty$); (2) *WiFi*; (3) *ZigBee*; (4) *Bluetooth*. Table 2.1 depicts the maximum communication distance adapted from a comparison between the key characteristics of each wireless protocol in [11]. The mean between the minimum and maximum range shown in [11] was considered as the maximum communication distance d_{max}.

The number of robots will vary from 3 robots to 33 robots with incremental steps of 6 robots, *i.e.*, $N = \{3, 9, 15, 21, 27, 33\}$ in order to understand the performance of the algorithm while changing the population size and the maximum communication distance.

Since these simulation experiments represent a search task, it is necessary to evaluate not only the completeness of the mission but also the speed. Therefore, the performance of the algorithm will be evaluated through the analysis of the final global solution of the population and the runtime of the simulation. If the group cannot find the optimal solution, the runtime is considered to be the simulation time (*i.e.*, 300 iterations).

A two-way *MANOVA* analysis was carried out to assess whether the factors on this study have a statistically significant effect on the team's performance. The significance of the maximum communication distance and the number of robots (independent variables) on the global solution and the runtime (dependent variables) was analyzed using a two-way *MANOVA* after checking the assumptions of multivariate normality and homogeneity of variance/covariance. The assumption of normality of each of the univariate dependent variables was examined using univariate tests of *Kolmogorov-Smirnov* (*p-value* < 0.05). Although the univariate normality of each dependent variable has not been verified, since $n \geqslant 30$ and using the Central Limit Theorem (*CLT*) [12] this statement was assumed [13]. Consequently, the assumption of multivariate normality was validated [13].

The assumption about the homogeneity of variance/covariance matrix in each group was examined with the *Box's M Test* (M $=$ 6465.13, F(69; 5368369.62) $=$ 92.98; *p*-value $=$ 0.001). Although the homogeneity of variance/covariance matrices has not been verified, the *MANOVA* technique is robust to this violation because all the samples have the same size [13].

When the *MANOVA* detected significant statistical differences, we proceeded to the commonly-used *ANOVA* for each dependent variable followed by the *Tukey's HSD Post Hoc*. The classification of the size effect (*i.e.*, measure of the proportion of the total variation in the dependent variable explained by the independent variable) was done according to Palant [14]. This analysis was performed using *IBM SPSS Statistics* for a significance level of 5%.

The *MANOVA* revealed that the maximum communication distance had a small effect and significant on the multivariate composite (*Pillai's Trace* = 0.75; F(6; 4,752) = 30.974; *p-value* = 0.001; *Partial Eta Squared* η_p^2 = 0.038; *Power* = 1.0). The number of robots also had a small effect and significant on the multivariate composite (*Pillai's Trace* = 0.080; F(10; 4,752) = 19.706; *p-value* = 0.001; η_p^2 = 0.04; *Power* = 1.0). Finally, the interaction between the two independent variables had a small statistically significant effect on the multivariate composite (*Pillai's Trace* = 0.032; F(30; 4,752) = 2.55; *p-value* = 0.001; η_p^2 = 0.016; *Power* = 1.0).

After observing the multivariate significance in the maximum communication distance and the number of robots, an univariate *ANOVA* for each dependent variable followed by the *Tukey's HSD Test* was carried out.

For the maximum communication distance, the dependent variable final global solution presents statistically significant differences (F(3; 2,376) = 45.185; *p-value* = 0.001; η_p^2 = 0.054; *Power* = 1.0) and the dependent variable runtime presents statistically significant differences (F(3; 2,376) = 53.683; *p-value* = 0.001; η_p^2 = 0.063; *Power* = 1.0).

For the number of robots, the dependent variable final global solution also presents statistically significant differences (F(5; 2,376) = 23.347; *p-value* = 0.001; η_p^2 = 0.047; *Power* = 1.0) and also the dependent variable runtime shows statistically significant differences (F(5; 2,376) = 39.816; *p-value* = 0.001; η_p^2 = 0.077, *Power* = 1.0).

Using the Tukey's *HSD* Post Hoc, it is possible to verify where the differences between maximum distances of communication lie.

Analyzing the team's final solution and the runtime variables, it appears that there are statistically significant differences between experiments without communication constraints and experiments using the *WiFi* protocol, the *ZigBee* protocol and the *Bluetooth* protocol.

It is noteworthy that without communication constraints the algorithm produces better solutions. Also, using *WiFi* protocol produces better solutions than using the *ZigBee* protocol and, on the other hand, this last one produces better solutions than the *Bluetooth* protocol as expected. In fact, using the *Bluetooth* protocol proves to be the "worse" communication protocol to employ.

Analyzing both the final global solution of the team and the runtime variables, it appears that there are statistically significant differences between a population inferior to 15 robots and a population superior to 21 robots, not showing statistically significant differences for a population between 3–15 robots and 21–33 robots.

Table 2.2 Tukey's *HSD* Post Hoc test to the maximum communication distance

d_{max}	Final solution	Runtime
No limit vs WiFi	0.002*	0.854
No limit vs ZigBee	0.001*	0.001*
No limit vs Bluetooth	0.001*	0.001*
WiFi vs ZigBee	0.207	0.019*
WiFi vs Bluetooth	0.001*	0.001*
ZigBee vs Bluetooth	0.001*	0.001*

*The corresponding *p*-value for mean difference when it is significant at the 0.05 level

Table 2.3 Tukey's *HSD* Post Hoc test to the number of robots

N	Final solution	Runtime
3vs9	1.000	0.861
3vs15	0.151	0.182
3vs21	0.001*	0.001*
3vs27	0.001*	0.001*
3vs33	0.001*	0.001*
9vs15	0.249	0.844
9vs21	0.001*	0.001*
9vs27	0.001*	0.001*
9vs33	0.001*	0.001*
15vs21	0.004*	0.001*
15vs27	0.001*	0.001*
15vs33	0.001*	0.001*
21vs27	0.842	0.654
21vs33	0.785	0.076
27vs33	1.000	0.845

*The corresponding *p*-value for mean difference when it is significant at the 0.05 level

Note that the "worst" result is obtained using 3 robots but may not be considered significantly "worse" than using 9 or even 15 robots. This may be relevant since the increase in the number of robots result in an increase in the cost of the solution.

As previously depicted in Tables 2.2 and 2.3, as Figs. 2.1 and 2.2 shows and using *Tukey's HSD Post Hoc* test, the conditions of the independent variables N and d_{max} can be divided into different homogeneous subsets.

For instance, since there are no statistically significant differences between teams of 3, 9 and 15 robots in the analysis of both the final global solution and the runtime, this can be considered as a subset of N, *i.e.*, $N^1 = \{3, 9, 15\}$ and $N^2 = \{21, 27, 33\}$. In other words, in an application were the cost of the solution needs to be considered, and since there are no significant advantages of having 15 robots instead of having just 3 or having 33 robots instead of 21, the choice would be using the minimum number of robots of each subset of N.

The same analysis can be conducted for the maximum communication distance. However, in this specific situation, three subsets can be considered analyzing

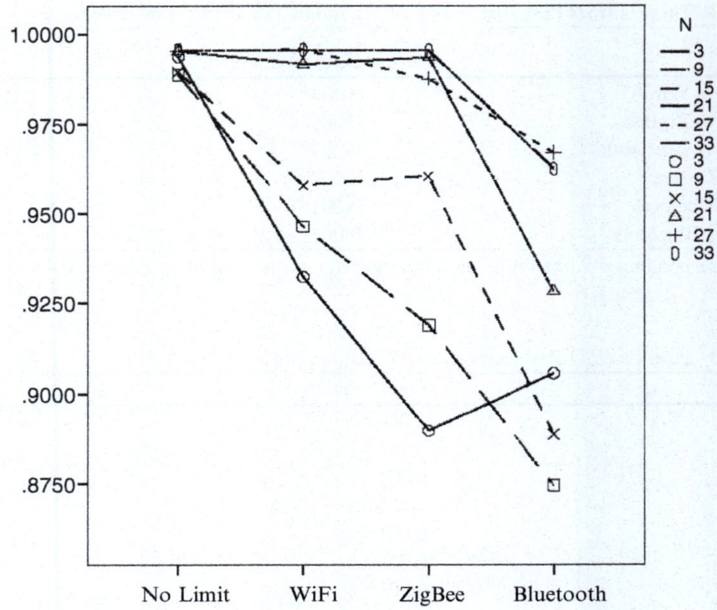

Fig. 2.1 Estimated marginal means of the final global solution

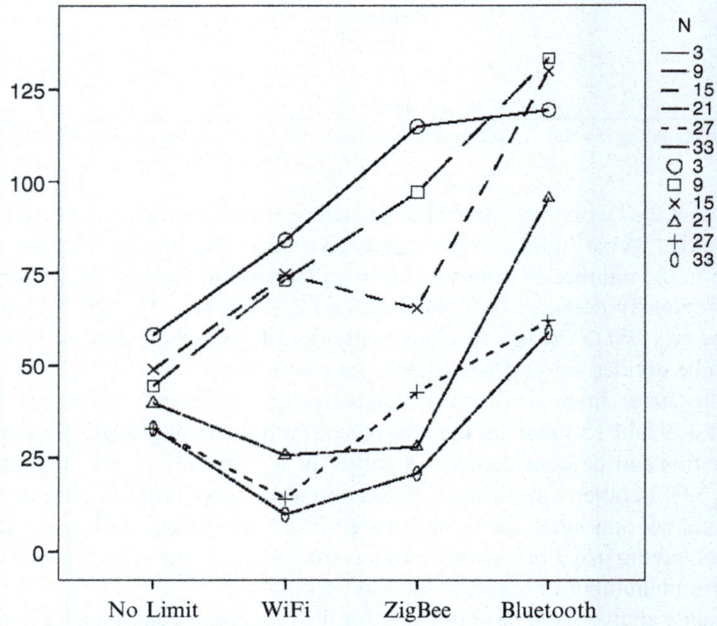

Fig. 2.2 Estimated marginal means of the runtime

the statistically significant differences between the different values of d_{max}, *i.e.*, $d_{max}{}^1 = NoLimit\}$, $d_{max}{}^2 = \{WiFi, ZigBee\}$ and $d_{max}{}^3 = \{Bluetooth\}$. Put differently, the choice between wireless technologies, if unable to have a preexistent infrastructure (*i.e.*, $d_{max}{}^1$), based on the maximum communication distance and ignoring other technical features, may be centered on the *WiFi* and *ZigBee* technologies.

2.4 Conclusions

This paper presented a methodology to evaluate the previously proposed algorithm, denoted as *RDPSO*, which takes into account real-world multi-robot systems characteristics.

Experimental results shows that the performance of the algorithm can be improved, thus decreasing the time needed to find the global optimum (*i.e.*, runtime), as the number of robots or the communication range increases. However, the choice on the number of robots and the wireless technology needs to take into account the global cost of the solution depending on the statistic significant differences between the independent variables.

As future work, a probabilistic model will be studied to optimize the swarm parameters such as the number of robots and the communication protocol in order to improve the overall runtime and find the optimal solution for a given scenario.

Acknowledgments This work was supported by a *PhD* scholarship (SFRH/BD /73382/2010) granted to the first author by the Portuguese Foundation for Science and Technology (*FCT*) and the Institute of Systems and Robotics (*ISR*) also under regular funding by *FCT*.

References

1. Floreano D, Mattiussi C (2008) Bio-inspired artificial intelligence: theories, methods, and technologies. MIT Press, Cambridge
2. Marques L, de Almeida AT (2004) Finding odours across large search spaces: a particle swarm-based approach. In: Proceedings of the 6th international conference on Climbing & Walking Robots (CLAWAR), Madrid, Spain
3. Marjovi A, Marques L, Penders J (2009) Guardians robot swarm exploration and firefighter assistance. In: Workshop on NRS in IEEE/RSJ international conference on Intelligent Robots and Systems (IROS), St Louis, USA
4. Bonabeau E, Dorigo M, Theraulaz G (1999) Swarm intelligence: from natural to artificial systems. Oxford University Press, New York
5. Beni G (2004) From swarm intelligence to swarm robotics. In: Proceedings of the swarm robotics workshop, Heidelberg, Germany, pp 1–9
6. Kennedy J, Eberhart R (1995) A new optimizer using particle swarm theory. In: Proceedings of the IEEE sixth international symposium on micro machine and human science, Nagoya, pp 39–43

7. Tillett J, Rao TM, Sahin F, Rao R, Brockport S (2005) Darwinian particle swarm optimization. In: Proceedings of the 2nd Indian international conference on artificial intelligence, Pune, pp 1474–1487
8. Rocha R, Dias J, Carvalho A (2005) Cooperative multi-robot systems: a study of visionbased 3-D mapping using information theory. Robot Auton Syst 53(3–4):282–311
9. Couceiro MS, Rocha RP, Ferreira NMF (2011) A novel multi-robot exploration approach based on particle swarm optimization algorithms. In: Proceedings of the IEEE international symposium on Safety, Security, and Rescue Robotics, SSRR2011, Kyoto, Japan, pp 327–332
10. Couceiro MS, Rocha RP, Ferreira NMF (2011) Ensuring Ad Hoc connectivity in distributed search with Robotic Darwinian Swarms. In: Proceedings of the IEEE international symposium on Safety, Security, and Rescue Robotics, SSRR2011, Kyoto, Japan, pp 284–289
11. Lee JS, Su YW, Shen CC (2007) A comparative study of wireless protocols: bluetooth, UWB, ZigBee, and Wi-Fi. In: Proceedings of the 33rd annual conference of the IEEE Industrial Electronics Society, Vigo, pp 46–51
12. Pedrosa AC, Gama SMA (2004) Introdução Computacional à Probabilidade e Estatística. Porto Editora, Portugal
13. Maroco J (2010) Análise Estatística com utilização do SPSS. Edições Silabo, Lisboa
14. Pallant J (2011) SPSS survival manual. Kindle Edition (ed) 4th edn. Open University Press, Maidenhead

Chapter 3
A Comparison Study Between Two Hyperspectral Clustering Methods: KFCM and PSO-FCM

Amin Alizadeh Naeini, Saeid Niazmardi, Shahin Rahmatollahi Namin, Farhad Samadzadegan, and Saeid Homayouni

Abstract Thanks to its high spectral resolution, hyperspectral imagery recently has been extremely considered in various remote sensing applications. A fundamental step in the processing of these data is image segmentation through a clustering process. One of the most widely used algorithms for clustering is fuzzy C-Means (FCM). However, the presence of spectrally overlapped classes in remote sensing data, and intrinsic sensitivity of FCM to initialized values and complex nonlinear patterns, affects the results of clustering. Particularly, this problem gets worse in case of hyperspectral data. To overcome the mentioned problems, two FCM approaches, i.e. clustering based on integration of particle swarm optimization (PSO) and FCM (i.e. PSO-FCM), and Clustering Based on Kernel-Based FCM (KFCM) are presented in this paper. The objective is an evaluation study of hyperspectral clustering methods. Experiments on the AVIRIS images taken over the northwest Indiana's Indian Pine, show that the PSO-FCM yields better performance in comparison with KFCM.

3.1 Introduction

Clustering is an unsupervised learning task that partitions data objects into a certain number of clusters, in a way that data in the same cluster should be similar to each other, while data in different clusters should be dissimilar [1]. By this definition, clustering can be very useful in remote sensing data analysis, because it can reveal useful information concerning the structure of the dataset. One of the most widely-used clustering algorithms is fuzzy clustering algorithm.

A.A. Naeini (✉) • S. Niazmardi • S. Rahmatollahi Namin • F. Samadzadegan • S. Homayouni
Department of Surveying Engineering, College of Engineering, University of Tehran, Tehran, Iran
e-mail: a.alizadeh@ut.ac.ir; s.niazmardi@ut.ac.ir; rnamin@ut.ac.ir; samadz@ut.ac.ir; homayounis@ut.ac.ir

A. Madureira et al. (eds.), *Computational Intelligence and Decision Making: Trends and Applications*, Intelligent Systems, Control and Automation: Science and Engineering 61,
DOI 10.1007/978-94-007-4722-7_3,

Fuzzy clustering algorithms aim to model fuzzy unsupervised patterns efficiently. One of the widely used fuzzy clustering algorithms is Fuzzy C-Means (FCM) algorithm [2]. The FCM algorithm is based on an iterative optimization of a fuzzy objective function. However, the main drawback of the FCM algorithm is that the results are highly sensitive to the selection of initial cluster centers and it might converge to the local optima. In order to solve this problem, one possibility is to use the swarm based methods such as PSO [3]. Moreover, the main problem of FCM is the nature of remote sensing data; as the information classes in images are usually overlapped in spatial and spectral domain [4]. However, by changing the characteristics of classifiers, one can overcome the mentioned problems, and achieve better results. One of the newest solutions is using the kernel based methods in clustering algorithms. This clustering technique is based on the FCM and the kernel concepts. Kernel function can transform the input data space into a new higher (possibly infinite) dimension space through some nonlinear mapping. Where the complex nonlinear problems in the original space can more likely be linearly treated and solved in the transformed space, according to the well-known Cover's theorem [5]. In recent years, some clustering techniques for remotely sensed image data has been proposed based on the two cited methods. In [6] after the preprocessing of data, the FCM clustering algorithm optimized by a particle swarm algorithm (PSO-FCM) and then utilized for the wetland extraction. In [7, 8] a kernel-based fuzzy C means clustering is used for clustering and recognition of multispectral remote sensing images.

Nonetheless, due to high dimensionality of hyperspectral data, stated problems are intensified. In the other word, regarding optimization and nonlinear complex problems, the number of local optima and nonlinear complexity are increased. If we modify the classifier to handle these issues, more accurate results are expectable. This can be done by using aforementioned algorithms. However, there is no study, which investigates the efficiency of kernel method and PSO in unsupervised classification of hyperspectral data.

The objective of this study is comparing two new FCM methods based on PSO and kernel for hyperspectral image clustering. In addition, a comparison has been done between these two methods and FCM and K-Means.

This paper is organized as the followings. In Sect. 3.2, a brief overview of the clustering is given. The methodology is described in Sect. 3.3, it includes two clustering methods based on PSO and kernel approaches. The results and discussion are presented in Sect. 3.4. Finally, the conclusion is given in Sect. 3.5.

3.2 Basic Concepts of Data Clustering

Clustering is the process of identifying natural groupings within the data, based on some similarity measure. Hence, similarity measures are fundamental components in most clustering algorithms. The most popular way to evaluate a similarity measure is the use of distance measures. The most widely used distance measure is the Euclidean distance [9].

Another important parameter is *Cluster Validity Index* which is used to evaluate of clustering methods. Cluster validity indices can be categorized in three different criteria: internal criteria, relative criteria, and external criteria. Indices based on internal criteria assess the fit between the structure imposed by the clustering algorithm and the data. Indices based on relative criteria compare multiple structures (generated by different algorithms, for example) and decide which of them is better in some sense. External indices measure the performance by matching cluster structure to the *a priori* information, namely the "true" class labels (often referred to as ground truth) [10]. Typically, clustering results are evaluated using the external criterion especially in remote sensing data that the goal is extraction of the specified classes [11, 12].

Clustering can be performed in two different modes: crisp and fuzzy clustering. In crisp clustering, the clusters are disjoint and non-overlapping in nature [13]. Any pattern may belong to one and only one class in this case. In case of fuzzy clustering, a pattern may belong to all the classes with a certain fuzzy membership grade [9]. The K-means (or hard c-means) algorithm starts with K cluster-centroids (these centroids are initially selected randomly or derived from some *a priori* information). Each pattern in the data set is then assigned to the closest cluster centre. Centroids are updated by using the mean of the associated patterns. The process is repeated until some stopping criterion is met. The FCM [2] seems to be the most popular algorithm in the field of fuzzy clustering. In the classical FCM algorithm, a within cluster sum function J_m is minimized to evolve the proper cluster centers as follows:

$$j_m = \sum_{i=1}^{c} \sum_{j=1}^{N} u_{ij}^m ||v_i - X_j||, m \geq 1 \tag{3.1}$$

Where $||v_i - x_j||$ is a distance measure between the center V_i of its cluster and the pattern X_j. Also u_{ij}^m is a fuzzy membership function and m is a constant known as the index of fuzziness. Given C clusters, we can determine their cluster centers v_j for j = 1 to C by means of the following expression:

$$V_j = \frac{\sum_{i=1}^{n} u_{ij}^m x_i}{\sum_{i=1}^{n} u_{ij}^m} \tag{3.2}$$

Now differentiating the performance criterion with respect to v_j (treating u_{ij}^m as constants) and with respect to u_{ij}^m (treating v_j as constants) and setting them to zero the following relation can be obtained:

$$u_{ik} = \sum_{j=1}^{C} \left(\frac{d_{ik}}{d_{jk}}\right)^{\frac{-2}{m-1}} = \sum_{j=1}^{C} \left(\frac{|x_k - v_i|}{|x_k - v_j|}\right)^{\frac{-2}{m-1}} \tag{3.3}$$

3.3 Clustering of Hyperspectral Data Based on PSO and Kernel Function

3.3.1 *FCM Clustering Based on PSO (PSO-FCM)*

3.3.1.1 Particle Swarm Optimization

PSO is a population based stochastic optimization technique inspired by the social behavior of bird flock (and fish school, etc.), and has been developed by Kennedy and Eberhart [14]. As a relatively new evolutionary paradigm, it has grown in the past decade and many studies related to PSO have been published. In PSO, each particle is an individual, and the swarm is composed of particles. The problem solution space is formulated as a search space. Each position in the search space is a candidate solution of the problem. Particles cooperate to find the best position (best solution) in the search space (solution space). Each particle moves according to its velocity which is computed as:

$$v_{id}(t+1) = wv_{id}(t) + c_1 r_1 (p_{id}(t) - x_{id}(t)) + c_2 r_2 (p_{gd}(t) - x_{id}(t)) \tag{3.4}$$

$$x_{id}(t+1) = x_{id}(t) + v_{id}(t+1) \tag{3.5}$$

In (3.4) and (3.5), $x_{id}(t)$ is the position of particle i at time t, $v_{id}(t)$ is the velocity of particle i at time t, $p_{id}(t)$ is the best position found by particle i itself so far, $p_{gd}(t)$ is the best position found by the whole swarm so far, ω is an inertia weight scaling the previous time step velocity, c_1 and c_2 are two acceleration coefficients that scale the influence of the best personal position of the particle ($p_{id}(t)$) and the best global position ($p_{gd}(t)$), r_1 and r_2 are random variables between 0 and 1 [15].

3.3.1.2 PSO-FCM

The FCM algorithm tends to converge faster than the PSO algorithm because it requires fewer function evaluations. But, it usually gets stuck in local optima. We integrate FCM with PSO to form a hybrid clustering algorithm called PSO-FCM, which maintains the merits of FCM and PSO. More specifically, PSO-FCM will apply FCM with four iterations to the particles in the swarm every eight generations such that the fitness value of each particle is improved [16]. A particle is a vector of real numbers of dimension $k \times d$, where k is the number of clusters and d is the dimension of data to be clustered. The objective function of the FCM algorithm defined in Eq. 3.1 is the fitness functions of the hybrid clustering algorithms. The hybrid PSO-FCM algorithm can be summarize as follows [17].

1. Randomly generation of particles.
2. Calculation of cluster centers using Eq. 3.2.
3. Calculation of fitness function using Eq. 3.1.
4. Updating of Pbest and Gbest according to fitness function of FCM
5. Updating velocity by Eq. 3.4.
6. Updating position by Eq. 3.5.
7. Repeat step 2–6 until the stopping criteria is met.

3.3.2 *Kernel-Based Fuzzy C-Means Algorithm*

The Fuzzy C-means algorithm use the Euclidian distance for calculating the similarities between pixels and cluster centers. However, this distance has some problems which affect the results of clustering. For example, it is sensitive to clusters shapes and outlier. In order to tackle these problems, a new modification of FCM, named kernel FCM was introduced [18]. The Basic idea of KFCM is to compute the Euclidian distance in another space of higher dimension via a non-linear map function φ, and by this map function, we expect simpler relation in new space (feature space) so the clusters can better be separated.

Nevertheless, mapping all data into feature map, can lead to expensive cost. In order to handle data in feature space, one can use their pair-wise scalar product and this scalar product can be computed directly by kernel function. Thus kernel function is a function, $K : X \times X \to \mathbb{R}$, such that [19]:

$$\forall x, y \in X, <\varphi(x), \varphi(y)> = k(x, y) \tag{3.6}$$

So we can rewrite the Euclidian distance between pixel i and cluster j, in feature space as follows:

$$d_{ij} = ||\varphi(x_i) - \varphi(v_j)||^2 = k(x_i, x_j) + k(v_j, v_j) - 2k(x_i, v_j) \tag{3.7}$$

There are many different kernels, but here we use radial base kernel due to its robustness [20].

$$K(x, y) = \exp\left(\frac{-(x-y)^2}{\sigma^2}\right) \tag{3.8}$$

Therefore, the Euclidian distance can be written as follows:

$$d_{ij} = ||\varphi(x_i) - \varphi(v_j)||^2 = 2(1 - k(x_i, v_j)) \tag{3.9}$$

By using this distance in FCM objective function, we can derive the objective function of KFCM.

$$J(X,U,C) = 2\sum_{j=1}^{c}\sum_{i=1}^{n} u_{ji}^{m}(1 - k(x_i, v_j)) \tag{3.10}$$

Where U is fuzzy partition matrix, Vj is the centroid of jth cluster and Xi is the vector ith pixel.

To optimize the KFCM objective function, an alternative optimization method is used. By this method the clusters centers and fuzzy partition matrix can be calculated by following equations in each iteration [21]:

$$u_{ji} = \frac{1}{\sum_{l=1}^{c}\left(\frac{(1-k(x_i,v_j))}{(1-k(x_i,v_l))}\right)^{\frac{1}{(m-1)}}} \tag{3.11}$$

$$v_j = \frac{\sum_{i=1}^{n} u_{ji}^{m} k(x_i, v_j) x_i}{\sum_{i=1}^{n} u_{ji}^{m} k(x_i, v_j)} \tag{3.12}$$

For using this method one should tune the kernel parameter (σ in 3.8). For tuning the parameter, we normalize the data and run the algorithm with ten different sigmas from 0.1 to 1, with a 0.1 increment, and the results have been compared with kappa coefficient, at last, the sigma of the best performance (1 in here) was chosen for comparison with other algorithms.

3.4 Results and Discussion

3.4.1 Dataset

The performances of two mentioned methods are evaluated using a sample hyperspectral image which is taken over northwest Indiana's Indian Pine test site in June 1992 [22]. This data set was chosen because its ground truth for evaluating algorithm is available. The data consists of 145 × 145 pixels with 220 bands. The 20 water absorption bands were removed from the original image. In addition, 15 noisy bands were also removed, resulting in a total of 185 bands [23]. The original ground truth has actually 16 classes, but in this study five classes of them are used. Also the ground truth map of five classes is shown in Figs. 3.1 and 3.2. These classes are selected because they have suitable spatial distribution.

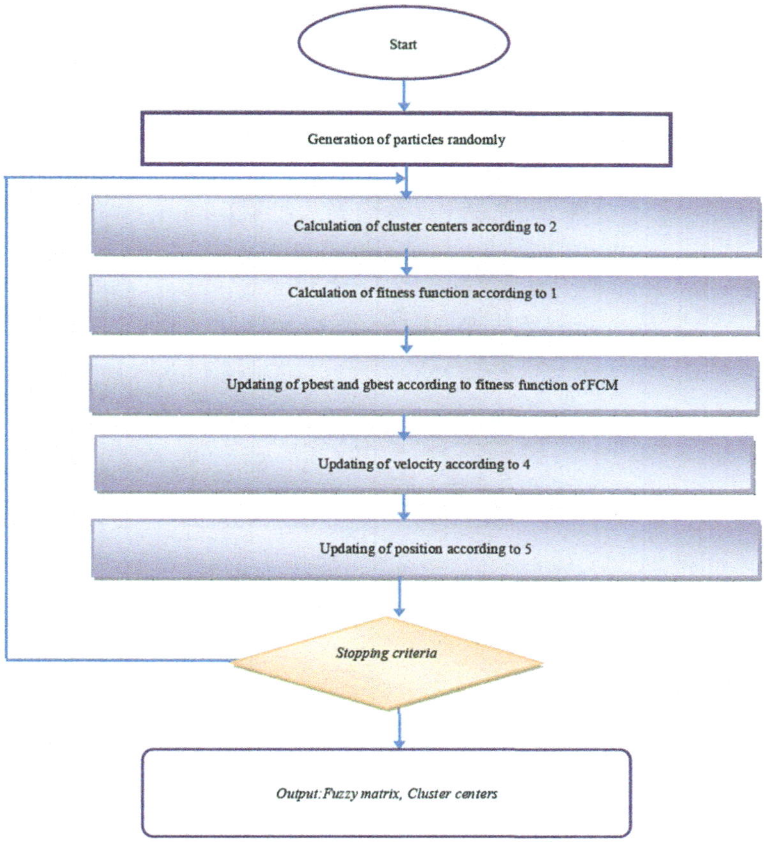

Fig. 3.1 Color composite of the image subset

3.4.2 *Performance Measure*

In this paper, confusion matrix was used to evaluate the true labels and the labels returned by the clustering algorithms as the quality assessment measure [12]. In addition, the Kappa coefficient of agreement is defined in Eq. 3.13 for individual classes, The Khat index [24] is calculated using Eq. 3.14.

$$K = \frac{N\sum_{i=1}^{r} x_{ii} - \sum_{i=1}^{r} (x_{i+} \times x_{+i})}{N^2 - \sum_{i=1}^{r} (x_{i+} \times x_{+i})} \tag{3.13}$$

Fig. 3.2 Ground truth of the area with five classes

$$K = \frac{N\sum_{i=1}^{r} x_{ii} - \sum_{i=1}^{r}(x_{i+} \times x_{+i})}{N^2 - \sum_{i=1}^{r}(x_{i+} \times x_{+i})} \tag{3.14}$$

In Eq. 3.13, K is the kappa coefficient and in Eq. 3.14, ki is Khat index for individual classes, r is the number of columns (and rows) in a confusion matrix, xii is entry (i,i) of the confusion matrix, xi+ and x + i are the marginal totals of row i and column j, respectively, and N is the total number of observations [24, 25].

In this study, four clustering methods, i.e. kmeans, FCM, PSO-FCM and KFCM, are compared to each other. These methods were developed based on the parameters listed in Table 3.1.

According to Fig. 3.3, it is clear that methods PSO-FCM, KFCM and FCM with kappa values 76.21, 67.28, 66.51 have better performance (accuracy) than kmeans method with kappa value 58.16

Because of the existence of overlap in information classes especially spectral domain in hyperspectral data, algorithms based on FCM obtains better results. Among the three methods based on FCM, two new presented methods i.e. PSO-FCM and KFCM have better results in contrast to FCM. It should be noted that PSO-FCM have global and local search, while KFCM is only able to do local search, it seems that transferring data to a space of high dimension can separate some clusters and enhance the FCM results. Therefore, it can be said, these two methods were efficient in hyperspectral clustering. In the other words, the mentioned methods can help reaching better performance for FCM.

Table 3.1 Parameters used in the clustering hyperspectral data sets

Algorithm	Parameters	Assigned value
kmeans	Iterations	50
FCM	Iterations	50
	m	2
PSO-FCM	Iterations	100
	FCM iterations	4
	PSO iterations	8
	P_{size}	35
	W	0.72
	C_1	0.49
	C_2	0.49
KFCM	Iterations	50
	m	2
	Sigma	1

'Assigned Value' refers to the number (value) of the parameters involved in the algorithms

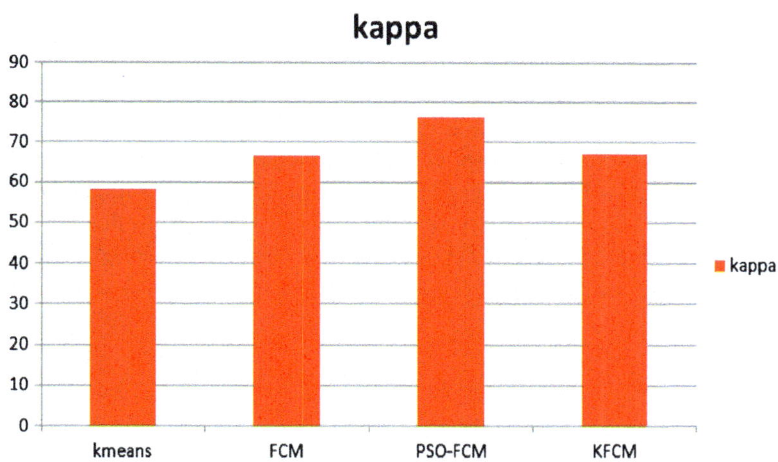

Fig. 3.3 Comparison of kappa coefficient in four clustering methods

Between two presented methods, PSO-FCM has better accuracy (about 12%). The reason is that PSO-FCM has both local ability search of FCM and global ability search of PSO, while KFCM method firstly is sensitive to both initialized values and sigma parameter, which means with either different initialized values or sigma parameter, KFCM converges to different values.

In order to have better comparison, the obtained results from clustering different methods are presented in Table 3.2. The results and their comparison with the thematic map, represents that in FCM and K-means, clusters are somehow combined, e.g. the Hay-windrowed class can also be seen in other clusters; but in KFCM

Table 3.2 Khat index obtained by the four investigated methods on the AVIRIS image

	Corn no-till	Grass\trees	Hay-windrowed	Soybean no-till	Woods
K-means	47.8	0	99.8	62.7	99.9
FCM	50.71	71.89	89.72	47.83	98.58
PSO-FCM	50.53	84.32	99.54	66.75	97.44
KFCM	68.76	67.07	74.49	40.87	93.65

and PSO-FCM, this issue is reduced. In addition, in all the methods, some parts were not clustered properly. As an example, in no-tilled corn and soybean lands, the KFCMs results show not necessarily all clusters can be better separated by using kernel method. For example, Wood class is completely clustered in FCM results but is mixed with other clusters in KFCM results.

3.5 Conclusion

In this article, two fuzzy clustering methods are evaluated. These methods are based on PSO and kernel approaches, respectively. Results show that the presented methods have better accuracy than standard FCM. Also, PSO-FCM yields better result than KFCM due to its inherent global and local search existing in this method. Also, there is an interesting ability of KFCM in data transformation which makes possible to introduce PSO in order to solve its problem. Accordingly, our future investigations will be dedicated to combining PSO and KFCM so that not only can find the optimum parameters of KFCM, but also will overcome the sensitivity to initialized values.

References

1. Wunsch D, Xu R (2008) Kernel-based clustering. In: Clustering, 1st edn. IEEE, New Jersey, pp 163–178
2. Bezdek J (1981) Pattern recognition with fuzzy objective function algorithms. Kluwer Academic Publishers, Norwell
3. Izakian H, Abraham A, Snasel V (2009) Fuzzy clustering using hybrid fuzzy c-means and fuzzy particle swarm optimization. Paper presented at the World Congress on Nature & Biologically Inspired Computing, Coimbatore, India, 9–11 December
4. Tso B, Mather PM (2009) Classification methods for remotely sensed data. CRC, Boca Raton
5. Cover T (1965) Geometrical and statistical properties of systems of linear compute. IEEE Trans Electron Comput 14:326–334
6. Liu H, Pei T, Zhou C, Zhu AX (2008) Multi-temporal MODIS-data-based PSO-FCM clustering applied to wetland extraction in the Sanjiang Plain. Paper presented at the International Conference on Earth Observation Data Processing and Analysis, Wuhan, China
7. Yun-song S, Yu-feng S (2010) Remote sensing image classification and recognition based on KFCM. In: Proceedings of the 5th international conference on Computer & Education, Hefei, 239 China, pp 1062–1065

8. Niazmardi S, Homayouni S, Safari A (2011) Remotely sensed image clustering based on kernel-based fuzzy C-means algorithm. SMPR, Tehran
9. Jain AK, Murty MN, Flynn PJ (1999) Data clustering: a review. ACM Comput Surv (CSUR) 24231:264–323
10. Jain AK (2010) Data clustering: 50 years beyond K-means. Pattern Recognit Lett 31:651–666
11. Saeedi S, Samadzadegan F, El-Sheimy N (2009) Object extraction from LIDAR data using an artificial Swarm Bee colony clustering algorithm. In: Stilla U, Rottensteiner F, Paparoditis N (eds) CMRT09 IAPRS 38 (Part 3)
12. Zhong S, Ghosh J (2003) A comparative study of generative models for document clustering. Paper presented at the SIAM international conference on data mining workshop on clustering high dimensional data and its applications, San Fransisco
13. Abraham A, Das S, Roy S (2008) Swarm intelligence algorithms for data clustering. In: Maimon O, Rokach L (eds) Soft computing for knowledge discovery and data mining. Springer, New York, pp 279–313. doi:10.1007/978-0-387-69935-6_12
14. Kennedy J, Eberhart R (1995) Particle swarm optimization. Paper presented at the IEEE International Conference Neural Network (ICNN), Perth, WA
15. Yang F, Zhang C, Sun T (2009) Particle swarm optimization and differential evolution in fuzzy clustering. In: Advances in neuro-information processing. Springer, Berlin, pp 501–508
16. Yang F, Sun T, Zhang C (2009) An efficient hybrid data clustering method based on K-harmonic means and particle swarm optimization. Expert Syst Appl 36:9847–9852
17. Li W, Yushu L, Xinxin Z, Yuanqing X (2006) Particle swarm optimization for fuzzy c-means clustering. Paper presented at the intelligent control and automation. WCICA 2006. The Sixth World Congress on Intelligent Control and Automation
18. Zhang DQ, Chen SC (2002) Fuzzy clustering using kernel method. In: International conference on control and automation (ICCA'02), Xiamen, China, pp 123–127
19. de Oliveria JV, Pedeycz W (eds) (2007) Advances in fuzzy clustering and its applications. Wiley, Chichester
20. Wang L, Jin Y, Du W, Inoue K, Urahama K (2005) Robust kernel fuzzy clustering. In: Fuzzy systems and knowledge discovery, vol 3613. Springer, Berlin/Heidelberg, pp 454–461
21. Graves D, Pedrycz W (2010) Kernel-based fuzzy clustering and fuzzy clustering: a comparative experimental study. Fuzzy Set Syst 161:522–543
22. Graves D, Pedrycz W (2007) Performance of kernel-based fuzzy clustering. Electron Lett 43:1445–1446
23. Mojaradi B, Emami H, Varshosaz M, Jamali S (2008) A novel band selection method for hyperspectral data analysis. Int Arch Photogramm Remote Sens Spat Inf Sci 37:447–454
24. Kumar M (2004) Feature selection for classification of hyperspectral remotely sensed data using NSGA-II. Water Resources Seminar CE 597D
25. Carletta J (1996) Assessing agreement on classification tasks: the kappa statistic. Comput Linguist 22:249–254

Chapter 4
Comparison of Classification Methods for Golf Putting Performance Analysis

J. Miguel A. Luz, Micael S. Couceiro, David Portugal, Rui P. Rocha, Hélder Araújo, and Gonçalo Dias

Abstract This paper presents a comparative case study on the classification accuracy between five methods for golf putting performance analysis. In a previous work, a digital camera was used to capture 30 trials of 6 expert golf players. The detection of the horizontal position of the golf club was performed using a computer vision technique followed by the estimation algorithm Darwinian Particle Swarm Optimization (DPSO) in order to obtain a kinematical model of each trial. In this paper, the estimated parameters of the models are used as sample and training data of five classification algorithms: (1) Linear Discriminant Analysis (LDA); (2) Quadratic Discriminant Analysis (QDA); (3) Naïve Bayes with Normal (Gaussian) distribution (NV); (4) Naïve Bayes with Kernel Smoothing Density Estimate (NVK) and (5) Least Squares Support Vector Machines with Radial Basis Function Kernel (LS-SVM). The five classification methods are then compared through the analysis of the confusion matrix and the area under the Receiver Operating Characteristic curve (AUC).

J.M.A. Luz (✉) • M.S. Couceiro
RoboCorp, Department of Electrical Engineering (DEE), Engineering Institute of Coimbra (ISEC), Coimbra, Portugal
e-mail: miguel.luz@isec.pt; micael@isec.pt

D. Portugal • R.P. Rocha • H. Araújo
Institute of Systems and Robotics (ISR), University of Coimbra (FCTUC), Coimbra, Portugal
e-mail: davidbsp@isr.uc.pt; rprocha@isr.uc.pt; helder@isr.uc.pt

G. Dias
RoboCorp, Faculty of Sport Sciences and Physical Education, University of Coimbra (FCDEF), Coimbra, Portugal
e-mail: goncalodias@fcdef.uc.pt

A. Madureira et al. (eds.), *Computational Intelligence and Decision Making: Trends and Applications*, Intelligent Systems, Control and Automation: Science and Engineering 61, DOI 10.1007/978-94-007-4722-7_4,

4.1 Introduction

The putt is a motor skill that combines art and science [1], representing about 43% of the strikes in a Golf game [2]. Coordination and control of this movement is an individual process that is different for each subject depending on his profile and characteristics [3]. Several studies [4–6] analyzed putting based on process measurements of motor execution, *i.e.*, movement action parameters like position, velocity and acceleration (Fig. 4.1). The same researchers assume that the aspects of stability and variability in the execution of this movement, observed at expert level players and beginners alike, may have significant differences in motor performance. Like other motor skills, the intra and inter-individual performance resulting from the execution of the putt is a "fingerprint" (*i.e.*, signature) that is unique to each individual [7].

A previous work [8] presented the experimental design and methodological aspects in the analysis of the effects of variability in golf putting performance of expert subjects based on previously detected and estimated data of the horizontal position of the putter during the execution. The process variables, such as the trajectory function were obtained by estimation of a sinusoidal model composed by the sum of three sine waves to fit the horizontal position of the putter during the execution using the Darwinian Particle Swarm Optimization (*DPSO*) method.

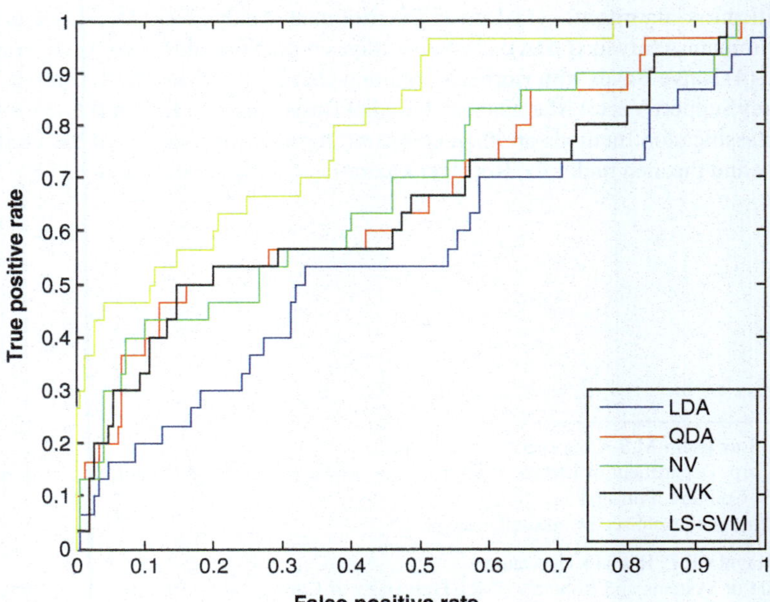

Fig. 4.1 Example of ROC curves for the third sine wave (Amplitude (a3) vs. Angular Frequency (b3)) of the fifth player (class 5). The best possible prediction method would yield a point in the *upper left* corner, i.e., coordinate (0,1), of the ROC space, hence the LS-SVM is the classifier which presents a superior performance for this trial

In this paper, the parameters of the kinematic model (*i.e.*, amplitudes, frequencies and phases of the sinusoids) of each expert player's putting trial are classified with five different classification methods in order to identify possible links between different executions of the same player, thus extracting putting signatures for every player. The several classification algorithms studied are described in Sect. 4.2, leading to the experimental results of the mentioned stages, that are represented and discussed in Sect. 4.3. Finally, the main conclusions are outlined in Sect. 4.4.

4.2 Survey of Classification Methods

In this section it is presented a review of the five classification algorithms used in this work. In particular, the Linear and Quadratic Discriminant Analysis (*LDA* and *QDA*), the Naïve Bayes with Normal (Gaussian) distribution and with Kernel Smoothing Density Estimate (*NV* and *NVK*) and the Least Squares Support Vector Machines with Radial Basis Function Kernel (*LS-SVM*). Despite the diversity of the methodologies to evaluate the performance of classifiers, the confusion matrix and the area under the Receiver Operating Characteristic curve (*AUC*) are some of the most well-known evaluation methods and will be used in this work. Further details about both methods can be found in [9].

A confusion matrix is a matrix containing information about actual and predicted classifications obtained with a classification system [10]. The confusion matrix lists errors and successes in the test set. The main diagonal represents the correctly classified samples while the other elements of the matrix correspond to samples that were incorrectly classified.

The Receiver Operating Characteristic (*ROC*) is a technique to visualize, evaluate, organize and select classifiers based on their performance (Fig. 4.1). The *ROC* graphs can show the line between the true positive and false positive rate of the classifiers [11].

To compare classifiers it is necessary to reduce the curve to a scalar value. A common method to achieve this reduction is to calculate the area under the *ROC* curve (*AUC*). The *AUC* is a way to measure classifiers performance. Since the *AUC* is a part of the area of the unit square, its value always varies between 0 and 1. An *AUC* value of 1 represents a perfect test while the *AUC* value of 0.5 represents a weak or worthless test.

4.2.1 *Linear Discriminant Analysis (LDA)*

There are many possible techniques for data classification. The Linear Discriminant Analysis (*LDA*) is one of the most commonly used techniques for data classification and dimensionality reduction in statistics, pattern recognition and machine learning, since it easily handles the case where the within-class frequencies are unequal and their performances have been examined on randomly generated test data [12]. This method maximizes the ratio of between-class variance to the within-class variance in any particular data set thereby guaranteeing maximal separability.

LDA is closely related to logistic regression, Principal Component Analysis (*PCA*) and Quadratic Discriminant Analysis (*QDA*), which also attempt to express one dependent variable as a combination of other features or measurements. Other methods are preferable in applications where it is not reasonable to assume that the independent variables are normally distributed, which is a fundamental assumption of the *LDA* method. Hence, a difference between independent variables and dependent variables must be made. *LDA* assumes that the conditional probability density functions (*PDF*) follow a normal distribution for all classes. In practice, the class means and covariances are usually not known and are estimated from the training set used; *e.g.*, using the maximum likelihood estimate or the maximum *a posteriori* estimate. Also, *LDA* assumes that all classes have a common covariance matrix and all covariances have full rank, this is called the homoscedastic assumption. The discriminant function is given by testing the probability that a sample x is contained in one class or another:

$$\delta_k(x) = x^{\mathrm{T}} \sum\nolimits^{-1} \mu_k - \frac{1}{2}\mu_k^{\mathrm{T}} \sum\nolimits^{-1} \mu_k + log\, P\,(C = k) \tag{4.1}$$

Where, Σ is the covariance matrix common to all classes and μ_k is the mean of class k. The value of x in each $\delta_k(x)$ is calculated and the classification of x is the class k that yields the largest value. When the number of observations of each sample exceeds the number of samples, the covariance estimates do not have full rank, and so cannot be inverted. One way to deal with this is to use a pseudo inverse instead of the usual inverse matrix or use a Shrinkage estimator of the covariance matrix. The resulting classifier implies that the decision boundary between pairs of classes is linear and a hyperplane when using more than two classes, this is verified through the comparison of classes using the log-ratio. In geometrical terms, it is clear that an input observation is in a given class if the multidimensional-space observation point is located on a certain side of a hyperplane, perpendicular to the normal to the discriminant hyperplane.

LDA finds other applications in areas like face recognition, marketing or financial prediction. For more details on the implementation of the method, one should refer to [12].

4.2.2 Quadratic Discriminant Analysis (QDA)

Although they differ in their derivation, Quadratic Discriminant Analysis (*QDA*) is similar to *LDA* [13]. The essential difference between them is in the way the linear function is fit to the training data. Also very popular, *QDA* separates measurements of classes of objects or events with a boundary between each pair of classes described by a quadratic equation. Normal distributions are assumed, but each class can have a different covariance matrix. Thus, the separate covariance matrices must be estimated for each class, which means that there are much more parameters than in *LDA*, that increase with the number of dimensions. Since the decision boundaries are functions of the parameters of the densities, counting the number of parameters must be carefully done. When the homoscedastic assumption is true, the best possible test for the hypothesis that a given measurement is from a given class is the likelihood ratio test, similarly to the *LDA*. The discriminant function is given by testing the probability that a sample x is contained in one class or another:

$$\delta_k(x) = -\frac{1}{2}\log\left|\sum\nolimits_{\mathrm{k}}\right| - \frac{1}{2}(\mathrm{x}-\mu_{\mathrm{k}})^{\tau}\sum\nolimits_{\mathrm{k}}^{-2}(\mathrm{x}-\mu_{\mathrm{k}}) + \log \mathrm{P}(\mathrm{C}=\mathrm{k}) \tag{4.2}$$

where Σ_k is the covariance matrix for class k, and μ_k is the mean of class k. After calculating each covariance matrix, estimating the mean, μ_k, and calculating $P(C = k)$, the classification of new samples is accomplished by calculating their discriminant function value for each class. The less rigid model underlying *QDA* may better approximate the situation in a real classification problem (compared to *LDA*). While the allowance of curved decision boundaries can lead to reduced bias in the estimation of the (unknown) optimal decision boundaries, having to estimate variance and covariance values with less data can lead to increased variance in the estimation of the optimal boundaries.

Literature show that both *LDA* and *QDA* perform well on large and diverse set of classification tasks.

4.2.3 Naïve Bayes with Normal (Gaussian) Distribution (NV)

Naïve Bayes (*NV*) is one of the most efficient and effective inductive learning algorithms for machine learning and data mining [14]. The *NV* classifier is designed for use when features are independent of one another within each class, which is a rather unrealistic assumption that is almost always violated in real-world applications. However, it has surprisingly good performance, performing well in practice even when that independence assumption is not valid. Furthermore, this assumption dramatically simplifies the estimation. The individual class-conditional marginal densities can be estimated separately; also if the variables are discrete, then an appropriate histogram estimate can be used. Assuming conditional independence among X_i's (vectors of observed random variables), Bayes Rules is given by:

$$P(Y = y_k|X_1 \ldots X_n) = \frac{P(Y = y_k)\Pi_i P(X_i|Y = y_k)}{\sum_j P(Y = y_j)\Pi_i P(X_i|Y = y_j)} \tag{4.3}$$

Hence, the Naïve Bayes classifier selects the class Y^{new} with maximum discriminant function for $X_{new} = \langle X_1, \ldots, X_n \rangle$:

$$Y^{new} \leftarrow \arg{}_{y_k}^{max} P(Y = y_k) \prod_i P\left(X_i^{new} | Y = y_k\right) \tag{4.4}$$

If X_i is continuous, a common approach is to assume that $P(X_i|Y = y_k)$ follows a normal (Gaussian) distribution:

$$P(X_i = x | Y = y_k) = \frac{1}{\sqrt{2\pi\sigma_{ik}^2}} e^{-\frac{1}{2}\left(\frac{x-\mu_{ik}}{\sigma_{ik}}\right)^2} \tag{4.5}$$

And the classification becomes:

$$Y^{new} \leftarrow \arg{}_{y_k}^{max} \pi_k \prod_i N\left(X_i^{new};\ \mu_{ik},\ \sigma_{ik}\right) \tag{4.6}$$

Where Π_k is estimated for each value of y_k by $\pi_k \equiv P(Y = y_k)$. Also, for each attribute X_i, it is necessary to estimate the class conditional mean μ_{ik} and variance σ_{ik}.

A well-known limitation of Naïve Bayes is in the case of binary features [15], where it can only learn linear discriminant functions, and thus it is always suboptimal for non-linearly separable concepts. Nonetheless, Naïve Bayes has proven effective in many practical applications, including text classification, medical diagnosis, and systems performance management [16]. It is also efficient using memory space and in terms of time complexity [17].

4.2.4 Naïve Bayes with Kernel Smoothing Density Estimate (NVK)

Kernel smoothing density estimation is an unsupervised learning procedure, which historically precedes kernel regression. This method fits a different but simple model separately at each query point x_0, using only observations close to the target point, in a way that the resulting estimated function becomes smooth. This is accomplished via a weighting function or kernel $K_\lambda(x_0;x_i)$ that assigns a weight to x_i based on its distance from x_0 [12]. Kernel methods use weights that decrease smoothly to zero with distance from the target point. In high-dimensional spaces the distance kernels are modified to emphasize some variable more than others.

The Kernel estimate is probably the most widely used nonparametric density estimation method. Kernels K_λ are typically indexed by a parameter λ that controls the width of the neighborhood. This results in a memory-based method that requires little or no training at all, where all the work gets done at evaluation time. The only parameter that needs to be determined from the training data is λ. The model is the entire training data set. Generally the kernel function is given by:

$$K_\lambda(x_0,\ x) = PDF\left(\frac{|x - x_0|}{h_\lambda(x_0)}\right) \tag{4.7}$$

Three popular probability density functions (*PDF*) used are Epanechnikov, Tri-Cube and Gaussian distribution function.

Bayesian classification and decision making are based on probabilities that a given set of measurements come from objects belonging to a certain class (probability theory) and statistical methods, based on class conditional probability density functions of features, are suitable in diverse classification tasks [18].

Estimated *PDF*s have been used for classification utilizing Bayes formula. The classification can be done based on the probability density function, instead of estimating posterior probability using Naïve Bayes. The attempt is to estimate the underlying density function from the training data, and the idea is that the more data in a region, the larger is the density function. Kernel smoothing density estimation leads naturally to a simple family of procedures for nonparametric density estimates for classification in a straightforward fashion using Bayes' theorem [12].

4.2.5 Least Squares Support Vector Machines with Radial Basis Function Kernel (LS-SVM)

Support Vector Machines (*SVM*) is a powerful methodology for solving problems in nonlinear classification, function estimation and density estimation which has also led to many developments in kernel based methods in general. This method solves convex optimization problems, typically by quadratic programming. *SVM* models are similar to multilayer perceptron neural networks. However, using a kernel function, *SVM*s are an alternative training method for polynomial, radial basis function and multi-layer perceptron classifiers in which the weights of the network are found by solving a quadratic programming problem with linear constraints, rather than by solving a non-convex, unconstrained minimization problem as in standard neural network training.

Many kernel mapping functions can be used but only a few have been found to work well in for a wide variety of applications. The default and recommended kernel function is the Radial Basis Function (*RBF*). According to [15], Kernel methods achieve flexibility by fitting simple models in a region local to the target point x_0. Localization is achieved via a weighting kernel K_λ, and individual observations receive weights $K_\lambda(x_0;\ x_i)$. Radial basis functions combine these ideas, by treating the kernel functions $K_\lambda(\mu;\ x)$ as base functions, where each basis element is indexed by a location and a scale parameter (μ_m and λ_m

respectively). Thus, Radial basis functions are symmetric p-dimensional kernels located at particular centroids:

$$f_\theta(x) = \sum_{m=1}^{M} K_{\lambda_m}(\mu_m, x)\theta_m \tag{4.8}$$

The centroids μ_m and scales λ_m have to be determined. A common choice for the probability density functions is the standard Gaussian density function. There are also several approaches for learning the parameters μ_m, λ_m and θ_m. For example, a popular method is estimating θ_m, given μ_m and λ_m by a simple least squares problem. Often the kernel parameters μ_m and λ_m are chosen in an unsupervised way using the X distribution alone. A way to do this, is to fit a Gaussian mixture density model to the training x_i, which provides both the centers μ_m and the scales λ_m.

The Least Squares Support Vector Machines (*LS-SVM*) is a reformulation to the standard *SVMs* [19] which was recently proposed. In fact, when the data points are linearly independent, *LS-SVM* is equivalent to Hard Marginal *SVM* [20]. *LS-SVM* involves the equality constraints only. Hence, the solution is obtained by solving a linear equations system.

4.3 Experimental Results

Experimental results present a comparative case study on the classification accuracy between five methods for the detection of signatures in the performance of the golf putting. In this stage, intensive *Matlab* simulation was performed using the detection algorithm and the *DPSO* as an estimation technique to obtain the putter's motion function that describes 30 putt executions of 6 different expert subjects (classes), in a total of 180 trials. After calculating all the estimation parameters, the five classification methods previously described were used and their performances compared.

Figures 4.2, 4.3, and 4.4 depict the confusion matrix and the area under the *ROC* curve (*AUC*) of the five classifiers, respectively. Based on these figures, it can be

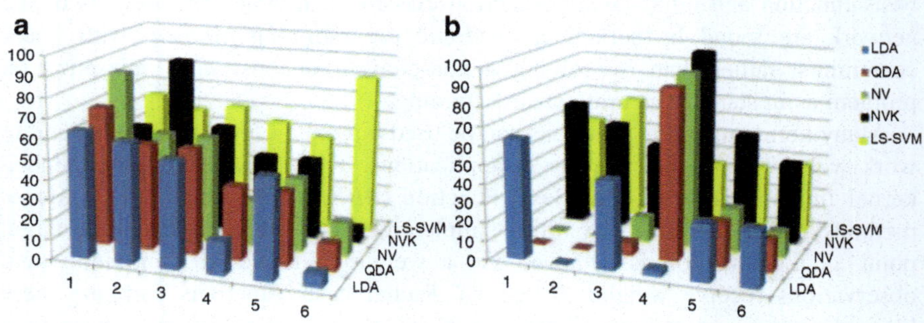

Fig. 4.2 Comparison of the percentual TP rate of the classifiers for the first sine wave. **(a)** Amplitude (a1) vs. Angular Frequency (b1); **(b)** Amplitude (a1) vs. Phase (c1)

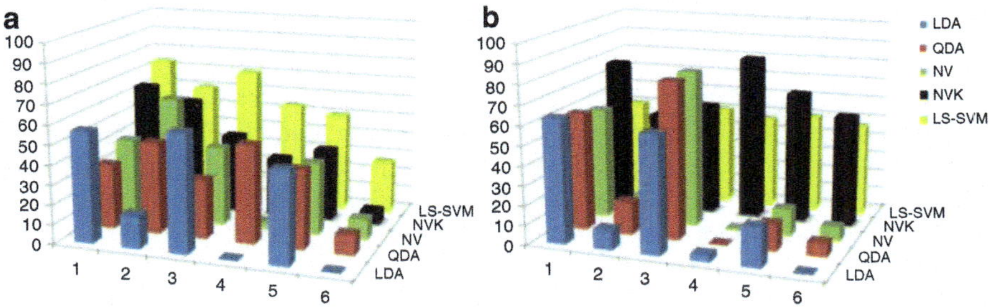

Fig. 4.3 Comparison of the percentual TP rate of the classifiers for the second sine wave. (**a**) Amplitude (a2) vs. Angular Frequency (b2); (**b**) Amplitude (a2) vs. Phase (c2)

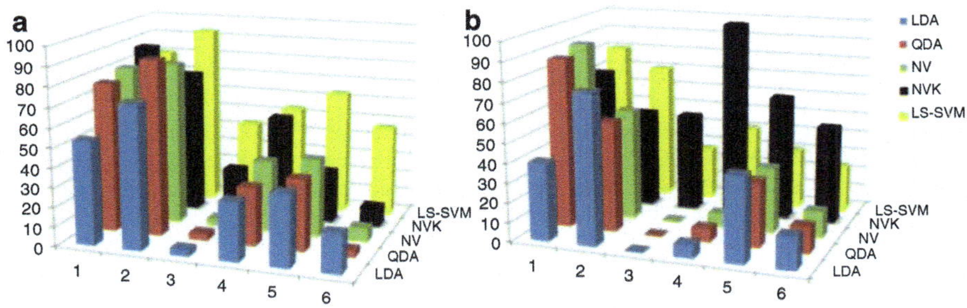

Fig. 4.4 Comparison of the percentual TP rate of the classifiers for the third sine wave. (**a**) Amplitude (a3) vs. Angular Frequency (b3); (**b**) Amplitude (a3) vs. Phase (c3)

Table 4.1 Average value of the AUC

Class	LDA	QDA	NV	NVK	SVM
1	0.619	0.601	0.671	0.680	0.744
2	0.650	0.623	0.692	0.685	0.737
3	0.566	0.582	0.634	0.761	0.734
4	0.507	0.585	0.574	0.675	0.690
5	0.622	0.651	0.692	0.766	0.797
6	0.493	0.602	0.650	0.718	0.745

confirmed the better classification accuracy the *LS-SVM* shows, since it presents a higher percentage of true-positives (*TP*) in nearly all situations, closely followed by the *NVK* method.

However, in order to confirm the superiority of the *LS-SVM* over the *NVK*, the *AUC* of each player's trials was determined for the five classification methods. In order to allow a straightforward comparison of the five classifiers, Table 4.1 depicts the average value of the *AUC* highlighting the maximum value for each player's trial. This table provides evidence about the superiority of the LS-SVM classifier which shows better results in the majority of the trials, closely followed by the NVK classifier.

4.4 Conclusions

Real-time automated analysis of sport games such as football, tennis or golf is a domain receiving increased attention. Nowadays, in many live broadcasts, computer vision analysis, with special attention to the ball's kinematic, is used for example to present the ball's velocity or checking the ball's relative position. Also, relevance is given to information computed offline, like player's statistics.

The adopted system for data retrieval, despite its complexity, is functional and allows retrieving different information simultaneously. To validate the work, towards using it in real situations, the putting performance of six expert golf players was evaluated. Several classification methods for golf putting performance analysis were used and compared, with the *LS-SVM* showing the most consistent results. For this reason, in the future we intend to apply *LS-SVM* to extract unique features related to each player when performing the putt, in order to obtain a putting signature for every single subject.

Acknowledgments This work was supported by PhD scholarships (SFRH/BD /73382/2010) and (SFRH/BD/64426/2009) by the Portuguese Foundation for Science and Technology (FCT), the Institute of Systems and Robotics (ISR) and RoboCorp at the Engineering Institute of Coimbra (ISEC) also under regular funding by FCT.

References

1. Pelz D (2000) Putting Bible: the complete guide to mastering the green. Publication Doubleday, New York
2. Alexander DL, Kern W (2005) Drive for show and putt for dough? J Sports Econ 6(1):46–60
3. Jonassen DH, Grabowski BL (1993) Handbook of individual differences, learning and instruction. Lawrence Erlbaum, Hillsdale
4. Delay D, Nougier V, Orliaguet JP, Coello Y (1997) Movement control in golf putting. Hum Mov Sci 16(5):597–619
5. Hume PA, Keogh J, Reid D (2005) The role of biomechanics in maximising distance and accuracy of golf shots. Sports Med 35(5):429–449
6. Karlsen J, Smith G, Nilsson J (2008) The stroke has only a minor influence on direction consistency in golf putting among elite players. J Sports Sci 26(3):243–250
7. Phillips E, Davids K, Renshaw I, Portus M (2010) Expert performance in sport and the dynamics of talent development. Sports Med 40(4):271–283
8. Couceiro MS, Dias G, Luz JMA, Figueiredo CM, Ferreira NMF (2010) Parameter estimation for a mathematical model of the golf putting. In: WACI'10 – 5th workshop on applications of computational intelligence, Portugal
9. Chan K, Lee TW (2002) Comparison of machine learning and traditional classifiers in glaucoma diagnosis. IEEE Trans Biomed Eng 49(9):963–974
10. Kaladhar DSVGK, Nageswara Rao PV, Ramesh Naidu Rajana BLV (2010) Confusion matrix analysis for evaluation of speech on Parkinson disease using Weka and MatLab. Int J Eng Sci Technol 2(7):2733–2736
11. Fawcett T (2006) An introduction to ROC analysis. Pattern Recognit Lett 27(8):861–874
12. Hastie T, Tibshirani R, Friedman J (2009) The elements of statistical learning: data mining, inference, and prediction, 2nd edn, Springer series in statistics. Springer, New York

13. Roushanzamir S, Valafar H, Valafar F (1999) A comparative study of linear and quadratic discriminant classifier techniques for variable selection: a case study in predicting the effectiveness of hydroxyurea treatment of sickle cell anemia. In: IJCNN '99 – international joint conference on neural networks, Washington, DC
14. Zhang H (2005) Exploring conditions for the optimality of naive Bayes. Int J Pattern Recognit Artif Intell 19(2):183–198
15. Rish I (2001) An empirical study of the naive Bayes classifier. In: IJCAI 2001 workshop on empirical methods in artificial intelligence, Seattle, USA
16. Domingos P, Pazzani M (1997) On the optimality of the simple Bayesian classifier under zero–one loss. Mach Learn 29:103–130
17. Zhang H, Ling CX, Zhao Z (2000) The learnability of naive Bayes. In: Hamilton H, Yang Q (eds) Advances in artificial intelligence (Proceedings of CAI2000), vol 1822/2000. Springer, Ottawa, Canada, pp 432–441
18. Rahimizadeh H, Marhaban MH, Kamil RM, Ismail NB (2009) Color image segmentation based on Bayesian theorem and kernel density estimation. Eur J Sci Res 26(3):430–436
19. Suykens JAK, Vandewalle J (1999) Least squares support vector machine classifiers. Neural Process Lett 9(3):293–300
20. Ye J, Xiong T (2007) SVM versus Least Squares SVM. In: Proceedings of the eleventh international conference on artificial intelligence and statistics (AISTATS 2007), San Juan, Puerto Rico, USA, 21–24 Mar, pp 640–647

Chapter 5
Switched Unfalsified Multicontroller Nonparametric Model Based Design

Fernando Coito, Luís Brito Palma, and Fernando Costa

Abstract In this paper, we present a controller design strategy for the implementation of a multicontroller structure for single-input single-output (SISO) plants. The overall control system can be viewed as a feedback interconnection of a SISO plant, a set of candidate controllers and a switched selection scheme that supervises the switching process among the candidate controllers. The switching scheme is designed without explicit assumptions on the plant model, based on the unfalsified control concept. A switched multicontroller structure is implemented and experimental results are presented.

5.1 Introduction

Dealing with nonlinear systems is an inherently difficult problem. As a consequence models and analysis of nonlinear systems will be less precise than for the simpler linear case. Thus, one should look for model representations and tools that utilize less precise system knowledge than the traditional approaches. This is indeed the trend in the area of intelligent control where a range of approaches, such as Fuzzy Logic, Neural Networks and Probabilistic Reasoning are being explored [1, 2]. The current paper uses operating regime decomposition for the partitioning of the operating range of the system in order to solve modeling and control problems.

The operating regime approach [3–6] leads to multiple-model or multiple controller synthesis, where different local models or controllers are used under different operating conditions (see Fig. 5.1). A supervisory mechanism selects the controller according to the current local region in which the process is operating.

F. Coito (✉) • L.B. Palma • F. Costa
Departamento de Engenharia Electrotécnica, Faculdade de Ciências e Tecnologia, UNL, 2829-516 Caparica, Portugal
e-mail: fjvc@fct.unl.pt; lbp@fct.unl.pt; f.martinscosta@hotmail.com

A. Madureira et al. (eds.), *Computational Intelligence and Decision Making: Trends and Applications*, Intelligent Systems, Control and Automation: Science and Engineering 61, DOI 10.1007/978-94-007-4722-7_5, © Springer Science+Business Media Dordrecht 2013

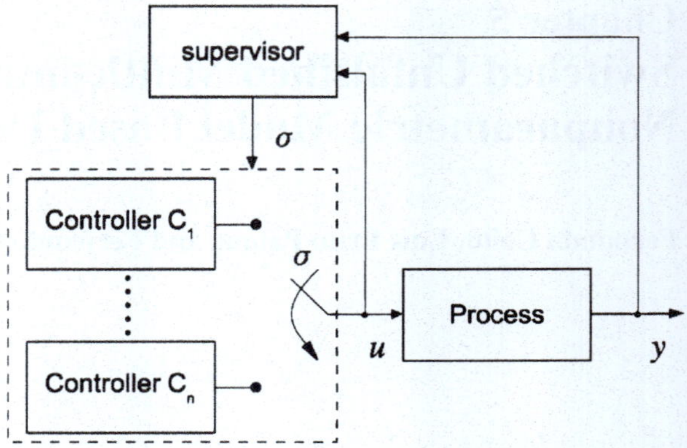

Fig. 5.1 Switching control. The switching decision between the controllers is performed trough the switching signal σ

In performance-based supervision the supervisor attempts to directly assess the potential performance of the candidate controllers, without estimating the process model [6–9].

Under the unfalsification concept the best controller among a set of candidate is selected straight from input/output data. The candidate controllers' performance is evaluated without actually applying them to the feedback loop. Controllers that do not yield the desired closed loop dynamics are entitled falsified. Only unfalsified controllers are candidate to actually control the process. The process is controlled by the controller presenting the best performance.

5.2 The Unfalsified Multicontroller

Control specifications are mapped into a closed-loop reference model W_m. Assume there is a set of controllers from which at least one is able to fulfill the control specifications. A controller is tagged as *falsified* if it is unable to comply with the specifications. Otherwise it is tagged *unfalsified*. The required information for controller evaluation consists in the past values from the set-point (r), the output (y) and the control action (u).

A further requirement is that each of the controllers (C_i) on the set must be "*causally-left-invertible*" – in the sense that the current value of $r_i(t)$ is uniquely determined by past values of $u(t)$ and $y(t)$. As proposed by Safonov [7], with the plant working under some stabilizing controller, controllers are evaluated according to the scheme presented in Fig. 5.2, where $\tilde{r}_i(t) = \tilde{r}_i(C_i, u_{\tau \leqslant t}, y_{\tau \leqslant t})$ is a fictitious set-point signal for controller C_i; $\tilde{y}_i(t)$ is a fictitious output signal corresponding to the output of model W_m, when the fictitious set-point signal is used. Proposed controllers are evaluated using performance index (5.1). If the performance index $V_i(t)$ is larger than a performance threshold γ, the controller is falsified by the available data at time t. Only unfalsified controllers are candidates to control the process.

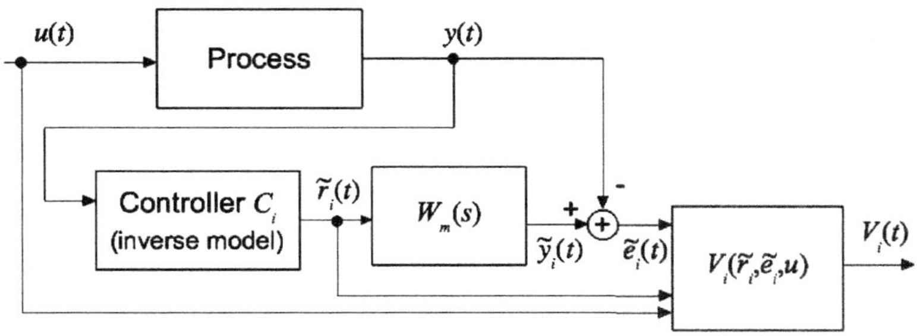

Fig. 5.2 Controller performance evaluation under unfalsified control framework

$$V(r,u,e,t) = \begin{cases} \dfrac{\|e(t)\|_t + \lambda\|u(t)\|_t}{\|r(t)\|_t} & \text{if} \quad \|r(t)\|_t \neq 0 \\ \infty & \text{if} \quad \|r(t)\|_t = 0 \end{cases} \tag{5.1}$$

The basic control selection approach is to choose the controller with the least performance index. It is necessary to implement a strategy limiting controller switch in order to avoid stability problems [10, 11].

5.3 Controller Design

The experimental results presented on this paper are obtained over a lab-scale heat/ventilation experiment (Fig. 5.3). The process consists of a fan that blows air through a tunnel. At the tunnel inlet an electrical resistor heats up the air. The air temperature is measured at the middle of the tunnel. Decreasing the fan speed leads to higher measured temperatures and slower dynamics. A higher fan speed has the opposite effect. Three operation regimes are considered according to fan speed: low, medium and high speed.

5.3.1 Controller Tuning

With low noise processes, by using experimental frequency response to characterize the dynamic behavior over selected operating points it is possible to obtain models (non-parametric) that are close to the process true behavior. From such models controllers may be designed by classical frequency based methods (Nyquist plot, lead-lag compensation, etc.), or by optimization over the frequency based algorithms [12]. The plant behavior is presented in Fig. 5.4, for each of the selected operating conditions.

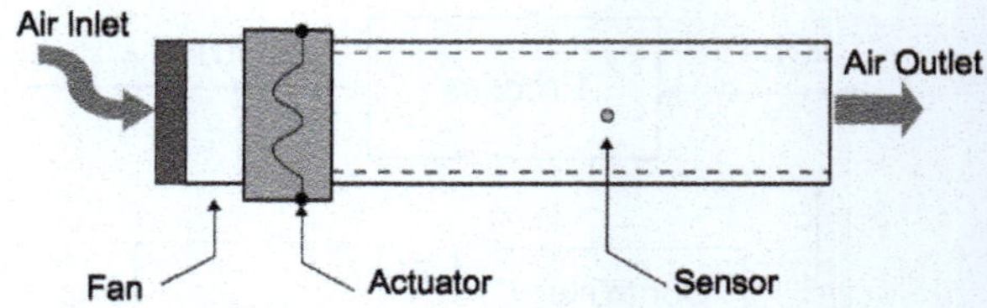

Fig. 5.3 Schematic of the lab-scale heat/ventilation experiment used for tests

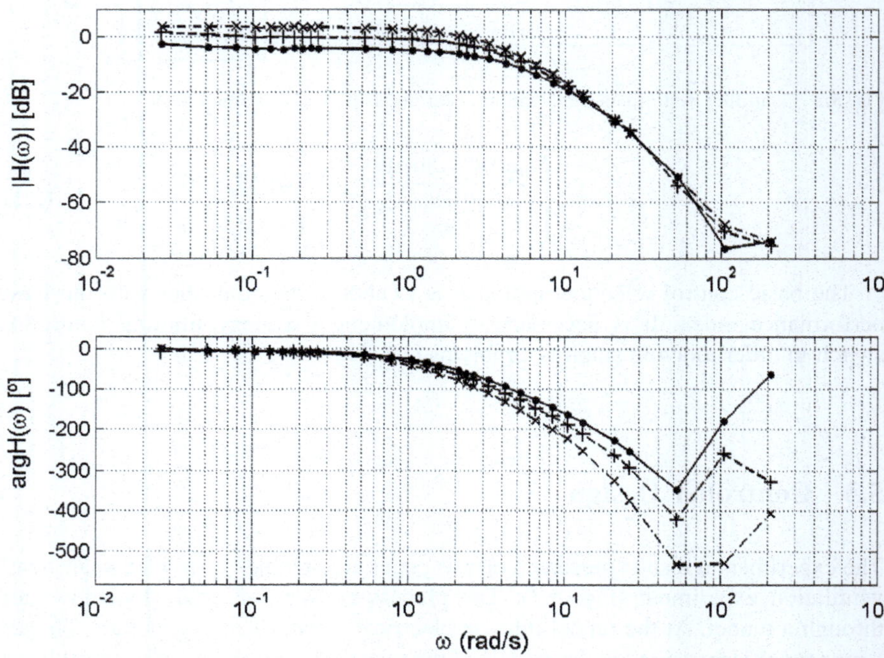

Fig. 5.4 Process experimental frequency response for low (-x-), medium (-+-) and high (-•-) fan speed

Each of the process models is used to design a closed-loop controller. The design goals are the same for all the pairs {controller, model}: (i) a closed-loop behavior yielding an overshoot of 3% and a rise-time under 1 s; (ii) small tracking error both in steady state and for frequencies under 0.3 rad/s; (iii) the control action is desired to have low high speed content.

To comply with these specifications, the controller is design through an optimization procedure. Let $H(\omega)$ be the process, $C(\omega)$ the controller, $G_{cl}(\omega)$ the designed closed-loop behavior and $G_{spec}(\omega)$ the desired closed-loop behavior. Functions (5.2), (5.3), and (5.4) are used to weight each of the design goals:

$$J_1 = \sum_{\omega_i < 20} \left| 1 - \frac{G_{cl}(\omega_i)}{G_{spec}(\omega_i)} \right| \tag{5.2}$$

Table 5.1 Controllers

Operation point	Controller
Low speed:	$\frac{39.2\left(1+\frac{s}{2.89}\right)^2}{\left(1+\frac{s}{0.04}\right)\left(1+\frac{s}{8.06}\right)}$
Medium speed:	$\frac{136\left(1+\frac{s}{2.29}\right)^2}{\left(1+\frac{s}{0.0154}\right)\left(1+\frac{s}{3.19}\right)}$
High speed:	$\frac{261\left(1+\frac{s}{3.02}\right)\left(1+\frac{s}{7.63}\right)}{\left(1+\frac{s}{0.016}\right)\left(1+\frac{s}{8.22}\right)}$

$$J_2 = \sum_{\omega_i<0.3} |1 - G_{cl}(\omega_i)| \tag{5.3}$$

$$J_3 = \sum_{\omega_i\in[1;5]} \left|\frac{H(\omega_i)}{1 + C(\omega_i)H(\omega_i)}\right| \tag{5.4}$$

The controller tuning optimizes the function $J = J_1 + J_2 + J_3$, with the constraint that the overall closed-loop system must be stable. Table 5.1 presents the results from the design procedure.

5.3.2 Single Controller Performance

In Fig. 5.5 experimental step responses for each of the operation points are compared with simulation results. The figure illustrates the performance achieved by applying the correct (tuned) controller. In each of the tests the use of a different fan speed leads to different process dynamic. Nevertheless, the use of an operation point tuned controller produces similar closed-loop performance.

To illustrate the need for multiple controllers a set of tests is performed, in which each of the designed controllers is applied to the process at one of the operating conditions, and the closed-loop step response is evaluated. This procedure is repeated for the three operation points.

Some relevant results from this set of tests are presented in Fig. 5.6. The tests show that using a controller tuned for a fan speed higher than the operating condition leads to increased overshoot and oscillations. In fact, the use of a high-speed tuned controller when the fan is working at low speed may even lead to an

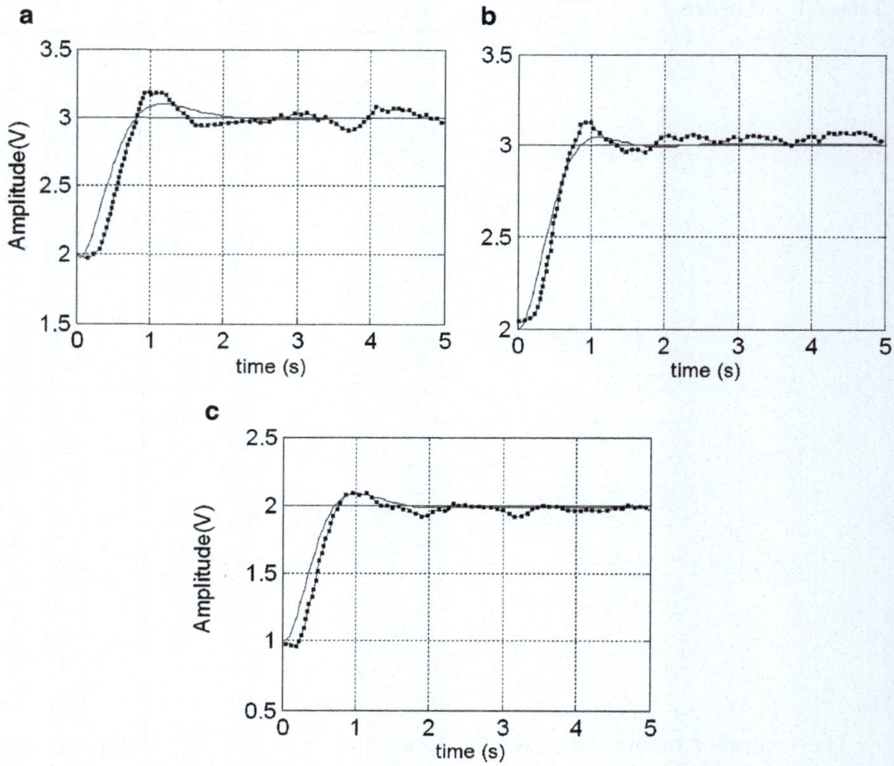

Fig. 5.5 Experimental closed-loop step response at different operating conditions, using the tuned controller: experimental result (-•-); expected behavior (—). (**a**) low fan speed; (**b**) middle fan speed; (**c**) high fan speed

unstable closed-loop behavior, as illustrated in Fig. 5.6b. The full set of tests performed on the plant shows that it is not possible to achieve a performance close to the desired with a single fixed-parameter linear controller.

5.4 Implementation and Test

A switched controller structure, combining the set of tuned controllers and an unfalsified based supervisor, is applied to the heat/ventilation experiment. Figure 5.7 shows the result from a test in which the fan speed is smoothly increased from the minimum to the middle speed range. The air temperature set-point corresponds to a 0.04 Hz square signal, with amplitude 1 V and offset 2.25 V.

While the temperature signal (Fig. 5.7a) follows the set-point, with no significant change on the performance, the control action is increasing (Fig. 5.7c). This stems

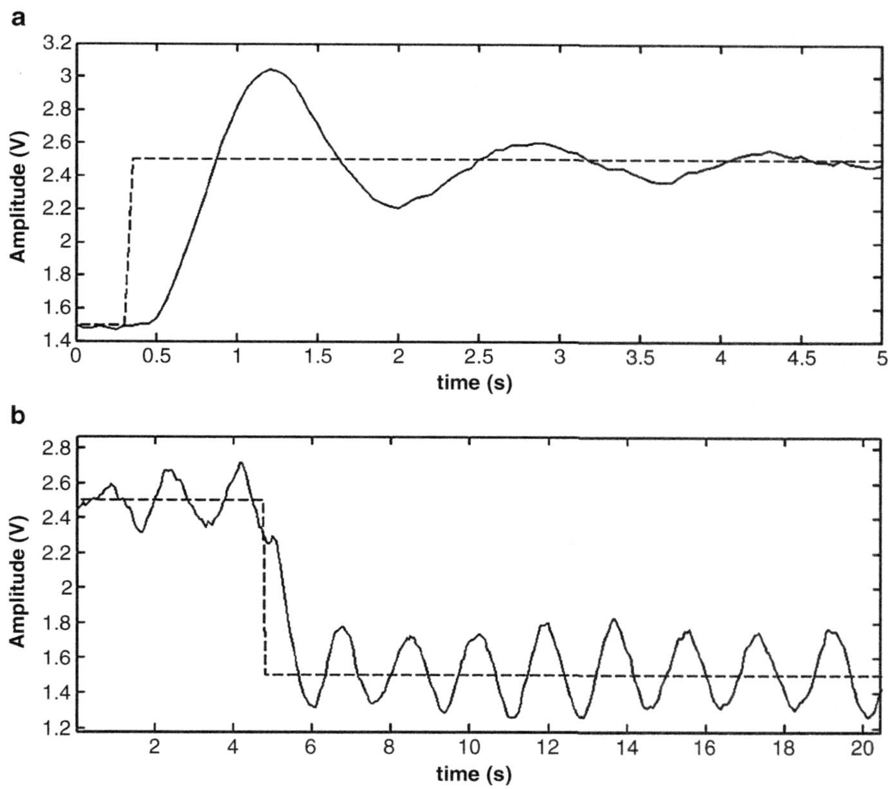

Fig. 5.6 Closed-loop step response at different operating conditions: (**a**) fan at middle speed with a high speed tuned controller; (**b**) fan at low speed with a high speed tuned controller

from the decrease of the process gain as the fan speed increases, requiring higher heating power (control action) for the same set-point. The test is stopped when the control action reaches the maximum value (5 V). To perform the test at a higher fan speed range the set-point offset needs to be reduced.

Figure 5.7b shows the three performance indexes (V_i), used for the selection of the active controller. As the fan speed increases the performance indexes relative positions change. At the beginning of the test the low fan speed index is the smallest and the high fan speed is the highest. Close to moment 40 s there is a change in the relative position and the middle fan speed becomes the smallest one. Shortly after the index crossing, the selected controller commutes from the one tuned for low fan speed to the controller tuned for middle fan speed, maintaining the systems closed-loop performance.

Figure 5.8 presents the results for a similar test, where the fan speed is slowly decreasing. The set-point signal is similar to the previous test, but its offset is 1.2 V. The fan starts at high speed and the test stops when the control action reaches the minimum allowed value (0 V). The controller selection changes at moment 48.8 s.

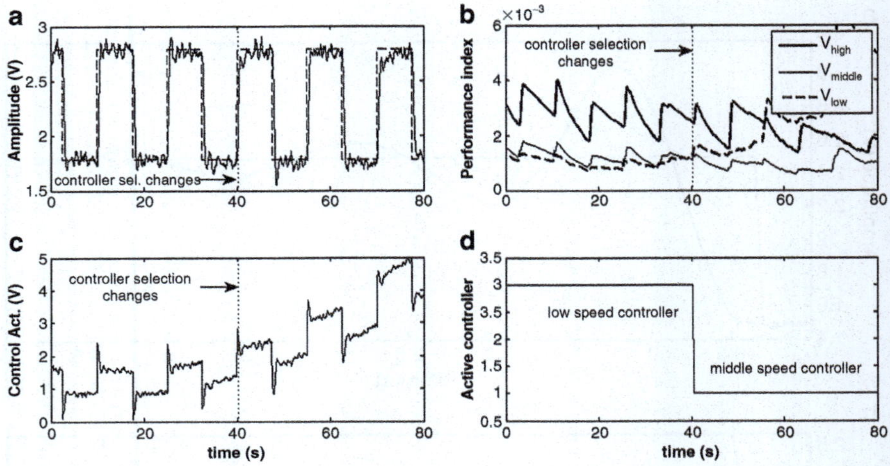

Fig. 5.7 Supervisory control test: (**a**) process output, (**b**) performance indexes, (**c**) control action and (**d**) controller selection. The fan speed is slowly increased from the minimum to the middle speed range

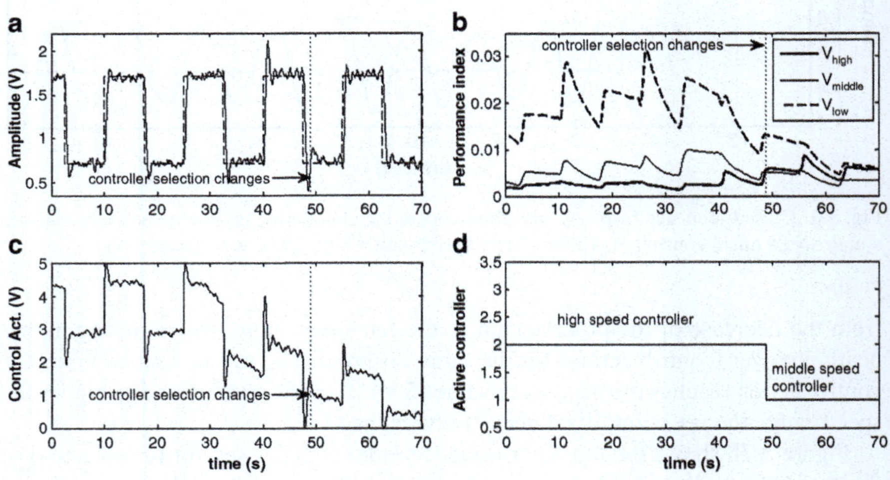

Fig. 5.8 Supervisory control: (**a**) process output, (**b**) performance indexes, (**c**) control action and (**d**) controller selection. The fan speed is slowly decreased from high to the middle speed range

The closed-loop performance is similar to the previous test except for the two steps closer to the selector change, which present significant overshoot. Observing Fig. 5.8b, it is noticed that, within the time period corresponding to these steps, the performance indexes V_{high} and V_{middle} are close to each other.

Figure 5.9 shows a test where the operation point suddenly changes from middle fan speed to high fan speed range. This change occurs at moment 21.3 s. This causes

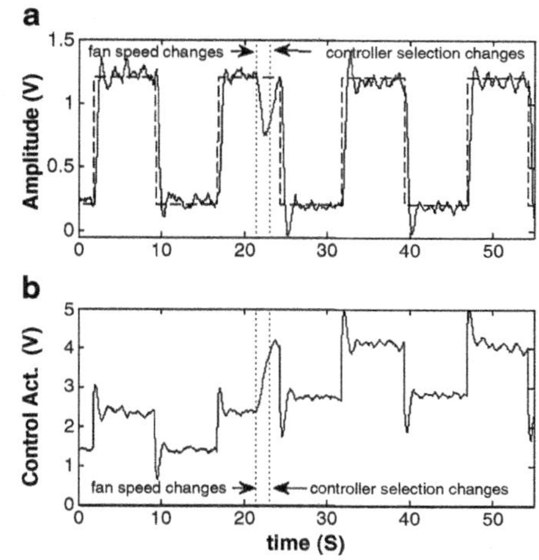

Fig. 5.9 Supervisory control: Process output and the control action. The fan speed is suddenly increased from the middle speed range into high speed

a large decrease on the process output. The controller reacts with a strong increase on the heating power signal. Within 3 s the output temperature signal is back to the value previous to the fan speed step change.

It takes 1.8 s for the supervisor to react and adjust the controller selection.

5.5 Conclusions

The switched multicontroller structure described shows to present good performance and fast reaction to modifications on the operating conditions.

An important feature is that no previous knowledge on the plant dynamics is required to implement the unfalsified control-switching scheme. As the performance evaluation algorithm does not require a specific controller structure, it can be used with a broad range of controllers, and its possible to combine controllers with different structures into a single switched multicontroller.

Once the control-switching scheme requires no process model, an interesting development is to design the controllers without using any process parametric models. In the paper, this is done through the use of experimental frequency response to characterize the dynamic behavior over the operating range.

References

1. Palma LB, Coito FJ, Neves-Silva RA (2005) Robust fault diagnosis approach using analytical and knowledge-based techniques applied to a water tank system. Int J Eng Intell Syst Electr Eng Commun 13:237–244
2. Feng G (2006) A survey on analysis and design of model-based fuzzy control systems. IEEE Trans Fuzzy Syst 14(5):676–697. doi:10.1109/TFUZZ.2006.883415

3. Hespanha JP (2001) Tutorial on supervisory control. In: Lecture notes for the workshop control using logic and switching, 40th conference on decision and control, Orlando, FL. http://www.ece.ucsb.edu/~hespanha/published
4. Deshpande AP, Patwardhan SC (2008) Online fault diagnosis in nonlinear systems using the multiple operating regime approach. Ind Eng Chem Res 47:6711–6726. doi:10.1021/ie071593q
5. Johanson TA, Murray-Smith R (1997) The operating regime approach. In: Johanson TA, Murray-Smith R (eds) Multiple model approaches to modelling and control. Taylor and Francis, London
6. Costa F, Coito F, Palma L (2011) Switched unfalsified multicontroller. IFIP Adv Inf Commun Technol 349:393–401, Technological Innovation for Sustainability
7. Safonov MG, Tsao TC (1997) The unfalsified control concept and learning. IEEE Trans Autom Control 42:843–847
8. Safonov MG, Cabral FB (2001) Fitting controllers to data. Syst Control Lett 43:299–308
9. Wang R, Safonov MG (2005) Stability of unfalsified adaptive control. In: Proceedings of the 2005 American control conference, Portland, pp 3162–3167
10. Hespanha JP, Morse AS (1999) Certainty equivalence implies detectability. Syst Control Lett 36:1–13
11. Morse AS (1997) Supervisory control of families of linear set-point controllers—part 2: robustness. IEEE Trans Autom Control 42:1500–1515. doi:10.1109/9.649687
12. Coito F, Ortigueira MD (2008) Fractional controller design trough multi-objective optimization. In: Proceedings of the 8th Portuguese conference on automatic control – CONTROLO'2008, Vila Real

Chapter 6
Evolving Fuzzy Uncalibrated Visual Servoing for Mobile Robots

P.J.S. Gonçalves, P.J.F. Lopes, P.M.B. Torres, and J.M.R. Sequeira

Abstract In this paper, fuzzy models obtained on-line and off-line for uncalibrated visual servoing, are proposed and validated for an unicycle mobile robot. This approach will recursively update the inverse fuzzy model based only on measurements, at a given time instant. The uncalibrated approach does not require calibrated kinematic and camera models, as needed in classical visual servoing to obtain the Jacobian. Experimental results obtained in an unicycle mobile robot performing eye-in-hand visual servoing are used to demonstrate the validity of the proposed approach, when compared to the previous developed off-line learning.

6.1 Introduction

Robots use sensor-based control to perform tasks, either in structured or unstructured environments. Vision sensors provide a vast amount of information on the environment in which robots move. Thus, vision is essential for robots working in unstructured environments. These types of sensor definitely enlarge the potential applications of the actual robots, mobile or manipulators, and are massively used. Visual servoing can be defined as a method to control dynamic systems using the information provided by visual sensors. In this paper, the control of a unicycle [1] mobile robot using a single camera attached to the robot (eye-in-hand) is addre-ssed [2].

P.J.S. Gonçalves (✉)
School of Technology, Polytechnic Institute of Castelo Branco,
Av. Empresário, 6000-767 Castelo Branco, Portugal

IDMEC/LAETA, Technical University of Lisbon (TU Lisbon),
Av. Rovisco Pais, 1049-001 Lisbon, Portugal
e-mail: paulo.goncalves@ipcb.pt

P.J.F. Lopes • P.M.B. Torres • J.M.R. Sequeira
School of Technology, Polytechnic Institute of Castelo Branco,
Av. Empresário, 6000-767 Castelo Branco, Portugal
e-mail: pedrotorres@ipcb.pt; sequeira@ipcb.pt

A. Madureira et al. (eds.), *Computational Intelligence and Decision Making: Trends and Applications*, Intelligent Systems, Control and Automation: Science and Engineering 61,
DOI 10.1007/978-94-007-4722-7_6,

Early approaches for visual servoing are based on a model already known, i.e. the robot-camera model, from which the relation between the image features and the robot kinematics is analytically obtained [2]. Apart from the stated approaches where the robot-camera model is already known, it can also be estimated [3]. This type of systems, *Uncalibrated Visual Servoing*, can deal with unknown robot kinematics and/or unknown camera parameters. By using this type of robot-camera models, the system becomes independent of robot type, camera type or even camera location. Several approaches claim to use uncalibrated visual control on a mobile robot, but the only uncalibrated part are the cameras as for example in [4].

In this paper, the robot-camera model estimation by learning is addressed both on-line and off-line, using fuzzy techniques to obtain a controller capable of controlling the robot. An inverse fuzzy model is used to derive the inverse robot-camera model, i.e. the Jacobian, in order to compute the wheel velocities in a straightforward manner. The inverse fuzzy model is applied directly as a controller, which is a simple way to implement a controller in real-time. Note that this feature is very important in robotic systems.

From the modeling techniques based on soft computing, fuzzy modeling is one of the most appealing. As the mobile robot together with the visual system is a nonlinear system, which is only partly known, it is advantageous to use fuzzy modeling as a way to derive a model (or an inverse model as in this case) based only on measurements. In this paper will be applied and compared two approaches, off-line [3], and on-line [5]. The first approach has already proven to perform well in visual servoing systems. Its main drawback is that the learning must be performed off-line. Since the environment in which the robot operates can vary importantly, new models should be derived very often, which take time and also the robot must stop the task that is performing. That is why an on-line modeling of the visual servoing system is preferable. This approach was not yet applied to control an unicycle mobile robot, using vision, which is the objective of the present paper.

The paper is organized as follows. Section 6.2 describes briefly the concept of visual servoing and presents the uncalibrated visual servoing approach. Section 6.3 presents very briefly fuzzy on-line and off-line modeling, and also discusses the problem of identifying fuzzy models directly from data measurements. Section 6.4 describes the experimental setup and presents the obtained results, where the identified fuzzy model is discussed. Finally, Sect. 6.5 presents conclusions and the future work.

6.2 Visual Servoing

Visual information is nowadays extensively used in robotic applications. The visual information from the work environment can be used to control robots, mobile and manipulators, performing a given task. In the following sub-sections, both classical and uncalibrated image-based visual servo control are presented.

Fig. 6.1 Coordinate frames of a generic unicycle robot

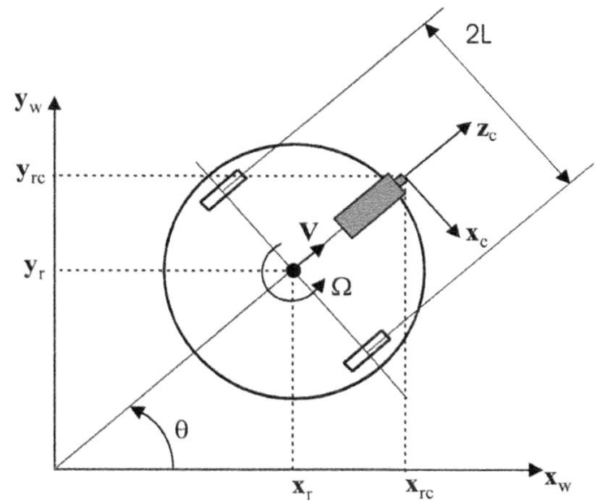

6.2.1 *Image-Based Visual Servoing*

In this paper, the image-based visual servoing system used is the eye-in-hand system [2], where the camera is attached to the mobile robot and looking the environment. The kinematic modeling of the transformation between the image features velocities and the wheel velocities must be found. In image-based visual servoing the image feature is a 2D point in the image. The kinematic modeling, according to Fig. 6.1 of the transformation between the image features velocities $\dot{S}$ and the wheel velocities V, for a generic unicycle robot, is defined as follows [6, 7]:

$$\dot{s} = J_i \cdot {}^{c}T_{rc} \cdot {}^{rc}T_r \cdot V \tag{6.1}$$

where J_i is the image Jacobian, defined as:

$$J_i = \begin{bmatrix} -\frac{1}{z} & \frac{x}{z} & -(1+x^2) \\ 0 & \frac{y}{z} & -x \cdot y \end{bmatrix} \tag{6.2}$$

x and y are the metric coordinates of the features in the 2D image space, Z is the depth between the camera and object frames, ${}^{c}T_{rc}$ is defined as the transformation between the camera frame world coordinates and the camera optical axis, and ${}^{rc}T_r$ is the transformation between the robot and the camera frame world coordinates. The robot wheel velocities V, needed to move the robot to a predefined point in the image, s^* are derived from the use of the *Visual Control Law*. To achieve this goal, the error between the desired and the actual image features position in the image, should be zero. The image features velocity at the desired position should also be zero, to guarantee that the mobile robot stops. Thus, the image features error velocity is related to the image features velocity as follows:

$$\mathrm{e} = s - s^* = \dot{s} \tag{6.3}$$

In this paper, an exponential decayment of the image features error during the servoing is chosen to be applied, as suggested in [2]:

$$\dot{e} = -K_p \cdot e \tag{6.4}$$

where K_p is a positive gain, that is used to increase or decrease the decayment of the velocity error. From (6.1), (6.3) and (6.4), the proportional visual control law is defined as [7]:

$$\dot{q} = -K_p \cdot (J_i \cdot {}^{c}T_{rc} \cdot {}^{rc}T_r \cdot)^{-1} \cdot (s - s*) \tag{6.5}$$

6.2.2 Uncalibrated Visual Servoing

To derive an accurate Jacobian, $J = J_i \cdot {}^{c}T_{rc} \cdot {}^{rc}T_r$, a perfect modeling of the camera, the chosen image features, the position of the camera related to the world, and the depth of the target related to the camera frame must be accurately determined. Even when a perfect model of the Jacobian is available, it can contain singularities, which hampers the application of a control law. Remind that the Jacobian must be inverted to send the camera wheel velocities to the robot. When the Jacobian is singular, the control cannot be correctly performed.

There are visual servo control systems that obviate the calibration step and estimate the robot-camera model. The robot-camera model may be estimated: *analytically*, using nonlinear least square optimization [8], and by *learning* or training, using fuzzy membership functions and neural networks [3, 9]. In addition, the control system may estimate an image Jacobian and use the known robot model, or a coupled robot-camera Jacobian may be estimated. To overcome the difficulties regarding the Jacobian, a differential relationship between the features and the camera, embarked in the robot, velocities was proposed in [9]. This approach states that the robot position P variation depends on the image features variation and the previous position of the robot:

$$\delta P(k) = F_k^{-1}(\delta s(k+1), P(k)) \tag{6.6}$$

In image-based visual servoing, the goal is to obtain a wheel velocity, $\delta P(k)$, capable of driving the robot according to a desired image feature position, $s(k+1)$, with an also desired image feature error, $\delta s(k+1)$, from any position in the feature space. This goal can be accomplished by modeling the inverse function F_k^{-1}, using inverse fuzzy modeling as proposed in this paper and presented in Sect. 6.3. This new approach to image-based visual servoing allows to overcome the problems stated previously regarding the Jacobian and the calibration of the robot-camera model. It was developed for a robot manipulator in [3], and it will be applied to an unicycle mobile robot in this paper.

6.3 Fuzzy Modeling

Fuzzy modeling often follows the approach of encoding expert knowledge expressed in a verbal form in a collection of if—then rules, creating the model structure. Parameters in this structure can be adapted using input–output data. When no prior knowledge about the system is available, a fuzzy model can be constructed entirely on the basis of systems measurements. In the following, we consider data-driven modeling based on fuzzy clustering. For the robot under control, the direct robot-camera model states that the image features move in the image space with variation $\delta s(k+1)$, given a predefined robot velocity $V(k)$ and knowing its actual position $P(k)$ in the world coordinate frame. In this paper, rule-based models of the Takagi-Sugeno (TS) type are considered.

6.3.1 *Off-Line Fuzzy Modeling*

In this paper is considered the rule-based model of the Takagi-Sugeno (TS) type [10]. The rules of a TS fuzzy model describe local input–output relations, typically in a linear form:

$$\begin{aligned} R_i:\ & \textbf{\textit{IF}}\, x_1 \text{ is } A_{i1} \textbf{\textit{ AND }} \cdots \textbf{\textit{ AND}}\, x_n \text{ is } A_{in} \\ & \textbf{\textit{THEN}}\, y_i = a_i x + b_i,\ i = 1, 2,\ \ldots, K \end{aligned} \tag{6.7}$$

Here R_i is the ith rule, $x = [x_1, \ldots, x_n]^T$ is the input (antecedent) variable, $A_{i1}, \ldots, A_{in}$, are fuzzy sets defined in the antecedent space, and y_i is the rule output variable. K denotes the number of rules in the rule base. The aggregated output of the model, $\hat{y}$, is calculated by taking the weighted average of the rule consequents:

$$\hat{y} = \frac{\sum_{k=1}^{K} \beta_i y_i}{\sum_{k=1}^{K} \beta_i} \tag{6.8}$$

where β_i is the degree of activation of the ith rule:

$$\beta_i = \prod_{j=1}^{n} \mu_{A_{ij}}(x_j),\ i = 1, 2, \ldots, K \tag{6.9}$$

and $\mu_{A_{ij}}(x_j) : R \rightarrow [0, 1]$ is the membership function of the fuzzy set A_{ij} in the antecedent of R_i. To identify the model in (6.7), the procedure presented in [3] is used to obtain the antecedents and consequents of the fuzzy rules.

6.3.2 On-Line Fuzzy Modeling

In On-Line Fuzzy Modeling and according to [5], also rule-based models of the TS type, are considered. Typically in the affine form, described in (6.7), where the input–output data is acquired continuously. The new data, arriving at some time instant, can bring new information from the system, which could indicate a change in its dynamics. This information may change an existing rule, by changing the spread of the membership functions, or even introduce a new one. To achieve this, the algorithm must be able to judge the informative potential and the importance of the new data. In the following are briefly presented the several steps of the algorithm used for on-line fuzzy modeling, proposed in [5], *evolving fuzzy systems*. The first step is based on the subtractive clustering algorithm, where the input–output data is partitioned. The procedure used must be initialized, i.e. the focal point of the first cluster is equal to the first data point and its potential is equal to one. Starting from the first data point, the potential of the next data point is calculated recursively using a Cauchy type function of first order:

$$P_k(Z_k) = \frac{1}{1 + \frac{1}{k-1}\sum_{i=1}^{k-1}\sum_{j=1}^{n+1}\left(d_{ik}^{j}\right)^2}, \quad k = 2, 3, \ \ldots \tag{6.10}$$

where $P_k(Z_k)$ denotes the potential of the data point z_k calculated at time k; $d_{ik}^{j} = z_i^j - z_k^j$, denotes projection of the distance between two data points (z_i^j and z_k^j) on the axis z^j.

When a new data point arrives it also influences the potential of the already defined center of the K clusters, $(z_i^*, i = 1, 2, \ldots, K)$.

A recursive formula for the update of the cluster centers potential is defined in [5]:

$$P_k(z_i^*) = \frac{(k-1)P_{(k-1)}(z_i^*)}{k - 2 + P_{(k-1)}(z_i^*) + P_{(k-1)}(z_i^*)\sum_{j=1}^{n+1}\left(d_{ik}^{j}\right)^2} \tag{6.11}$$

where $P_k(z_i^*)$ is the potential at time k of the cluster center, related to the rule i.

The next step of the algorithm is to compare the potential of the actual data point to the potentials of the existing cluster centers. A crucial part of the evolving fuzzy systems is the rule creation and modification. In this approach, if the potential of a new data point is higher than the potential of the existing cluster centers, then the new data point is accepted as a new cluster center and a new rule is formed. If in addition to the previous condition the new data point is close to an old cluster center, the old cluster center is replaced. This last condition is defined by:

$$\frac{P_k(z_k)}{max_K\, P_k(z_i^*)} - \frac{\delta_{min}}{r} \geq 1 \tag{6.12}$$

where $r \in [0.3; 0.5]$ is the spread of the antecedent [5], and δ_{min} is the shortest distance between the new candidate Z_k and all the existing cluster centers z_i^*. The fuzzy sets in the antecedent of the rules are Gaussian, with the form:

$$\mu_{ij} = e^{-r\|x_j - x_{ij}^*\|^2} \tag{6.13}$$

The consequents of the fuzzy rules are obtained using the global parameter estimation procedure based on the weighted recursive least squares, presented in [5].

6.3.3 *Visual Servo Fuzzy Modeling*

For the robotic application in this paper, the inverse model is identified using input–output data from the inputs V(k) and the outputs δs(k + 1), adapting the procedure described in [3]. In this paper, the training set is obtained while moving the robot in the track of the Portuguese Robotics Open competition, Autonomous Driving, Fig. 6.2a. Note that we are interested in the identification of an inverse model as in (6.6).

6.4 Experimental Results

To validate the proposed approach, the KIKA robot [7] is used with a standard webcam in eye-in-hand configuration. The experimental apparatus is presented in Fig. 6.2b.

6.4.1 *Fuzzy Modelling Results*

To obtain the data for model identification, the robot must move in the competition track, Fig. 6.2a. The goal of the robot was to follow the right lane of the track. For that, the image feature chosen was the middle of the line segment identified in the right lane. This line segment was obtained from the image taken with the webcam, after binarization, in the row with height 300, following the work in [11].

The variables needed for identification, $V(k)$ and δs(k + 1), are directly obtained from the wheel velocity sensors and the image, respectively. δs(k + 1), is determined using (6.14), where the desired position of the right lane in the images was set to $s^* = (470, 300)$.

$$\delta s(k+1) = s * - s(k+1) \tag{6.14}$$

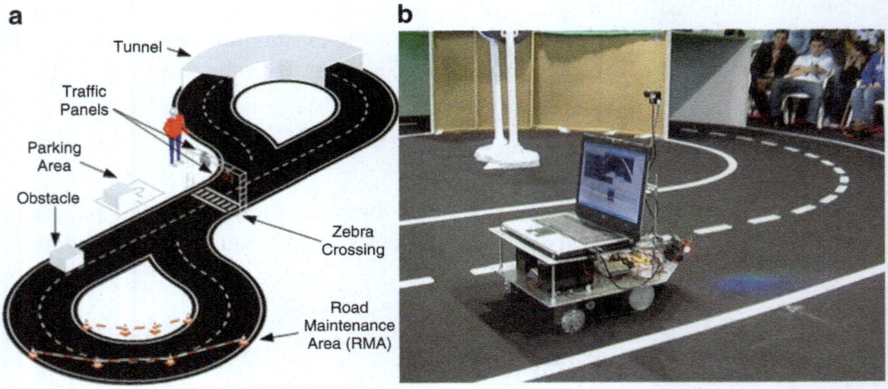

Fig. 6.2 (**a**) Portuguese Robotics Open – Autonomous Driving Track. (**b**) The KIKA robot at the 2010 Portuguese Robotics Open

Table 6.1 Results obtained for each wheel velocity, both for off-line and on-line fuzzy models

	Wheel	Rules	VAF (%)	MSE [mm^2]
Off-line	Right	6	99.876	0.029
	Left	6	99.121	0.006
On-line	Right	28	99.535	0.105
	Left	29	93.990	0.041

To estimate the modeling accuracy VAF (Variance Accounted For) [3] was used, and MSE (Mean Squared Error). A perfect match occurs, when VAF is 100\% and MSE is 0.

Table 6.1 presents VAF and MSE for the fuzzy modeling, on-line and off-line. The off-line fuzzy model obtained with six clusters showed the best results of VAF and MSE, meaning that the best model for estimating the wheel velocities, right V_r and left V_l, is obtained from the off-line approach. The on-line approach presented worst results because the model is being updated at each sample time, i.e., it uses less information to construct the model at each time step, not coping with all the data as the off-line approach does. This induces wrong rules, despite the fact that the on-line system constructed 28/29 rules, a large number compared to the off-line approach. The input data, 2D point, is depicted in Fig. 6.3a. Note that the y coordinate has always the same value, because only line 300 is used to detect the line segment on the right lane, as previously defined. Figures 6.3b and 6.4 presents the results of the modelling, the output of the models follows closely the real data used to train the model.

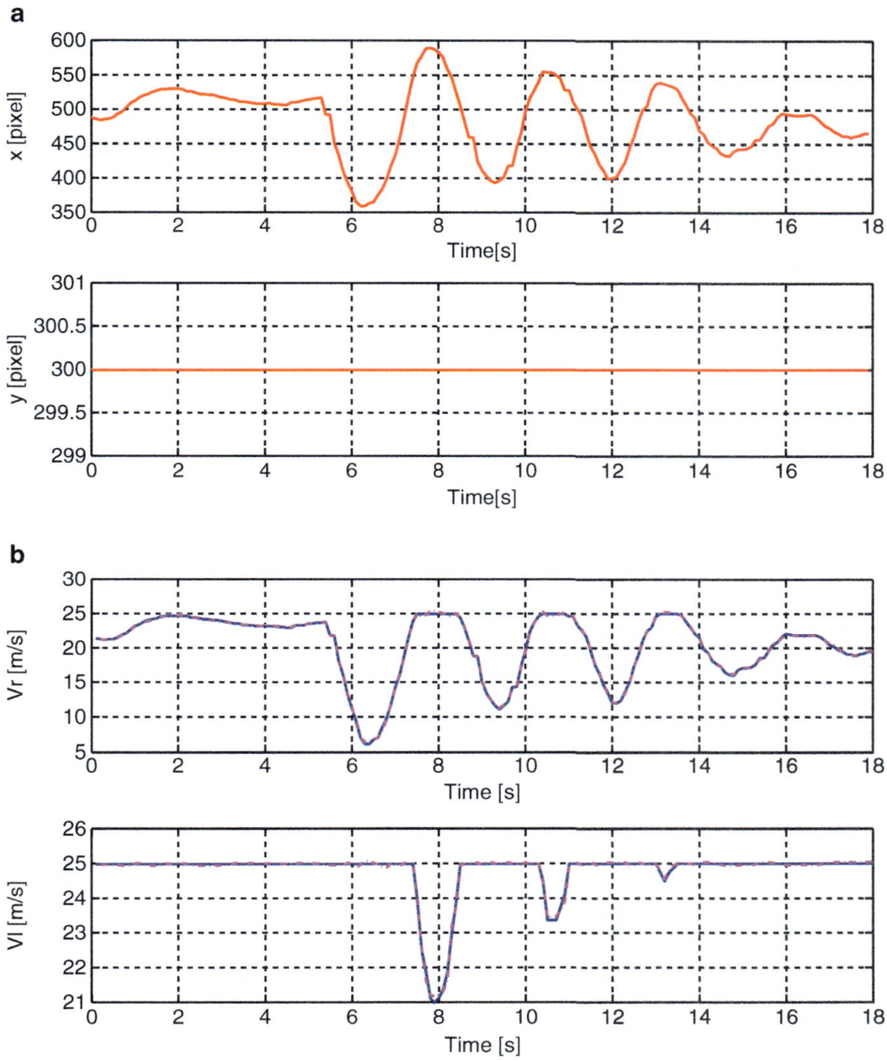

Fig. 6.3 Input – Output data for fuzzy modeling. (**a**) Input data for fuzzy modeling – 2D point. x and y coordinate. (**b**) Fuzzy modeling results – wheel velocities Vr and Vl. Solid: real output data, and *dash-dotted*: output of the inverse fuzzy model

6.4.2 Control Results

This section presents the obtained control results, using the fuzzy model based control, i.e., using the obtained inverse model directly as the controller. In this section it is also presented the output of the classical image-based visual servoing approach, for comparison with the off-line fuzzy approach.

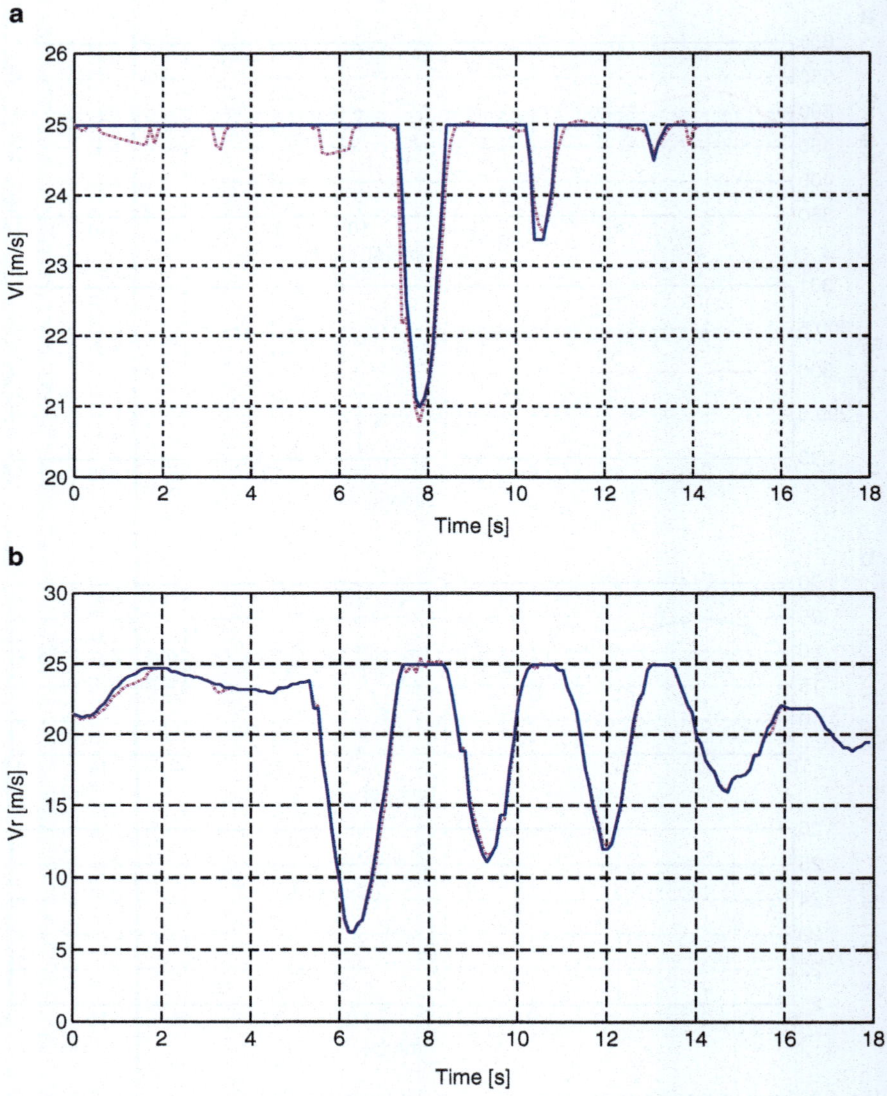

Fig. 6.4 Evolving fuzzy modeling results. *Solid*: real output data, and *dash-dotted*: output of the inverse fuzzy model. (**a**) wheel velocities V_l. (**b**) wheel velocities V_r

Figure 6.5a depicts the error of the x coordinate of the feature point, for the image-based and for the fuzzy uncalibrated visual servoing. Both approaches show an oscillatory behavior but the fuzzy approach presents a MSE = 36.2056 [pixel^2] better than the MSE = 45.7667[pixel^2] for the classical visual servoing. The velocities, in both approaches, also present oscillatory behavior, as depicted in Fig. 6.5b. The oscillatory behavior is due to the high vision sample time present in the experimental setup.

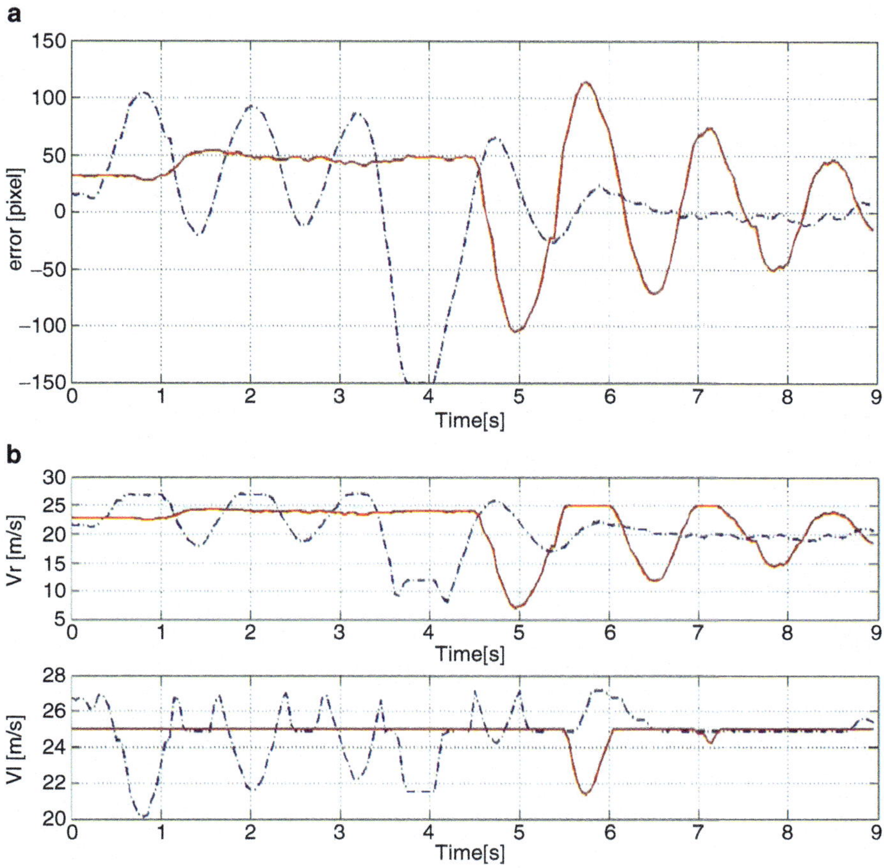

Fig. 6.5 Fuzzy uncalibrated visual servoing control results. *Solid*: image-based visual servoing, and *dash-dotted*: Fuzzy uncalibrated visual servoing. (**a**) Feature error, 2D point x coordinate. (**b**) Wheel velocities

6.5 Conclusions and Future Work

This paper introduces an evolving uncalibrated inverse controller to eye-in-hand image-based visual servoing, based on fuzzy modelling, to unyclicle mobile robots. The fuzzy modelling approach was applied to obtain two inverse models, one for each wheel, of the mapping between image features variations and wheel velocities, the first obtained with off-line data and the second with on-line data arriving at each time step. The off-line inverse model is directly used as controller of an unyclicle mobile robot performing visual servoing, due to the six rules of the model when compared with 28 and 29 rules, for each on-line model. The mapping between image features variations and wheel velocities obtained by the inverse fuzzy modelling promotes a new approach to the classic Jacobian modelling. This approach is independent from the robot's kinematic model or camera calibration, and also avoids the necessity of inverting on-line the Jacobian. The obtained experimental results showed that the fuzzy controller can successfully drive the robot in the track, which demonstrates the usefulness of the uncalibrated approach. This approach has also the great advantage of the needless calibration of the overall system. The obtained results show that the proposed approach outperforms the classical image-based visual servoing.

As future work, the proposed evolving inverse fuzzy model based control scheme need to decrease the number of rules generated, to allow its usage in real-time control of the robot.

Acknowledgments This work was partly supported by the Strategic Project, PEst-OE/EME/LA0022/2011, through FCT (under IDMEC-IST, Research Group: IDMEC/LAETA/CSI), FCT project PTDC/EME-CRO/099333/2008 and FP7 project ECHORD.

References

1. Mariottini GL, Prattichizzo D, Oriolo G (2007) Image-based visual servoing for nonholonomic mobile robots using epipolar geometry. IEEE Trans Robot 23(1):87–100
2. Hutchinson S, Hager G, Corke P (1996) A tutorial on visual servo control. IEEE Trans Robot Autom 12(5):651–670
3. Gonçalves PJS, Mendonça LF, Sousa JMC, Caldas Pinto JR (2008) Uncalibrated eye-to-hand visual servoing using inverse fuzzy models. IEEE Trans Fuzzy Syst 16(2):341–353
4. Dixon WE, Dawson DM, Zergeroglu E, Behal A (2001) Adaptive tracking control of a wheeled mobile robot via an uncalibrated camera system. IEEE Trans Syst Man Cybern B Cybern 31(3):341–352
5. Angelov P, Filev D (2004) An approach to online identification of takagi-sugeno fuzzy models. IEEE Trans Syst Man Cybern B Cybern 34(1):484–498
6. Gonçalves PJS, Correia LMPM (2005) Controlo em tempo-real, baseado em visão, de um robô móvel. In: Proceedings of Engenharia' 2005, Covilhã, Portugal, 21–23 Nov 2005
7. Torres PMB, Gonçalves PJS, Simões R, Santos V, Albardeiro D, Afonso N (2009) KIKA – Keep it Intelligent and Keep it Autonomous. In: Proceedings of Engenharia' 2009, Covilhã, Portugal, 25–27 Nov 2009
8. Peipmeier J, McMurray G, Lipkin H (2004) Uncalibrated dynamic visual servoing. IEEE Trans Robot Autom 20(1):143–147
9. Suh IH, Kim TW (1994) Fuzzy membership function based neural networks with applications to the visual servoing of robot manipulators. IEEE Trans Fuzzy Syst 2(3):203–220
10. Sousa JMC, Kaymak U (2002) Fuzzy decision making in modeling and control. World Scientific Pub. Co., River Edge, Sinagapore
11. Coulaud JB, Campion G, Bastin G, De Wan M (2006) Stability analysis of a vision-based control design for an autonomous mobile robot. EEE Trans Robot Autom 22(5):1062–1069

Chapter 7
Evaluating the Potential of Particle Swarm Optimization for Hyperspectral Image Clustering in Minimum Noise Fraction Feature Space

Shahin Rahmatollahi Namin, Amin Alizadeh Naeini, and Farhad Samadzadegan

Abstract Particle Swarm Optimization methods are the optimization techniques inspired by the social movements of animals. In these methods, the particles movements are based on simple rules, but make complex overall behavior and search of the space. The clustering is the process of dividing the existing data to diverse groups based on the inherent characteristics and similarity of them and can also be seen as an optimization problem. Due to the complexity of this problem in hyperspectral remotely sensed data, different feature space and clustering techniques are applied and evaluated in this field. In this paper, a PSO clustering algorithm is evaluated in Minimum noise fraction space for hyperspectral AVIRIS image taken over the northwest indiana's indian pine agricultural land. The comparison of the results with the K-means clustering method shows better obtained performance for the PSO clustering in minimum noise fraction feature space.

7.1 Introduction

The Particle Swarm Optimizations (PSO) are population-based tools with high potentiality to solve complex optimization applications [1]. These algorithms have global and local search mechanisms to reach to the optimum solutions iteratively. Currently there exists a great amount of literature investigating different aspects of these algorithms and evaluating their potential to solve complex optimization applications.

One of the complex applications that the optimization algorithms can be applied is clustering. Clustering is a method of partitioning a set of data into subsets, in a way that the data grouped together, are similar to each other and are diverse from the data in

S. Rahmatollahi Namin (✉) • A.A. Naeini • F. Samadzadegan
Department of Surveying Engineering, College of Engineering,
University of Tehran, Tehran, Iran
e-mail: rnamin@ut.ac.ir; a.alizadeh@ut.ac.ir; samadz@ut.ac.ir

A. Madureira et al. (eds.), *Computational Intelligence and Decision Making: Trends and Applications*, Intelligent Systems, Control and Automation: Science and Engineering 61,
DOI 10.1007/978-94-007-4722-7_7,

other clusters [2]. In this study, a PSO optimization algorithm is used in order to find the best cluster centers in a way that the accuracy of the clusters are optimized.

Despite the inherent complexities of the clustering that the algorithms should be dealing with, the clustering of the hyperspectral images bring other necessities which are the result of the high dimensionality and correlation of these data. Although the high spectral resolution in these images allows recognition of spectrally similar objects, the high dimensionality of these data and the resulting correlation between them decreases the performance and accuracy of the common clustering algorithms, because a great number of local optimum solutions exist. Hence, new clustering algorithms and different feature space selection are considered as an active research in this field.

In order to overcome the mentioned problems, the Minimum Noise Fraction (MNF) transformation is utilized in this research. This transformation, tries to decrease the dimensionality of the image space and minimize the noise influence in the data by maximizing the signal to noise ratio simultaneously [3]. This transformation is widely used for hyperspectral data; as they obtained results representing better performance in handling these images [4].

There exist different studies that have investigated the PSO algorithms. Some of these algorithms have verified different interaction models between the particles [5–7]. Based on these interaction models, the PSO methods can be divided into GBest and LBest methods. The GBest methods consider sharing the information to be available to all the particles. On the other hand, the LBest methods share the data available to neighbor particles. The comparison of these two approaches reveals that the higher connection between the particles in GBest methods causes faster divergence and better global results as compared to LBest methods [7, 8].

There are some literatures that utilize PSO to cluster images. In Omran et al. [9], the number of clusters is considered constant and the particles are made up of all the cluster centers. This method has been evaluated based on three different fitness functions. Omran et al. have also proposed a dynamic PSO clustering which estimates the number and centers of clustering methods [10]. Hyperspectral image clustering based on PSO to solve the number and parameters of clusters is also presented by Paoli et al. [11].

In this paper a PSO clustering algorithm will be applied to cluster hyperspectral data in band and MNF feature spaces for hyperspectral data. Section 7.2 introduces clustering concepts. Section 7.3 introduces PSO and PSO clustering algorithm. Section 7.4 is the proposal of the MNF transform and the experimental results are given in Sect. 7.5. Finally the conclusions and future works are given is Sect. 7.6.

7.2 Data Clustering Concepts

Clustering is an automated unsupervised classification method which divides the set of data into subsets of clusters or classes. The objective of this process is the separation of these data in a way that the data in one cluster are similar, and the data in different clusters are dissimilar to each other [12].

Based on these definitions, a similarity measure is a necessity in data clustering. The most well-known similarity measure is the use of distance measures and the most widely used distance measure is the Euclidian distance [13]. Also clustering validity index is the evaluation criterion for clustering methods and comparison.

Cluster validity indices are defined based on internal, relative and external criteria. The internal criteria use the data itself to assess the clustering method; the relative indices compare multiple structures, e.g. obtained from different clustering algorithms, and choose the best among them. Also the external indices measure performance by comparing the results to a priori information named true class labels [14]. If the priori information is available, the clustering is usually evaluated based on external indices as they are the most reliable assessments.

In this paper, the Euclidian distance is considered as the similarity measure. Also the kappa coefficient as an external index is utilized to assess the clustering performance.

7.3 Particle Swarm Optimization and PSO Clustering

In this section, the Particle Swarm Optimization (PSO) is explained first. Then the clustering method based on PSO is proposed.

7.3.1 Particle Swarm Optimization (PSO)

PSO is a population-based optimization which solves the optimization problem by an inspirational model of the social behavior of birds and fish, and is proposed by Kennedy and Eberhart [5, 15]. In this algorithm, each particle is able to solve the problem completely and is a candidate solution to it. The solution space is formulated as a search space in a way that each location in this space is a solution to the problem.

In PSO, the particles are in communication with each other in order to find the best solution location, which optimizes the fitness function. In this algorithm, each particle's movement is based on the velocity and location equations as follows.

$$V_{id}(t+1) = wv_{id}(t) + c_1 r_1 (p_{id}(t) - x_{id}(t)) + c_2 r_2 (p_{gd}(t) - x_{id}(t)) \tag{7.1}$$

$$x_{id}(t+1) = x_{id}(t) + v_{id}(t+1) \tag{7.2}$$

In these equations, $x_{id}(t)$ is the ith particle location in time t, $v_i(t)$ is the ith particle velocity in time t, $p_{id}(t)$ also labeled as *pbest*, is the best solution found so far for i^{th} particle, and $p_{gd}(t)$ is the best solution found for the population until time t, which is also called *gbest*. It is noticeable that in all variables, d indicates the dimension of data. Also w is the inertia weight that scales the previous time velocity; c_1 and c_2 are two acceleration coefficients for the *gbest* and *pbest* velocity elements, and r_1 and r_2 are random variables in range 0–1.

In optimization based on PSO, two main components are population and fitness functions. Population consists of a number of particles which are the candidate solutions of the problem. Also fitness function is the fitness criterion that represents the favorability of each location in search space of the problem.

7.3.2 Clustering Based on PSO

The solution of an optimization problem based on PSO requires determination of population and fitness function. In clustering based on PSO, each particle is composed of N_c cluster centers. Therefore each article can be represented by Eq. (7.3).

$$\mathrm{m_i} = (\mathrm{m_{i1}}, \ \ldots, \ \mathrm{m_{ij}}, \ \ldots, \ \mathrm{m_{iNc}}) \tag{7.3}$$

In this equation, m_{ij} is the jth candidate cluster center vector suggested by ith particle. This definition yields a candidate solution population for clustering centers.

Also as the clustering is solved by an optimization approach in PSO, the fitness function is considered as an external index based on Eq. (7.4).

$$\mathrm{F} = (1 - \mathrm{kappa}) \tag{7.4}$$

In this equation, Kappa is the kappa coefficient which is computed based on a priori knowledge of cluster labels for data. Based on these concepts, the flow chart of PSO clustering algorithm is illustrated in Fig. 7.1 in the next page. In this process, the input is the number of clusters and data to be clustered in band dimension or other feature spaces like Minimum Noise Fraction (MNF) feature spaces, and the output is the optimum cluster centers.

7.4 Minimum Noise Fraction Transform

The great number of spectrally near dimensions in hyperspectral images allows objects with subtle spectral diversity to be recognizable. But, despite the high information content in these images, the high dimensional data comes with new problems like intensive computation in its clustering, high inband correlation and great redundancy [3].

These problems make utilizing data compression and reduction techniques necessary. The common technique to yield this purpose in remote sensing field is the Principal Component Analysis (PCA) which comes by interesting characteristics.

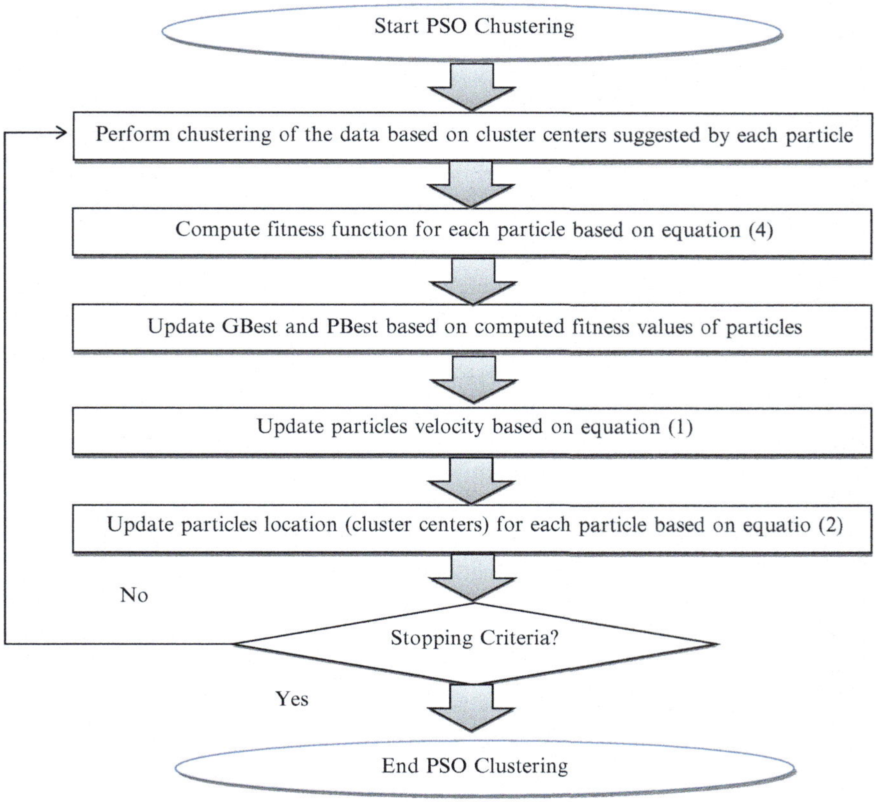

Fig. 7.1 Flowchart of PSO for clustering

This technique preserves the total variance in data, minimizes the mean square approximations and generates uncorrelated coefficients [3].

The PCA transforms the data into a new orthogonal space by rotating the coordinate system in such a way that the newly mapped data features are uncorrelated and the features variances are sorted in descending order. The computation of this new feature space is done by the aid of the Eigen-values and vectors and the covariance matrix of the data, also the summation of the variances in new feature space and dimension space are equal. After the PCA is computed, a lower number of features with highest variances are selected and the other features are discarded.

Although the PCA comes with computationally interesting properties, it considers the variances to be completely internal. But the hyperspectral remote sensing data is affected by external conditions that result in change in variances, also labeled external variances. The atmospheric transformations, angle of sun incidence, and approximate calibration of detector parameters of sensors are the cause of this phenomenon [3].

On the other hand, the subtle spectral features detection gets harder with the existence of noise and external influences. The existence of noisy, water absorption, and highly correlated bands cause PCA transform to result in less accurate results. Therefore, the MNF transform that tries to deal with the existing noise is suggested and usually results in better consequences.

The MNF transform is a linear transformation that maximizes the signal to noise ratio represented by each component and results in better image quality in comparison with PCA. It determines the inherent dimension of data, segregates noise, and reduces computational requirements [3]. The process of computing MNF consists of two PCA rotations and a noise whitening step. it uses noise covariance matrix to rescale the noise in data. This process causes a transformed feature space with unit noise variance and no band to band correlations. After this step, the PCA transformation of the new transformed space data is performed.

In order to compute MNF transform, computation of the noise and signal covariance matrices are essential. The signal covariance computation is similar to the PCA covariance matrix computation, but the noise covariance matrix should be estimated based on other methods like flat-field image, or if not possible, based on statistical properties of the satellite image.

As the flat-field image does not usually exist for the satellite hyperspectral images, the corresponding in-scene statistical estimation is used to compute the noise covariance matrix [16]. The assumption of this method is that the noise adheres to multidimensional normal distribution and is uncorrelated with signal. It also considers a high spatial dependency among nearby signals but a week spatial dependency for noise. Considering these, the noise differences can be estimated with the following equation.

$$\mathrm{p}_1 - \mathrm{p}_2 = \mathrm{s}_1 + \mathrm{n}_1 - \mathrm{s}_2 - \mathrm{n}_2 \approx \mathrm{n}_1 - \mathrm{n}_2 \tag{7.5}$$

In this equation, p_1 and p_2 are two adjacent pixel values of the image, $s_1 \approx s_2$ are the signal values of these pixels and n_1 and n_2 are the noise of these two pixels.

After estimation of the noise values, a PCA transform is performed on these values. After this transformation, the orthogonal noise space Q is generated. By dividing the newly transformed space by the standard deviation of the feature spaces u_i, a white noise space is generated as Eq. (7.6).

$$R'(x) = \left[\frac{Q_1(x)}{\sqrt{u_1}}, \frac{Q_2(x)}{\sqrt{u_2}}, \ldots, \frac{QN_c(x)}{\sqrt{u_{N_c}}}\right] \tag{7.6}$$

In Eq. (7.6), $R'(x)$ is the transformed image data normalized by standard deviation of noise in transformed space. This feature space is the white noise space.

After computation of the white noise space, the transformed image data is orthogolized once more by the PCA transform.

After the computation of this feature space, the first several MNF bands are eliminated. This is equal to eliminating the last PCA bands, as the MNF bands emphasize the noise content. Then the noise is smoothed in the next several bands in order to minimize the noise while retaining the information. Then the inverse transformation can be performed to acquire noise free image data.

7.5 Experimental Results

In order to evaluate the PSO algorithm for clustering hyperspectral images in MNF and band dimension spaces, the results of this algorithm is compared with the Kmeans algorithm in these two clustering feature spaces.

7.5.1 *Data Set*

The evaluation of the clustering methods are performed on a hyperspectral image which is taken over northwest Indiana's Indian Pine test site in June 1992 [17–19]. The existence of the ground truth for this dataset allows using an external validity index to evaluate the performed algorithms. This data consists 220 spectral bands and contains 145 × 145 pixels. The known 20 water absorption bands and 15 noisy bands were removed from the original image, resulting in a total of 185 bands [20].The ground truth contains 16 classes, but only five classes are used in this study. The ground truth map of five classes is shown in Figs. 7.2 and 7.3 below. The suitable spatial distribution of these classes is the cause of selecting them for this investigation.

7.5.2 *Performance Measure*

In order to evaluate the clustering results, the confusion matrix is computed for each method. Comparison of the true labels and the labels returned by the clustering algorithms as the quality assessment measure are acquired based on the confusion matrices [21]. The Overall accuracy, Kappa coefficient for individual classes and the Khat index [22] are used to compare the results. The equations of these three comparison criteria are given in Eqs. (7.7), (7.8) and (7.9) respectively.

$$OA = \frac{\sum_{i=1}^{n} x_{ii}}{N} \tag{7.7}$$

$$K = \frac{N\sum_{i=1}^{r} x_{ii} - \sum_{i=1}^{r}(x_{i+} * x_{+i})}{N^2 - \sum_{i=1}^{r}(x_{i+} * x_{+i})} \tag{7.8}$$

$$K_i = \frac{Nx_{ii} - (x_{i+} * x_{+i})}{N * x_{i+} - x_{i+} * x_{+i}} \tag{7.9}$$

In Eq. (7.7), OA is the overall accuracy, in Eq. (7.8) K is the kappa coefficient and in Eq. (7.9), K_i is Khat index for individual classes, r is the number of row or columns in confusion matrix, x_{ii} is the entry (i,i) of the confusion matrix, x_{i+} and x_{+i} are the marginal totals of row i and column j, respectively, and N is the total number of observations [22].

Fig. 7.2 Image subset of Aviris hyperspectral data

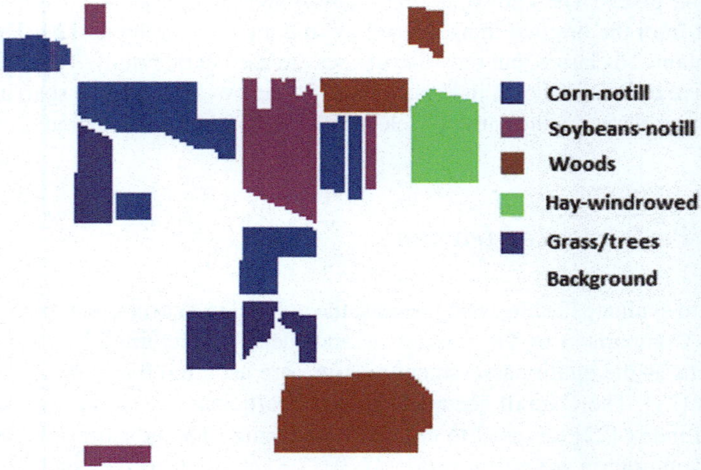

Fig. 7.3 Ground truth of Aviris hyperspectral data with five classes

In this study the kmeans and PSO clustering methods are compared to each other in MNF and band dimension feature classes. The parameters of these methods are represented in Table 7.1.

According to Fig. 7.4, PSO and Kmeans in MNF feature space have gained better accuracy results in comparison with the corresponding methods in band dimension spaces. The overall accuracy values for the Kmeans, PSO, Kmeans +MNF and PSO+MNF are 0.56, 0.79, 0.70 and 0.81 respectively. Also the kappa coefficient values for these methods are 0.44, 0.73, 0.61 and 0.75 respectively. The results show that the best performance is obtained for PSO in MNF space in comparison with the other three methods.

The clustering methods comparison for PSO and Kmeans in band dimension space or MNF space represented in Fig. 7.4 reveals that the PSO algorithm has

Table 7.1 Algorithm parameters used for hyperspectral data clustering

Algorithm	Parameter	Value
Kmeans	IteCount	100
PSO	IterCount	100
	PopulationSize	20
	W	1
	C1	0.5
	C2	0.5
MNF transform params for Kmeans and PSO	NoBands	12

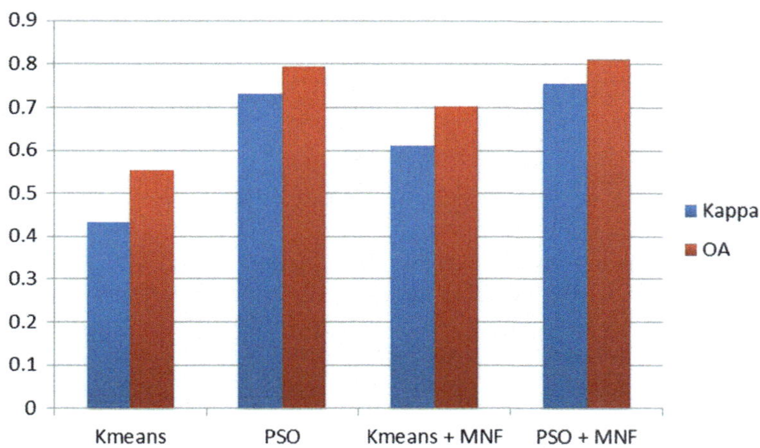

Fig. 7.4 Comparison of overall accuracy and Kappa coefficient in four clustering methods

better performance results in comparison with Kmeans. It is noticeable that the global and local search ability of PSO is the cause of better performance when compared with Kmeans which is merely able to do local search. Also the comparative study of feature spaces suggest that using MNF transform results in better clustering performance, as the accuracies for both methods have been increased while computing in MNF space. It seems that transferring data to a space of uncorrelated and noise-free space with lower dimensions causes better identification and differentiation of classes.

A quantitative comparison of the performance measures suggest that PSO has about 10% better kappa accuracy in band dimension space and about 14% in MNF space when compared to Kmeans. The reason is the global search ability of PSO. This ability causes more stability and reliability of the PSO in comparison with Kmeans, which means lower dependency to initial cluster centers.

Figure 7.5 and Table 7.2 are given to have a visual inspection of the results. According to these outputs, it is noticeable that Kmeans clusters are somehow combined and in some cases these classes are falsely clustered, e.g. the green area on right side of image (Hay-windrowed class) is not clustered correctly in Kmeans

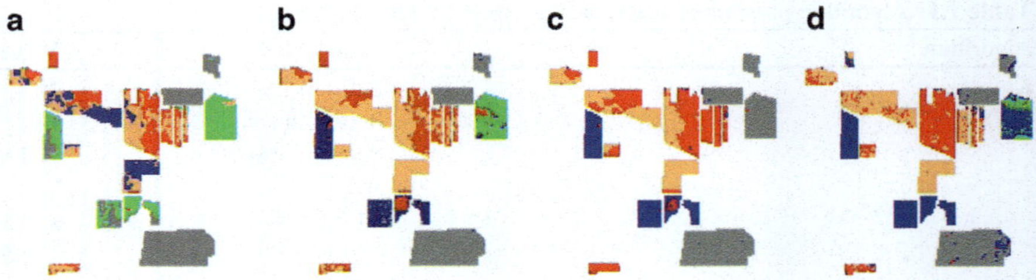

Fig. 7.5 Clustering results of Kmeans (**a**), PSO (**b**), Kmeans+MNF (**c**), and PSO+MNS (**d**)

Table 7.2 Khat index in present yielded by the four investigated methods on the AVIRIS image

	Corn no-till	Grass\trees	Hay-windrowed	Soybean no-till	Woods
Kmeans	0.22	−0.14	0.94	0.34	0.997
PSO	0.67	0.81	0.80	0.39	0.99
Kmeans+MNF	0.45	0.94	0	0.56	0.99
PSO+MNF	0.74	1	0.29	0.72	0.89

+MNF method, but in PSO methods this issue is reduced. Also, although the overall performance is increased by using the MNF transform, the Hay-windowed class cluster result is hardly distinguishable in this feature space, but the clustering accuracy for other classes are highly increased.

7.6 Conclusion

In this paper, the PSO method is evaluated in band dimension and MNF feature spaces. In order to compare the potential of the PSO algorithm in these feature spaces, the results are compared to the common Kmeans clustering method. Results show that PSO methods have better accuracy than Kmeans, because of the global search capability of this method. Also, computation in noise-free uncorrelated MNF feature space results in better performance in comparison with band dimension space. Although using PSO algorithm for clustering results in better performance, but there are some PSO and Kmeans fusion and combination algorithms which are said to result in better performance than the PSO clustering. Also the simultaneous dimension reduction with clustering is possible using the PSO clustering algorithms. Accordingly, our later investigations will be to combine PSO and Kmeans, to reach to better results and also propose methods to be capable of select optimum bands simultaneously with the cluster centers.

References

1. Kennedy J, Eberhart R (1995) Particle swarm optimization. In: Proceedings of the 1995 I.E. international, vol 4. Perth, pp 1942–1948
2. Hopper F, Klawonn F, Kruse R, Runkler T (1999) Fuzzy cluster analysis: methods for classification, data analysis and image recognition. Wiley, Chichester
3. Cheng CM (2000) Comparison of principal components analysis and minimum noise fraction transformation for reducing the dimensionality of hyperspectral imagery. Geogr Res 33(1):163–178
4. Lee JB, Woodyatt AS, Berman M (1999) Enhancements of high spectral resolution remote-sensing data by a noise-adjusted principal components transform. IEEE Trans Geosci Remote Sens 28(3):295–304
5. Engelbrecht AM (2007) Computational intelligence: an introduction. Wiley, Chichester
6. Mendes R, Cortez P, Rocha M, Neves J (2002) Particle swarms for feedforward neural network training. Learning 6:1
7. Eberhart R, Simpson P, Dobbins R (1996) Computational intelligence PC tools. Academic Press Professional, Inc, San Diego
8. Krink T, Vesterstrom J, Riget J (2002) Particle swarm optimization with spatial particle extension. In: Evolutionary computation, CEC'02. Proceedings of the 2002 congress, vol 2, pp 1474–1479
9. Omran M, Salman A, Engelbrecht A (2002) Image classification using particle swarm optimization. In: Proceedings of the 4th Asia-Pacific conference on simulated evolution and learning, vol 2002. Singapore, pp 370–374
10. Omran MGH, Salman A, Engelbrecht AP (2005) Dynamic clustering using particle swarm optimization with application in unsupervised image classification. Trans Eng Comput Technol 9:199–204
11. Paoli A, Melgani F, Pasolli E (2010) Swarm intelligence for unsupervised classification of hyperspectral images. IEEE 5:V96–V99
12. Ezzatabadipur H (2010) Study and evaluation of the hyperspectral remote sensing image clustering. University of Tehran, Farsi
13. Jain AK, Murty MN, Flynn PJ (1999) Data clustering: a review. ACM Comput Surv (CSUR) 31:264–323
14. Jain AK (2010) Data clustering: 50 years beyond K-means. Pattern Recognit Lett 31:651–666
15. Poli R, Kennedy J, Blackwell T (2007) Particle swarm optimization. Swarm Intell 1:33–57
16. Green AA, Switzer P (1984) Min/max autocorrelation factors for multivariate spatial imagery. Technical report 6. Department of Statistics, Stanford University, Stanford
17. Yun-song S, Yu-feng S (2010) Remote sensing image classification and recognition based on KFCM. In: Proceedings of the 5th international conference on Computer & Education, Hefei, China, 2010, pp 1062–1065
18. Graves D, Pedrycz W (2010) Kernel-based fuzzy clustering and fuzzy clustering. A Comp Exp Study Fuzzy Set Syst 161:522–543
19. AVIRIS image Indian Pine Test Site (June 12, 1992) Retrieved from: http://cobweb.ecn.purdue.edu/~biehl/MultiSpec/hyperspectral.html
20. Mojaradi B et al (2008) A novel band selection method for hyperspectral data analysis. The international archives of the photogrammetry, remote sensing and spatial information sciences, vol XXXVII. Part B7. Beijing
21. Zhong S, Ghosh J (2003) A comparative study of generative models for document clustering. In: Proceedings of the workshop on clustering high dimensional data and its applications in SIAM data mining conference
22. Kumar M (2004) Feature selection for classification of hyperspectral remotely sensed data using NSGAII. In: CE 597D water resources seminar
23. Tso B, Mather PM (2009) Classification methods for remotely sensed data. CRC, Boca Raton

Chapter 8
On a Ball's Trajectory Model for Putting's Evaluation

Gonçalo Dias, Rui Mendes, Micael S. Couceiro, Carlos M. Figueiredo, and J. Miguel A. Luz

Abstract The main goal of the present work is to draw conclusions about the influence of constraints such as different striking spots on the presence of a ramp, in the performance of golf putting while determining if the player's performance is indeed unique and presents a signature for each of the 3 expert players in study. Each of them performed 30 trials at 2 distinct spots (in a total of 60 trials) and a standard curve for their performance was obtained with the Robust Linear Least Squares (LLSQ) method. The obtained results show that the putting performance is a "fingerprint", unique to each player.

8.1 Introduction

The putt is a motor skill that combines art and science [1, 2], representing about 43% of a Golf game [3, 4]. Coordination and control is an individual process of this gesture that is different from subject to subject depending on their profiles and characteristics [5–7]. Just like other sports (e.g. football, basketball or rugby), the resulting variability and biological "noise" from putting performance when considering different constraints can be considered as a "fingerprint" (i.e., signature)

G. Dias (✉)
Faculty of Sport Sciences and Physical Education, University of Coimbra (FCDEF UC), Research Center for Sport and Physical Activity (CIDAF)/RoboCorp, Coimbra, Portugal
e-mail: goncalodias@fcdef.uc.pt

R. Mendes
Coimbra College of Education (ESEC/IPC)/CIPER/RoboCorp/Coimbra, Portugal
e-mail: rmendes@esec.pt

M.S. Couceiro • C.M. Figueiredo • J.M.A. Luz
RoboCorp, Department of Electrical Engineering (DEE), Engineering Institute of Coimbra (ISEC), Coimbra, Portugal
e-mail: micael@isec.pt; cfigueiredo@isec.pt; miguel.luz@isec.pt

A. Madureira et al. (eds.), *Computational Intelligence and Decision Making: Trends and Applications*, Intelligent Systems, Control and Automation: Science and Engineering 61, DOI 10.1007/978-94-007-4722-7_8, © Springer Science+Business Media Dordrecht 2013

that is unique to each player [8]. Considering that every player has different morphological and functional features representing a certain individual profile, it is hard to perform a variability study, solely based on statistics (e.g., mean value, standard deviation, variation coefficient, etc.). This means that performance curves commonly used in cognitive system's researches may not be enough to explain differences in performance [9].

The main goal of the present work is draw conclusions about the influence of constraints such as different striking spots or the presence of a ramp, in the performance of golf putting while determining if the player's performance is indeed unique.

8.2 Methodology

8.2.1 Sample

In this study 3 expert golf players participated, with handicap between 5 and 15 and over 10 years of golf practice. All players were adult, volunteer, male and right-handed.

8.2.2 Experimental Design and Procedures

The adopted task under examination was the golf putting, meaning the strike of a ball (Titleist; model Pro V1) with a putter (Putter Jumbo Black Beauty; size 35; standard) on a horizontal and still surface, placed on the ground over a ramp.

The players performed 30 trials at 2 m from the hole, with the constraint of a ramp and a deviation of 25° for both the left and the right side of the hole as in Fig. 8.1.

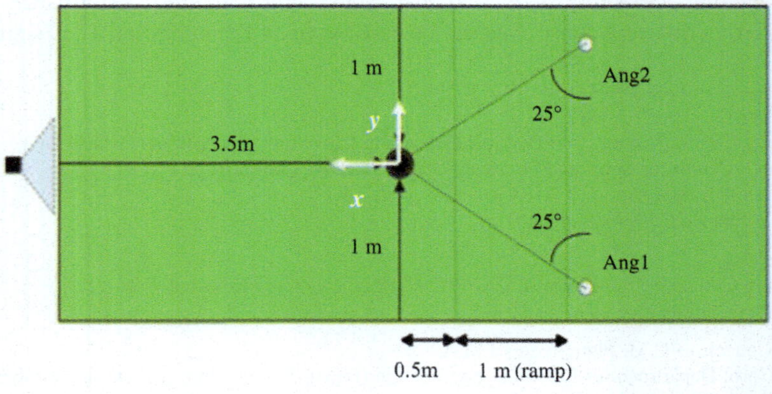

Fig. 8.1 Experimental design

To perform this study, a digital camera Casio Exilim High Speed EX-FH25 was used. The autonomy of the digital camera was also considered and in order to smoothly record the entire session without further problems, rechargeable 2,700 mA batteries and 16 Gb memory cards were used.

This camera was placed 2 m after the apparatus ending, capturing the entire experimental device, in order to retrieve the ball's trajectory and eventual deviation from the hole. The camera was static, at 1.55 m high, with an inclination of 22° pointing down, working still in its tripod guaranteeing reliability for later data analysis. It was shooting at 30 fps at a resolution of 1,280 x 720 pixels and its lens had a focal length of 26 mm. Both apparatus and digital camera were always in the exact same place, so that everything was recorded in the same conditions.

8.3 Data Analysis

Algorithms for object detection are one of the fundamental issues in several fields such as robotics [10, 11]. The vision-based techniques can be classified into two categories, the stereo vision [12] approach and the motion-based approach [13]. In the latter one, the motion field is computed from consecutive images obtained from the same camera and dynamic objects are then detected when their motion is dominant in the scene (Fig. 8.2).

Fig. 8.2 Frame from a video sequence of a player striking at Ang1

Table 8.1 Real coordinates of the experimental device

Ref #	x-axis [mm]	y-axis [mm]
1	0	0
2	2,000	1,000
3	0	1,000
4	−500	1,000
5	−1,500	1,000
6	−2,000	1,000
7	2,000	−1,000
8	0	−1,000
9	−500	−1,000
10	−1,500	−1,000
11	−2,000	−1,000

The most commonly used calibration method is perhaps the direct linear transformation (DLT) method originally reported by Abdel-Aziz and Karara [14]. The DLT method uses a set of control points whose object space/plane coordinates are already known. The control points are normally fixed to a rigid frame (i.e., calibration frame). The flexibility of the DLT-based calibration often depends on how easy it is to handle the calibration frame. Since a 2D scene was captured and the ball is hit towards the camera position (Fig. 8.1), several yellow markers with known location were placed in the experimental device so the conversion to real coordinates could be made (Table 8.1). The center of the hole is the reference of this coordinate system being the point (0,0).

A simple algorithm for ball detection throughout the video sequence was created, detecting this round object (with a high form factor value) based on image differences and noise reduction as the remaining scenario keeps stable as the ball travels through it.

Another feature included in the algorithm, as it is based on a video sequence analysis, is the fact that the ball can only be detected within a small region of interest based on the previous detection, in order to prevent false detections and lower the computational cost. Exception being made for the first position, as it can only be either on Ang1 or Ang2, it is easy to define the first region of interest.

8.4 Results

Here are presented the obtained standard curves based on the Robust Linear Least Squares (LLSQ) [15], run iteratively until it was achieved a determination coefficient (R2) higher than 0.9.

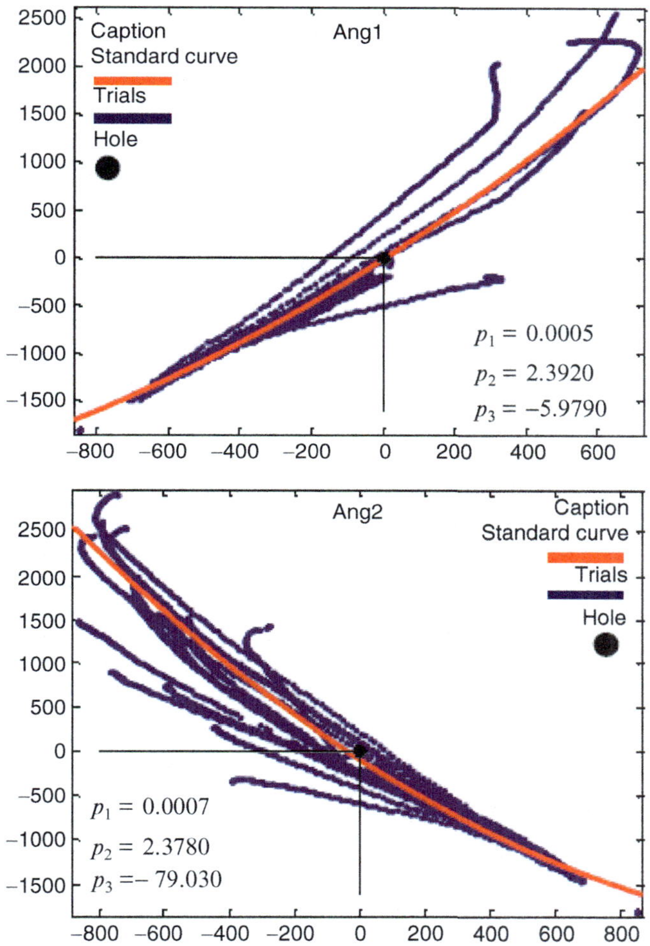

Fig. 8.3 Entire session of trials and standard curve for player 1 based on the LLSQ method

As fitness model (8.1), we considered the second degree polynomial function:

$$y = p_1x^2 + p_2x + p_3y \tag{8.1}$$

In Fig. 8.3 are presented the entire session of trials performed by player 1 and the standard curve for each set (i.e., Ang1 and Ang2) based on LLSQ method.

In Fig. 8.4 are presented the entire session of trials performed by player 2 and the standard curve for each set (i.e., Ang1 and Ang2) based on LLSQ method.

In Fig. 8.5 are presented the entire session of trials performed by player 3 and the standard curve for each set (i.e., Ang1 and Ang2) based on LLSQ method.

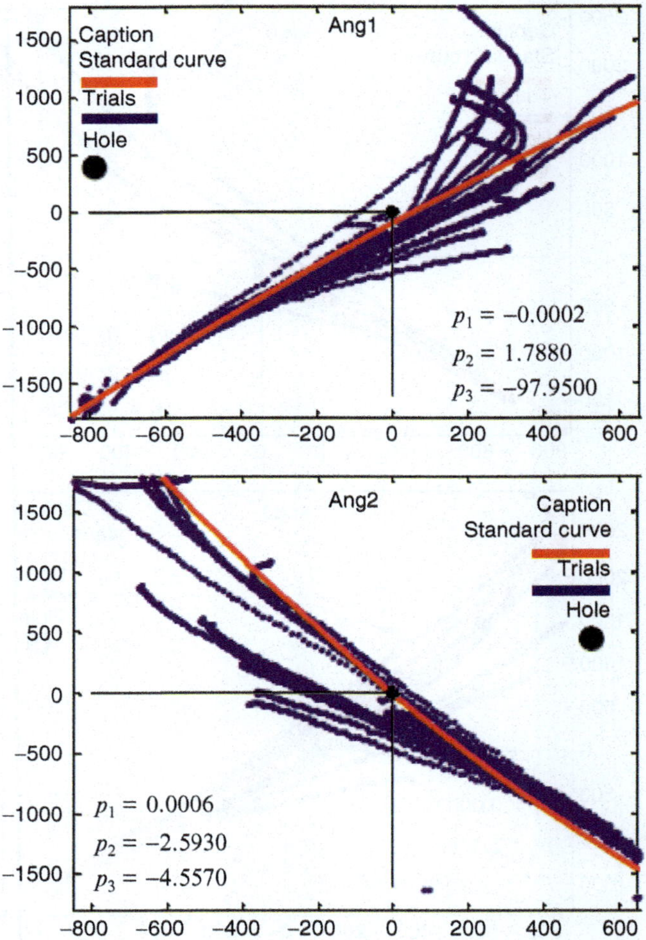

Fig. 8.4 Entire session of trials and standard curve for player 2 based on the LLSQ method

By analysing Figs. 8.3, 8.4 and 8.5, where the trajectory of the entire session of trials is presented plus the standard curve for each set (i.e., 30 trials at Ang1 and 30 trials at Ang2), it can be verified that player 1 strikes harder than players 2 or 3. Furthermore, the standard curve of this player intercepts the hole, always with concave up, meaning that the quadratic coefficient p1 is positive. Player 1 performance is quite similar to player 2 at Ang2.

For player 2, as opposite to the other players, has lighter strikes, clearly showing the ramp influence in ball's trajectory (Fig. 8.4). This player presents a standard curve similar to player 3 when considering the strike at Ang1 and a standard curve similar to player 1 when considering the strike at Ang2.

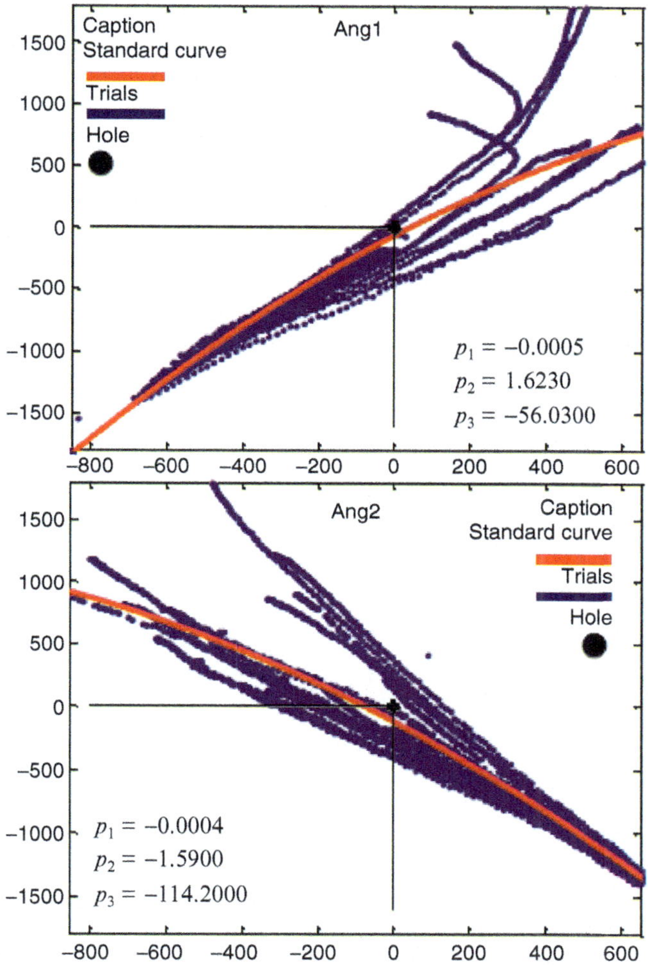

Fig. 8.5 Entire session of trials and standard curve for player 3 based on the LLSQ method

8.5 Discussion and Conclusions

The obtained results with the Robust Linear Least Squares (LLSQ) method, when the players deal with the constraints (i.e., different striking spots and the ramp), show that their performance is different from player to player [8]. Anyway, the individual profiles of the participant players in this study, show similarities in their standard curves, as all of them are closer to the hole, as expected for expert players.

It was also noted that for the players, striking spot Ang2 is a more difficult place to perform the putt than Ang1. This has to do with the field of view they have to the hole. When they strike at Ang2 their "stereo vision" is compromised and it results in erroneous information processing at central nervous system.

Given the results of empirical findings and the factors mentioned above, it can be concluded that the putt gesture needs be analyzed not only from the quantitative point of view but also from a qualitative point of view that privileges the process variables of motor execution. Like other motor skills, the inter-individual differences resulting from the execution of putt is a "fingerprint" that is unique to each individual [9].

Acknowledgments This work was supported by a PhD scholarship (SFRH/BD/73382/2010) by the Portuguese Foundation for Science and Technology (FCT). The authors would also like to thank RoboCorp (ISEC) for providing the facilities and equipment for this study.

References

1. Pelz D (1989) Putting like the Pro´s. Harpercollins, New York
2. Pelz D (1990) The long putter. Pelz Rep 1(3):1–5
3. Pelz D (2000) Putting Bible: the complete guide to mastering the green. Publication Doubleday, New York
4. Alexander DL, Kern W (2005) Drive for show and putting for dough? J Sports Econ 6(1):46–60
5. Pelz D, Mastroni A (1989) Puft like the Pro´s. Penguin, Englewood Cliffs
6. Jonassen DH, Grabowski BL (1993) Handbook of individual differences, learning and instruction. Lawrence Erlbaum, Hillsdale
7. Pelz D (1993) How's your eye sight? Golf Austr 11:96–99
8. Newell KM (1986) Constraints on the development of coordination. In: Wade MG, Whiting HTA (eds) Motor development in children: aspects of coordination and control. Martinus Nijhoff, Boston, pp 341–360
9. Davids K, Button C, Bennett S (2008) Dynamics of skill acquisition – a constraints-led approach. Human Kinetics Publishers, Champaign
10. Ohya A, Kosaka A, Kak A (1997) Vision-based navigation of mobile robot with obstacle avoidance by single camera vision and ultrasonic sensing. In: Proceedings of the 1997 IEEE/RSJ international conference on Intelligent Robots and Systems. IROS '97, Grenoble
11. Wang H, Yuan K, Zou W, Peng Y (2005) Real-time obstacle detection with a single camera. In: IEEE international conference on industrial technology, Hong Kong, 5–18 Dec 2005
12. Salinas RM, Aguirre E, Silvente MG, Gonzales A (2007) People detection and tracking through stereo vision for human-robot interaction. Image Vis Comput 25(6):995–1007, Butterworth-Heinemann
13. Kim YG, Kim H (2004) Layered ground floor detection for vision-based mobile robot navigation. In: Proceedings of the IEEE international conference on robotics and automation. ICRA '04, New Orleans Louisiana, 26 April–1 May 2004
14. Abdel-Aziz YI, Karara HM (1971) Direct linear transformation into object space coordinates in close-range photogrammetry. In: Proceedings of the symposium on close-range photogram merry. University of Illinois at Urbana-Champaign, Urbana, pp 1–18
15. Bjorck Ake (1996) Numerical methods for least squares problems. Society for Industrial and Applied Mathematics (SIAM), Philadelphia. ISBN 0-89871-360-9

Chapter 9
Efficient Discriminative Models for Proteomics with Simple and Optimized Features

Lionel Morgado, Carlos Pereira, Paula Veríssimo, and António Dourado

Abstract The broad diversity in biology offers a panoply of interesting categorization problems for the machine learning community. New challenges arise in modern subjects such as protein classification, where huge and complex datasets are common, and demand the most accurate and fast classifiers to retrieve meaningful biological traits in acceptable time. Although the Support Vector Machine algorithm has been playing a significant role by offering the most precise solutions in diverse domains, the problem of protein classification is far from being solved. Other successful Kernel Methods such as the Relevance Vector Machine and extensions that combine Recursive Feature Elimination in formulations capable of performing feature selection like SVM-RFE and RVM-RFE, were tested in a benchmark environment and compared to other popular statistical models such as Nearest Neighbor, Random Forest, Artificial Neural Networks and Logistics Regression. The results show that SVM-RFE can create classifiers with the highest recognition ability even using a simple compact feature set easily computable from protein primary structure. Plus, these models allow getting predictions in a time scale reduced by orders of magnitude when compared with the standardly used PSI-BLAST.

L. Morgado (✉) • A. Dourado
Center for Informatics and Systems of the University of Coimbra, Polo II – University of Coimbra, Coimbra, Portugal
e-mail: lionel@dei.uc.pt; dourado@dei.uc.pt

C. Pereira
Center for Informatics and Systems of the University of Coimbra, Polo II – University of Coimbra, Coimbra, Portugal

Department of Informatics Engineering and Systems, Coimbra Institute of Engineering – ISEC, Quinta da Nora, 3030-199 Coimbra, Portugal
e-mail: cpereira@dei.uc.pt; cpereira@isec.pt

P. Veríssimo
Department of Biochemistry and Center for Neuroscience and Cell Biology, University of Coimbra, 3004-517 Coimbra, Portugal
e-mail: paulav@ibili.uc.pt

A. Madureira et al. (eds.), *Computational Intelligence and Decision Making: Trends and Applications*, Intelligent Systems, Control and Automation: Science and Engineering 61, DOI 10.1007/978-94-007-4722-7_9,

Keywords Protein Family Classification • Kernel Machines • Support Vector Machine • Relevance Vector Machine • Feature Selection • Recursive Feature Elimination

Abbreviations

AUC	Area Under the Curve
FN	False Negative
FP	False Positive
FPR	False Positive Rate
KM	Kernel Machine
RFE	Recursive Feature Elimination
ROC	Receiver Operating Characteritic
RVM	Relevance Vector Machine
SVM	Support Vector Machine
TN	True Negative
TP	True Positive
TPR	True Positive Rate

9.1 Introduction

Proteins are known to be crucial for the survival of living organisms. These biological polymers made of small units called amino acids, are important in structural and mechanical functions, intervening as well in processes such as cell signaling, immune response and digestion, among others. The shape of a protein unravels its function, and that's the reason why so much attention has been given to the study of protein spatial conformation. The present theories defend that similar proteins have analogous activity, and therefore the scientific community has been putting much effort to create catalogues with all identified proteins. This knowledge eases the study of the biological mechanisms in which they intervene.

If on one hand, traditional laboratorial techniques used to obtain the structure and function of peptides are very expensive and sometimes even unfeasible, on the other hand, getting the primary structure representation is much more affordable and fast. Since protein folding is dependent on local and global properties of the amino acid chain, it comes with no surprise that many researchers invest their time in the development of computational approaches to retrieve the missing traits from the sequence alone. The computational methodologies for protein classification based on sequence similarity can be divided into three big families: generative models, pairwise sequence comparison approaches and discriminative classifiers. Nowadays, sequence comparison methods such as BLAST and PSI-BLAST [1] are the most established and accepted by the community since they are accurate and easy to update. The last property is very important once proteomics is a fast evolving field where numerous proteins are discovered in a daily basis. However, a prediction system based on sequence alignment techniques can become very slow when very large repositories are involved, since a query needs to be made over the database. Moreover, jobs can consist in the massive categorization of millions of unknown instances against these very extensive databases and therefore they can take weeks or even last for months using common computers.

Supervised Kernel Machines such as the Support Vector Machine [2] and the Relevance Vector Machine [3] belong to the group of discriminative classifiers. These learning algorithms can build compact discriminative models that proved their efficiency finding accurate decision hyperplanes in numerous domains. In proteomics, considerable work was already developed using this technology [4–16], but since this is an area in constant change, new algorithms still under study.

Protein family categorization is a typical problem related to protein structure and function analysis. It adequately illustrates the present challenges faced in computational biology. The numerous proteomes that started being deciphered few years ago by means of high-throughput procedures, make today part of huge repositories that still waiting for characterization [16]. Because no perfect algorithm was yet found for protein classification, new benchmarks still being created to evaluate the performance of emerging methods and establish standards for exempt algorithm comparison. Many collections gather proteins in the primary structure form and segregate the samples according to groups of interest. Using the sequence of amino acids it is possible to create numerous features. Some can eventually highlight patterns hidden in the original data that can guide the learning algorithm toward better decision frontiers. Nevertheless, uninformative and redundant attributes can introduce an overhead in the final models. It is also known that during the training phase, they can also mislead the learning algorithm, promoting suboptimal solutions with increased memorization of the training samples and consequently an inferior capacity of generalization. Eliminating redundant and uninformative features is therefore desirable. This can be done by choosing a subset of characteristics or by projecting the original attributes to a new reduced representation like it is done in popular techniques like Principal Component Analysis. Projection techniques show disadvantages such as the loss of the original data representation which compromises the interpretability of the solutions, and always require the presence of the initial features before executing the projection to a lower dimension space. Feature selection approaches like Recursive Feature Elimination don't suffer from these weaknesses. In [17] the state-of-the-art learning algorithm SVM was adapted to the RFE scheme. Feature selection became possible using information from the decision frontier to weight the features. The fundaments of the technique which was called SVM-RFE were later extended to other KMs, namely the RVM, originating the RVM-RFE algorithm [18]. Here both formulations are studied in the context of protein family recognition using a benchmark dataset from [19]. The main mathematical aspects of the approaches are exposed in the next section. The experimental procedures and the results can be found in Sect. 9.3. In the last section a reference to future work ends the paper.

9.2 Kernel Machines with Recursive Feature Elimination

The SVM and RVM are examples of KMs. The expression KM refers to a broader class of supervised and unsupervised learning algorithms rooted in statistical learning theory [24].

In the case of supervised learning, a KM tries to find an estimator $f : X \rightarrow Y$ that minimizes the average error or expected risk $R[f] = \iint^{L} (y, f(x)) dP(y|x) dP(x)$, where the distribution of the examples $P(x)$ and their classifications $P(x|y)$ are unknown, and where L is a loss function that measures the error of predicting y with $f(x)$. For that, a dataset $D_n \equiv \{(x_i, y_i) \in X \times Y\}_{i=1}^{N}$ composed of two variables X and Y statistically related is needed. Because in practice the precise value of $R[f]$ cannot be found, it is used a stochastic approximation that minimizes the training error (also known as empirical risk):

$$R_{emp}[f] = \frac{1}{N} \sum\nolimits_{i=1}^{N} \frac{1}{2} |f(x_i) - y_i| \tag{9.1}$$

Using these fundaments, a classifier $f(x) = w.\phi(x) + b$ can be constructed from a training set $\{(x_i, y_i) \in \Re^n \times \{-1, 1\} \forall i \in \{1, \ldots, N\}\}$ for the binary case. Initially, a projection is done from an eventually non-linear input problem representation to a linear feature space ϕ. This new space is implicit in a kernel function $K(x_i, y_i)$ given by the dot product:

$$K(x_i, y_i) = \boldsymbol{\phi}(x_i), \boldsymbol{\phi}(y_i). \tag{9.2}$$

The kernel encloses the similarity between data instances which is exploited to search the decision hyperplane. The search process is oriented by an objective function that can be represented by the expected error (the error of the classifier in a finite test set). Instead, for convenience and efficiency, during the training phase the objective function can be determined only on the train data, estimating the importance of a feature in the decision through the changes in cost function $DJ(i)$ caused by removing it, or equivalently, by making its weight zero.

For a large number of features it is desirable to be able to eliminate several features at every step. However, because a feature ranking criterion may not be a good feature subset ranking criterion, it is necessary to use an iterative procedure such as RFE. RFE is a backward feature elimination method that can be defined by the following three simple steps:

1. Train a classifier (optimization of weights w_i considering the function J)
2. Determine the ranking criteria for all features ($DJ(i)$ or ${w_i}^2$)
3. Remove the feature for which the ranking criterion is the smallest

The fusion between SVM and RVM learning with RFE generates a feature selection algorithm, which allows eliminating iteratively unnecessary data.

Several features can be eradicated each step to speed up the process. In this case we can't consider a ranking for all features but instead a set rank. There is no formal optimum for the number of features that must be eliminated by the end of a given iteration, but a tradeoff between the processing time and the effectiveness of the method must be considered when defining this value. Removing too many features at a single iteration may degrade the classifiers, but if too few are depleted the analysis can be extended for prohibitive time periods.

9.3 Experiments and Results

The aim of the present study was to compare the performance of SVM, RVM, SVM-RFE and RVM-RFE to other data mining algorithms previously tested in a benchmark environment for protein family classification. The other previously tested algorithms are: Nearest Neighbor, Random Forest, Artificial Neural Networks and Logistics Regression. Details about the tests with these methods are not exposed here, but can be consulted in [20].

In the new tests the amino acid composition was introduced as an easily and fast computable feature set, which allows the creation of compact models from which fast and accurate predictions can be obtained. All experiments were performed with the protein benchmark *SCOP40_Sequence_Minidatabase*. This subset of the SCOP40 protein database [21] is composed by 1375 proteins and was designed to simulate 55 binary classification problems, where the objective is to recognize if a given protein belongs or not to a given family. The partition of the benchmark was respected. Both the dataset and the partition instructions can be obtained from the internet at [22].

The effectiveness of learning was accessed for all the tested algorithms by measuring for each model created the accuracy, sensitivity, specificity and the area under the curve of the receiver operating characteristic plot. For some particular cases the TPR and the FPR were also calculated.

A ROC curve [25, 26] consists in the plot of the true positive rate as a function of false positive rate at varying decision thresholds, and expresses the ability of a discriminative model to correctly rank examples and separate distinct classes. A score can be obtained by calculating the area under the ROC curve through the trapezoidal integration:

$$AUC = \sum \left\{ (1 - \varphi_i, \Delta\phi) + \frac{1}{2}[\Delta(1 - \varphi_i)\Delta\phi] \right\}, \tag{9.3}$$

where $TPR = sensitivy = 1 - \varphi$ and $1 - FPR = specificity = 1 - \phi$. A perfect model has AUC $= 1$, a random classifier is expressed by AUC ~ 0.5 and the worst case comes when AUC $= 0$.

To study the complexity of the decision models, the number of train examples retained as support/relevance vectors was examined. Although this kind of analysis is often forgotten, it is very important since it allows a first insight on the extension of data overfitting. It can happen that some discriminative models with similar accuracy scores may use a different number of support/relevant vectors. In this situation, using fewer vectors implies a reduced memorization of training examples, which can eventually be advantageous to recognize new unseen instances. Moreover, when fewer samples are kept in a model it becomes computationally more efficient.

In all experiments the samples were projected to a gaussian space. The kernel parameters and the SVM cost value were tuned using a combination of a grid search to find the C parameter with a hill-climbing scheme to find the gamma parameter for the gaussian kernel. For the RVM only one parameter for the projection function needed to be tweaked, and a hill climbing strategy was enough to tackle the problem.

A quad-core machine running at 2.4 GHz Matlab version 2008a in a Windows XP environment was used to run all the experiments.

To what concerns the learning algorithms, the SVM employed was LIBSVM 2.91 [23] and the RVM used was version 2.

The initial feature extraction stage consisted in calculating the amino acid composition for each protein using the 20 standard units. Other amino acids or symbols we simply not considered.

Studies concerning the normalization of the instances, the attributes and both at the same time were done. Since no improvements were observed for any kind of normalization (see Table 9.1), the consecutive tests were executed without performing this step.

The high accuracy values shown in Table 9.1 may be misleading and not reflect their true capacity of the classifiers to distinguish between classes. The highly unbalanced classes present in the current problem deviate learning in such a way that the models produced tend to classify all test instances as belonging to the major group (in this case the class with examples not belonging to the family to recognize). The prevalence of high false negatives is explained by the formulation of the learning algorithms used that privileges decision frontiers that minimize the total number of wrongly classified instances even if that comes at the cost of misclassifying a complete smaller class that may overlap another much larger. Because sensitivity and specificity are also based in the total amount of examples correctly and wrongly classified, the convenience of these metrics is also questionable for the severe class imbalanced problems here faced. This is the main reason why receiver operating characteristics and the area under the curve were computed. The area under the ROC curve gives us an important measure of the capacity of a classifier to rank examples for varying thresholds. Because a prediction is made applying a threshold to a decision value, the ROC-AUC can subsequently be used to adapt the bias of a decision function to optimize the trade-off between the rate of positive examples correctly classified (also known as true positive rate), and the rate of negatives wrongly classified (also known as false positive rate).

According to the ROC-AUC records the SVM was able to find better decision hyperplanes than the RVM, and the fusion with RFE further improved the performance of both learning formulations. The mean number of vectors used by the RVM to define the decision hyperplanes was less than 7% of the mean number of SVs, even though the RVM achieved high ROC-AUC scores. In any case, the SVM

Table 9.1 Main results for the experiments executed

	Quality metric						
Method	ROC-AUC	Sensitivity	Specificity	Accuracy	#SVs/RVs	TPR	FPR
SVM_i	92.00	2.40	99.60	97.49	168	–	–
SVM_{ii}	91.70	1.80	99.70	97.55	154	–	–
SVMiii	88.20	0.5	99.90	97.74	178	–	–
SVMiv	88.60	1.30	99.90	97.73	196	–	–
RVM	83.30	29.90	97.60	96.22	11	–	–
SVM-RFE	94.45	4.20	99.33	97.23	167	4.20	0.70
RVM-RFE	89.50	40.2	97.20	96.06	5	–	–
SVM-RFEii	94.45	97.70	86.10	86.43	167	97.70	13.70

The table contains the following mean values for the 55 protein families from the benchmark: ROC-AUC, sensibility (sens), specificity (spec), accuracy (acc), the number of support vectors (SVs) or relevance vectors (RVs) used by the model to describe the decision frontier, TPR and the FPR. SVM_i , SVM_{ii} , SVM_{iii} and SVM_{iv} stand for SVM: without any kind of normalization, with features normalized, with both features and instances normalized, and with just instances normalized. $SVM\text{-}RFE_{ii}$ refers to the best model obtained from SVM-RFE after optimizing the threshold of the classifier

models reached 9% higher ROC-AUC than the RVM when no feature selection was applied, and 5% more using only the features that RFE found has being the most informative.

ROC-AUC values express an important measure of the capacity of a classifier to rank examples, but it may be necessary to adapt the threshold of individual models to fully exploit their potential. An intuitive solution to diminish the high false negative problem due to imbalanced classes can be constructed after finding the trade-off point that maximizes the TPR and at the same time minimizes the FPR. By giving a weight to the misclassified test examples proportional to the size of the classes they belong, searching the optimum trade-off point can be done by seeking the threshold for which the maximum of the difference between the true positive rate and the false positive rate in a ROC curve is attained. This point is equivalent to a rough estimate of the tangent between the ROC curve and a straight line with 45° slope. Adjusting the threshold of a classifier, the recognition rate of the class with the reduced number of elements can be augmented, but frequently at the cost of misclassifying some elements from the larger class that takes to a decrease in the global accuracy if the classes cannot be perfectly separated. This procedure was applied to the best models found by SVM-RFE, resulting in an increment of the TPR from a mean value for the 55 binary models of approximately 0 to approximately 1, but unfortunately accompanied by an increase of the FPR from near 0–0.137. Yet, in this situation it is important to remember that the emphasis must be given to the ability to distinguish between two classes. Despite the risk of a slight accuracy loss, more information can be retrieved from an adjusted classifier rather than using a version that classifies correctly a larger number of test examples but instead cannot or can roughly detect correctly instances from the minor class. Because the ROC curve is independent of the threshold of a classifier, ROC-AUC remains untouched after a model adaptation.

The learning time was recorded to give a perspective of the practical implications of using kernel methods here analyzed. LIBSVM-RFE took around 4 h and 30 min to execute, much longer than the 13 min needed to run LIBSVM without the RFE scheme. The RVM implementation showed to be slower than LIBSVM taking 23 min in the experiments. The exact processing time for the RVM-RFE was not measured but the proportion between the times for the RVM and RVM-RFE was analogous to the one between LIBSVM and LIBSVM-RFE, thus making the RVM-RFE from far the most time consuming methodology. Despite this extensive period for learning, in the end the time gain can be significant case the models will be used repeatedly in numerous jobs (Table 9.3).

The time needed to get predictions for the whole protein collection was measured using the best SVM-RFE models with the sequence amino acid composition and using PSI-BLAST. In the end an average speedup of nearly 70 was observed (see Table 9.3).

Finally, it was recognized that some combinations of features can promote the construction of models with increased capability to discern between two families of proteins, but no tendency was observed for the general case. Therefore, no universal rule can be formulated about the importance of particular amino acids for the problem of protein family recognition.

9.4 Conclusions and Future Work

The SVM and RVM algorithms are able to learn discriminative patterns from protein data using very simple features such as the amino acid composition. The SVM was the best performer tested from all algorithms (see Tables 9.2 and 9.3). Just by using

Table 9.2 ROC-AUC for the benchmark studied

	Features					
Method	i	ii	iii	iv	v	vi
1nn	75,77	81,54	82,52	73,43	71,74	56,44
RF	69,65	82,30	80,30	83,44	73,96	72,53
ANN	79,88	88,75	88,34	90,22	83,46	82,54
LogReg	87,15	90,63	91,75	87,66	74,86	83,08
SVM	90,47	93,14	93,76	93,96	82,88	85,51

Results obtained from [22]

Algorithms include: *1nn* nearest neighbor, *RF* Random forest, *ANN* artificial Neural Networks, *LogReg* logistics regression, *SVM* Support Vector Machine. As features the distance metric from several algorithms was extracted, namely from BLAST, *SW* Smith-Waterman, *NW* Needleman-Wunsch, *LA* Local Alignment Kernel, *LZW* Lempel-Ziv-Welch, *PPMZ* Prediction by Partial Match

Table 9.3 Processing time needed to get a prediction (values in seconds)

	Processing time			
Algorithm	Mean	Max	Min	Std
SVM	0.003	0.055	0.002	0.001
BLAST	0.211	0.988	0.060	0.203

the accounting of amino acids for a protein sequence it was possible to train competitive classifiers in terms of recognition, showing that this simple feature set is informative enough to create models that can recognize correctly the family of most proteins. The simplicity of the procedure comparatively to other classifiers or methods such as PSI-BLAST opens a new view on the possibility to create faster prediction systems based on SVMs that without loss in the recognition can operate over unclassified sequences at a proteome scale in a small fraction of the time needed by the techniques employed currently. To confirm this hypothesis it is still necessary to perform new studies with multiclass SVM architectures, and is advisable to extend the tests to other protein datasets.

The RVM also showed an interesting reduction in the complexity of the models preserving a high capacity of discrimination. Future work could be done on ways to use the RVMs to filter the data and find compact models to represent homologues or specific patterns inside proteins. This could bring a significant reduction in data storage and a deeper understanding of the biological diversity.

The large training time needed by the kernel methods here tested is an inconvenience that must be attenuated with improved formulations or parallelized forms. Additionally, incremental SVM formulations could be further developed in order to bring to reality updatable models.

Acknowledgments This work was executed under the project FCOMP-01-0124-FEDER-010160 (PTDC/EIA/71770/2006), designated BIOINK – Incremental Kernel Learning for Biological Data Analysis, supported by Fundação para a Ciência e Tecnologia and FEDER through Program COMPETE (QREN).

References

1. Altschul S, Madden T, Schaffer A, Zhang J, Zhang Z, Miller W, Lipman D (1997) Gapped BLAST and PSI-BLAST: a new generation of protein database search programs. Nucleic Acids Res 25:3389–3402
2. Vapnik V (1998) Statistical learning theory. Adaptive and learning systems for signal processing, communications, and control. Wiley, New York
3. Tipping M (2001) Sparse Bayesian learning and the relevance vector machine. J Mach Learn Res I:211–214
4. Jaakkola T, Diekhans M, Haussler D (1999) Using the Fisher Kernel Method to detect remote protein homologies. In: Proceedings of the international conference on intelligent systems for molecular biology, Heidelberg
5. Krogh A, Brown M, Mian I, Sjolander K, Haussler D (1994) Hidden Markov models in computational biology: applications to protein modeling. J Mol Biol 235:1501–1531. doi:10.1006/jmbi.1994.1104
6. Kuang R, Ie E, Wang K, Wang K, Siddiqi M, Freund Y, Leslie C (2005) Profile-based string kernels for remote homology detection and motif extraction. J Bioinform Comput Biol 3:527–550. doi:10.1142/S021972000500120X
7. Leslie C, Eskin E, Noble W (2002) The spectrum kernel: a string kernel for SVM protein classification. In: Proceedings of the Pacific symposium on biocomputing, vol 7, pp 564–575
8. Leslie C, Eskin E, Cohen A, Weston J, Noble W (2004) Mismatch string kernels for discriminative protein classification. Bioinformatics 20:467–476. doi:10.1093/bioinformatics/btg431

9. Melvin I, Ie E, Kuang R, Weston J, Noble W, Leslie C (2007) Svm-fold: a tool for discriminative multi-class protein fold and superfamily recognition. BMC Bioinform 8(4). doi:10.1186/1471-2105-8-S4-S2
10. Aydin Z, Altunbasak Y, Pakatci I, Erdogan H (2007) Training set reduction methods for protein secondary structure prediction in single-sequence condition. In: Proceedings of the 29th annual international conference IEEE EMBS, Lyon
11. Kurgan L, Chen K (2007) Prediction of protein structural class for the twilight zone sequences. Biochem Biophys Res Commun 357(2):453–460
12. Cheng J, Baldi P (2006) A machine learning information retrieval approach to protein fold recognition. Bioinformatics 22(12):1456–1463
13. Mei S, Fei W (2010) Amino acid classification based spectrum kernel fusion for protein subnuclear localization. BMC Bioinform 11(Suppl 1):S17. doi:10.1186/1471-2105-11-S1-S17
14. Du P, Li Y (2006) Prediction of protein submitochondria locations by hybridizing pseudo-amino acid composition with various physicochemical features of segmented sequence. BMC Bioinform 7:518. doi:10.1186/1471-2105-7-518
15. Lanckriet G, Deng M, Cristianini N, Jordan M, Noble W (2004) Kernel-based data fusion and its application to protein function prediction in yeast. In: Pacific symposium on biocomputing, pp 300–311
16. Kuang R, Gu J, Cai H, Wang Y (2009) Improved prediction of malaria degradomes by supervised learning with SVM and profile kernel. Genetica 36(1):189–209
17. Guyon I, Weston J, Barnhill S, Vapnik V (2002) Gene selection for cancer classification using support vector machines. Mach Learn 46(1–3):389–422. [Online]. Available: citeseer.ist.psu.edu/guyon02gene.html
18. Zhang W, Liu J (2007) Gene selection for cancer classification using relevance vector machine. In: The first international conference on bioinformatics and biomedical engineering, pp 184–187. doi:10.1109/ICBBE.2007.50
19. Webpage dedicated to Protein classification benchmark collection of the international center for genetic engineering and biotechnology: http://net.icgeb.org/benchmark
20. ICGEB/EMBNet Protein classification benchmark collection webpage 2: http://hydra.icgeb.trieste.it/benchmark_previous/index.php?page=33
21. Murzin A, Brenner S, Hubbard T, Chothia C (1995) SCOP: a structural classification of proteins database for the investigation of sequences and structure. J Mol Biol 247:536–540. doi:10.1006/jmbi.1995.0159
22. ICGEB/EMBNet Protein classification benchmark collection webpage: http://hydra.icgeb.trieste.it/benchmark_previous/index.php?experiment=33
23. Chang C, Lin C (2004) LIBSVM: a Library for Support Vector Machines. Software available at http://www.csie.ntu.edu.tw/~cjlin/libsvm
24. Cristianini N, Shawe-Taylor J (1999) An introduction to support vector machines. Cambridge University Press, Cambridge
25. Bradley AP (1997) The use of the area under the ROC curve in the evaluation of machine learning algorithms. Pattern Recognit 30(7):1145–1159. doi:10.1016/S0031-3203(96)00142-2
26. Fawcett T (2006) An introduction to ROC analysis. Pattern Recognit Lett 27:861–874

Chapter 10
Meta-heuristics Self-Parameterization in a Multi-agent Scheduling System Using Case-Based Reasoning

Ivo Pereira, Ana Madureira, and Paulo de Moura Oliveira

Abstract This paper proposes a novel agent-based approach to Meta-Heuristics self-configuration. Meta-heuristics are algorithms with parameters which need to be set up as efficient as possible in order to unsure its performance. A learning module for self-parameterization of Meta-heuristics (MH) in a Multi-Agent System (MAS) for resolution of scheduling problems is proposed in this work. The learning module is based on Case-based Reasoning (CBR) and two different integration approaches are proposed. A computational study is made for comparing the two CBR integration perspectives. Finally, some conclusions are reached and future work outlined.

10.1 Introduction

The Scheduling problem can be defined as *"a decision-making process that is used on a regular basis in many manufacturing and services industries. It deals with the allocation of resources to tasks over given time periods and its goal is to optimize one or more objectives"* [1], and several approaches have been developed to its resolution. However, many of those approaches are impractical in real manufacturing environments, which are inherently dynamic incorporating complex constraints and where a variety of unexpected disruptions may occur.

I. Pereira (✉) • A. Madureira
GECAD – Knowledge Engineering and Decision Support Research Center,
Institute of Engineering – Polytechnic of Porto (ISEP/IPP), Porto, Portugal
e-mail: iasp@isep.ipp.pt; amd@isep.ipp.pt

P. de Moura Oliveira
Department of Engineering, University of Trás-os-Montes e Alto Douro, Vila Real, Portugal
e-mail: oliveira@utad.pt

A. Madureira et al. (eds.), *Computational Intelligence and Decision Making: Trends and Applications*, Intelligent Systems, Control and Automation: Science and Engineering 61,
DOI 10.1007/978-94-007-4722-7_10,

In most real-world environments, scheduling is a progressive reactive process where the presence of real-time information requires continuous reconsideration and review of the pre-established plans. Research on scheduling has not fully addressed this dynamic issue, mostly focusing on the optimization of static scheduling plans [2].

In spite of all research made so far, the scheduling problem is still known to be NP-complete, even in static environments [1]. This fact presents several challenges to conventional algorithms and stimulate researchers to explore new directions. Multi-Agent technology has been considered an important approach for developing industrial distributed systems [3–5] motivating its use in this work.

Learning is a crucial component of intelligence, autonomy and pro-activeness which must be a study target of agents and MAS [6]. Panait and Luke [7] described two different approaches of learning in MAS: team learning and concurrent learning. In the first, there is only one apprentice involved, with the objective to learn about the environment and improve a set of agents. In concurrent learning, there are multiple apprentices trying to improve parts from the team.

The parameterization of MH is a known hard task, requiring knowledge about the domain in use and which parameters are more suitable for that domain. For this reason, systems should be capable of parameterizing themselves autonomously. In order to outfit the system with self-parameterization capability, Case-based Reasoning (CBR) is proposed, to implement learning through past experience. With CBR, our proposed system can remember previously solved cases with the objective to autonomously decide which MH should be used and self-parameterize itself, for the resolution of new similar problems.

The paper is organized as follows: Sect. 10.2 describes the dynamic scheduling problem used in this work. CBR is explained along Sect. 10.3 and some applications to scheduling problem resolution are presented. In Sect. 10.4, the implemented prototype is described detailing the MAS and the different perspectives of CBR integration. The computational study is presented in Sect. 10.5 and, finally, in Sect. 10.6 some conclusions and some future work are presented.

10.2 Dynamic Scheduling Problem

The scheduling problem treated in this work is named Extended Job-Shop Scheduling Problem (EJSSP) [8] and has some major extensions and differences compared to the classic Job-Shop Scheduling Problem (JSSP).

JSSP have a set of tasks processing in a set of machines, with each job having an ordered list of operations, each one characterized by the respective processing time and machine where is processed [9].

The main elements of JSSP problem are:

- a set of jobs $J_1,\ldots,J_n$ to be scheduled on a set of machines $M_1,\ldots,M_n$;
- d_j represents the due date of job J_j;

- t_j represents the initial processing time of job J_j;
- r_j represents the release time of job J_j;

EJSSP problems consist in JSSP problems with additional restrictions, with the objective to better represent reality [8]. Those restrictions are:

- Different release and due dates for each job;
- Different priorities for each job;
- Possibility that not every machines are used for all jobs;
- A task can have more than one operation being processed in the same machine (recirculation);
- Two or more operations of the same job can be processed simultaneously;
- Possibility of existence of alternatives machines, identical or not, for operations' processing.

Scheduling Problem is classified as NP-hard combinatorial optimization problem. The methods for their resolution can be divided in exact and approximation algorithms [8]. In the first, it is made an exhaustive solutions space search and it is ensured that optimal solution are found. In the other hand, approximation algorithms [10], including heuristics and MH, do not guarantee the optimal solution but have the objective to find a good solution in an acceptable period of time. For this reason, they are used in this work together with MAS.

10.3 Case-Based Reasoning

Case-Based Reasoning (CBR) is an Artificial Intelligence methodology aiming to solve new problems by using information about previous similar problems resolution [11]. CBR operates under the premise that similar problems can require similar solutions [12].

CBR roots are found in the work of Roger Schank [13] about dynamic memory and how the memory of previous situations can affect problems' resolution and learning processes. There are also references about the study of analogical reasoning [14].

Previous CBR solved cases and respective solutions are stored in a repository named casebase, in order to reuse them in the future [12]. A CBR system solves a new problem by reusing similar previously solved cases [15].

The CBR cycle (known as the '4 REs' cycle) is described in Fig. 10.1 and consists in four main phases [12, 16]:

1. Retrieve the most similar case or cases;
2. Reuse the retrieved information and knowledge;
3. Revise the proposed solution;
4. Retain the revised solution to be useful for future problem solving.

In the first phase, the initial description of a new case is used to retrieve a old case from the casebase. In the Reusing phase that retrieved case is analysed with the

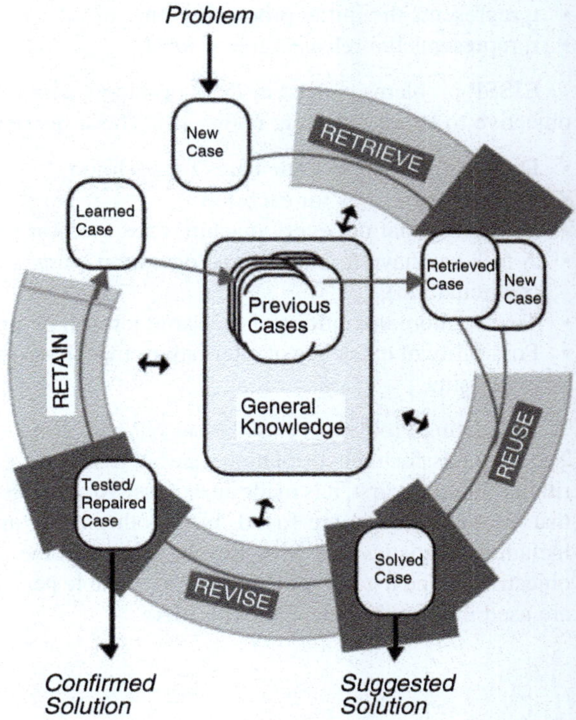

Fig. 10.1 The CBR cycle [16]

objective to recommend a solution for the resolution of the new case. In the next phase (Revising) the suggested solution is tested, for example by executing it in the system. Finally, in the Retaining phase, the useful information is stored for future use and the casebase is updated with the new learned case.

10.4 CBR-Based Module Prototype

Meta-heuristics are very valuable in achieving good solutions in reasonable execution times. Sometimes they even obtain optimal solutions. However, they require the most appropriate tuning of the parameters, which has revealed to be a hard task, since it need some expertise knowledge from the MH to use and from the problem to solve [17].

The proposed system adopts and provides a self-parameterization of MH through CBR, according to the problem to solve, with the possibility that parameters can change in run-time. In accordance to the situation being treated, the system must be able to decide which MH and parameters should be used. It is even possible to change from one MH to another, according to current state and previous information.

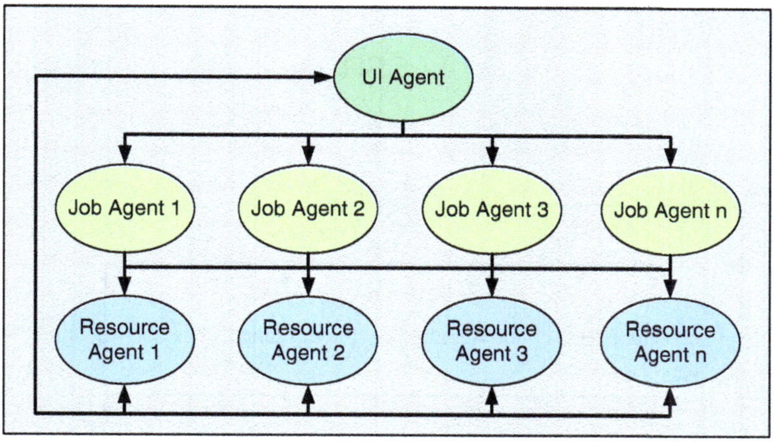

Fig. 10.2 Multi-agent scheduling aystem

In the proposed MAS, there are agents representing jobs and agents representing resources/machines. The system is able to find optimal or near optimal solutions through the use of MH, deal with dynamism (arriving of new jobs, cancelled jobs, changing jobs attributes, etc.), adapt the parameters of the algorithm according to the current situation, switch from one MH to another, and perform a coordination between agents through cooperation or negotiation mechanisms.

Since it is impossible to predict each problem to treat, the system should be capable of learning about its experience during lifetime, as humans do. To perform this learning mechanism, it is proposed the use of CBR for its implementation.

The next three subsections describe the implemented MAS and also the two different perspectives of CBR integration on the MAS.

10.4.1 Multi-agent Scheduling System

The developed MAS for Scheduling problem resolution is a hybrid autonomous architecture represented by hierarchical and team models (if using cooperation) or market models (if using negotiation), as defined by Horling and Lesser [18]. It is composed by three kinds of agents as shown in Fig. 10.2.

In order to provide a reliable communication with the user, a User Interface Agent (UI Agent) was implemented. This agent defines the user interface and dynamically generates the necessary Job and Resource agents, respectively to the number of jobs and machines in the scheduling problem. UI Agent is also responsible for assigning each task to the respective Job Agent and, in the end, applying repair and coordination mechanisms.

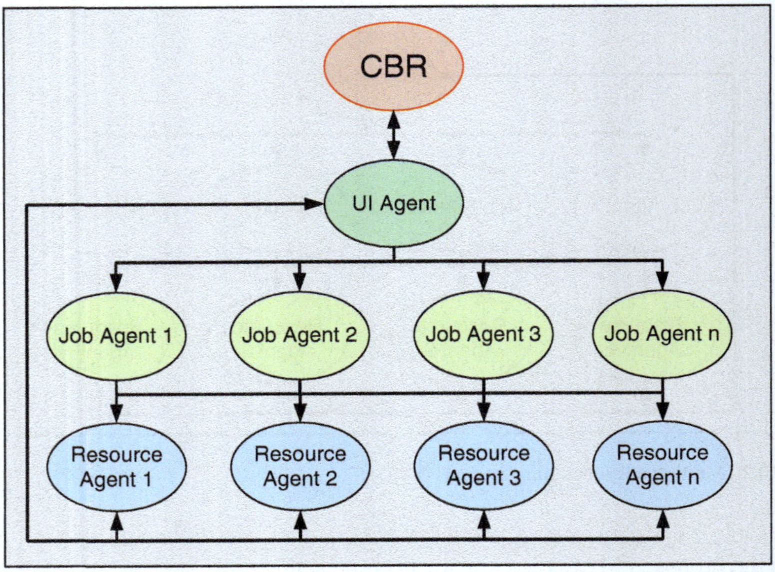

Fig. 10.3 CBR global integration

Job agents are responsible for processing the necessary information about each job. They generate the earliest and latest processing times of the respective job and automatically separate each operation for the respective Resource Agent.

Resource agents are responsible for scheduling the operations of the respective single-machine scheduling problems. These agents apply MH in order to find the best possible single-machine schedules/plans of operations. In the end, Resource agents communicate their solutions to the UI Agent, for later feasibility validation.

10.4.2 Global CBR Perspective

In our system, it is possible to use CBR through two different approaches. In the first, one the most similar cases to the new problem is retrieved, apart from the MH to use. The retrieved case contains the MH and respective parameters to use. The second approach consists in an *'a priori'* selection of the MH and then retrieving the most similar case, only with the tuning parameters.

Both approaches represent a Solution Reuse, but the first one is more appropriate to the Scheduling problem resolution, since the system should have the capacity to decide which MH to use and defining the parameters. This is important because not all MH are effective for the resolution of any kind of problem.

In a global perspective of CBR integration on the system (Fig. 10.3), there is an external module (agent) responsible for executing the CBR. This module communicates with UI Agent in order to suggest a MH to use and configure the respective parameters according to problem characteristics. This communication

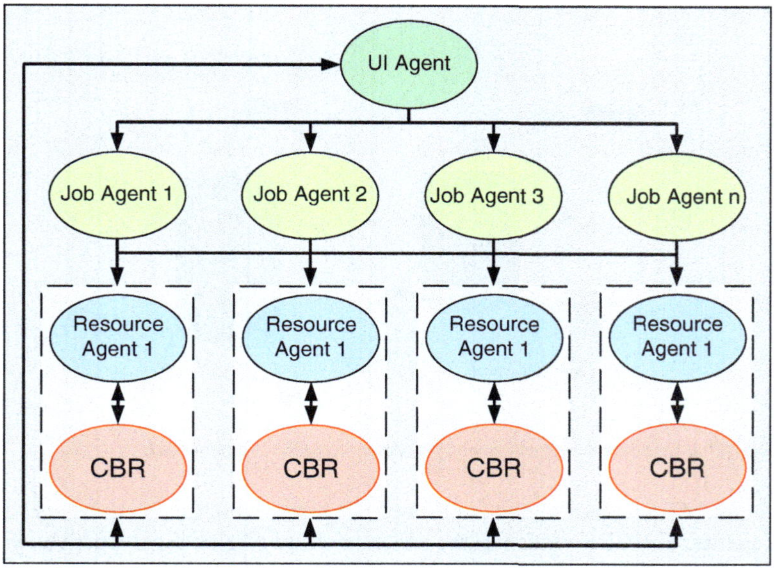

Fig. 10.4 CBR resource agents' integration

occurs after loading the problem to solve and before communicating with Job and Resource agents. In this perspective, it is possible to assume that the system performs a team learning approach, since the CBR module performs self-parameterization for the whole system.

As previously illustrated in Fig. 10.1, every new problem, or perturbations occurred, leads to a new case in the system, with the previous most similar cases being retrieved from the casebase. Then the better case is reused, becoming a suggested solution. After the solution revision, the case is executed in the MAS. This revision is performed in order to allow the MH to escape from local optimal solutions and stagnation, since it applies some disturbance in the parameters of the proposed solution. After the conclusion of the MAS execution, the case is confirmed as a good solution, being retained on the casebase as a new learned case, for future use.

10.4.3 Resource Agents' CBR Perspective (Multi-apprentice)

Another possible approach of CBR integration is in a resource agents' perspective, where a concurrent learning is made, since each resource agent has his own CBR module for performing its self-parameterization of MH (see Fig. 10.4).

In this perspective, the problem is quite different, since each resource/machine has to solve his own Single Machine Scheduling Problem, instead of EJSSP. However, in this concurrent approach, each resource agent, through CBR module, uses his own and independent technique with self-tuned parameters. This is an approach opposing with global perspective, since each resource agent can use

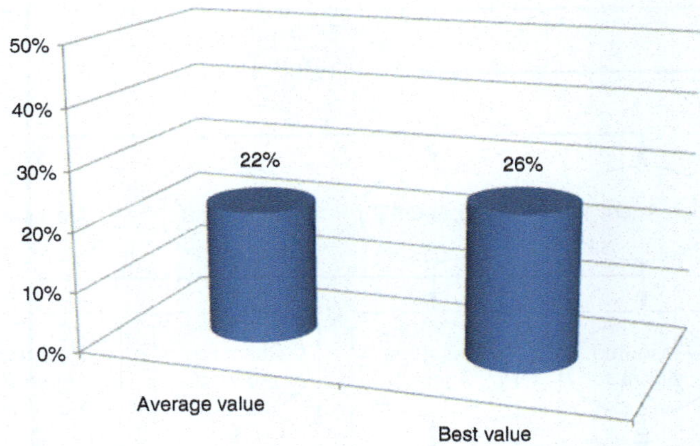

Fig. 10.5 Obtained results improvement percentage on global perspective

a different MH and parameters, and does not know which techniques are used by other agents. The CBR cycle has the same structure of global perspective, working in the same way, like previously described.

10.5 Computational Study

The objective of the computational study is to analyze the CBR integration and reach some conclusions about which perspective presents advantages when compared with previous obtained results (without the integration of CBR based module) and with the two proposed perspectives together.

For the computational study, all instances available in OR-Library Job-Shop Scheduling problems (82 instances) were used, proposed by Adams, Balas and Zawack [9], Fisher and Thompson [19], Lawrence [20], Applegate and Cook [21], Storer, Wu and Vaccari [22], and Yamada and Nakano [23]. These instances cover problems with 10, 20, 30, and 50 jobs, processed by 5, 10, 15, and 20 machines. All instances were executed five times (before and after CBR integration). From the obtained results we retrieve the best and the average values of obtained solutions. The machine used for the computational study is a HP Z400 Workstation, with the following main characteristics: Intel® Xeon® CPU W3565 @ 3.20 GHz, 6 GB RAM, Samsung HD103SJ disk with 1 TB, and Windows 7 64-bit.

10.5.1 Global Application

The comparison between previous obtained results (without self-optimization through the introduction of CBR in the system) and global perspective's obtained results is shown in Fig. 10.5.

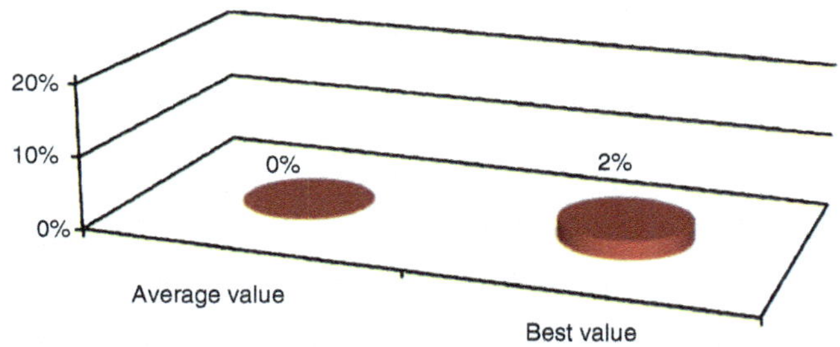

Fig. 10.6 Obtained results improvement percentage on resource agents' perspective

Analyzing this figure, average obtained results were improved by 22 % and best obtained results were improved by 26 %. We consider that system performance was improved significantly with CBR-based module since previous behaviour and performance was already quite effective.

10.5.2 Resource Agents' Application

The comparison of obtained results before the introduction of CBR in the system with resource agents perspective is shown in Fig. 10.6. Analyzing this figure, it is possible to conclude that the results improvement was very poor, since average results were not improved at all and best results were improved only by 2 %. This behavior indicates that this perspective is not adequate for the system.

10.5.3 Global Perspective vs Resource Agents' Perspective

When comparing the two perspectives, it is possible to analyze and conclude which one is more effective in order to integrate the system.

Figure 10.7 compares the average and best obtained results between both proposed perspectives. The average results from a global perspective were all better when comparing with resource agents perspective. Comparing the best obtained results, it is possible to conclude that best results from global perspective were 87 % better and only 11 % were better when executing resource agents' perspective. The obtained results were the same in 2 % of instances. Thus, it is possible to conclude that global perspective, in which the learning process is based in one apprentice, is more effective for integrating the system, and, for this system, a team learning approach revealed better results than a concurrent learning approach.

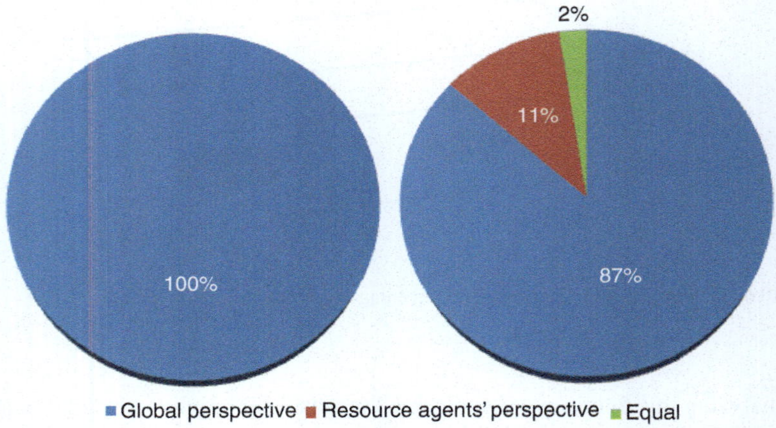

Fig. 10.7 Comparison of average (*left*) and best (*right*) obtained results between perspectives

10.6 Conclusions and Future Work

This paper addressed the use of CBR to self-parameterize MH to solve Scheduling problems, integrated on a MAS. With CBR, the system is able to choose and parameterize MH autonomously, for the resolution of static or dynamic scheduling problems.

Two perspectives of CBR integration were proposed. The first one was based on a team learning architecture, with CBR being integrated in a global perspective, choosing and self-parameterizing the same MH for every resource agents . The second perspective was based on a concurrent learning approach, where every resource agent has his own CBR module, choosing and parameterizing MH for the resolution of his own local SMSP problem.

From the computational study it was possible to conclude that global perspective revealed better results when comparing with resource agents' perspective. With this, we conclude that, in this system, a team learning approach is more effective than a concurrent learning approach.

Future work includes improvements on CBR global perspective and the implementation of other learning techniques, to compare with CBR.

Acknowledgments This work is supported by a PhD scholarship (SFRH/BD/63404/2009) by FCT "Fundação para a Ciência e a Tecnologia", FEDER Funds through the "Programa Operacional Factores de Competitividade – COMPETE" program and by National Funds through FCT "Fundação para a Ciência e a Tecnologia" under the project: FCOMP-01-0124-FEDER-PEst-OE/EEI/UI0760/2011.

References

1. Pinedo M (2012) Scheduling: theory, algorithms, and systems, 4th edn. Springer, Boston
2. Ouelhadj D, Petrovic S (2008) A survey of dynamic scheduling in manufacturing systems. J Sched 12:417–431
3. Madureira A, Santos J, Fernandes N, Ramos C (2007) Proposal of a cooperation mechanism for team-work based multi-agent system in dynamic scheduling through meta-heuristics. In: IEEE international symposium on assembly and manufacturing, Ann Arbor, pp 233–238
4. Monostori L, Váncza J, Kumara S (2006) Agent based systems for manufacturing. CIRP Ann Manuf Technol 55(2):697–720
5. Wellner J, Dilger W (1999) Job shop scheduling with multiagents. Workshop planen und Konfigurieren, Universität Würzburg, German, 3–5 March
6. Alonso E, D'inverno M, Kudenko D, Luch M, Noble J (2001) Learning in multi-agent systems. Knowl Eng Rev 16(3):277–284
7. Panait L, Luke S (2005) Cooperative multi-agent learning: the state of the Art. Auton Agents Multi-Agent Syst 11:387–434
8. Madureira A (2003) Meta-heuristics application to scheduling in dynamic environments of discrete manufacturing. PhD dissertation, University of Minho, Braga, (in portuguese)
9. Adams J, Balas E, Zawack D (1988) The shifting bottleneck procedure for job shop scheduling. Manag Sci 34:391–401
10. Gonzalez T (2007) Handbook of approximation algorithms and metaheuristics, Chapman&hall/Crc computer and information science series. Chapman&Hall/Crc, Boca Raton
11. Kolodner J (1993) Case-based reasoning. Morgan Kaufmann Publishers Inc, San Mateo
12. Beddoe G, Petrovic S, Li J (2009) A hybrid metaheuristic case-based reasoning system for nurse rostering. J Sched 12(2):99–119
13. Schank R (1982) Dynamic memory; a theory of reminding and learning in computers and people. Cambridge University Press, New York
14. Gentner D (1983) Structure mapping – a theorical framework for analogy. Cognit Sci 7:155–170
15. Petrovic S, Yang Y, Dror M (2007) Case-based selection of initialisation heuristics for metaheuristic examination timetabling. Expert Syst Appl 33:772–785
16. Aamodt A, Plaza E (1994) Case-based reasoning: foundational issues, methodological variations, and system approaches. Artif Intell Commun 7:39–52
17. Madureira A, Pereira I (2010) Self-optimization for dynamic scheduling in manufacturing systems. In: Technological developments in networking, education and automation. Springer, Dordrecht, pp 421–426
18. Horling B, Lesser V (2004) A survey of multi-agent organizational paradigms. Knowl Eng Rev 19(4):281–316
19. Fisher H, Thompson GL (1963) Probabilistic learning combinations of local job-shop scheduling rules. In: Muth JF, Thompson GL (eds) Industrial scheduling. Prentice-Hall, Englewood Cliffs, pp 225–251
20. Lawrence S (1984) Resource constrained project scheduling: an experimental investigation of heuristic scheduling techniques (Supplement), Carnegie-Mellon University
21. Applegate D, Cook W (1991) A computational study of the job-shop scheduling instance. ORSA J Comput 3:149–156
22. Storer RH, Wu SD, Vaccari R (1992) New search spaces for sequencing instances with application to job shop scheduling. Manag Sci 38:1495–1509
23. Yamada T, Nakano R (1992) A genetic algorithm applicable to large-scale job-shop instances. In: Manner R, Manderick B (eds) Parallel instance solving from nature. North-Holland, Amsterdam, pp 281–290

Chapter 11
Haptic-Based Robot Teleoperation: Interacting with Real Environments

Pedro Neto, Nélio Mourato, and J. Norberto Pires

Abstract Robot teleoperation *via* haptic interface has been increasingly studied in recent years. In fact, the capacity to remotely feel the forces being exerted on a robot end-effector is a major advantage for many modern robot applications. This study aims to explore the capabilities of a low-cost haptic device, a force/torque sensor and a commercially available industrial robot, both together for haptic-based robot teleoperation purposes. A ballistic approach is implemented to make the mapping between the motion produced on the haptic device by the operator and the resulting robot motion. Inspired by some studies in the field of robot haptics, the robot and haptic device are connected by a virtual spring, which allows the operator to feel the sensation of pull and push the robot at the same time that teleoperates it. Results related to the process of robot teleoperation and contact with different objects are analyzed and discussed.

11.1 Introduction

Over the last few years, robots have gain new markets and have been used for new applications. In some of these new applications, haptic-based systems can have a fundamental role and in many cases they are the starting point for the evolution of robotics in all the fields where force feedback from the robot is an important issue for the success of the robotic tasks. However, most of the "sophisticated" haptic devices are relatively expensive.

This paper discusses a haptic-based robotic system that allows a user to teleoperate a robot and at the same time receives force feedback from the same robot, all of this with a high level of abstraction from the specific robot commands. The proposed

P. Neto (✉) • N. Mourato • J.N. Pires
Department of Mechanical Engineering (CEMUC) – POLO II,
University of Coimbra, 3030-788 Coimbra, Portugal
e-mail: pedro.neto@dem.uc.pt

A. Madureira et al. (eds.), *Computational Intelligence and Decision Making: Trends and Applications*, Intelligent Systems, Control and Automation: Science and Engineering 61,
DOI 10.1007/978-94-007-4722-7_11,

platform is composed by a low-cost haptic device (the Falcon), a commercially available industrial robot, and attached to the robot wrist a 6 DOF force/torque (F/T) sensor which reports the contact forces from the interaction of the robot end-effector with its surrounding environment. The appearance of the Falcon has allowed an increase of use and diffusion of haptic technology in many different areas. It is possible to say that the great advantage of the Falcon is its price, and in fact, the Falcon has limited control capabilities when compared to other and more professional devices. Nevertheless, as a research tool, the Falcon is a relatively good device that allows to explore most of the capabilities of modern haptic devices.

One of the major concerns of today's robotics has to do with the need for intuitive ways for robot control and programming [1, 2]. This is the key point to remember, through the proposed haptic system the operator can interact with a robot without the need to be a specialist in robotics, on the contrary, the operator can focus only on the correct definition of the robotic task. Experimental tests involving robot teleoperation and contact were carried out to verify the reliability and effectiveness of the proposed platform.

11.2 Related Work

Haptic technology has experienced a tremendous growth in recent years. Applications of haptic technology have been spreading rapidly in several areas such as video game industry, arts and creation, manufacturing, telerobotics, education and training, as well as in the medical field [3].

Through the physical simulation of a real environment it is possible to calculate in real time the contact forces between the various virtual objects and by this way interact with this virtual environment through a haptic device. This creates in the user the sensation of being in contact with a real object [4]. In fact, when combined with computer graphics, haptic technology can be used to train people for tasks requiring hand-eye coordination such as surgery or spaceship manoeuvres' [4]. The Falcon was initially used as a computer game controller, however, its applications was quickly expanded to other areas [5].

Haptic technology has been widely used for interaction with virtual environments, however, there is a growing trend for applications involving interaction with real scenarios. Several control techniques have been proposed to cope with uncertainty, workspace mapping, time delay, master/slave models and telepresence [6–9].

11.3 Haptic-Based Robotic Platform

11.3.1 Architecture

Motion data from the Falcon grip is received in a computer running a software interface that manages the entire platform. These data are analyzed and converted

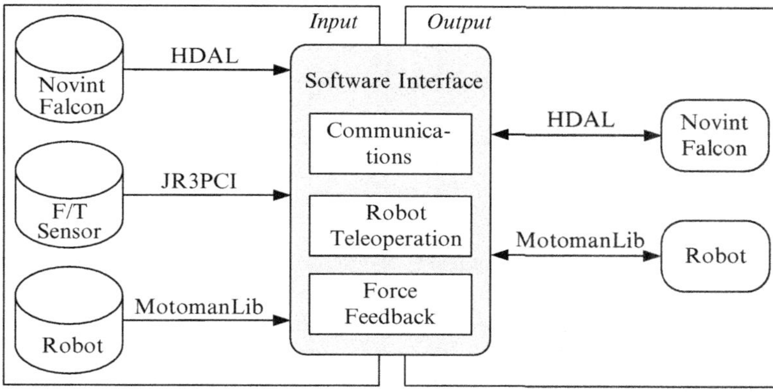

Fig. 11.1 Interaction between different system components

into robot commands. Simultaneously, data reported by the F/T sensor are received by the software interface, treated, and then through the Falcon are used as force feedback to the operator.

The Falcon is a relatively inexpensive 3 DOF haptic device. It provides accurate tracking and realistic force feedback sensation, allowing a user to interact with it and learn muscle memory. It is also important to refer that one of the most common criticisms to the Falcon is its limited workspace. The robot is a commercially available industrial robot Motoman HP6 equipped with a Motoman NX100 controller. The 6 DOF F/T sensor is a 85M35A from JR3.

The way the equipments of the platform communicate and interact with each other is here exposed. The software interface (coded in C++) controls/manages the whole system, in other words, it acquires data from the Falcon, robot and the F/T sensor, interprets such data and sends control commands to the robot and Falcon (Fig. 11.1). Motion data from the Falcon is acquired using a free tool named Haptic Device Abstraction Layer (HDAL). Moreover, the HDAL is also used to send feedback forces to the Falcon. The data from the F/T sensor are acquired using an ActiveX named JR3PCI [10]. Adding the MotomanLib component to the application interface it is possible to control and monitor robot state.

11.4 Haptic-Based Robot Teleoperation

Haptic teleoperation provides telepresence as it allows a user to remotely operate a slave robot through a master device and at the same time "feel" the remote environment. However, reliable telepresence is difficult to achieve [7]. The potential of haptic teleoperation is enormous, but the connection of master/slave stations in a coherent way is still a challenging task, especially due to some aspects such as workspace mapping, communications delay and the way the force feedback is transmitted to the operator. Whereas the master station is controlled by a human operator, the slave station often interacts with an unknown and dynamic environment (collisions can occur). The nature of this interaction greatly influences overall system performance.

11.4.1 Ballistic Approach

Concerning to haptics, position control is one of the most common control paradigms. Usually, it refers to a mapping in which the displacement of an object in physical space directly dictates the displacement of another different object in real or virtual space. For this particular study, the displacement of the Falcon grip will produce a displacement on the robot end-effector. Note that the workspace of the Falcon is much smaller than the workspace of the robot.

Robot positioning accuracy depends on the robot motion which in turn is controlled by the speed with which the user's hand moves the Falcon grip (ballistic approach). The ballistic approach makes the mapping between the motion (displacement and velocity) produced on the haptic device and the resulting robot motion [8]. Making an analogy with the way a human being performs a simple positioning task, videlicet, when a human searches for accuracy he/she moves the arm or any other part of the body slowly and, on the other hand, if he/she searches for large displacements the arm is moved quickly (accuracy is more difficult to achieve).

11.4.2 Workspace Mapping and Scaling Control

Motion data from the haptic device (master device) is used to control the robot end-effector (slave device), creating the concept of robot teleoperation. Both the master and slave workspaces are significantly different in size, so that a certain type of workspace mapping is needed. The displacement produced on the Falcon grip in two consecutive time intervals, t_n and t_{n-1}, is given by:

$$X^*_{novH2} = X^*_{nov}(t_n) - X^*_{nov}(t_{n-1}) \quad (11.1)$$

where X^*_{nov} is the absolute position of the Falcon grip in relation to its reference frame H2, Fig. 11.2. The displacement to send to the robot (relative to workspace H1) will depend on the Falcon grip displacement X^*_{novH2} and the velocity with which the operator moves the grip. A rotation matrix, R, makes the relation between workspace H2 and H1. Workspace H1 is a dynamic workspace, in other words, H1 has a fixed orientation and a dynamic origin which is coincident with the tool center point (TCP) of the end-effector in each instant of time. So, the displacement to send to the robot can be calculated as:

$$X_a = R \cdot \left(k_s \cdot X^*_{novH2}\right), \quad (11.2)$$

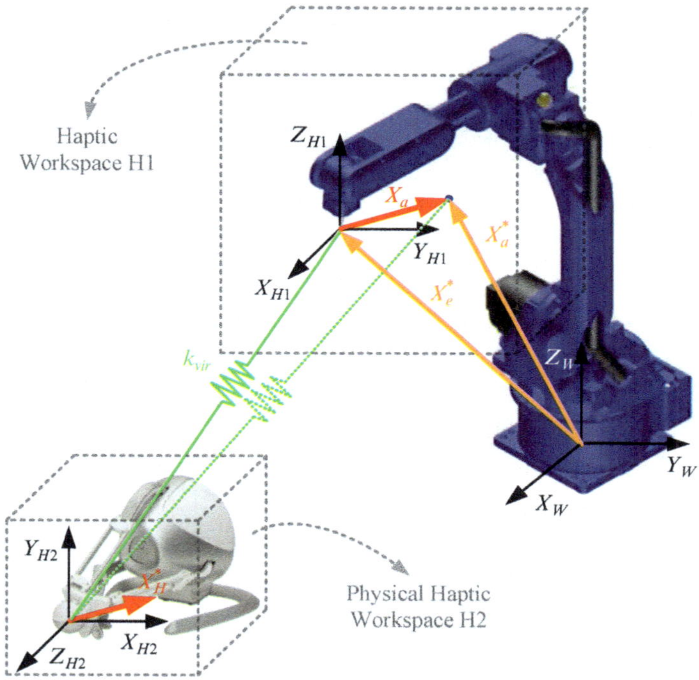

Fig. 11.2 Workspace mapping

where k_s is a scale factor used to make the mapping between H1 and H2. It depends on the velocity $\left\|\dot{X}_{novH2}\right\|$ with which the operator moves the Falcon grip:

$$\dot{X}_{novH2} = \frac{X^*_{novH2}}{\Delta t}, \tag{11.3}$$

where $\Delta t = t_n - t_{n-1}$. This means that for the same grip displacement the motion to send to the robot will directly change with the velocity with which the operator moves the haptic cursor; if the velocity increases the displacement for the slave robot will also increase, and *vice versa*. So, k_s can be calculated as:

$$k_s = k_r \cdot \Delta t, \tag{11.4}$$

where k_r is a scaling factor. If we are moving the Falcon grip at relatively high speed, in this way we should have that $k_r > 0$. In the other hand, if we are moving the grip at relatively low speed, in this case k_r should have a value near zero.

$$k_r = \begin{cases} k_{sp} \cdot \left\|\dot{X}_{novH2}\right\|, & \left\|\dot{X}_{novH2}\right\| > v_t \\ k_{sn}, & \left\|\dot{X}_{novH2}\right\| \leq v_t \end{cases}, \tag{11.5}$$

where k_{sp} is a positive scaling factor, k_{sn} is a negative scaling factor and v_t is the velocity threshold.

The Falcon is a 3 DOF haptic device; however, we need 3 DOF to control the end-effector translations and 3 DOF to control the end-effector orientation. Since the Falcon's grip has four buttons, one of the buttons is used to "transform" the motion extracted from the Falcon not in robot end-effector displacements but in orientations. From the displacement produced on the Falcon grip X^*_{novH2}, the rotations for the end-effector about each one of the Cartesian axes can be established:

$$X_r^* = \begin{cases} R_x = \arctan(x^*_{novH2}) \\ R_y = \arctan(y^*_{novH2}) \\ R_z = \arctan(z^*_{novH2}) \end{cases}, \tag{11.6}$$

where the rotation angles to send to the robot (relative to H2) are $X_r = R \cdot X_r^*$.

11.4.3 Force Feedback

Force feedback from the F/T sensor is used not only to create the reactive force that the Falcon transmits to the operator, but also to act on the displacements to send to the robot. It is important to start by establishing that we have two types of feedback forces to send to the Falcon: the force due to effective contact and a virtual force. The concept of virtual force associated with haptics-robotics has been studied by a number of authors [6, 11]. In this paper, and taking as inspiration the studies mentioned above, an adapted virtual force concept is implemented.

Considering that the grip of the master device is linked to the robot end-effector by a virtual spring with a constant elastic modulus k_{vir}, the virtual force can be considered to be proportional to the distance between the actual end-effector position X_e^* and the end-effector position after applied the displacement X_a in workspace H1, X_a^* (Fig. 11.2).

$$F_{vir} = k_{vir} \cdot (X_a^* - X_e^*) \tag{11.7}$$

From Fig. 11.2, $X_a^* = X_e^* + X_a$, and thus (11.7) becomes $F_{vir} = k_{vir} \times X_a$. The virtual force can not be too high because it can be confused with the force due to effective contact. To solve this situation, it was applied a smooth function to the virtual force: $f(F_{vir}) = F_{virN} = (F_{vir})^{1/6}/2$. From the contact forces measured by the F/T sensor, F_c, the force feedback to send to the master device can be calculated as:

$$F_{cmd} = k_{cmd} \cdot F_c, \tag{11.8}$$

where k_{cmd} is a scale factor that makes the mapping between the forces measured by the F/T sensor and the forces to send to the Falcon. In this way, the force feedback

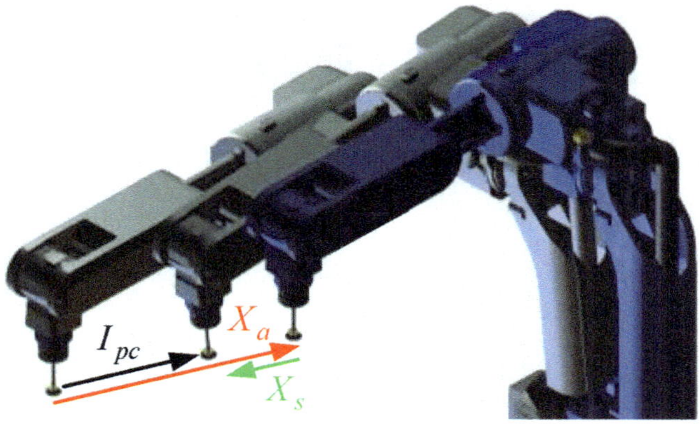

Fig. 11.3 The effect of the contact force on the displacement sent to the robot

to send to the Falcon, F_d, can be described as the sum of the force feedback due to effective contact and the virtual force:

$$F_d = F_{cmd} + F_{virN} \tag{11.9}$$

The measured contact forces also produce effect on the displacement vector sent to the robot. Thus, when the robot end-effector becomes in contact with an object and the contact forces exceed a predefined value, it is intended that the displacement to send to the robot is adjusted to avoid excessive contact forces. It means that if F_c is bigger than zero, the displacement to send to the robot will be adjusted by the factor X_s:

$$X_s = \frac{F_c}{F_{\max}} \cdot X_a, \tag{11.10}$$

where $F_{\max}$ is the maximum force value predefined by the operator. The adjusted displacement to send to the robot is (Fig. 11.3):

$$I_{pc} = X_a - X_s. \tag{11.11}$$

We can see that I_{pc} decreases with the augment of F_c, and when $F_c = F_{\max}$, X_s becomes equal to X_a and $I_{pc} = 0$. In practice it was established that when $F_c \geq F_{\max}$, the following condition is applied: $|X_s| \leq 0.9 \cdot |X_a|$; in this case I_{pc} can be a small value, but never zero. A similar reasoning can be done for the end-effector rotations but using torques instead forces and the factor X_{rs} instead X_s.

11.4.4 Summary

The block diagram on Fig. 11.4 summarizes the control methods applied to the proposed platform. This diagram contains contributions from the studies above

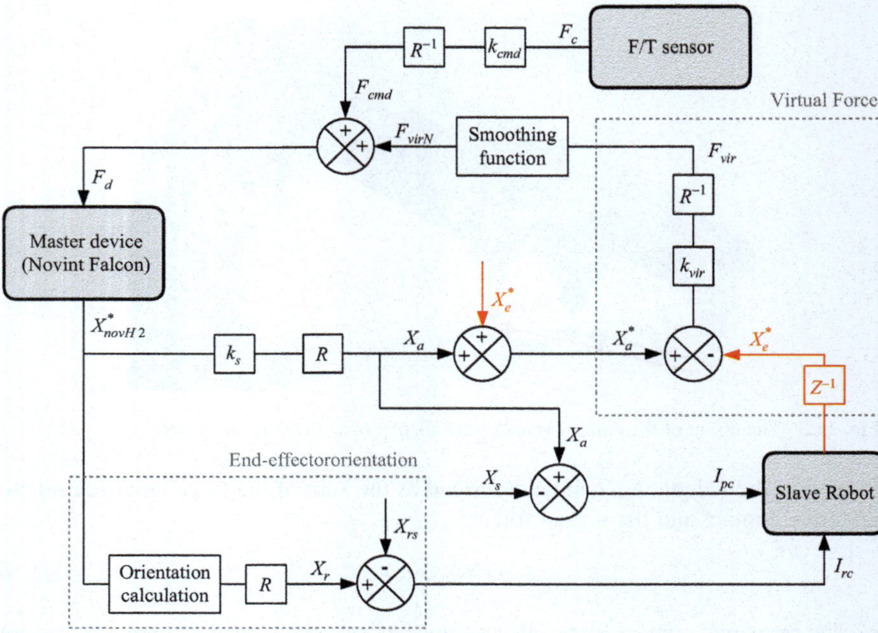

Fig. 11.4 Block diagram for the proposed system

mentioned, which were combined and adapted to this specific platform. Note that it is not necessary to acquire the position of the robot end-effector X_e^* (at red) to achieve our goals, since it was assumed that the origin of workspace H1 is dynamic, coincident in each instant with the TCP of the robot end-effector.

11.5 Experiments

Two different experiments involving contact with two different objects with different stiffness values (a sponge and a wood top table) were carried out. These experiments aim to validate the proposed system, giving to the reader a better understanding about the way the system works and reacts to contact.

11.5.1 Results and Discussion

Qualitatively analyzing the proposed platform, we can say that in terms of robot teleoperation it is really easy and intuitive to move the robot for precision positioning or for large robot displacements. The force feedback delivered to the operator helps him/her to understand the magnitude of the contact forces and in which

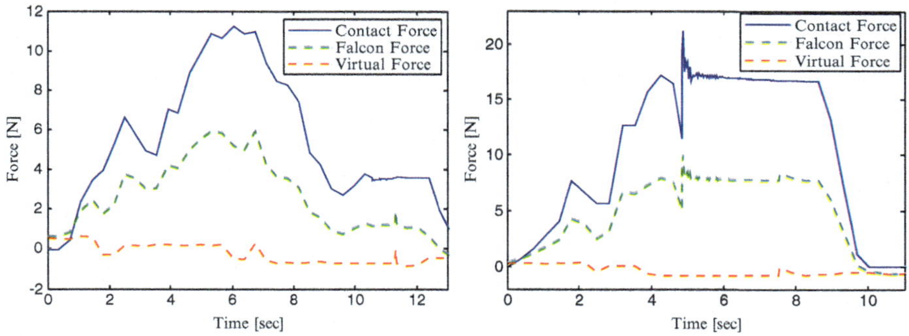

Fig. 11.5 Contact with a sponge (*left*) and contact with a wood top table (*right*)

direction these forces are acting. However, the system presents some important negative points. When the operator wants to produce a large displacement on the robot, he/she needs to constantly release the grip button and repositioning the grip. This is due to the relative small workspace of the Falcon when compared with the robot workspace.

The forces in vertical direction reporting the contact between the end-effector and a sponge are represented in Fig. 11.5 (left). As expected, whereas the operator moves the end-effector against the sponge the contact forces increase. Owing to the force feedback transmitted to the Falcon, the operator feels a significant push up in the hand that holds the grip. On contrary, when the end-effector is moved up the contact forces alleviate. The virtual force is limited to relatively small values when compared with the other forces. Through the graphs (Fig. 11.5) it is possible to see that when the end-effector is moved down the virtual forces have a positive value. In the other hand, when the end-effector is moved up the virtual forces have a negative value. This means that at the moment of the contact, the effect of the virtual forces is not perceptible. Nevertheless, in free space the virtual forces allow the operator feels the sensation of being pushing or pulling the end-effector.

Maximum forces are achieved at the moment of the contact with the wood top table and not with the sponge, as expected. The operator feels that the table presents a greater stiffness than the sponge because the force feedback increases quickly (6–18 N in 1.5 s) and the operators feels a hard push up in the grip (Fig. 11.5 (right)). The "disturbance" (Time $\approx$ 5 s) in graph of Fig. 11.5 can be justified by the effect of the spring embedded into the robot tool, due to vibrations of the robot in conjugation with the spring effect, or due to the reaction of the operator that when feels a hard push has tendency to react and counteract the situation.

11.6 Conclusions

A haptic-based robotic platform for interaction with real environments was proposed. This platform allows to teleoperate an industrial robot whereas the operator feels the forces that the robot wrist is subjected. Moreover, the platform showed to be intuitive to operate, even for non-specialists in robotics. The ballistic approach allows relatively fine and stable positioning of the robot end-effector. The virtual spring effect is visible and useful when the robot is teleoperated in free space.

References

1. Neto P, Pires JN, Moreira AP (2010) High-level programming and control for industrial robotics: using a hand-held accelerometer-based input device for gesture and posture recognition. Ind Robot Int J 37(2):137–147
2. Ferreira M, Moreira AP, Neto P (2011) A Low-cost laser scanning solution for flexible robotic cells: spray coating. Int J Adv Manuf Technol 58:1031–1041
3. Eid M, Orozco M, El Saddik A (2007) A guided tour in haptic audio visual environments and applications. Int J Adv Media Commun 1:265–297
4. Salisbury K, Barbagli F, Conti F (2004) Haptic rendering: introductory concepts. IEEE Comput Graph Appl 24:24–32
5. Schill F, Mahony R, Corke P, Cole L (2008) Virtual force feedback teleoperation of the insectbot using optic flow. In: Australian conference on robotics and automation, Canberra
6. Park J, Khatib O (2006) A haptic teleoperation approach based on contact force control. Int J Robot Res 25:575–591
7. Cortesao R, Park J, Khatib O (2006) Real-time adaptive control for haptic telemanipulation with kalman active observers. IEEE Trans Robot 22:987–999
8. Conti F, Khatib O (2005) Spanning large workspaces using small haptic devices. In: Proceedings of the 1st world haptics conference, Pisa, pp 183–188
9. Horie T, Tanaka K, Abe N, Taki H (2001) Generation of haptic model and remote control of robot using force sensor and phantom. In: International conference on virtual systems and multimedia, Berkeley, pp 774–756
10. Mendes N, Neto P, Pires JN, Moreira AP (2010) Fuzzy-PI force control for industrial robotics. In: Vadakkepat P et al (eds) Trends in intelligent robotics, CCIS 103. Springer, Heidelberg, pp 322–329
11. Son W, Kim K, Amato NM, Trinkle JC (2000) Interactive dynamic simulation using haptic interaction. In: IEEE/RSJ international conference on intelligent robots and systems. IEEE Press, New York, pp 145–150

Chapter 12
Multi-agent Predictive Control with Application in Intelligent Infrastructures

J.M. Igreja, S.J. Costa, J.M. Lemos, and F.M. Cadete

Abstract This paper describes a stabilizing decentralized predictive controller for multivariable systems with application in intelligent infrastructures. Cooperative distributed control systems can be seen as a set of computational agents that cooperate with their neighbors, in a open information interchange infrastructure, to achieve local performance indexes, suitable to globally control large scale, geographically expansive, systems. An agent behavioral induction approach is introduced and discussed in the model predictive control framework. Cooperative iteration convergence, with information interchange, is proved. The algorithm is fully developed for a serially chained systems and an application to a real water delivery canal is presented.

12.1 Introduction

This paper describes a stabilizing decentralized predictive control for multivariable systems with application to intelligent infrastructures. Cooperative distributed control systems can be seen as a set of computational agents that cooperate with their neighbors, in an open information interchange infrastructure, to achieve local performance indexes, suitable to globally control in closed-loop large scale, geographically expansive, systems. A behavioral induction approach is introduced and discussed in the model predictive control framework. The predictive control algorithm is fully developed for generic serially chained systems.

J.M. Igreja (✉)
Automation and Power Systems Engineering Department, ISEL,
R. Conselheiro Emdio Navarro, 1, Lisbon, Portugal
e-mail: jigreja@deea.isel.pt

S.J. Costa
Chemical Engineering Department, ISEL, R. Conselheiro Emdio Navarro, 1, Lisbon, Portugal

J.M. Lemos • F.M. Cadete
Control of Dynamic Systems Group, INESC-ID/IST, R. Alves Redol, 9, Lisbon, Portugal
e-mail: jlml@inesc-id.pt

A. Madureira et al. (eds.), *Computational Intelligence and Decision Making: Trends and Applications*, Intelligent Systems, Control and Automation: Science and Engineering 61,
DOI 10.1007/978-94-007-4722-7_12,

Decentralized control systems arise from the need to deal with a large number of interconnected geographical distributed subsystems where information exchange between local subsystems is a priori somehow constrained. This physical inability decentralized framework since each system is often simpler to understand than a large composite MIMO system. Fault tolerance, growing geographical expansive systems, large scale networked systems, such as power systems grids and water distribution canal pools, are other good reasons to decentralize control.

David Fogel's book [3] dedicates one chapter to closed-loop control and model prediction recognizing that "*Intelligence requires iterative prediction and control*" as two basic features. In this paper computational agents include, beside these two, cooperative behavior for obtaining the same common goals.

The paper's main contributions consist, first, in the derivation of a novel decentralized control algorithm as a decentralized version of the original MIMO SIORCH. In second, the developed algorithm is included in a computational multi-agent scenario where cooperation and behavior induction are key features suitable for application in intelligent infrastructures [8]. It has also been proved that, considering all the interactions between all subsystems, that the overall decentralized controlled system is stable assuring its straightforward application to stable and unstable systems. The proof is omitted for lack of space.

12.2 MPC Framework

Model Predictive Control (MPC) is a well understood control methodology [6, 10]. Nonlinear Model Predictive Control has important advantages for advanced process control [2]. This is due to the fact that it allows an easy constraints inclusion, making it a decisive advantage for industrial applications when compared to other methods. It can also be very easily extended to the multivariable case. Other key advantages are the explicit open-loop prediction calculations and the direct design of controllers for systems with delays. However, a major issue consists in ensuring closed-loop stability for finite control horizons that implies the introduction of a dedicated constraint, in the optimization underlying problem.

In the centralized MPC regulation problem (12.1), (12.2), (12.3), and (12.4), the following feasible optimization problem is solved to find the subsystems inputs $u_i(k)$ at each instant in discrete time k, see [13] for instance. The discrete global system model is considered to be linear [11] and the cost function $J(k)$ is quadratic with symmetric weighting matrices Q_i and R_i.

$$\min_u J(x, u; x(k)) = \sum_i w_i J_i(x_i, u_i; x_i(k)) \tag{12.1}$$

$$s.t.\, x(k+1) = Ax(k) + Bu(k) \tag{12.2}$$

$$u_i \in \Omega_i \tag{12.3}$$

where $w_i > 0$, $\Sigma_i w_i = 1$ and Ω_i is an admissible control set.

$$J_i(k) = \sum_{j=1}^{N} \left(x_{i,k+j}^T Q_i x_{i,k+j} + u_{i,k+j-1}^T R_i u_{i,k+j-1} \right) \tag{12.4}$$

where $Q_i \geq 0$, $R_i > 0$, subsystem state at k, $x_{i,k}$, is given and the pair $(A_i, Q_i{}^{1/2})$ is detectable [11].

Centralized MPC achieves the best attainable performance, Pareto Optimal, as the effects of interconnections among subsystems are accounted for exactly. Any conflicts between control objectives are resolved optimally. The controller is defined by implementing the first input of the corresponding solution for the optimization problem (receding horizon).

In a decentralized version of MPC (12.5), (12.6), (12.7), (12.8), and (12.9) the global model is partitioned and the global cost function is replaced by local ones that measure the impact of local control action in a communication based strategy due to information exchange constrains [7, 12]. For each subsystem a local cost function is iterated L times before the resulting control input u_i^L is found. Remark that at each iteration u_i^l the optimization problem is solved for $x_{j\neq i}^{l-1}$ and $u_{j\neq i}^{l-1}$ which are broadcasted by the other subsystem controllers. If the communication based iterates converge, different types of equilibrium can be achieved by adjusting the weights α_h, where h indexes the set that includes all subsystems directly connect to subsystem i and includes itself. If $\alpha_{h\neq i} = \varepsilon$, with $\varepsilon \ll 1$, except for $\alpha_{h=i} \gg \varepsilon$ (inducing egocentric behavior) then at limit ($\varepsilon \rightarrow 0$) a Nash equilibrium is achieved, resulting in potentially poor closed-loop performance or even instability because conflicts between different controllers are solved in a competitive iteration rather than in a cooperative one. If $\alpha_h \neq 0$ (inducing altruistic behavior) then all iterates generated by a cooperation algorithm are strictly feasible, satisfying all the local constraints, and resulting in a nominal distributed closed-loop stable control law, as it will be shown for the particular algorithm developed in this paper.

$$\min_{u_i^L} \bar{J}_i^L(x_h, u_i; x_h(k)) = \min_{u_i^L} \sum_h \alpha_h J_{h,i}(k) \tag{12.5}$$

$$s.t.\, x_i(k+1) = A_{ii} x_i(k) + B_{ii} u_i(k) \tag{12.6}$$

$$+ \sum_{j\neq 1} \left(A_{ij} x_j^{l-1}(k) + B_{ij} u_j^{l-1}(k) \right) \tag{12.7}$$

$$u_i \in \Omega_i \tag{12.8}$$

where $l = 1, \ldots, L$ and

$$J_{h,i}(k) = \sum_{j=1}^{N} \left(x_{h,k+j}^T Q_i x_{h,k+j} \right) + \sum_{j=0}^{N-1} u_{i,k+j}^T R_i u_{i,k+j} \tag{12.9}$$

In a pure altruist scenario α_h will be all equal. General conditions for cooperative iterates convergence and solution uniqueness, by the computational agents, is discussed in [1].

12.3 Decentralized SIORHC

SIORHC is a receding tracking horizon control algorithm equipped with stabilizing zero terminal constraints for moderate values of the prediction horizon dimension, Fig. 12.2. Details about the centralized original algorithm, early results and extensions to MIMO systems can be found respectively in [6], [5] and [4]. In this section a decentralized version of the algorithm is derived for serially chained systems based on neighborhood optimization, see [14], as a particular case of the decentralized MPC discussed above when there are no input constraints, $\Omega_i = \mathbf{R}^m$, and when zero equality terminal constraints at the end of the prediction horizon, forcing the outputs to take a particular value, are introduced. This constraint will play a major role in guaranteeing global closed-loop stability, see Sect. 12.5. In fact, in this case the assumption of a linear model and unbound inputs yields, by solving the stated optimization problem for each local control agent, an analytical iterative solution that converges to the unique optimal distributed solution, providing that a given condition can be verified *a priori*. Cooperative behavior among the control agents can be induced by adjusting weights on local cost functions yielding different overall performances. Figure 12.1 depicts generic service oriented architecture, an open computational infrastructure, where the agents can cooperate by subscribing services. This figure stresses the shared and broadcasted information during the iterative cooperation by each control agent. About cooperative service oriented open process control technological issues see, for instance, [9].

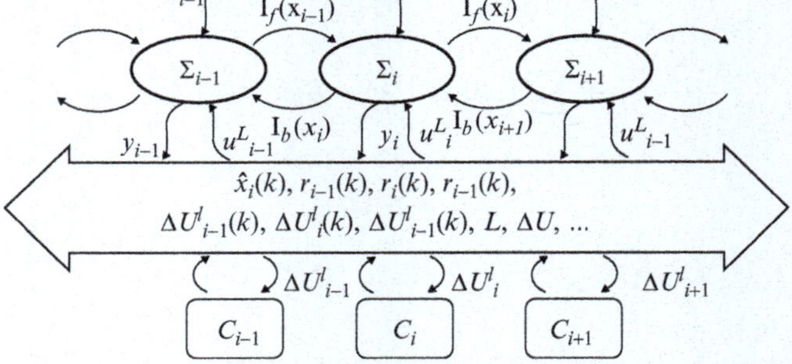

Fig. 12.1 Agent shared information service, C_i; Information interchanged by system i, Σ_i, is depicted in the service bus. $I_{f(b)}$ denotes forward (backward) interactions and the d_i denotes disturbances

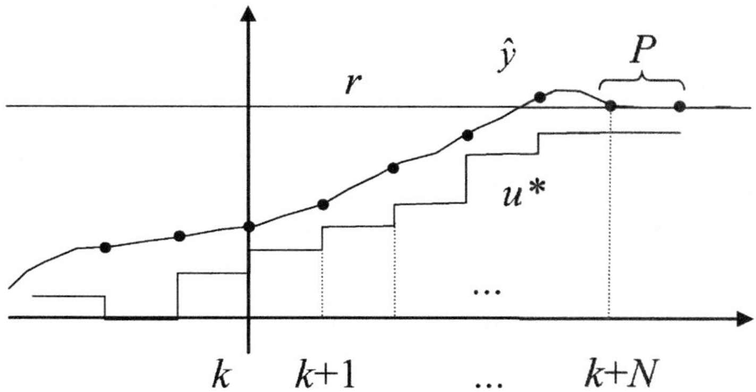

Fig. 12.2 SIORHC prediction horizon with constraints

Consider the extended or augmented linear state model

$$\begin{bmatrix} x_{k+1} \\ y_k \end{bmatrix} = \begin{bmatrix} A & B \\ C & 0 \end{bmatrix} \begin{bmatrix} x_k \\ \Delta u_k \end{bmatrix} \tag{12.10}$$

For this system the predicted outputs at $k + j$ can be obtained from

$$\begin{aligned} \hat{y}(k+j) &= \sum_{i=0}^{j-1} CA^{j-i-1} B\Delta u(k+i) + \hat{y}_0(k+j) \\ \hat{y}_0(k+j) &= CA^j x(k) = \Gamma x(k) \end{aligned} \tag{12.11}$$

where $\hat{y}_0$ is the output predicted value without moves. For $j = 1 \ldots N, N{+}1, \ldots, N{+}P$, (12.11) yields the following predictive equations

$$\begin{aligned} \hat{Y}_N &= G_N \Delta U + \Gamma_N x(k) = G_N \Delta U + \hat{Y}_{0N} \\ \hat{Y}_P &= G_P \Delta U + \Gamma_P x(k) = G_P \Delta U + \hat{Y}_{0P}, \end{aligned} \tag{12.12}$$

with

$$\begin{aligned} \hat{Y}_N &= [y_{k+1} \ldots y_{k+N}]^T \\ \hat{Y}_P &= [y_{k+N+1} \ldots y_{k+N+P}]^T \\ \Delta U &= [\Delta u_k \ldots \Delta u_{k+N-1}]^T. \end{aligned} \tag{12.13}$$

Equation (12.12) can be rearranged by considering only input related interactions between neighboring serially connected systems with $i = 1, 2, \ldots$, see Fig. 12.1,

$$\hat{Y}_{iN} = \sum_{h=-1}^{1} G_{Ni(i+h)} \Delta U_{i+h} + \hat{Y}_{0Ni}$$
$$\hat{Y}_{iP} = \sum_{h=-1}^{1} G_{Pi(i+h)} \Delta U_{i+h} + \hat{Y}_{0Pi}. \tag{12.14}$$

Local controllers are obtained by minimizing the local cost function for the i subsystem

$$\bar{J}_i = \sum_{g=-1}^{1} \left(\alpha_{i+g} \sum_{j=1}^{N} e_{i+g,k+j}^T Q_i e_{i+g,k+j} \right) + \sum_{j=0}^{N-1} \Delta u_{i,k+j}^T R_i \Delta u_{i,k+j} \tag{12.15}$$

with zero terminal horizon constrain given by, see Fig. 12.2,

$$\left[y_{1,k+N+1} \dots y_{1,k+N+P}\right]^T = \left[r_{1,k+N+1} \dots r_{1,k+N+P}\right]^T \tag{12.16}$$

where $e_{i,k} = r_{i,k} - y_{i,k}$, is the subsystem i error, at instant k, in relation with the reference sequence, $r_{i,k}$, and $Q_i \geq 0$ and $R_i > 0$ are symmetric weighting matrices. Weights α_i can change agents' behavior as explained before.

Equivalently minimizing $\bar{J}_i$with respect to ΔU_i can be written as

$$\min_{\Delta U_i} \bar{J}_i = \sum_{g=-1}^{1} \alpha_{(i+g)} \left\| Y_{RN(i+g)} - \hat{Y}_{N(i+g)} \right\|_{Q_{i+g}}^2 + \left\| \Delta U_i \right\|_{R_i}^2$$
$$s.t. \hat{Y}_{Pi} = Y_{RPi} \tag{12.17}$$

The stated QP optimization problem with constraints can now be solved by finding the vector ΔU_i that minimizes the Lagrangian

$$\mathrm{L_i} := \sum_{g=-1}^{1} \alpha_{(i+g)} \left\| E_{N(i+g)} \right. \tag{12.18}$$

$$- \sum_{h=-1}^{1} G_{N(i+g)(i+g+h)} \Delta U_{(i+g+h)} \Big\|_{Q_{(i+g)}}^2 + \tag{12.19}$$

$$\left\| \Delta U_1 \right\|_{R_i}^2 + \left[E_{Ni} + \sum_{h=-1}^{1} G_{Pi(i+h)} \Delta U_{(i+h)} \right]^T \lambda_i$$

where

$$E_{Ni} = Y_{RNi} = \hat{Y}_{0Ni}$$
$$E_{Pi} = Y_{RPi} = \hat{Y}_{0Pi} \tag{12.20}$$

and λ_i is a $P.p_i$ column vector of Lagrange multipliers. The conditions to find the minimum are

$$\begin{aligned} \frac{\partial L_i}{\partial \Delta U_i} &= 0 \\ \frac{\partial L_i}{\partial \lambda_i} &= 0. \end{aligned} \tag{12.21}$$

Solving (12.21), for all the subsystems, yields the building block matrices for the decentralized SIORHC serially connected systems solution, *viz.*,

$$\Phi = \left[block\, pentadiagonal\left\{\cdots \Phi_{i(i-2)}\Phi_{i(i-1)}\Phi_{ii}\Phi_{i(i+1)}\Phi_{i(i+2)}\cdots\right\}\right] \tag{12.22}$$

$$\Psi = \begin{bmatrix} \vdots \\ \Psi_i \\ \vdots \end{bmatrix} \tag{12.23}$$

$$\begin{aligned} \Phi_{i(i-2)} &= S_i\left(\alpha_{i-1}G_{N(i-1)i}^T Q_{i-1}G_{N(i-1)(i-2)}\right) \\ \Phi_{i(i-1)} &= S_i\left(\alpha_{i-1}G_{N(i-1)i}^T Q_{i-1}G_{N(i-1)(i-1)} + \alpha_i G_{Nii}^T Q_i G_{Ni(i-1)}\right) + G_{Pii}^T W_i G_{Pi(i-1)} \\ \Phi_{ii} &= M_i \\ \Phi_{i(i+1)} &= S_i\left(\alpha_i G_{Nii}^T Q_i G_{Ni(i+1)} + \alpha_{i+1}G_{N(i+1)i}^T Q_{i+1}G_{N(i+1)(i+1)}\right) + G_{Pii}^T W_i G_{Pi(i+1)} \\ \Phi_{i(i+2)} &= S_i\left(G_{N(i+1)i}^T Q_{i+1}G_{N(i+1)(i+2)}\right), \end{aligned} \tag{12.24}$$

$$\Psi_i = S_i \sum_{h=-1}^{1} \alpha_{i+h}G_{N(i+h)i}^T Q_{i+h}E_{N(i+h)} + G_{Pii}^T W_i E_{Pi}, \tag{12.25}$$

where

$$S_i = I - G_{Pii}^T W_i G_{Pii} M_i^{-1} \tag{12.26}$$

$$Wi = \left(G_{Pii}M_i^{-1}G_{Pii}^T\right)^{-1} \tag{12.27}$$

and

$$M_i = \sum_{h=-1}^{1} \alpha_{i+h}G_{N(i+h)i}^T Q_{i+h}G_{N(i+h)i} + R_i. \tag{12.28}$$

Remark that for the solution by the Lagrangian multiplier method to exist $Pp_i < Nm_i$ and the rank of G_{Pii} should be equal to Pp_i. Remark also that if simultaneous calculation was possible then $\Delta U = \Phi^{-1}\Psi$.

12.4 Iterated Solution

Each decentralized SIORHC control agent $C_i(k)$ must iterate during the sampling time period k the following difference equation

$$\Delta U_i(l) = \Phi_{ii}^{-1}\left(\Psi_i \sum_{j \neq i} \Phi_{ij}\Delta U_j(l-1)\right) \tag{12.29}$$

where $l = 1, 2, \ldots, L$, indexes the number of iterations during each sampling time period, for simplicity of notation the time index has been omitted in (12.29). Starting with the initial condition

$$\Delta U_j(0) = \left[\Delta u_j^L(k+1|k-1)\ldots\Delta u_j^L(k+N-1|k-1) \quad 0\right]^T \tag{12.30}$$

that is the control input sequence obtained in the last time $k - 1$ for future instants $k + 1, \ldots, k + N - 1$ and because there is no prediction at $k + N|k - 1$ the last array position is filled with a zero row block, meaning no moves at the horizon end.

The iterates will converge to the algebraic fully-iterated solution if the following spectrum radius condition hold: $\forall i \ |\lambda_i| < 1$, where λ_i are the difference equation matrix eigenvalues given by

$$\left|\lambda_i I - \Phi_d^{-1}\Phi_{nd}\right| = 0 \tag{12.31}$$

where

$$\Phi_d + \Phi_{nd} = \Phi, \tag{12.32}$$

with $\Phi_d = [diag\{\ldots\Phi_{ii}\ldots\}]$.

12.5 Water Delivery Canal

The experimental canal of *Núcleo de Hidráulica e Controlo de Canais* (Universidade de Évora, Portugal) is taken hereafter as a prototype. The canal has four pools with a length of 35 m, separated by three undershoot gates, with the last pool ended by an overshoot gate. Four water off-takes, each one placed downstream from each branch

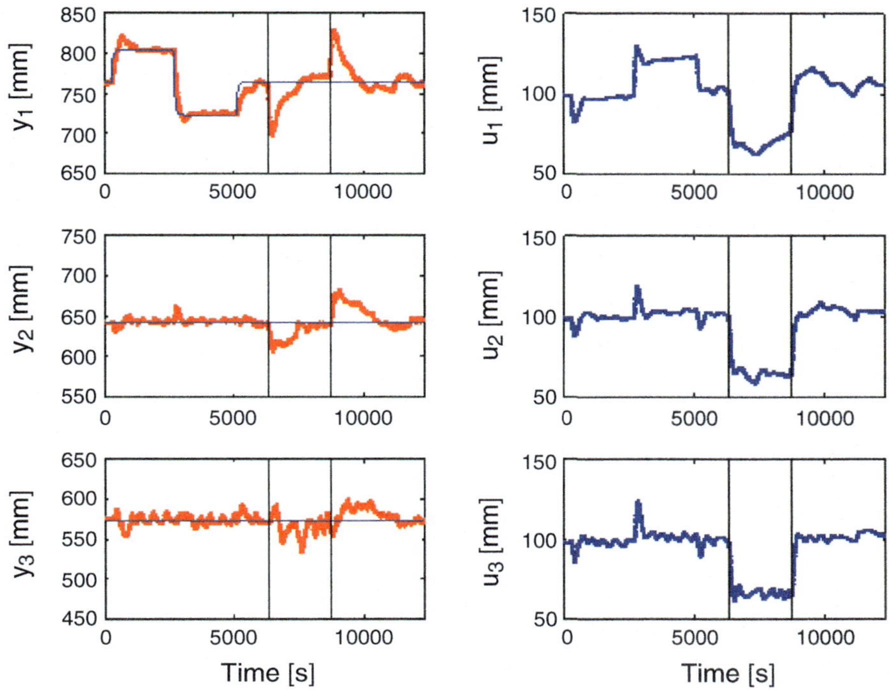

Fig. 12.3 Control of a water delivery canal with Decentralized SIORHC

and equipped with a valve, allow to mimic water use for irrigation. Water level sensors are installed at the end of each pool. The level sensors allow to measure values between 0 and 900 mm that corresponds to the canal border. The nomenclature is as follows: For pool number i, $i = 1,\ldots,4$, the downstream level is denoted y_i. Pool number i ends with gate number i. The opening degree of gate i is denoted u_i. All these quantities are measured in [mm].

Figure 12.3 shows experimental results obtained for upstream control of the first 3 pools using Decentralized SIORHC. The vertical lines indicate the opening and closing of the off-take of pool 1.

12.6 Conclusion

A decentralized version of SIORHC, a model predictive controller for linear systems with constraints on the terminal state that ensures closed-loop stability, has been presented. This results in a set of multiple local agents that act in a cooperative way to achieve a common goal. The distributed control agents predict their control actions in a finite horizon relying in a local model that assumes only interaction with neighboring subsystems, via the input signals. Cooperation is accomplished by an iterative procedure that converges to the optimal decentralized (*i.e.* fully-iterated) solution. Different type of closed-loop performance can be achieved, adjusting weights in each subsystem cost function, ranging from a so-called egocentric behavior to a pure altruistic one. The minimal requirements of a generic open computational infrastructure, where the agents can cooperate for sharing and broadcasting information, are given. Closed-loop stability results show that it can be directly applied to stable and unstable large scale linear systems. Application results obtained in an experimental water delivery canal validate this new approach to decentralized control.

Acknowledgments This work was supported by FCT under project *AQUANET – Decentralized and reconfigurable control for water delivery multipurpose canal systems*, contract PTDC/EEA-CRO/102102/2008, and INESC-ID multi-annual funding through the PIDDAC Program funds.

References

1. Camponogara E, Jia D, Krogh BH, Talukar S (2002) Distributed model predictive control. IEEE Control Syst Mag 22:44–52
2. Findeisen R, Allgöwer F (2002) An introduction to nonlinear model predictive control. In: Proceedings of the 21st Benelux meeting on systems and control, Veidhoven
3. Fogel DB (1995) Evolutionary computation. Willey, New York
4. Igreja JM (2002) Mixed constrained stabilizing predictive control: an extension to multivariable processes. In: Proceedings of the 5th Portuguese conference on automatic control, Aveiro
5. Mosca E, Lemos JM, Zhang J (1990) Stabilizing I/O receding horizon control. In: Proceedings of the 29th decision and control conference, Honolulu, Hawaii, Dec 1990
6. Mosca E (1995) Optimal, predictive and adaptive control, Information and systems sciences series. Prentice Hall, Englewood Cliffs
7. Negenborn RR (2007) Multi-agent model predictive control with applications to power networks. PhD thesis, Delft University of Technology
8. Negenborn RR, Lukszo Z, Hellendoorn H (eds) (2010) Intelligent infrastructures, vol 42, 1st edn, Intelligent infrastructures, control and automation: science and engineering. Springer, Dordrecht
9. Osório A, Afsarmanesh H, Camarinha-Matos L (2010) Open services ecosystem supporting collaborative networks. In: Ortiz A, Franco R, Gasquet P (eds) Proceedings of the balanced automation systems for future manufacturing networks. Springer, Boston
10. Rawlings JB, Mayne DQ (2009) Model predictive control: theory and design. NobeHill Publishing, Madison
11. Rugh WJ (2006) Linear system theory, Information and systems sciences series. Prentice Hall, Upper Saddle River
12. Vaccarini M, Longhi S, Katebi MR (2009) Unconstrained network decentralized model predicitve control. J Process Control 19:328–339
13. Ventak AS, Hisken IA, Rawlings JB, Wright SJ (2008) Distributed MPC strategies with application to power system automatic generation control. IEEE Trans Control Syst Technol 16(6):1192–1206
14. Zhang Y, Li S (2007) Networked model predictive control based on neighbourhood optimization for serially connected large-scale processes. J Process Control 17:37–50

Chapter 13
Single-Objective Spreading Algorithm

E.J. Solteiro Pires, Luís Mendes, António M. Lopes, P.B. de Moura Oliveira, and J.A. Tenreiro Machado

Abstract This paper addresses the problem of finding several different solutions with the same optimum performance in single objective real-world engineering problems. In this paper a parallel robot design is proposed. Thereby, this paper presents a genetic algorithm to optimize uni-objective problems with an infinite number of optimal solutions. The algorithm uses the maximin concept and ε-dominance to promote diversity over the admissible space. The performance of the proposed algorithm is analyzed with three well-known test functions and a function obtained from practical real-world engineering optimization problems. A spreading analysis is performed showing that the solutions drawn by the algorithm are well dispersed.

E.J. Solteiro Pires (✉) • P.B. de Moura Oliveira
INESC TEC - INESC Technology and Science (formerly INESC Porto),
Escola de Ciências e Tecnologia,
Universidade de Tras-os-Montes e Alto Douro, Vila Real, Portugal
e-mail: epires@utad.pt; oliveira@utad.pt

L. Mendes
Departamento de Engenharia Electrotécnica, Escola Superior de Tecnologia e Gestão,
Instituto Politécnico de Leiria, Portugal

Instituto de Telecomunicações, Lisboa, Portugal
e-mail: lmendes@ipleiria.pt

A.M. Lopes
Instituto de Engenharia Mecânica, Faculdade de Engenharia,
da Universidade do Porto, Porto, Portugal
e-mail: aml@fe.up.pt

J.A. Tenreiro Machado
Grupo de Investigacao em Engenharia do Conhecimento e Apoio à Decisão,
Departamento de Engenharia Electrotécnica, Instituto Superior de Engenharia do Porto,
Instituto Politecnico do Porto, Porto, Portugal
e-mail: jtm@isep.ipp.pt

A. Madureira et al. (eds.), *Computational Intelligence and Decision Making: Trends and Applications*, Intelligent Systems, Control and Automation: Science and Engineering 61,
DOI 10.1007/978-94-007-4722-7_13,

13.1 Introduction

Some single-objective problems have several optimal solutions. A finite number of solutions can be found at different peaks of the objective function or an unlimited set of solutions exist along one or more regions in the parameter space. Therefore, achieving a well-spread and non-dominated parameter front is of paramount importance to the decision maker, since he can choose from a set of optimum solutions the one best suited to be developed or implemented in the problem context.

Evolutionary algorithms use different approaches to promote the solutions diversity, such as the sharing model, crowding and maximin techniques. The sharing model, originally suggested by Goldberg and Richardson [7], is used to obtain a set of optimal solutions distributed by several optimal peaks. To promote solutions in less crowded regions of the parameter space and, therefore, to *force* the population to be well distributed, the sharing model degrades the fitness values of similar solutions according to the distance of their closest neighbors. In this method, the number of optimum solutions under each peak is proportional to its value.

The crowding technique, introduced by De Jong [3], replaces offspring by *similar* solutions existing in the population in order to maintain its diversity. *Similarity* is measured, like in the sharing model, by evaluating the distance between solutions. In some crowding technique variants, the offspring compete against their parents, giving the winners a chance to breed in the next population. This method tends to spread solutions over the most prominent search space peaks, achieving a number of solutions in each peak proportional to its base size.

Deb and Tiwari [5] proposed an algorithm to solve several types of optimization: single-objective uni-optimal problems, single-objective multi-optima problems, multi-objective uni-optimal and multi-objective multi-optima problems. The Deb and Tiwari algorithm is capable of solving the problem without knowing at the outset the type of optimization to perform. In single-objective multi-optima problems they [5] use the ε-dominance and the crowding distance in the objective and the parameter space, respectively. These authors apply the algorithm in multi- modal problems which have several optimum solutions in different peaks.

In 2003, Balling [1] proposed a multi-objective optimization technique based on a fitness function derived from the maximin strategy and Li [9] used the maximin fitness in a particle swarm multi-objective optimizer. Moreover, Pires et al. [12] adopted the maximin technique to find a well distributed non-dominated Pareto front in multi-objective optimization algorithms. The maximin strategy can be adopted by genetic algorithms (GAs) to obtain a set of solutions well distributed along the parameter space. The maximin technique is mainly used to obtain a set of points from a continuous front instead of finding the peaks of multi-modal functions.

The method proposed in this paper uses the maximin concept together with the ε-dominance [8] in order to find a front in the parameter space. This method presents a better performance than the one reported in the literature [10], which adopts the sharing scheme, because it reduces significantly the solutions dispersion along the optimum front and, consequently, leading a better defined front.

The paper is organized as follow. Section 13.2 describes the maximin spreading algorithm. This optimization method promotes solutions diversity in the parameters space. Section 13.3 presents the simulation results of four optimization problems. The first three problems follow two well-known and widely used functions: the Himmelblau's, squared sine and double pulse functions. These equations are used by the algorithm as the fitness function in three different optimizations. The last function presented in Sect. 13.3 is obtained from real-world engineering applications. It optimizes the kinematics of parallel robots. Section 13.4 studies the spreading of the solutions drawn by the proposed algorithm. Finally, Sect. 13.5 outlines the main conclusions.

13.2 Maximin Spreading Algorithm

In multi-objective problems there are several ways to find the Pareto front. One of them is based on the ε-dominance concept in which solutions with slightly inferior fitness values are allowed to remain in the population as non-dominated ones. This technique is used to get a set of solutions with good spread and diversity over the multi-objective space [8].

The proposed single-objective algorithm adopts the above-mentioned concept and the maximin scheme as the main techniques to achieve good diversity in the parameters space. Initially, the algorithm divides the objective space using virtual horizontal parallel hyperplanes (straight lines or planes for a 1-D or 2-D dimensional parameter space, respectively), that are separated from each other by a ε-distance (see Fig. 13.1a). Two consecutive hyperplanes define a ε-rank, where all the solutions have the same preference, even if their objective values are different, as illustrated in Fig. 13.1a. After the division of the objective space, the algorithm selects the solutions, starting from the best ε-ranks (r_1), until a ε-rank is found with more solutions than the empty slots available in the new population. However, if the best ε-rank (rank r_1) has more solutions than the number of individuals of the new population, then the solutions of the other ε-ranks are not considered. In both cases, the best distributed solutions in the parameter space are selected according to the maximin scheme. The basic idea behind the maximin sorting scheme is the choice of the solutions that decrease the large gap areas where no solution exists in the already selected population. For example, consider the population solutions of one ε-rank depicted in Fig. 13.1b. In this case, two parameters $\{x_1, x_2\}$ are considered. Initially, two extreme solutions for each parameter are selected, $\{a, b\}$ and $\{d, c\}$ for x_1 and x_2, respectively. Through this selection the set $S \equiv \{a, b, c, d\}$ is initialized. Then, the solution e is selected because it has the larger distance to the set S. After that, solution f is chosen for inclusion into the set $S \equiv \{a, b, c, d, e\}$, for the same reasons. The process is repeated until population S is completed. The maximin technique boosts its performance as the number of iterations increases, since all the solutions, which are already in the first ε-rank, tend to spread along a non-dominated front.

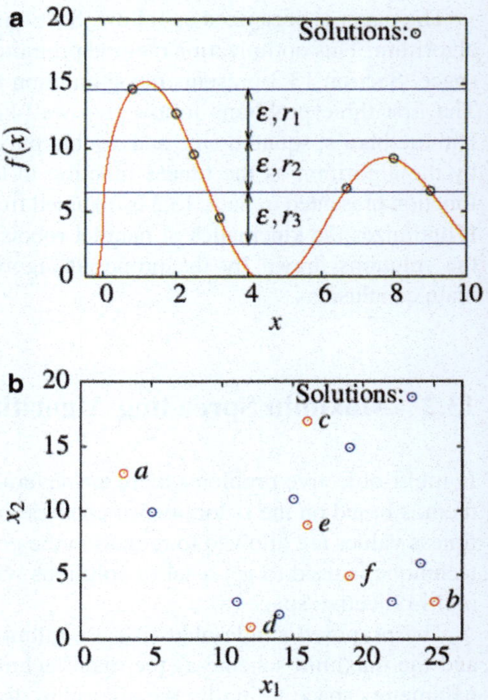

Fig. 13.1 (**a**) Problem with one objective and one parameter $\{x\}$, where r_m represents the rank m = {1,…,3} and ε is the rank length. (**b**) Solutions in a bi-dimensional parameter space $\{x_1, x_2\}$ represented by the *dot-points*. Solutions $\{a, b, \ldots, f\}$ are the most distant from each other

The maximin sorting scheme is depicted in Fig. 13.2. In each generation new offspring (set *D*) are merged with their progenitors (set *P*), according with Fig. 13.2, leading to the new set *R* (line 1). After that, the algorithm may select, for each one of the optimization parameters (n_{par}), the extreme solutions (getMin and getMax functions) from rank r_1 (lines 5–7) and introduces them into the final population (set *S*). Then, the individuals of lower rank are removed (getRankMin function) from the auxiliary population *A* and inserted into the set *S* until the number of solutions of the current rank surpass the allowed number of solutions of set *S* (lines 9–12). Next, the squared distance, c_{aj} (13.1a), between each solution, a_j, and the solutions already selected, s_i, is evaluated (lines 13–15). After that, the solution a_j, whose squared distance to the set *S* is the larger (*k* solution), is selected (13.1b) (getMaxCi function, line 17). Each time a solution enters into the set *S* (line 18), the cost c_{al} of the set *A* is reevaluated (lines 19–21). This process ends when the set *S* is completed, requiring, at most, $O(pop_{\mathrm{dim}}{}^2)$ computations, where pop_{dim} represents the number of population elements.

$$\boldsymbol{c_{aj}} = \mathbf{min}_{s_i \in S, a_j \in A} \left\| \boldsymbol{a_j} - \boldsymbol{s_i} \right\|^2 \tag{13.1a}$$

$$\boldsymbol{S} = \boldsymbol{S} \cup \left\{ \boldsymbol{a_j} : \boldsymbol{c_{aj}} = \mathbf{max}_{ai \in A} \boldsymbol{c_{ai}} \right\} \tag{13.1b}$$

Fig. 13.2 Maximin spreading algorithm pseudo-code

```
R = P ∪ D
S = ∅
A = getRankMin(R)
if #A > pop_dim then
    for i = 1 to n_par do
        S = S ∪ getMin(A, i) ∪ getMax(A, i)
    end for
end if
while #S + #A ≤ pop_dim do
    S = S ∪ A
    A = getRankMin(R)
end while
for j = 1 to #A do
    c_{a_j} = min_{s_i ∈ S} {||a_j − s_i||^2}
end for
while #S < pop_dim do
    k = getMaxCi(A)
    S = S ∪ k
    for l = 1 to #A do
        c_{a_l} = min{||a_l − k||^2, c_{a_l}}
    end for
end while
```

The proposed algorithm can be used with a static constant or a dynamic variable ε-parameter across all the iterations. In the dynamic technique, as the number of algorithm iterations increases, the ε-length decreases so that the algorithm favors the search of new solutions in regions near the optimal front. Therefore, the dynamic ε-technique leads to better results than the static ε-method. In fact, a more well defined (*i.e.*, low dispersed) front is obtained at the end of the algorithm, while maintaining a good exploration at the beginning.

13.3 Simulation Results

This section presents the simulation results of the proposed algorithm for several problems. The first three functions, which have well-known optima values, are used to validate the presented algorithm. The Himmelblau's and the squared sine functions are selected directly from the literature and the third one, the double pulse function, is chosen because it has two continuous parameters fronts. The last function is obtained from a real-world engineering problems, namely, the kinematic optimization of parallel robots. For each considered problems there is only one optimization function. However, each problem has several different optimums solutions differing between them by the parameters values. In all the optimization problems parameters are encoded directly into real-value vectors.

For all the above-mentioned optimization problems the successive generations of new solutions are reproduced based on a linear ranking scheme and simulated

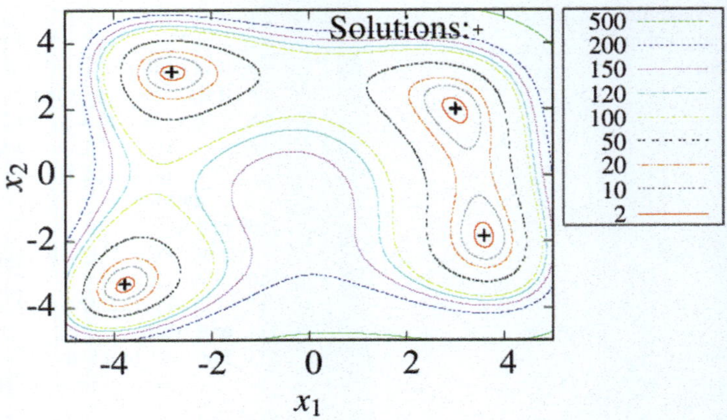

Fig. 13.3 Solutions found for the Himmelblau's function

binary crossover [4, 6] (with $\eta_c = 20$). When mutation occurs the operator replaces the value of one design parameter according to a uniform distribution function. The probabilities of crossover and mutation are $p_c = 0.6$ and $p_m = 0.05$, respectively.

At the end of each GA cycle, it is used a $m\varepsilon$ $-(\lambda + \mu)$ strategy to select the solutions which survive for the next iteration. This means that the best solutions among parents (λ) and offspring (μ) are chosen based on ε rank sort. To fill the last gap of the population the maximin (m) technique is used.

For the first three optimization functions, it is used a dynamic ε that is decreased during the optimization. The ε begins with the value of 5 and stops when reaches the final value 0.01. The ε is decreased by 1 % every 10 iterations, or when all the solutions belong to the same rank.

13.3.1 Benchmark Functions

The first optimization function used is the well-known Himmelblau's function [5]:

$$f_1(x_1, x_2) = \left(x_1^2 + x_2 - 11\right)^2 + \left(x_1 + x_2^2 + -7\right)^2 \tag{13.2}$$

where $-5 \leq x \leq 5, i = \{1, 2\}$. It has two parameters and four isolated minima, all having the optimal function value of $f_1^* = 0$. In this experiment a set of 100 solutions is used and the algorithm is executed during 1,000 iterations.

Figure 13.3 shows the obtained solutions for the Himmelblau's function. It is clear that the algorithm converges and is able to find the four minima in only one simulation. Moreover, the solutions are well distributed by all the four optimal as depicted by Table 13.1.

Table 13.1 Solution distribution over the optimums of $f_1(x_1, x_2)$

Optimal position $\{x_1, x_2\}$	$\{3.000, 2.000\}$	$\{-2.805, 3.131\}$	$\{-3.779, -3.283\}$	$\{3.584, -1.848\}$
Solution percentage	23%	18%	32%	27%

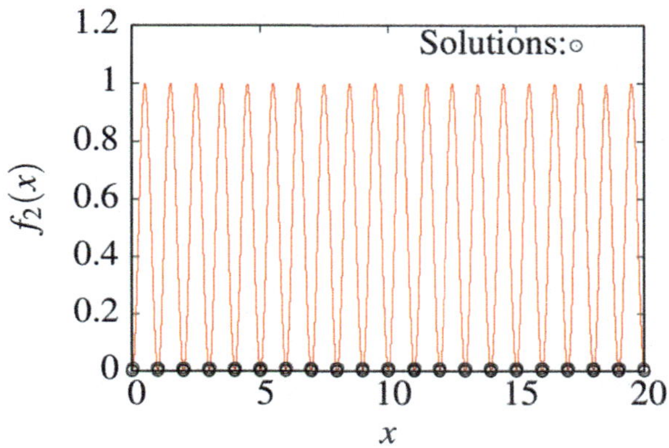

Fig. 13.4 Solutions found for the sine function

The second optimization function is a squared sine function:

$$f_2(\boldsymbol{x}) = \boldsymbol{sin}^2(\pi \boldsymbol{x}) \tag{13.3}$$

where $\mathbf{0} \leq \boldsymbol{x} \leq \mathbf{20}$. It has one parameter and 21 isolated minima, all with the optimal value of$\boldsymbol{f}_2^* = \mathbf{0}$. In this experiment a set of 100 solutions is used and the algorithm is executed during 1,000 iterations. Figure 13.4 depicts the final solution set obtained for the sine function. It could be seen that the algorithm find all the isolated optimal.

At the end of the simulation, the solutions are well dispersed by all the optimal minimal. This phenomena can be observed examining the number of solutions in the neighborhood of each minima. The percentage of solutions for each minimum can be seen in Table 13.2.

The third function (13.4) is used with one parameter and two continuous optimal fronts in the parameter space where $\mathbf{0} \leq \boldsymbol{x} \leq \mathbf{20}$.

$$f_3(\boldsymbol{x}) = \begin{cases} \mathbf{1}, & \mathbf{3.5} < x < 4.0 \ \textbf{or} \ \mathbf{16.0} < x < 16.5 \\ \mathbf{0}, & \textbf{Otherwise} \end{cases} \tag{13.4}$$

Figure 13.5 shows that the algorithm finds a representative solution set of the non-dominated front in a single algorithm execution. Moreover, the final solutions are spread over the optimal parameter front. In this experiment, it is also considered a population with 100 individuals and 1,000 iterations. In this case the solutions found by the algorithm are divided between the pulses (see Table 13.3).

Table 13.2 Solution distribution over the optimums of $f_1(x_1, x_2)$

Optimal position $\{x\}$	Solution percentage (%)
{0}	3
{1}	4
{2}	5
{3}	5
{4}	5
{5}	5
{6}	4
{7}	5
{8}	5
{9}	5
{10}	5
{11}	6
{12}	5
{13}	5
{14}	5
{15}	5
{16}	5
{17}	5
{18}	5
{19}	5
{20}	3

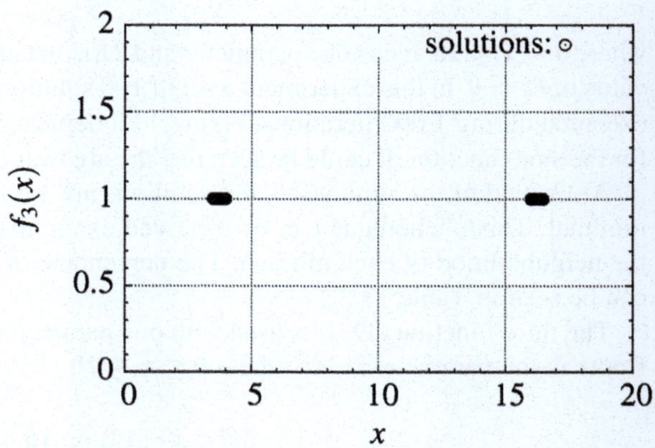

Fig. 13.5 Solutions found for the double pulse function

Table 13.3 Solution distribution over the optimums of $f_3(x)$

Optimal position $\{x\}$	[3.5, 4.0]	[16.0, 16.5]
Number of solutions (%)	53	47

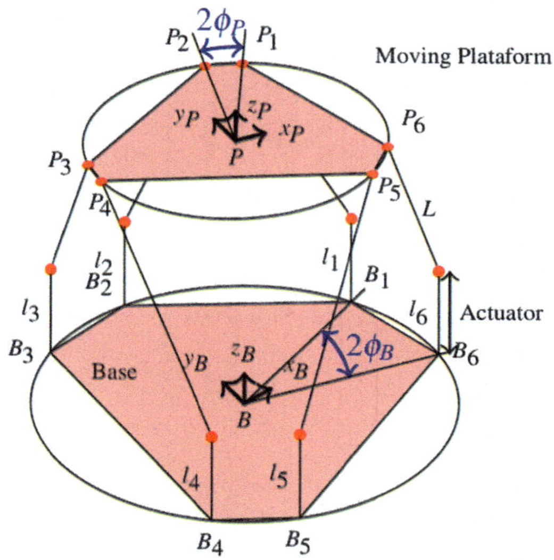

Fig. 13.6 Schematic representation of the parallel manipulator structure

13.3.2 Real World Problem Function

The last function describes a real problem, where, the kinematic optimization of a Stewart-type parallel manipulator, based on maximum dexterity [2], is considered (Fig. 13.6). The manipulator's mechanical structure comprises a fixed (base) platform and a moving platform, linked together by six independent, identical, open kinematic chains. Each chain comprises two links: the first link (linear actuator) is always normal to the base and has a variable length, l_i ($i = \{1, \ldots, 6\}$), with one of its ends fixed to the base and the other one attached, by a universal joint, to the second link; the second link (arm) has a fixed length, L, and is attached to the moving platform by a spherical joint.

The function to maximize is given by expression (13.5).

$$f_4(r_B, \phi_P, \phi_B, L) = \frac{\sigma_{\max}(\mathbf{J}_c)}{\sigma_{\min}(\mathbf{J}_c)} \tag{13.5}$$

where $\sigma_{\max}(\mathbf{J}c)$ and $\sigma_{\min}(\mathbf{J}c)$ represent the maximum and minimum singular values of the inverse kinematic jacobian matrix, $\mathbf{J}c$ (13.6).

$$\mathbf{J}_c = \begin{bmatrix} \frac{(\mathbf{e}_1 - l_1 \mathbf{z}_B)^T}{\mathbf{z}_B^T \mathbf{e}_1 - l_1} & \frac{{}^P\mathbf{p}_{1|_B} \times (\mathbf{e}_1 - l_1 \mathbf{z}_B)^T}{\mathbf{z}_B^T \mathbf{e}_1 - l_1} \\ \vdots & \vdots \\ \frac{(\mathbf{e}_6 - l_6 \mathbf{z}_B)^T}{\mathbf{z}_B^T \mathbf{e}_6 - l_6} & \frac{{}^P\mathbf{p}_{6|_B} \times (\mathbf{e}_6 - l_6 \mathbf{z}_B)^T}{\mathbf{z}_B^T \mathbf{e}_6 - l_6} \end{bmatrix} \tag{13.6}$$

This is a 6×6 matrix in which all vectors can be obtained using vector algebra applied to each parallel manipulator's kinematic chain. Vectors $\mathbf{e}_i$ are given by

$$\mathbf{e}_i = \mathbf{x}_{P(pos)} - \mathbf{b}_i + {}^P\mathbf{p}_{i|_B} \tag{13.7}$$

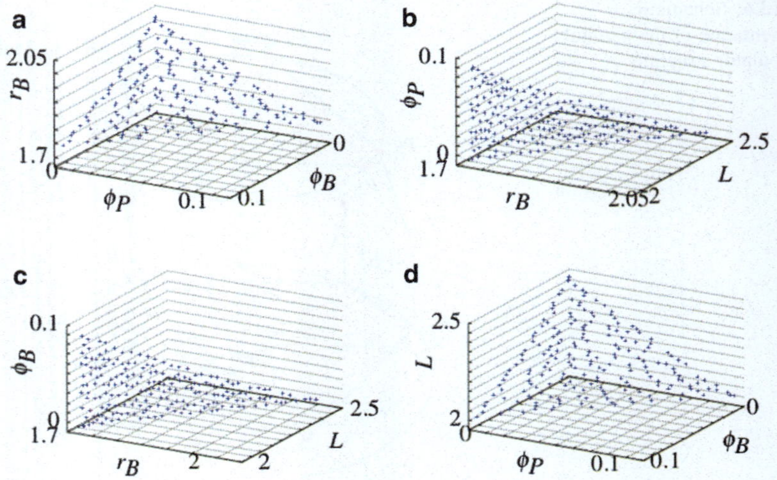

Fig. 13.7 Optimal solutions set of kinematic parameters. Projections of the solutions in the plane: (**a**) $\phi_P \times \phi_B \times r_B$ (**b**) $r_B \times L \times \phi_P$, (**c**) $r_B \times L \times \phi_B$ and (**d**) $\phi_P \times \phi_B \times L$. $\phi_P[°], \phi_B[°], r_B[r_B]$ units

where $\mathbf{x}_{P(pos)}$ is the position of the moving platform expressed in the base frame, $\mathbf{b}_i$ represents the positions of the connecting points between kinematic chains and base and ${}^P\mathbf{p}_{i|_B}$ are the positions of the points P_i connecting the kinematic chains to the moving platform, expressed in the base frame.

The scalars l_i ($i = \{1,...,6\}$) are the displacements of the actuators, given by

$$l_i = e_{iz} + \sqrt{L^2 - e_{ix}^2 - e_{iy}^2} \tag{13.8}$$

This work considers a particular manipulator pose (position and orientation), corresponding to the centre of the manipulator workspace namely, $[0\ 0\ 2\ 0\ 0\ 0]^T$ (units in r_P and degrees, respectively). Thus, for this pose, the jacobian matrix will be a function of the four kinematic parameters, that are sufficient to define the parallel manipulator's kinematic structure.

The parameters are codified by real values through the vector $p = [r_B\ \phi_P\ \phi_B\ L]^T$. The solutions are randomly initialized in the range $\mathbf{1.0 \leq r_B \leq 2.5}$, $\mathbf{0^0 \leq \phi_P, \phi_B \leq 25^0}$, and $\mathbf{2.0 \leq L \leq 3.5}$.

The ε is initialized with the value 20.0 and, during the evolution, its value is sucessively decreased until it reaches 0.03. This decrement is done at a rate of 10% every time the best 200 solutions are in the first rank and when ε has not changed in the last 100 generations.

The solutions are classified by the fitness function given by equation (13.5), in case the solution is admissible, otherwise it takes a very high value (1×10^{20}).

The global results (Fig. 13.7) show that there are multiple sets of solutions that are optimal. Moreover, the algorithm draws a representative solution set of the optimal parameters front.

Table 13.4 Performance indices results

Index	f_1	f_2	f_3	f_4
SP	0.001	0.002	0.003	0.005
MDG	0.948	0.371	1.205	0.003

Table 13.5 Analysis results

Index	f_1	f_2	f_3	f_4
Median	0.007	0.003	1.000	1.416
Average	0.006	0.005	1.000	1.416
Std dev	0.004	0.004	0.000	0.001
Max	0.010	0.010	1.000	1.414
Min	0.000	0.000	1.000	1.417
Max – Min	0.010	0.010	0.000	0.003
ε	0.010	0.010	0.010	0.003

13.4 Performance Indices

To study the solution diversity the spacing index (*SP*) [11] and the minimal distance graph (*MDG*) [12] are employed. In this work, both functions adopt the distances in parameter space instead of objective space as was initially proposed by their authors.

The resultant metric values, for the functions considered in Sect. 13.3, are shown in Table 13.4. The indices present small values indicating that the algorithm obtains a good solution spread. When the index *MDG* is used in the functions f_1, f_2 and f_3, the obtained values are greater than those evaluated with the *SP* index. This happens since the *MDG* index uses, necessarily, the distances between the different niches.

Table 13.5 shows the median, the average, the standard deviation, the maximum, the minimum, the objective range and the ε value for the considered functions. In all the optimization problems, the final solution set are spread in the best rank. This could be confirmed through the maximum and minimum values (Max – Min $\leq \varepsilon$). For function f_3 all the solutions have the value 1 since it is a binary function and the rank value is 0.01. The solutions of the other functions are spread along the interval rank. The rank interval, ε, is fundamental to allow the movement of solutions along the front. Otherwise, they remain trapped in an optimal of the front.

The algorithm finds all the minimum for three functions (f_1, f_2, f_3) which have been commonly adopted as a benchmark in multi-objective evolutionary community. For the practical problem, and since its optimums values are not known, is inferred that algorithm finds the function optimal.

13.5 Conclusions

A new spreading technique was developed with the purpose of solving a wide range of different single-objective optimization problems. To access the algorithm performance, it was firstly applied in benchmark functions. For these functions the algorithm found a correct representative solution set. Then, it was used in a function obtained from a practical engineering problems, namely, in kinematic optimization of parallel robots. In all the test cases, the algorithm has found, in only one execution, several solutions with the same maximum performance but with different parameters values. The results show that the proposed algorithm solves different types of single-objective problems, maintaining a good spreading in parameter space.

References

1. Balling R (2003) The maximin fitness function; multi-objective city and regional planning. In: Evolutionary multi-criterion optimization, second international conference, Portugal, 8–11 April 2003. Proceedings, Lecture notes in computer science, vol 2632, pp 1–15
2. Barbosa MR, Solteiro Pires EJ, Lopes AM (2010) Optimization of parallel manipulators using evolutionary algorithms. In: Soft computing models in industrial and environmental applications, 5th international workshop, Portugal, pp 79–86
3. De Jong KA (1975) An analysis of the behavior of a class of genetic adaptive systems. PhD thesis, University of Michigan
4. Deb K (2001) Multi-objective optimization using evolutionary algorithms, Interscience series in systems and optimization. Wiley, Chichester
5. Deb K, Tiwari S (2005) Omni-optimizer: a procedure for single and multi-objective optimization. In: Evolutionary multi-criterion ptimization, third international conference, Mexico, 9–11 March. Proceedings, Lecture notes in computer science, vol 3410, pp 47–61
6. Goldberg DE (1989) Genetic algorithms in search, optimization, and machine learning. Addison – Wesley, Reading, MA, USA
7. Goldberg DE, Richardson J (1987) Genetic algorithms with sharing for multi-modal function optimisation. In: Proceedings of the 2nd international conference on genetic algorithms and their applications, Mahwah, NJ, USA, pp 41–49
8. Laumanns M, Thiele L, Deb K, Zitzler E (2002) Archiving with guaranteed convergence and diversity in multi-objective optimization. In: Proceedings of the genetic and evolutionary computation conference. Morgan Kaufmann Publishers, New York, USA, pp 439–447
9. Li X (2004) Better spread and convergence: particle swarm multiobjective optimization using the maximin fitness function. In: Proceeding of genetic and evolutionary computation conference. Lecture notes in computer science, Seattle, USA, vol 3102, pp 117–128
10. Mendes LM, Solteiro Pires EJ, Vaz J, Rosário MJ (2007) Automated design of radio-frequency single-ended switched capacitor arrays using genetic algorithms. In: Proceedings of the Midwest symposium on circuits and systems, Montreal. IEEE Computer Society, Montreal, Canada, pp 453–456
11. Schott JR (1995) Fault tolerant design using single and multicriteria genetic algorithm optimization. Master's thesis, Massachusetts Institute of Technology, Department of Aeronautics and Astronautics, Cambridge, MA
12. Solteiro Pires EJ, de Moura Oliveira PB, Tenreiro Machado JA (2005) Multi-objective maximin sorting scheme. In: Conference on evolutionary multi-criterion optimization 2005, México. Lecture notes in computer science, vol 3410. Springer, pp 165–175

Chapter 14
Fault Tolerant Control Based on Adaptive LQG and Fuzzy Controllers

Carla Viveiros, Luis Brito Palma, and José Manuel Igreja

Abstract This paper describes a fault tolerant switching loop control study case for a three water tank laboratory setup with sensors and actuators redundancies. Fault diagnosis and tolerance are two main characteristics of intelligent setups when closed-loop control is used for large complex systems. Discrete adaptive LQG (linear quadratic) and Fuzzy fault tolerant controllers, for nominal situation, are presented and tested in the laboratory setup. For abrupt fault situation (structural fault), switching from one controller to another based on controller performance evaluation using PCA (principal component analysis) technique is implemented. The main contributions are: (a) proposed approach based on PCA to detect structural faults; (b) performance assessment based on performance index through *Mahalanobis* distance using T^2 statistics and (c) control switching between LQG and Fuzzy controllers based on the performance index. Experimental results applied to a three-tank laboratory setup in the presence of a structural fault are shown to evaluate the performance of the proposed approach.

Keywords Structural faults • Principal component analysis • Controller performance monitoring • Switching control

---

C. Viveiros (✉) • J.M. Igreja
Automation & Power Systems Engineering, Instituto Superior de Engenharia de Lisboa, 1959-007 Lisbon, Portugal
e-mail: cviveiros@deea.isel.ipl.pt; jigreja@deea.isel.ipl.pt

L.B. Palma
Electrical Engineering, Universidade Nova de Lisboa,
2829-516 Caparica, Portugal
e-mail: lbp@fct.unl.pt

A. Madureira et al. (eds.), *Computational Intelligence and Decision Making: Trends and Applications*, Intelligent Systems, Control and Automation: Science and Engineering 61,
DOI 10.1007/978-94-007-4722-7_14,

14.1 Introduction

Our present-day society is strongly dependent on the availability and correct functioning of control systems. One of the many benefits of Fault Tolerant Control (FTC) approach is to maintain overall system stability and acceptable performance in the presence of faults. FTC approaches [1] can be classified in two categories: passive (e.g. robust control) and active approach (e.g. adaptative control). To implement active fault tolerant control approaches two tasks are needed: fault detection and isolation (FDI) and controller reconfiguration or accommodation. In the area of controller performance there are several techniques proposed, e.g. [2–4], in this work the principal component analysis (PCA) technique will be used. PCA is a multivariate statistical method, within the class of linear methods, which uses linear correlations while reducing data dimension and is suitable for control loop performance assessment and fault detection. In [5], PCA and non linear PCA were presented and the performance index is based on Q statistics. In [6], a benchmark to monitor model predictive control was proposed and performance index was based on T^2 statistics. In [7] a technique called Auto Regressive Moving Average (ARMA) – PCA is proposed. The ARMA filter is introduced to remove autocorrelation for the score variable and the performance index was based on T^2 and Q statistics. In this work the performance index is based on *Mahalanobis* distance between PCA scores. When a fault occurs it is necessary to accommodate such fault or reconfigure the system to maintain overall system stability. Structural faults [1] are those which affect the system structure, thus breaking control loops in some situations. In order to achieve fault accommodation two approaches can be used: adaptative control and switched control. Adaptative control [8] is used to modify the control law to handle the situation where the system parameters are changing over time, on the other hand, switched control is determined by a bank of controllers designed for specific purposes (nominal or faulty situations) that switches from one controller to another [9]. Meanwhile, controller reconfiguration [10] can be achieved, for instance, by controller re-design technique using an optimization approach such as LQG or by learning control techniques using fuzzy logic approach [11]. Various switching control approaches have been presented in the literature, e.g. in [12] a switching control method between PID controllers based on performance assessment is proposed, methods based on direct search optimization [13] or based on time-delay in the input channel [14] are other examples. A supervisor must deal with faults that may occur during process operation, therefore it must include functions such as [15]: monitoring, automatic protection and supervision with fault diagnosis. The paper is organized as follow: Sect. 14.1 presents a brief state-of-art. In Sect. 14.2 the proposed architecture as well as LQ polynomial and Fuzzy control techniques is described. Controller performance evaluation is presented in Sect. 14.3. A brief description of DTS200 system and experimental results are presented in Sect. 14.4 followed by conclusions in Sect. 14.5.

14.2 Proposed Approach for Fault Tolerant Switching Control Between LQG and Fuzzy Controllers

In this section an approach for fault tolerant switching control based on the controller performance evaluation will be presented. From this approach and based on performance index, the supervisor decides which controller LQG or Fuzzy has the best performance.

14.2.1 Supervisory Control Architecture

Emerging fault tolerant control techniques proposed new controller designs capable to tolerate system malfunctions, faults in order to maintain overall stability and desirable performances. Two main tasks have to be considered: fault detection and diagnosis to detect and diagnose faults and controller reconfiguration to accommodate such faults. Re-design controller, set-point changes or switching from one controller to another are various possibilities in order to tackle the fault effects. Several schemes of FTCS have been presented, e.g. Blanke [1] introduced an approach which included execution level and supervisory level, Karsai [16] introduced a scheme of fault adaptive control technology centered on model-based approaches for fault detection and isolation, controller selection and reconfiguration for hybrid systems. The architecture proposed here, is based on Blanke [1] scheme and it includes a control loop with actuators, sensors, and a switching control block that switches from LQG to Fuzzy controllers in the execution level, based on the performance index. In Fig. 14.1 the proposed architecture to detect structural faults and reconfigure the system to guarantee an acceptable closed-loop system performance is presented. The methodology proposed in this paper is applied to the three water tank system DTS200 [17], Fig. 14.6. In nominal and faulty situations the main goal is to maintain h_1 level around a desirable set-point. In faulty conditions, the structural fault f_1, consisting in total leakage from V_{1L} was applied to evaluate the system performance.

In the supervision level there is a supervisor that receives all signals. The supervisor tasks consist in: (a) parameter estimation through recursive least squares $\hat{\theta}(k)$; (b) switching from one controller to another based on a performance index I_{switch} and (c) fault detection through PCA and *Mahalanobis* distance [18]. One method to ensure system fault tolerance, in a nominal situation, is adaptive control which is more appropriate to deal with structural faults and it is not restricted to a set of faults. When a fault f occurs, supervision must reconfigure the system switching from one controller to another, based on controller performance to guarantee an acceptable system performance in closed loop. LQG adaptive control has the ability to adapt to some unknown type of faults, but it depends on parameter estimation $\hat{\theta}(k)$. If there is a sudden change on reference or if there is an abrupt fault leading to long period of time of control saturation, parameter estimation through recursive

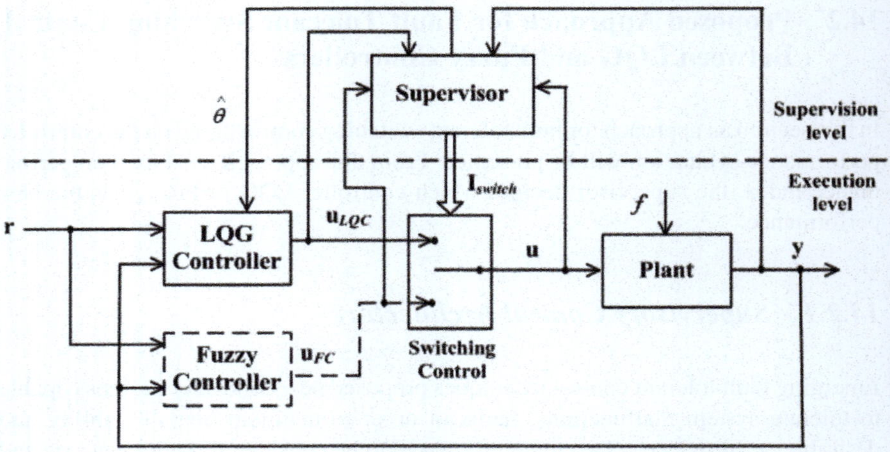

Fig. 14.1 Supervisory control architecture

least squares can lead to system instability. Fuzzy controllers do not depend on parameter estimation, thus, is suitable in those situations and can be used as an alternative controller.

14.2.2 Linear Quadratic Polynomial Controller

Below is presented a brief description of LQ polynomial controller [8]. To model the DTS200 around a pre-defined set-point *r(k)*, an ARX (n = 2, m = 2, d = 1) model is used to represent the dynamical system, the corresponding transfer function is:

$$\frac{Y(z)}{U(z)} = \frac{b_1 z^{-1} + b_2 z^{-2}}{1 + a_1 z^{-1} + a_2 z^{-2}} \tag{14.1}$$

with *Y*(z) and *U*(z) as the Z-transform of the system output and control input. The performance index is:

$$J(k) = (P\,y(k+d) - Q\,r(k))^2 + (\rho u(k))^2 \tag{14.2}$$

Where $P = Q = 1$ and the scalar $\rho = 0.01$ is a design parameter used to tune the closed-loop performance at the end of the design. The optimal control *u*(k) that minimizes performance index is straightforward to determine. In the minimum-phase core one solves the Diophantine equation, for the intermediate polynomials $F(z^{-1})$ and $G(z^{-1})$. The structure of the optimal LQ polynomial controller is presented in Fig. 14.2,

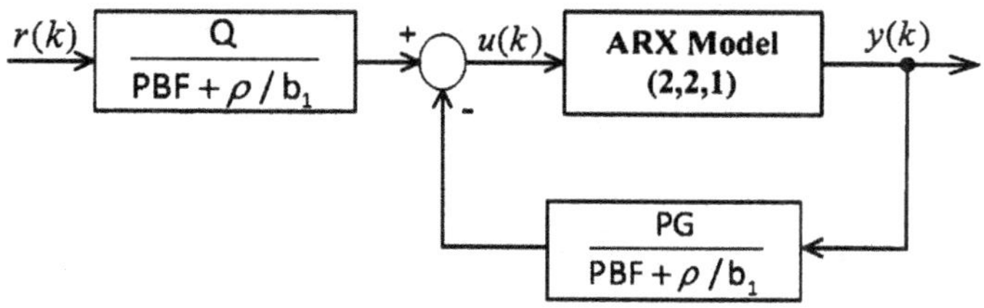

Fig. 14.2 Optimal LQG controller structure

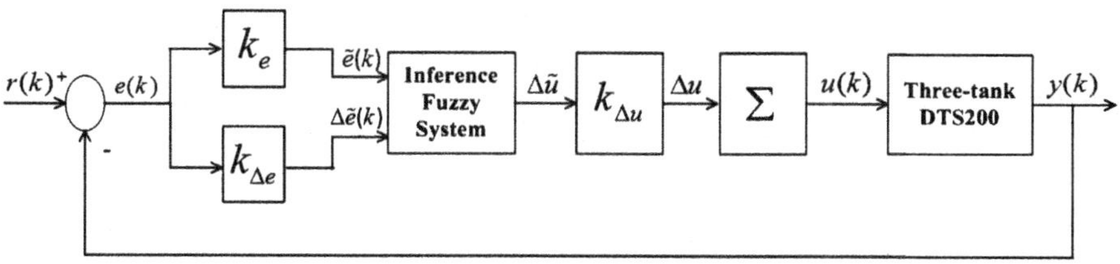

Fig. 14.3 Fuzzy controller structure

Finally, the optimal control equation is given by:

$$u(k) = \frac{\hat{b}_1}{\hat{b}_1^{\,2} + \rho^2}\left[\left(\frac{\rho^2 - \hat{b}_2\,\hat{b}_1}{\hat{b}_1}\right)u(k-1) + \hat{a}_1\,y(k) + \hat{a}_2\,y(k-1) + r(k)\right] \quad (14.3)$$

Parameters from polynomials $A(z^{-1})$ and $B(z^{-1})$ are estimated using recursive least squares (RLS) algorithm, thus obtaining parameters $\hat{\theta}(k) = [\hat{a}_1\ \hat{a}_2\ \hat{b}_1\ \hat{b}_2]$.

14.2.3 Fuzzy PI Controller

The fuzzy proportional integral (PI) controller [19] structure is presented below, where *r(k)* is the desired reference, control action *u(k)* is calculated considering error *e(k)* and scaling factor (k_e, $k_{\Delta e}$, $k_{\Delta u}$), *y(k)* is the system output (Fig. 14.3).

The fuzzy PI control action equation can be described as:

$$\begin{aligned} u(k) &= u(k-1) + k_{\Delta u} f_{NL}(e(k), ke, \Delta e(k), k_{\Delta e}) \\ e(k) &= r(k) - y(k) \end{aligned} \tag{14.4}$$

where f_{NL} is a non linear function.

14.3 Controller Performance Evaluation

A closed loop controller performance assessment approach is based on PCA and the performance index is computed using *Mahalanobis* distance between the projections of matrix **X**. This matrix **X** contains the control action vectors (u) from each controller and the output vectors (y) obtained through ARX models.

14.3.1 Nominal PCA Model

Principal components analysis (PCA) divides original data into the significant patterns, such as linear directions in scores subspace, and the uncertainties, such as noises located in residual subspace. The nominal data is stored into a data matrix $X \in \mathbb{R}^{nxm}$ with n samples (rows) and m variables (columns). The nominal loading and score matrix (P_0, T_0) are obtained when closed loop performance is considered to be acceptable and consistent with user specification. For this particular case, it was used an optimal LQG controller to maintain the desired level in a fixed set-point.

Using singular value decomposition (SVD) algorithm applied to the covariance matrix **S** we can obtain the matrix **V** and **Λ**,

$$\mathbf{S} = \frac{1}{n-1}\mathbf{X}^{\mathrm{T}}\mathbf{X} = \mathbf{V}\Lambda\mathbf{V}^{\mathrm{T}} \tag{14.5}$$

where $V \in \mathbb{R}^{mxm}$ is the eigenvector matrix of the nominal data and Λ the eigenvalue matrix. The scaled matrix **X** is decomposed as follows:

$$\mathbf{X} = \mathbf{T}\mathbf{P}^{\mathrm{T}} \tag{14.6}$$

where $\mathbf{T} \in \mathbb{R}^{nxa}$ is called the *score matrix* containing the projections of matrix **X** into a lower dimension space, $\underline{a}$ is the number of principal components and $\mathbf{P} \in \mathbb{R}^{mxa}$ is the *loading matrix*. The columns of **P** matrix are equal to the first $\underline{a}$ columns of the **V** matrix, corresponding to the $\underline{a}$ largest eigenvalue of the covariance matrix **S**. If $\mathbf{P}^{\mathrm{T}}\mathbf{P} = \mathbf{I}$, the linear mapping of PCA is given by:

$$\mathbf{T} = \mathbf{X}\mathbf{P} \tag{14.7}$$

The projection of matrix T into the m-dimensional space is given by:

$$\hat{\mathbf{X}} = \mathbf{TP}^{\mathrm{T}} \tag{14.8}$$

The residual space is obtained by the following equation:

$$\mathbf{E} = \mathbf{X} - \hat{\mathbf{X}} \tag{14.9}$$

The following steps were taken to obtain the nominal PCA model (P_O, T_O),

$$x(k) = [\,y(k)\ y(k-1)\ y(k-2)\ u(k-1)\ u(k-2)\,] \tag{14.10}$$

From Eqs. (14.5) and (14.10) we can obtain:

$$\mathbf{P}_0 = \mathbf{V}(:, 1:a), \quad \mathbf{T}_0 = x(k)\mathbf{P}_0 \tag{14.11}$$

14.3.2 Performance Index

Performance index is calculated using *Mahalanobis* distance between nominal score,t_0, and the mean of score variable,t. The score variable,t, includes a sliding window with d samples. The minimum distance corresponds to better closed loop performance. In this work the following values were used: n = 300 (samples); m = 5 (variables); $T_s = 1$ s (sampling time); a = 2 (number of principal components). The following figure represents the performance index methodology:

We can see from Fig. 14.4 that PCA scores, t, are calculated using ARX models. To determine the performance index between PCA scores, the followings steps were taken for both controllers, LQG and Fuzzy:

$$x(k) = [\,\hat{y}(k)\ \hat{y}(k-1)\ \hat{y}(k-2)\ u(k-1)\ u(k-2)\,] \tag{14.12}$$

$$t(k) = x(k)\mathbf{P}_0 \tag{14.13}$$

$$\hat{x}(k) = t(k)\mathbf{P}_0{}' \tag{14.14}$$

$$t_m(k) = mean(t(k-d+1:k)) \tag{14.15}$$

$$I(k) = mahal(t_m(k), \mathbf{T}_0) \tag{14.16}$$

$$I_{switch}(k) = \min\left(I_{LQG}(k), I_{Fuzz}(k)\right) \tag{14.17}$$

This method can be applied to several types of controllers and not just for LQG or Fuzzy controllers. The following figure illustrates the criteria of switching control

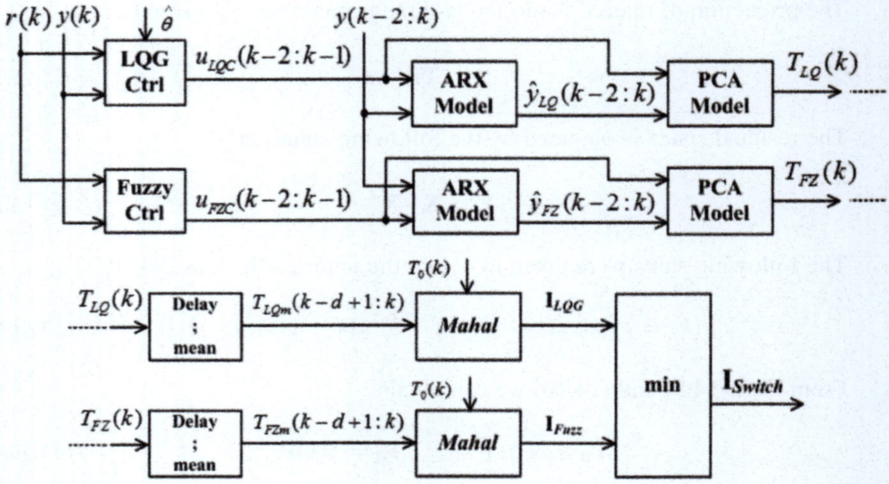

Fig. 14.4 Approach to compute performance index methodology

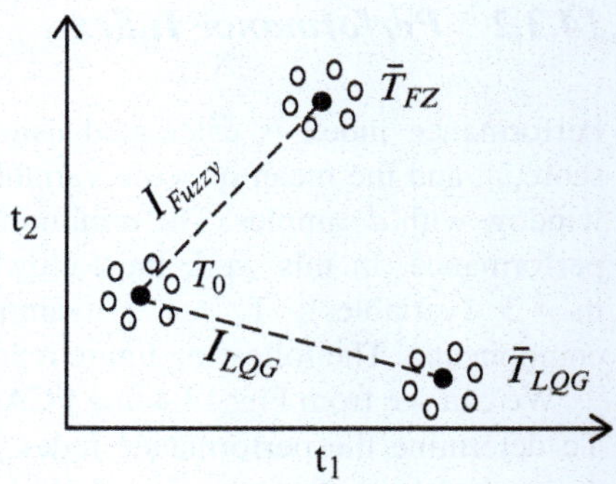

Fig. 14.5 Switching control criteria

between LQG and Fuzzy controllers based on minimum distance between PCA scores from both controllers (Fig. 14.5), where $t(k) = [t_1(k) \;\; t_2(k)]$.

14.4 Experimental Results

This section presents obtained experimental results using the *Matlab* environment, with LQG and Fuzzy controllers, in the presence of structural faults, f_1.

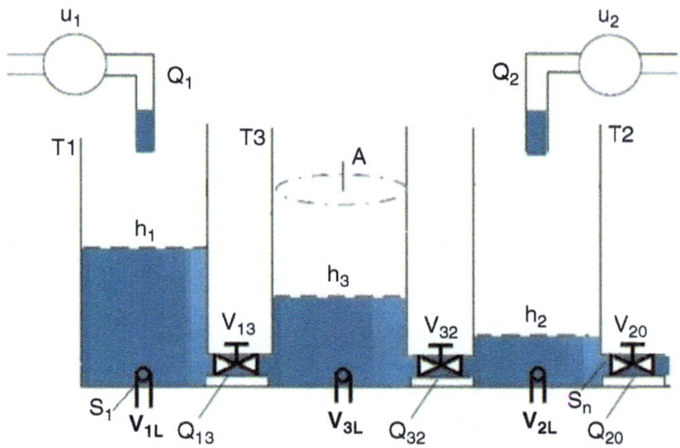

Fig. 14.6 Three-water tank DTS200 schematic diagram

14.4.1 Three-Tank System DTS200

The three-tank system *DTS200* [17], Fig. 14.6, is an equipment used as a benchmark for implementation and realization of different control and fault detection algorithms in real-time environment (Fig. 14.7).

14.4.2 Experimental Results in Nominal and Faulty Situations

All values are normalized between [0,1]. The experiment begins with *Fuzzy* controller until 60[s], after that switching between controllers is based on *Mahalanobis* distance. Structural fault is applied during 15[s], from 120 to 135 [s]. Although the system is nonlinear, it is under assumption that the process data are linear since set-point is fixed, $r(\mathrm{k}) = 0.3$.

In the first subplot, we can see the reference (–.), output (–) and fault f_1(– –) which was applied for 15 [s] to evaluate system performance. In the second subplot, switching from one controller to another was conducted based on the performance index, the minimum *Mahalanobis* distance. The variable "act" represents the chosen controller and "u" is the control action from range [0,1]. In the third subplot the distances from controllers I_{LQG}(– –), I_{Fuzz}(–.), the minimum distance I_{switch}(–) and a fixed threshold (5%) is presented. When I_{switch} is higher than the threshold, switching from one controller to another occurs. In the last subplot the error between the original data $X$and the reconstructed $\hat{X}$, Eq. (14.9) is presented.

In Fig. 14.8 the *Mahalanobis* distances between PCA scores, t, and nominal score, t_O, is presented. It can be seen that in the time period from 135 to 137 [s] the minimal distance is Ifz, leading to Fuzzy controller action, while in the time period from 165 to 167 [s] the minimal distance is Ilq, leading to LQG controller action.

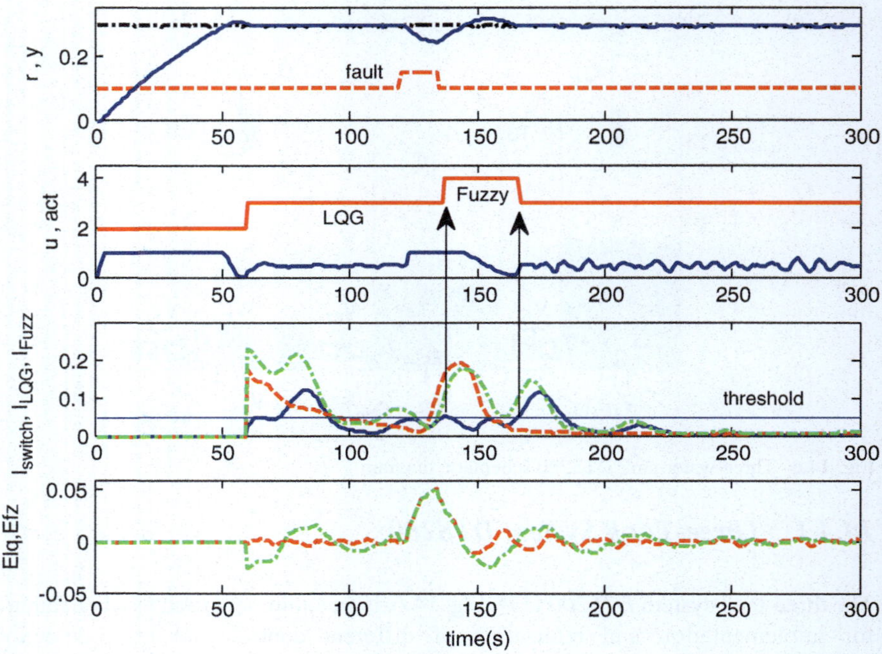

Fig. 14.7 Switching control based on the performance index

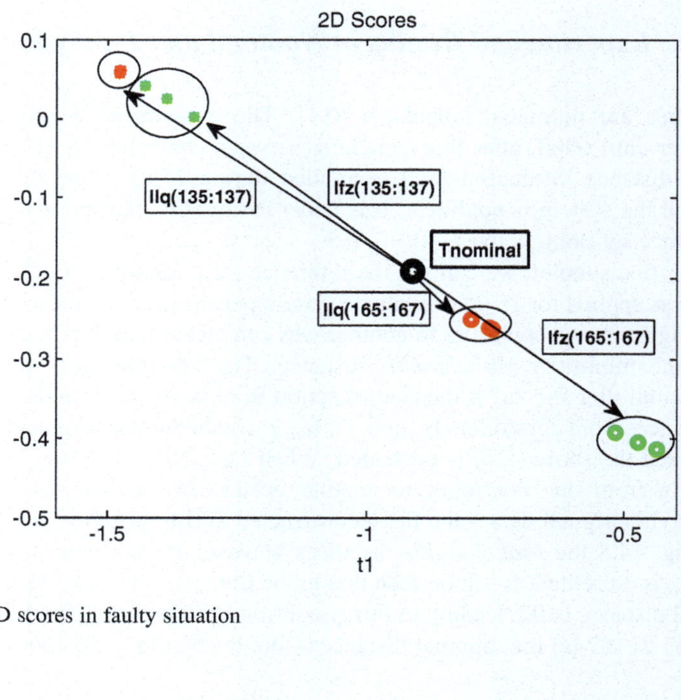

Fig. 14.8 2D scores in faulty situation

14.5 Conclusions and Further Work

A closed loop controller performance assessment approach based on PCA was proposed. It is under assumption that the process data are linear, thus principal component analysis can be used. Performance index was calculated based on *Mahalanobis* distance between PCA scores. The minimum distance implies better performance. A structural fault was applied to evaluate system performance and based on the performance index, switching control performance assessment presented good results.

Future work will be extended to nonlinear principal component (NLPCA) methods to deal with nonlinear systems. For some nonlinear processes, PCA performs poorly due to its assumption that the process data are linear. To handle the problem caused by nonlinear data, several nonlinear PCA (NLPCA) methods based on auto-associative neural networks, nonlinear PCA approach based on principal curves and neural networks had been developed in recent years.

References

1. Blanke M, Kinnaert M, Lunze J, Staroswiecki M (2006) Diagnosis and fault-tolerant control. Springer, Berlin
2. Palma LB, Coito F, Gil PS (2011) Real-time detection of oscillations in control loops. In: Proceedings of the III international conference on power engineering, energy and electrical drives. Malaga, Spain
3. Jelali M (2007) Performance assessment of control systems in rolling mills – application to strip thickness and flatness control. J Process Control 17:805–816
4. AlGhazzawi A, Lennox B (2009) Model predictive control monitoring using multivariate statistics. J Process Control 19:314–327
5. Zabiri H, Thao T (2007) A principal component approach in diagnosing poor control loop performance. In: Proceedings of the world congress on engineering and computer science, San Francisco, 24–26 Oct 2007
6. Qiang Z, Li S (2006) Performance monitoring and diagnosing of multivariable model predictive control using statistical analysis. Chin J Chem Eng 14(2):207–215
7. Chen J, Wang W (2009) Performance assessment of multivariable control systems using PCA control charts. In: Proceedings of the 4th IEEE conference on industrial electronics and applications, ICIEA, China, pp 936–941
8. William LS (2000) The control handbook. CRC Press Inc, Florida, ISBN: 8172247850
9. Lunze J, Richter J (2006) Control reconfiguration: survey of methods and open problems. Ruhr – University, Bochum
10. Steffen T (2005) Control reconfiguration of dynamic systems: linear approaches and structural tests. Springer, Berlin. ISBN 3540257306
11. Diao Y, Passino K (2001) Stable fault tolerant adaptive fuzzy/neural control for turbine engine. IEEE Trans Control Syst Technol 9(3):494–509
12. Ohnishi Y (2009) A switching method of PID controllers based on performance assessment. In: Proceedings of the ICROS-SICE international joint conference, Japan
13. Luus R, Chen Y (2004) Optimal switching control via direct search optimization. Asian J Control 6(2):302–306
14. Shields B, Barth E, Goldfarb M (2006) Predictive control for time-delayed switching control systems. ASME J Dyn Syst Meas Control 128:999–1004

15. Isermann I (1997) Supervision, fault-detection and fault-diagnosis methods – an introduction. Control Eng Pract 5(5):639–652
16. Karsai G, Biswas G, Narasimhan S, Szemethy T, Peceli G, Simon G, Kovacshazy T (2002) Towards fault-adaptative control of complex dynamic systems. In: Software-enabled control. Wiley/IEEE, Hoboken/Piscataway, pp 347–368. ISBN 9780471234364
17. Amira (2009) Lab manual DTS200 three tank system, Duisburg, Germany
18. Mahalanobis PC (1936) On the generalized distance in statistics. Proc Natl Inst Sci India 12:49–55
19. Driankov D, Hellendoorn H, Reinfrank M (1996) An introduction to fuzzy control, 2nd edn. Springer, Berlin Heidelberg New York, ISBN: 3540606912

Chapter 15
P2P Web Service Based System for Supporting Decision-Making in Cellular Manufacturing Scheduling

Maria Leonilde R. Varela, Rui Barbosa, and Susana Costa

Abstract With the increase of the Internet and Virtual Enterprises (VEs), interfaces for web systems and automated services are becoming an emergent necessity. In this paper we propose a Peer-to-peer (P2P) web-based decision-support system for enabling access to different manufacturing scheduling methods, which can be remotely available and accessible from a distributed knowledge base. The XML-based modeling and communication is applied to manufacturing scheduling. Therefore, manufacturing scheduling problems and methods are modeled using XML. The proposed P2P web-based system works as web services, under the SOAP protocol. The system's distributed knowledge base enables sharing information about scheduling problems and corresponding solving methods in a widened search space, through a scheduling community, integrating a VE. Running several methods enables different results for a given problem, consequently, contributing for a better decision-making. An important aspect is that this knowledge base can be easily and continuously updated by any contributor through the VE. Moreover, through this system once suitable available methods, for a given problem, are identified, it enables running one or more of them, for enabling a better manufacturing scheduling support, enhanced though incorporated fuzzy decision-making procedures.

Keywords Cellular Manufacturing Scheduling • Web Services • Distributed Knowledge Base • Virtual Enterprise • Fuzzy Decision-making

---

M.L.R. Varela (✉) • R. Barbosa • S. Costa
Department of Production and Systems, School of Engineering, University of Minho, Azurém Campus, 4800-058 Guimarães, Portugal
e-mail: leonilde@dps.uminho.pt; rui_barbosa_1980@hotmail.com; surp@gmail.com

A. Madureira et al. (eds.), *Computational Intelligence and Decision Making: Trends and Applications*, Intelligent Systems, Control and Automation: Science and Engineering 61, DOI 10.1007/978-94-007-4722-7_15,

15.1 Introduction

In today's knowledge-based economy, the competitiveness of enterprises and the quality of work life are directly tied to the ability of effective creation and share of knowledge both, within and across organizations.

Manufacturing scheduling is a complex task that involves a wide range of knowledge. Slight differences on the manufacturing environment originate distinct problems, which even though being closely related, require different solving methods to be applied.

The effective and efficient resolution of those problems begins with the identification of suitable scheduling methods for solving them. When there are alternative methods for solving a problem we can obtain alternative solutions, which should be evaluated against specified criteria or objectives to be reached. Thus, users are able to properly solve a problem, through the execution of one or more scheduling methods, local or remotely accessible through the web, and, subsequently, select de most suited solution obtained.

This work attempts to offer new possibilities to scheduling, giving a contribution in terms of manufacturing scheduling problems solving through a P2P web-based decision support system. The system follows a networked-based computing model, which permits sharing scheduling knowledge by means of a distributed knowledge base. This distributed scheduling repository enables accessing knowledge arising from an extended range of contributors and, therefore, a potentially more useful search, through a widened search space, based on the principles of Virtual Enterprises (VE) [1, 2].

The system permits the characterization of each problem to be solved and, then, the access to corresponding solving methods. For problems identification, a problem classification model that includes a set of parameters is used. The data representation model for scheduling problems and related concepts is based on XML (extensible markup language), used as a specification language for scheduling data representation and processing on the web [3–5].

Moreover, fuzzy decision-making procedures are incorporated in the proposed P2P web-based scheduling decision-support system, in order to enable a better manufacturing scheduling activity, by the users, namely regarding cellular manufacturing problems solving.

This paper is organized as follows. The next section briefly presents the state of the art, regarding web-based scheduling approaches. Section 15.3 describes the nature of cellular manufacturing scheduling. Section 15.4 presents the proposed P2P web-based system, including a brief description of its architecture and main functionalities. Moreover, an example is given to illustrate the system's use, for solving a cellular manufacturing scheduling problem. Finally, in Sect. 15.5, a conclusion is presented.

15.2 State of the Art

Recently we notice a tendency for an increase in scheduling systems accessible through the Internet. The systems that we came across usually involve solvers or a set of solvers, addressing each one the resolution of a restricted range of scheduling problems, using specific techniques or tools, such as mathematical programming. Moreover, they are not usually designed to easily incorporate new methods' implementations by users.

An example of a web system that can be used for scheduling purposed is the NEOS Server, developed under the auspices of the Optimization Technology Center of Northwestern University and Argonne National Laboratory, generally used for optimization problems solving. It makes nearly 50 solvers available through a broad variety of network-based interfaces. According to the authors, although having evolved along with the web and the Internet, it is limited to some degree by early design decisions (http://www-neos.mcs.anl.gov/).

We can also refer the BBN's Vishnu scheduling system, as another interesting web-based scheduling system (http://vishnu.bbn.com) and the FortMP, a Mathematical Programming Solver, by Mitra's Group, at Brunel University (http://www.brunel.ac.uk/depts/ma/research/com).

The e-OCEA, a portal for scheduling intends to support identifying scheduling problems, to help development of new algorithms and to conduct benchmarks through the Internet. However, one requirement of this system is to only consider elements (algorithms, data sets, schedules and modules) that are e-OCEA compatible ones (http://www.ocea.li.univ-tours.fr/eocea).

The LekiNET, a prototype Internet scheduling environment, by Benjamin P. C. et al. (http://www.stern.nyu.edu/om/software/lekin/index.htm), which has migrated from LEKIN, a flexible job shop system, is a system that has some similarities to our proposed system, but focusing more on cost effective choices, based on scheduling agents for problems solving. The authors propose a migration scheme to transform existing standalone scheduling systems to Internet scheduling agents that can communicate with each other and solve problems beyond individual capabilities. They treat each system as an agent and build the relations between the systems. So, wrappers need to be specifically designed for each system [6, 7].

15.3 Cellular Manufacturings Scheduling

Manufacturing scheduling consists on a temporal assignment of tasks to manufacturing resources, where one or more goals and constraints have to be considered. It is one of the significant tasks to be performed to achieve competitive manufacturing levels, which usually means to deliver products on time and to use resources efficiently.

Moreover, low flow times, few ongoing jobs in the system, low inventory levels, high resource utilization and, certainly, low production costs are also important objectives, which can be better satisfied through the execution of the most suitable scheduling methods available for solving each particular problem.

In order to perform the scheduling process it becomes necessary to clearly specify the problem to be solved. Manufacturing scheduling problems have a set of characteristics that must be clearly defined.

Due to the existence of a great variety of scheduling problems there is a need for a formal and systematic manner of problem representation that can serve as a basis for their classification. A manufacturing scheduling problems notation for achieving this was developed by Varela et al. [8, 9], based on existing notations available in the literature [10–13]. This notation allows identifying the characteristics of each problem to be solved, and it is used as a basis for an XML-based problem specification model used in this work.

After clearly specifying a manufacturing scheduling problem it turns necessary to select a suitable method for solving it. Among the variety of scheduling methods available it is possible to identify three fundamental categories [14]:

- Calculus-based (analytical, based on derivatives);
- Enumerative (brute-force);
- Heuristic guided random search (e.g. SA, Genetic Algorithms and Neighborhood search approaches).

The first two categories deal, mainly, with algorithms that are not appropriate for solving problems of great dimension or lack analytical structure because of the computational time to find a solution [14–16].

The third category includes heuristic and approximate algorithms [17–19] that are more appropriate for the solution of NP-hard or NP-complete, known as "difficult" problems [10, 11]. Although these kinds of algorithms do not guarantee optimal solutions, they are usually faster and they allow obtaining near-optimal solutions or usually at least acceptable ones. Sometimes they could even reach an optimal solution. Most of these algorithms are relatively recent and have been increasingly explored [6]. Some well-known heuristic algorithms refer to Simulated Annealing, Tabu Search and Genetic Algorithms and other are neighborhood search based approaches [6, 10, 11, 18].

Scheduling methods for solving cellular manufacturing problems have to concerns about ordering jobs to be performed on a set of machines or work centers, in a given manufacturing environment, which is usually characterized by its possibility to process a range of different operations integrating a set of related work, generally grouped together, based on some similarities, which to some extent, enable to consider them to be more or less closely related to each other, in terms of processing requisites. In this context machines or work centers are usually equipped with flexible machines, in order to enable to process some quite different operations, among a set of more or less closely related jobs, usually referred as a family of jobs.

Therefore, in cellular manufacturing usually arise more or less complex manufacturing scenarios, even when processing simple jobs, including only one or two operations and in this sense it may become an even more complex task to select appropriate solving methods, for each given manufacturing scheduling problem.

15.4 Scheduling Web Services Based System

In our proposed system any method which is accessible through the Internet, once its signature and location are specified within our web system, can be used, for solving problems put by users. No further requirements are necessary for being able to remotely use available methods. The association of scheduling problems to resolution methods is done using the information available in a DKB, about both problems and solving methods [8]. We did not come across to systems with identical architecture and underlying approach, i.e. oriented for solving a large variety of manufacturing scheduling problems based on a continuously updatable distributed knowledge base, which allows a network of peers to provide scheduling services to users and the dynamic enlargement of the number of methods that can be accessed.

15.4.1 General Web Service Description

A web service enables a client to remotely invoke a method on a given machine, connected through a network, upon receiving a description of a problem instance to be solved, through a selected method, by sending a request with the data for running the method for the given problem instance. Afterwards the results are returned to the client in order to be transformed and presented to the final user in a certain format. Those web services can be identified and described over the web, by using UDDI registry and WSDL descriptions.

Moreover, the web services may be implemented and accessible for remote executions by using different platforms, programming languages and communication protocols. Therefore, web services enable to remotely invoke methods, namely manufacturing scheduling ones, which can be performed, for instance, through SOAP (Simple Object Access Protocol, http://www.w3.org/TR/soap/).

15.4.2 Web System Architecture

Web systems are systems that emerged from previous traditional systems, which are systems that beyond other advantages are able to implement web services that can be easily used and shared through the web [20].

Moreover, the last tendencies show that computing environments are characterized by increasing heterogeneity, distribution and cooperation, where P2P (peer-to-peer) based distributed knowledge bases also play an important role [20].

In networked environments, distributed knowledge bases and intelligent brokers, for information retrieval from specialized servers and knowledge repositories distributed across the Internet, enable to establish high-quality problem solving, through knowledge and resources sharing, forming virtual enterprises (VE) [1, 2].

In VEs several business partners contribute with a local knowledge base (KB) component within networked environments. In these environments each member can be seen as a VE business partner interested on solving manufacturing scheduling problems.

One of the greatest benefits of these networked environments, in the context of this work, is to easily support the concept of community. Consequently, it is possible for users to organize themselves into groups that can collaborate with each other in order to achieve certain goals. One of the main goals aimed at in this work is the collaborative improvement of the resolution of manufacturing scheduling problems. This is possible through the DKB for manufacturing scheduling by accessing several different scheduling approaches. This is achieved by providing a mechanism that allows the members of the organization's network to share their scheduling knowledge and scheduling methods.

Whenever a member stores knowledge in its KB component, he/she is automatically contributing to the enrichment of the whole distributed knowledge repository, which is available to all members of the VE.

Different manufacturing scheduling methods' implementations may provide results in different formats; consequently the system must have a description of them in order to format them according to the problem output to be returned to the client as the very last step of the service.

Moreover, the proposed system integrates an interface, for problem results presentation (as well as for problem data input), which is automatically generated by the system according to the corresponding implemented method's signature, previously specified.

Therefore, for the system to be able to use such implementations in a programmatic way, they must be described within the system. This description must include, among other things, the address to the running method or program and its signature, which, in turn, includes the definition of the parameters that are necessary for its execution (inputs) and its output format.

All the scheduling information is stored in XML documents, which are validated according to associated DTDs, before being put in the corresponding knowledge base component [8, 20].

The result from running an implemented method on a given problem instance can be delivered to the client as an XML file and/or can be transformed into some more expressive output, like a Gantt chart or other charts, namely obtained through excel files processing, which is another input and output format enabled through the proposed system, among others.

As previously referred the DKB scheduling system is based on the principles of VE. In this VE, each business partner contributes with a local knowledge base (KB) component in the organization's network. Each one can be seen as a VE member or partner interested on solving scheduling problems. At each business partner the knowledge base can be continuously improved with new problem descriptions and available solving methods.

Whenever a member stores knowledge in its KB component, he/she is automatically contributing to the enrichment of the whole distributed knowledge repository, which is available to all members of the VE.

The methods are usually available in the knowledge base of the VE partners but they can also be found in other sites not belonging to the community.

15.4.3 Web System Functionalities

The system has been designed and implemented as a set of web services (http://www.w3.org) using the SOAP protocol (http://www.w3.org/TR/soap/).

Each web service accepts as input a problem definition, i.e. data for running a method, and returns outputs from a method's invocation.

The SOAP protocol (Simple Object Access Protocol), like other communication protocols, consist on a possible way that can be used for remote methods invocation and can be defined as the sequence and structure of requests and responses required to invoke communications on a remote machine [20, 21].

The XML specification language provides a vocabulary for describing remote procedure calls, which are transmitted between computers using the Hyper Text Transfer Protocol (HTTP). SOAP clients make procedure requests to a server, which returns results to the clients. Clients use the same HTTP facilities as web browser clients, and servers use the same HTTP facilities as web servers. The SOAP protocol requires a minimal number of HTTP headers to be sent along with the XML method request for solving a given problem instance.

Upon receiving a request, a server must deliver a response to the client. The response may take one of two forms: the result of processing the method or a fault report, indicating that something has gone wrong in handling the request from the client. As with a request, the response consists of HTTP headers and an XML payload.

By using the SOAP protocol we are able to invoke scheduling methods implemented on different programming languages. Moreover, these methods, local or remotely available, may be running on different platforms.

As referred previously, our web scheduling system includes a Knowledge Base component that encompasses all knowledge and modules necessary for remote methods invocation and for performing a set of other system functionalities. This component is controlled by ASP (Active Server Pages) and corresponding server-side components.

On the other hand, there are also correspondent components for each of the methods servers waiting for requests.

This environment is heterogeneous as servers can use their own technology, i.e. use different implementation languages or/and different operating systems.

Several other protocols that could be used exist, namely XML-RPC (eXtensible Markup Language – Remote Procedure Call), CORBA (Common Object Request Broker Architecture) or RMI (Remote Method Invocation) [20, 21].

15.4.4 Fuzzy Decision-Making

An important aspect for improving the decision-making process consists on enabling the use of fuzzy parameters on the scheduling problems specification. For this purpose, fuzzy sets are being used.

The main goal of a fuzzy approach is to find the solution or the alternative decisions that may better satisfy and suite to a given situation that occurs in a fuzzy environment [22].

The proposed system can be used as a support tool, for the decision-making process, by providing, in real-time, a range of alternative solutions to the decision maker, indicating the advantages and disadvantages (trade-offs) of each evaluated alternative. With such information the decision maker can take an enlightened choice of the best alternative for each specific application scenario.

Great part of the decision-making process occurs in an environment in which the objectives, the constraints and the consequences of possible actions are not accurately known. To deal quantitatively with imprecision it becomes sometimes necessary to appeal to tools provided by the fuzzy sets theory [23].

With this type of approach we can easily represent any fuzzified parameter of our optimization problem. For instance, we may represent fuzzy coefficients, as for example, "the deadline of a job x is about 7 time units". This meaning that the preference of the decision maker is 7 time units but deviations are allowed to the left and right of the preferred value, by using an appropriate fuzzy function [22]. Moreover, other situations occur when we do not have access to precise information regarding some other kind of problem data, namely regarding a set of parameters that is necessary to be specified for each problem instance to be solved, according to a nomenclature used [24, 25], which includes a wide range of parameters related to the manufacturing environment and concrete problem characteristics to be defined.

15.4.5 Web System Illustration

The following image, in Fig. 15.1 illustrates de developed system's interface for supporting the user to solve a given cellular manufacturing scheduling problem.

Though the system's interface the user is able to choose among a number of alternatives ways for problem data insertion, namely by loading a file in xml, excel, pdf or txt format, or by directly data insertion through the web browser interface.

Once the user inserted the problem data a set of alternative methods is put available through the system's interface, for solving the given problem, for single up to multiple machines problem cases, varying from simple sequencing rules or procedures up to more complex methods, including meta-heuristics as, for example, based on neighborhood search techniques.

Moreover, the system enables to apply fuzzy logic to some specific problem parameters, as described above, which also enriches the range of results that can be obtained, for a given problem instance resolution, therefore, enabling a better decision-making, for each specific scenario considered.

In terms of problem results, several alternatives are also available through the system, for results presentation as well as for results storage, in xml, excel txt or pdf formats.

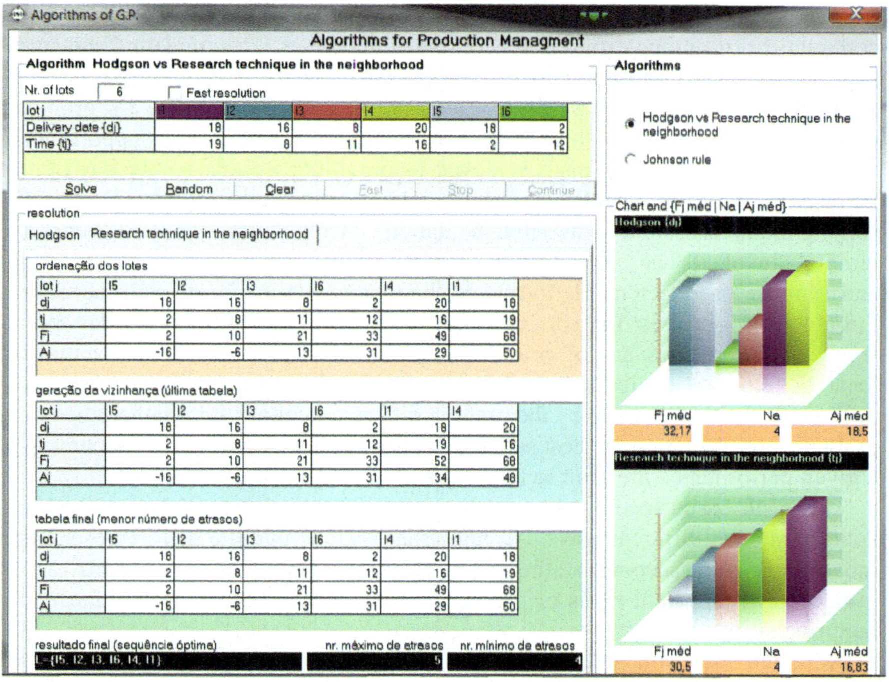

Fig. 15.1 Web system interface illustration

Particularly interesting results presentations arise from the use of Gantt chart, tables or other graphics, as shown if Fig. 4.4, which enable even more clear results interpretation.

15.5 Conclusion

In today's knowledge-based economy, the competitiveness of enterprises and the quality of work life are directly tied to the ability of effective creation and share of knowledge both, within and across organizations.

In production enterprises, it is important nowadays, as a competitive strategy, to explore and use software applications, now becoming available through the Internet and Intranets, for solving scheduling problems, which can be achieved in an easily way by using web service technology.

This work is based on an XML-based specification framework for manufacturing scheduling concepts modeling, which is used as specification language for better supporting manufacturing scheduling decision-making, based on web services. Some of the important functions of the proposed system include the ability to represent scheduling problems and the identification of appropriate available methods for solving them in the context of a P2P network and underlying distributed knowledge base about scheduling problems and solving methods.

The XML-based data modeling is used in order to make possible flexible communication among different scheduling applications. This modeling and specification language contributes to the improvement of the scheduling process, by allowing an easy selection of several alternative methods available for problems solving, as well as an easy maintenance of the knowledge base, by supporting the scheduling web services. This is achieved by providing a user-friendly way of new knowledge insertion, which includes knowledge about scheduling problems and solving methods as well as its implementations, available through the Internet and remotely invoked. These may be implemented on different programming languages and running on different platforms. Such implementations are easily accessible through web services for solving scheduling problems, where cellular manufacturing based ones are of primordial importance, through the invocation of local or remotely available scheduling methods. The system allows comparing different solutions obtained by the running different methods for a same scheduling problem, and to choose the best solution found to solve each problem, according to a given performance measure to be reached.

The XML based specification can be generated and visualized by computers in appropriate and different ways. An important issue is that the data representation model is general, accommodating a large variety of scheduling problems, which may occur in different types of production environments, besides cellular manufacturing systems.

Although the main goal is the service for scheduling problems solving, the system can also be used for teaching purposes, and from this point of view, several different features can be put forward by the system. These include historical and referencing information about each problem class and solving methods. The system will grow as more knowledge is included in the distributed knowledge repository.

The web system is based on an Internet-based infrastructure, associated to a distributed knowledge base (DKB). This DKB is spread through a set of members (partners) forming a virtual enterprise. These members can store information about methods for solving different kinds of manufacturing scheduling problems, as well as its implementations. Each partner, in a decentralized computing environment, is able to directly interact with each other as well as with any other external user. This web-based communication and direct collaboration is enabled through a dynamic web interface, which permits executing diverse scheduling functions, including the ability to represent different scheduling problems, search for appropriate solving methods and running methods that are available and accessible through the DKB, within the VE's network. As running different methods enables obtaining different solutions for a given problem, we believe that the proposed system may contribute to a better decision-making process, enhanced by means of fuzzy based support and collaboration of partners organized as a virtual enterprise.

Acknowledgments The authors wish to acknowledge the support of: (1) The Foundation for Science and Technology – FCT, under the scope of the financed Project on "Ubiquitous oriented embedded systems for globally distributed factories of manufacturing enterprises" – PTDC/EME-GIN/102143/2008, and (2) EUREKA, under the Project E!4177-Pro-Factory UES.

References

1. Putnik GD (2000) BM_virtual enterprise architecture reference model. Technical report RT-CESP-GIS-2000-<GP-01>, Universidade do Minho, Portugal
2. Putnik GD (2001) BM_virtual enterprise architecture reference model. In: Gunasekaran A (ed) Agile manufacturing: 21st century manufacturing strategy. Elsevier Science Publ, New York, pp 73–93
3. Abiteboul S et al (2000) Data on the Web - from relations to semistructured data and XML. Morgan Kaufmann Publishers, San Francisco
4. Ceponkus A, Hoodbhoy F (1999) Applied XML. Wiley Computer Publishing, New York
5. Harper F (2001) XML standards and tools. eXcelon Corporation, Chicago
6. Pinedo M (2002) Scheduling theory, algorithms and systems. Prentice-Hall Inc, Englewood Cliffs
7. Yen B et al (2004) Internet Scheduling Environment with Market-Driven Agents. IEEE Transactions on Systems, Man and Cybernetics – Part A: Systems and Humans Vol. 34, No. 2, March, 2004
8. Varela L, Aparício J, Silva S (2002a) An XML knowledge base system for scheduling problems. In: IICS'02 – Proceedings of the innovative internet computing system. Springer-Verlag in the lecture notes in computer science series, Springer, Berlin, Germany
9. Varela L, Aparício J, Silva S (2002b) Scheduling problems modeling with XML. In: The 4th international meeting for research in logistics, proceedings of research in logistics, Lisbon, Portugal
10. Blazewicz J et al (1996) Scheduling computer and manufacturing processes. Springer, Berlin
11. Brucker P (1995) Scheduling algorithms. Springer, Berlin
12. Graham RL, Lawler EL, Lenstra JK, Rinnooy Kan AHG (1979) Optimization and approximation in deterministic sequencing and scheduling: a survey. Ann Discret Math 5:287–326
13. Jordan C (1996) Batching and scheduling. Springer, Berlin
14. Artiba A, Elmaghraby S (1997) The planning and scheduling of production systems. Chapman & Hall, London
15. Conway RW, Maxwell WL, Miller LW (1967) Theory of scheduling. Addison-Wesley Publishing Company, Inc, Reading
16. French S (1982) Sequencing and scheduling – an introduction to mathematics of the job-shop. Wiley, New York
17. Arts E, Lenstra JK (1997) Local search in combinatorial optimization. Chichester, London
18. Morton T, Pentico D (1993) Heuristic scheduling systems. Wiley, New York
19. Osman IH, Kelly JP (1996) Meta-heuristics: theory and applications. Kluwer Academic Publishers, Boston
20. Varela MLR, Aparício JN, Carmo-Silva S (2005) A scheduling web service. In: Kendall G, Burke E, Petrovic S, Gendreau M (eds) Multidisciplinary scheduling – theory and applications. Springer, Nottingham. ISBN 0-387-25266-5
21. Laurent S et al (2001) Programming Web services with XML-RPC. O'Reilly and Associates Inc, O'Reilly Media, Beijing
22. Varela MLR, Ribeiro RA (2002) Evaluation of simulated annealing to solve fuzzy optimization problems. J Intell Fuzzy Syst 14(2):59–71
23. Zadeh L (1965) Fuzzy sets. Inform Control 18:338–353
24. Varela MLR, Carmo-Silva S (2008) An ontology for a model of manufacturing scheduling problems to be solved on the web. In: Azevedo A (ed) Innovation in manufacturing networks, 8th IFIP international conference on information technology for balanced automation systems. Springer, Porto, Portugal, 23–25 June 2008, pp 197–204. ISSN: 1571–5736 / 1861–2288 (Internet), ISBN: 978-0-387-09491-5; eISBN: 978-0-387-09492-2
25. Carmo-Silva S, Varela MLR (2011) An ontology for production systems environments of scgeduling problems. In: Putnik GD, Ávila P (eds) Business sustainability II, management technology and learning for individuals, organisations and society in turbulent environments. University of Minho, Guimarães & ISEP, Porto, Portugal

Chapter 16
Web-Based Decision Support System for Orders Planning

António Arrais-Castro, Maria Leonilde R. Varela, and Sílvio Carmo-Silva

Abstract Competition in the global economy is intensifying the implementation of the mass customization paradigm. This requires a substantial increase in the personalized interaction between client and producer. Agility in the order-production-delivery cycle optimization is a key element for industrial enterprises as a means to meet requirements of this paradigm. The use of agile methodologies for this purpose requires improving processes of product specification and product data management. Under today's technology this tends to be predominantly carried out with the aid of the Internet using mostly web services. For a company, this also requires better integration of front-office processes (interaction with the outside) and back office (including production processes). Mass Customization scenarios are characterized by large product variety dependent on product specific requirements of individual customers. In this process there is a need to provide the customer with tools for easy product specification, selection or configuration. Web-based configurators can provide an opportunity to both producer and customer, through an interactive process, for a more formal, faster, effective and better product and orders specification. In this paper, we propose an architecture and describe functionalities of a web-based system for interactive products and customer orders configuration. The proposed system can also be a valuable tool to support production and delivery.

Keywords Mass configuration • Web services • Product configurator

A. Arrais-Castro • M.L.R. Varela (✉) • S. Carmo-Silva
Department of Production and Systems, School of Engineering, University of Minho, Azurém Campus, 4800-058 Guimarães, Portugal
e-mail: arraiscastro@gmail.com; leonilde@dps.uminho.pt; scarmo@dps.uminho.pt

A. Madureira et al. (eds.), *Computational Intelligence and Decision Making: Trends and Applications*, Intelligent Systems, Control and Automation: Science and Engineering 61, DOI 10.1007/978-94-007-4722-7_16,

16.1 Introduction

Globalization introduces new challenges for companies that have to face a potentially large number of business competitors and new requirements that have to be fulfilled in order to maintain market share and improve customer retention. The implementation of competitive strategies and process optimization procedures is a requirement for survival. The integration of front-office and back-office processes needs to be efficient, cost-effective and error-free. Besides, companies need to adopt a strategy of differentiation. Having the lowest prices is no longer enough: companies need to adapt their products to the specific needs of customers. Modern consumers are particularly exigent, demanding higher levels of quality, functionality and correspondence to their specific needs. Traditionally, the inclusion of optional configurations and product variants had more to do with charging "extras" and gaining financial profits than with offering the customer a richer buying experience.

The "experience economy" [1] is a result of the growing demand for individualization. Companies need to fulfill customers' individual needs while maintaining prices as low as possible, which is a permanent challenge for modern enterprises. Therefore, companies are forced to create product families with a growing number of variants, adopting a strategy of differentiation based on variety [2].

The concept of Mass Customization was introduced by Davis [3] and is defined by Tseng and Jiao [4] as "producing goods and services to meet individual customers' needs with near mass production efficiency". Besides offering differentiation opportunities and cost optimization, Mass Customization allows for a richer interaction with customers, maximizing customer retention. Frank Piller [2] enumerates three distinct and alternative levels of Mass Customization, connecting them with associated benefits: customized products (differentiation option), efficiency (cost option) and lock-on (relationship option).

A Mass Customization strategy can be achieved designing and building modular products, so that each module can be integrated with other modules to build a final product. This generates a larger variety of products to better fit customer preferences. It requires processing information agility and a high level of integration with existing information systems, particularly at production management and product development processes. These requirements imply efficient management of information flows, from customer interaction to production.

Industrial companies can efficiently focus on customers only if they use information intensively [5]. With Mass Customization, the individual wishes and needs of each customer have to be transformed into a unique product specification, according to Piller. The author considers that costs associated with product customization can be determined by quantifying the efforts associated with specifying customers' needs, individual product configuration, transfer of user preferences to the production stages, additional complexity at production management and interactions with suppliers and product distribution.

16.2 Motivation

The implementation of a Web Service based product configuration system will allow creation of a central channel for the distribution of information about simple and complex products, supporting customers and business partners with their configuration and product selection needs. Implementing product design using a set of configurable modules the system will support the implementation of Mass Customization strategies [4], generating value from an optimized interaction with customers and business partners.

This communication channel can be used to automatically feed distinct point of sale (POS) systems, including websites, interactive product configurators for physical points of sale and promotional interactive media. Additionally the system will connect these POS systems with management back-ends, including CRM and Production Management Systems.

The main component of the product configuration system is a Web service layer that connects with the company's information systems (data import) and feeds external systems (data export). This layer will expose configuration engine functionality, facilitating the order capture process and its conversion to production orders. It will be a valuable tool to support the implementation of methodologies that require agile product and order information processing.

The system will also help the implementation of agile production methods and approaches, like Lean Manufacturing. This approach is derived from TPS – Toyota Production System, developed at Toyota after World War II. It aims to eliminate waste, targeting for elimination any activities that don't create value for the end customer. The method has two pillar concepts: automation and JIT – Just In Time. JIT was introduced by Taiichi Ohno, chief engineer at Toyota in the 1950s, as a method for the fulfillment of customers' needs with a minimum of delays while reducing finished and in-process inventory [6]. With JIT, inventory is seen as incurring cost and waste, instead of stored value. Nothing should be produced, transported or bought before the required time. It all comes to having "the right material, at the right time, at the right place, and in the right amount".

The make-to-order philosophy that JIT implies minimizes risks of having unsold products. Web based product configurators improve agility while capturing sales orders and converting them into production orders, thus helping to implement JIT.

Agile manufacturing [7] was designed to make businesses more flexible, requiring greater agility from production operations, setup and lead-time reduction and more efficiency while managing product information. An Agile company needs to determine customer needs quickly and continuously reposition itself against its competitors [8]. This means that it also needs to design products quickly based on customers' individual needs.

Product configuration support systems are being developed over the years and some interesting examples of such kind of systems can be found through the following papers [9–22]. These existing systems are based on the use of several distinct approaches and technologies and with the growing of web-based utilities, web-service oriented systems intend to be more and more important and useful and this is why we decided to give our contribution in this domain [23], expecting that such a kind of system will help capture customers' needs and preferences continuously, thus contributing to agile product definition and management processes.

16.3 Proposed Architecture

The product configuration system (WBPC – Web Based Product Configurator) includes two main components: An Integration Module, which is responsible for integration with existing information systems, and a Service Module, which interacts with external systems (client and server applications).

Integration will be built using connection modules called connectors. Several connectors will be available for reading and writing information from/ into external systems. Each read connector is responsible for reading the information available in a backend system and transform this information into XML. The write connectors receive XML data and convert it to a format that is adequate for the associated system. XML Schemas (XSD) will be used to help validate the information and check its completeness and integrity, as shown through Fig. 16.1.

The Data Broker component is the central data processing module and controls how data import and export is managed. It includes two sub-components: Data Validator and Data Transformer. The Data Broker is isolated from the details of each external system, accepting only XML information. Data Validator maps the

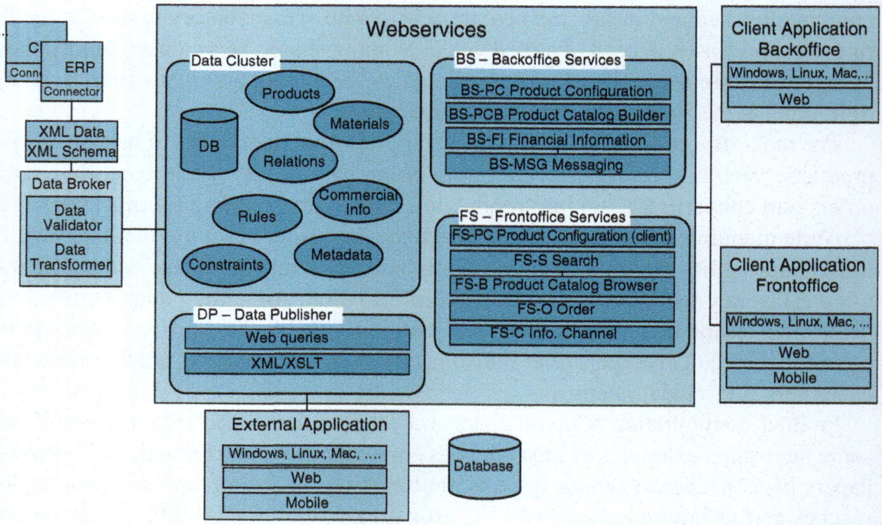

Fig. 16.1 System architecture

information received by the broker to the corresponding XML Schema. Any validation errors cause the rejection of the underlying data.

The Data Transformer module is responsible for making the imported information compatible with the structure the service broker expects to receive. It uses XSLT style sheets (XSL Transformation), applying them to the previously validated XML data, and forwarding the transformation results to the Web Service layer. The Data Broker component can be replaced by a third party middleware solution, like TIBCO (www.tibco.com) or BizTalk Server (www.microsoft.com/biztalk). These solutions will help modeling business rules at the integration level.

The web services layer is responsible for the implementation of functionalities that support product configuration and information publishing. It features four main components:

Client applications' only requirement is the ability to call Web methods using HTTP and SOAP, and process the returning data (XML formatted).

16.4 Product Planning

During product planning the modules and product constraints are defined. The outputs control how products can be built, according to constraints and business policies. A set of pre-approved configurations can be published, although it is not mandatory. Later the list of configurations can be extended with configurations captured from user interaction.

Valid product configurations comply with business regulation and fulfill production and business requirements.

A valid configuration can be built from inside the company, using a product configurator (for example, while generating a base product catalog) and from outside the company, by customers and business partners, when structuring a product or service for a given scenario.

Two distinct configuration levels are associated with customization processes:

- Controlled configuration
  - Customer can chose among a given set of components and attributes. Component or attribute value *n* depends on previous choices made until configuration step n − 1. It also depends on system rules, that determine compatibility among components and attributes.
- Free configuration
  - Customer is free to choose components or attributes (sizes, shapes, materials, base components). Customer may inclusively submit an initial drawing that will be evaluated by the company for budgeting. Typically a visual 2D or 3D editor is used to support customer order editing.

The proposed system architecture supports controlled configuration. A product consists on a set of modules or parts that are inter-related according to compatibility

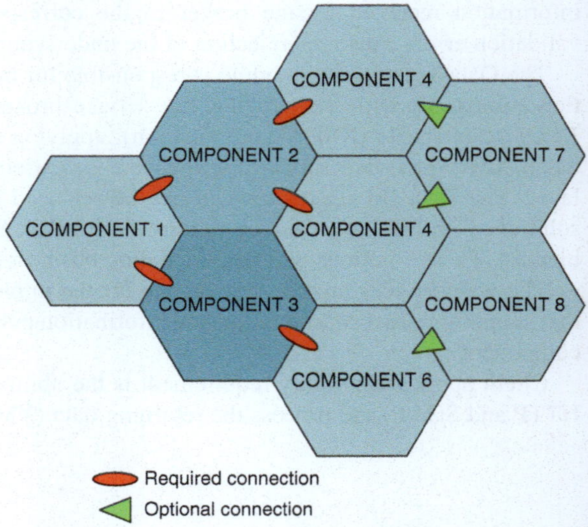

Fig. 16.2 Structure of a component-based product

rules, commercial rules and affinity levels. A module can be an undividable fabricated component, i.e., a part, or it may consist on sub-products made of parts or components, i.e. assemblies, as illustrated on Fig. 16.2.

Starting from a chosen initial component there will be several distinct possibilities for choosing a product or service. The choice of a new component is influenced by:

– Production constraints or requirements
  - Only valid combinations are allowed;
– Compatibility constraints
  - Compatibility between component A and component B in the context of the configuration of product P means that a valid configuration can be built using P = (A,B) or P = (A,C), if B and C are configuration alternatives; P = (A,B, C) if B and C are configuration alternatives.
– Structural rules associated with the product itself
  - Minimum, maximum and recommended quantities of components belonging to a specific class or type, for example;
– Business policies
  - If B and C are component alternatives in the context of the configuration of product P, we may have two different alternatives with distinct costs: P = (A,B) and P = (A,C);
  - On the other hand, commercial policies may define how optional components are combined with base components. Assuming P1 is based on base components (A,B,C) and that the accessories A1, A2 and A3 are available, different

combinations may be possible, each introducing packaging and delivering constraints, although not having a relevant impact on P1 production:

– P1 = (A,B,C) + A1
– P1 = (A,B,C) + A2
– P1 = (A,B,C) + A3
– P1 = (A,B,C) + A1 + A2
– P1 = (A,B,C) + A1 + A3
– P1 = (A,B,C) + A2 + A3
– P1 = (A,B,C) + A1 + A2 + A3

- Finally, commercial constraints may impose the unavailability of certain combinations due to insufficient profit margins or lack of strategic interest.

To build a base set of product configurations the company needs to define the technical structure of the product, listing the main components that integrate it and associated constraints. Next the company defines a set of attributes for each component, including the allowed range of values and data type. Values are based on production constraints and requirements and on product characteristics.

Once all attributes and possible values have been listed, the company builds a compatibility matrix, identifying which attributes are compatible with different components. If the company desires to help the customer choose configurations based on recommendations and component selection suggestions, it may create an affinity matrix. The matrix can be based on company product information and also on customer feedback. It may be automatically updated by the system (based on system usage) and manually (by operators).

With both compatibility and affinity matrixes the system has enough information to support product configuration. The last step is to assign values to components and options, to allow for alternative quantification and comparison. This is done by filling the commercial matrix. The step is optional since it is not required if the customer is building a product configuration for RFPS (Request for Proposals).

After defining the product configuration parameters, information is stored centrally in the system and publication and access rules are created. According to publication parameters the configuration is then available through the web service layer.

The following steps illustrate the workflow associated with simple product configuration scenario, expressed through Tables 16.1, 16.2, and 16.3, regarding a company that wants to allow customers to configure their pool table orders.

The company starts by identifying the main components of the product. Some components are part of the product itself and are complemented by sets of accessories or additional optional components. Since these are sold as part of the final product they are included in the product configurator. The following table includes a list of general components:

For each component the company registers minimum, maximum and typical quantities. Minimum and maximum values generate constraints, typical values generate configuration suggestions.

After building the list of general components the classification phase is performed. Each component is classified according to the associated level of pre-fabrication

Table 16.1 Sample list of product components

Component	Minimum quantity	Typical quantity	Maximum quantity	Mandatory
Cabinet and slate	1	1	1	Y
Table legs	4	4	6	Y
Table felt	1	1	1	Y
Table rail, including ruber cushion and wooden frame	1	1	1	Y
Cue holder	1	1	1	N
Cue sticks	0	4	8	N
Plastic triangle	1	1	1	N
Cue balls kit	1	1	1	N
Clue chalk box	1	1	*	N
Lamp	1	1	1	N

*stands any [maximum] quantity

Table 16.2 Sample list of compatibility matrix

			Existing models/configurations		
Item	Attributes	Allowed values	Paris	London	Lisbon
Cabinet	Materials	Wood – Maple	X		X
		Wood – Mahogany	X	X	X
		Wood – Poplar/tulipwood	X		X
		Chipboard	X		X
	Finishing	Matte	X	X	X
		Glossy	X	X	
	Color	Black	X	X	X
		Darkened wood	X		X
		Wengue	X		X
		Pine		X	
Cue sticks	Model	Precise LV1	X		X
		Precise LV2	X		X
		Precise LV3	X	X	X
		Precise LV4	X	X	X
		Precise LV5	X	X	X
		Precise LV6	X	X	X
	Accessories	Bag	X	X	X
		Maintenance kit	X	X	X

(finished components, which need to be integrated; partially pre-fabricated components, which require finishing; kit based components that need to be fabricated according to specifications). For each component attribute a list of valid values is defined. These values are derived from production constraints and commercial policies.

The next step is to build compatibility matrix mapping Items, attributes and values. These can also be matched with existing configurations (Table 16.2).

The company may allow the user to completely customize a product from scratch. Alternatively, it may choose to guide him offering a base set of options associated with a particular model or configuration domain.

Table 16.3 Sample list of affinity matrix

Item	Attributes	Allowed values	Existing models/configurations	
			Paris	London
Cabinet	Materials	Wood – Maple	0	
		Wood – Mahogany	300	500
		Wood – Poplar/tulipwood	500	1,100
		Chipboard	100	
	Finishing	Matte	0	0
		Glossy	0	0
	Color	Black	0	
		Darkened wood	50	0
		Wengue	50	50
		Pine	25	
Cue Sticks	Model	Precise LV1	0	
		Precise LV2	20	
		Precise LV3	30	0
		Precise LV4	50	50
		Precise LV5	75	75
		Precise LV6	100	100
	Accessories	Bag	100	100
		Maintenance kit	60	60

The compatibility matrix may be complemented by the affinity matrix, which connects components grading their relation according to compatibility levels (look and feel, other customer preferences and functional affinity, among others). This matrix includes static values (manually fed using the system back-office components) and dynamic values (automatically updated by the system based on user customer preferences).

When a customer has chosen a given set of options for product P, the system presents him a set of components and suggested options (O1, O2 and O3). These options will be shown according to associated affinity levels. Levels are obtained from the affinity matrix and may be weighted according to the company selling strategies.

Finally, the compatibility and affinity matrices may be complemented by a commercial matrix, which is used to:

Associate options with costs (some options are obviously cheaper than others);

Define pricing policies (a component may have different prices depending on the configuration it is used in; this may reflect positive cost differentiation or stock management strategies).

Table 16.3 shows a sample list of affinity matrix for the current scenario described.

16.4.1 Conclusion

The proposed system will be a valuable tool to support the adoption of Mass Customization with minimum impact on existing information systems. The companies will obviously need to adopt agile production methods in order to quickly respond to customers' orders and configurations. Promoting increased individualized interaction with customers and business partners will allow companies to obtain higher levels of customer retention.

Integrating Web Service Based product configuration systems with their backend systems, including Production and Enterprise Resource Planning, will allow for greater cost savings and more agile processes. This level of integration will contribute to an efficient make-to-order strategy minimizing inventory costs. Adopting integration mechanisms, like a Middleware layer, will support a richer level of integration with external information systems, minimizing information processing costs.

An effective Mass Customization strategy will only be possible if the company builds a successful information flow, covering the configuration process, product knowledge base, marketing and production restrictions and business rules and constraints.

Adopting recommendation techniques will help customers without significant knowledge about product details to express their needs and obtain a configuration that fulfills them. This way personalized interaction and individualization will not be exclusive for effective clients (who have sufficient knowledge about the products) and business partners, supporting new clients that know a lot about their needs and frequently know very little about product details and associated constraints.

The adoption of Web standards widely used and supported will facilitate the evolution of the model and an easier implementation in different scenarios. The impact on existing information systems and IT environment will be minimal if the system architecture includes specialized connectors that exclusively know the details on how the information is stored in associated systems.

A modular approach to the configuration system will support the adoption of different configuration engines, depending on the product complexities and associated constraints. Using adequate authentication and authorization mechanisms, the producer will support Web service consumption from a wide range of heterogeneous clients, supporting the development of multi-channel configuration platforms.

Acknowledgments This work was developed under the scope of project PEst-OE/EME/UI0252/2011, financed by The Foundation for Science and Technology – FCT.

References

1. Pine J, Gilmore J (1999) The experience economy. Harvard Business School Press, Boston
2. Piller F (2002) Customer interaction and digitizability – a structural approach to mass customization. In: Rautenstrauch C (ed) Moving towards mass customization. Springer, Heidelberg/Berlin/New York, pp 119–138
3. Davis SM (1987) Future perfect. Addison-Wesley, New York
4. Tseng M, Jiao J (2001) Mass customization. In: Gavriel Salvendy (ed) Handbook of industrial engineering, technology and operation management, 3rd edn. New York
5. Blattberg RC, Glazer R (1994) In: Blattberg RC (ed) Marketing in the information revolution, the marketing information revolution. Harvard Business School Press, Boston, pp 9–29

6. Goddard WE (1986) Just-in-time: surviving by breaking tradition. Oliver Wight Ltd. Pub, USA
7. Kidd PT (1994) Agile manufacturing forging new frontiers. Addison-Wesley, Reading
8. Nagel R, Dove R (1991) Twenty-first century manufacturing enterprise strategy – an industry led review, vol 1 and 2. Iacocca Institute, Leigh University, Bethlehem
9. Hiramatsu A, Naito A, Ikkai Y, Ohkawa T, Komoda N (1997) Case based function tree generator for client–server systems configuration design. In: Proceedings of the IEEE international conference on systems, man and cybernetics, Orlando, FL, USA
10. Hong G, Hu L, Xue D, Tu Y, Xiong Y (2008) Identification of the optimal product configuration and parameters based on individual customer requirements on performance and costs in one-of-a-kind production. Int J Prod Res, Taylor & Francis, Ablingdon, ROYAUME-UNI, 46(12):3297–3326
11. Jiao J, Helander MG (2006) Development of an electronic configure-to-order platform for customized product development. Comput Ind 57(3 (Abr)):231–244
12. Lee HJ, Lee JK (2005) An effective customization procedure with configurable standard models. Decis Support Syst 41(1):262–278
13. Li B, Chen L, Huang Z, Zhong Y (2006) Product configuration optimization using a multiobjective genetic algorithm. Int J Adv Manuf Technol 30(1–2):20–29
14. Luo X, Tu Y, Tang J, Kwong C (2008) Optimizing customer's selection for configurable product in B2C e-commerce application. Comput Ind 59(8 (Out)):767–776
15. Mailharro D (1998) A classification and constraint-based framework for configuration. Artif Intell Eng Des Anal Manuf 12(4):383–395
16. McGuinness D, Wright J (1998) An industrial-strength description logic-based configurator platform. IEEE Intell Syst 13(4):69–77
17. Mittal S, Frayman F (1989) Towards a generic model of configuration tasks. In: Proceedings of the 11th international joint conference on artificial intelligence. IJCAI '89. San Mateo, CA, Morgan Kaufmann, pp 1395–1401
18. Sabin D, Weigel R (1998) Product configuration frameworks – a survey. IEEE Intell Syst 13(4):42–49
19. Tseng H, Chang C, Chang S (2005) Applying case-based reasoning for product configuration in mass customization environments. Expert Syst Appl 29:913–925
20. Yang D, Dong M, Miao E (2008) Development of a product configuration system with an ontology-based approach. Comput Aided Des 40(8 (Ago)):863–878
21. Yeh J, Wu T, Chang J (2007) Parallel genetic algorithms for product configuration management on PC cluster systems. Int J Adv Manuf Technol 31(11–12)
22. Zhou C, Lin Z, Liu C (2008) Customer-driven product configuration optimization for assemble-to-order manufacturing enterprises. Int J Adv Manuf Technol, Springer, 38(1–2): 185–194
23. Castro AA, Varela MLR, Carmo-Silva S (2009) Products and orders configurator based on web services. In: Proceedings of the iDEMi09 – international conference on integration of design, engineering and management for innovation, Porto, Portugal

[illegible]

Chapter 17
Product Documentation Management Through REST-Based Web Service

Filipe Rocha, Maria Leonilde R. Varela, and Sílvio Carmo-Silva

Abstract Innovation is a creative process strongly associated with development and deployment of new products. This is essential in the global economy of today for the success and sustainability of companies. Collaboration between those involved in new product development (NPD) is an important requisite for product development and innovation. New product development involves not only the company and customers but also suppliers, free lancers and other stakeholders. Strong collaboration and agile communication among all are critical requirements for product development success. This is due to the highly competitive world economy we live in nowadays and also because data and relevant information that needs to be shared is distributed over networks of resources and users. Hence, web-based applications, decentralized repositories and databases are used to store and manage product and process development information as well as NPD stakeholders' interaction. Thus, for meeting today's new product development requirements Internet based collaborative tools and services must be applied. Web services are important assets in product development, helping the integration of data and knowledge bases and also for process and application interactions. This paper reports on a web services based research work focused on the development of a document-oriented web-based application for product development using Apache CouchDB technology and REST web services. The application seeks the managing of product and process information, as well as documentation generation, throughout the product development process cycle in an Internet based collaborative environment. An additional important function addressed deals with product development history.

Keywords Web services • Product information • Document • Management • CouchDB • REST

F. Rocha • M.L.R. Varela (✉) • S. Carmo-Silva
Department of Production and Systems, School of Engineering, University of Minho, Azurém Campus, 4800-058 Guimarães, Portugal
e-mail: filiperocha@gmail.com; leonilde@dps.uminho.pt; scarmo@dps.uminho.pt

A. Madureira et al. (eds.), *Computational Intelligence and Decision Making: Trends and Applications*, Intelligent Systems, Control and Automation: Science and Engineering 61, DOI 10.1007/978-94-007-4722-7_17,

17.1 Introduction

Innovation is closely associated with new product development (NPD). This is an activity that can benefit very much from collaboration between stakeholders involved. This can be enhanced through the use of agile tools and technology for efficient and fast information sharing and by carrying out of collaboration tasks. Such tools and technology are mostly available through the Internet. In fact, nowadays the Internet can be considered the best channel for collaboration and knowledge exchange and the path for companies' sustainability and competitiveness [1]. This is not only true inside a corporation, where actual new product development happens, but also for suppliers and customers innovating in a cooperative environment [1–4].

Zhang et al. [5] enhances and describes a web-based collaborative design platform. This platform provides a product catalogue, a communication system between peers, knowledge access restrictions and project management.

The use of web services [6] can be seen as a feasible interface between involved parties in product innovation, helping the integration of data and knowledge bases, processes and application interactions [1, 3, 5, 7].

One important aspect of this work is to access product and process information, in the collaborative distributed context of NPD.

Access to information for product design and development is not an easy task when using relational databases (RDB). Tables are referencing other tables and to retrieve little information, extreme computational work needs to be done by selecting and joining data from this structure.

RDB are designed to store and report on highly structured, interrelated data. Moreover, they include schemas and storage of the existing data that need frequent update. This often causes problems not anticipated in the initial database design. Each change in the initial database schema is a challenge for local and distributed upgrades. When dealing with paperwork, this is common and happens when we need to adapt predefined paper forms by adding or removing information and fitting its strict form to the needs of the moment.

In this paper, we propose a new approach to manage new product development information. This new concept is based on a document-oriented database (DoDB) with native distributed properties, as an alternative to the Relational Database Managements System (RDBMS). Hence, instead of the strict schema provided by tables we get to work with this new NoSQL (Not only SQL) next generation databases: schema-free, with easy replication support and simple API and capable of dealing with huge amounts of data [8].

For reasons described later in this paper we chose an open source Apache project named Apache CouchDB, commonly referred just as CouchDB [9].

This paper, in addition to this introduction follows with a literature review and the description of a proposed approach for managing product information. Based on this a system architecture is described and illustrated. Finally a conclusion is presented.

17.2 Literature Review

Companies use proprietary software, most of the times centralized, having high costs with licenses and high risk with information losses. Aziz et al. [10] talk about the shortcomings of centralized architectures and proposes a decentralized based on open standards and open source tools, stating that systems like this are implementable at low cost.

Sharma et al. [3] referring collaborative product innovation presents a theoretical framework identifying tools and processes for people and teams to collaborate on product development.

Zhang et al. [5] reviewed the state of the art on Internet-based product information, concluding that despite several Internet web-based technologies that had been applied to product development still none had been applied to real industrial applications. They also found a number of practical problems remaining unsolved, including dynamic product information collection and update, real-time collaboration, scalability and interoperability. These problems occur because there are no real applications yet available, probably due to the existence of bottlenecks on the database back-ends. Relational databases have very strict schemas and aggravate those problems with dynamic information, i.e., a change on the initial schema represents a change in the whole system. Moreover, dynamic product information is not what we get with schemas because there is no predefined schema for each product or service.

Major problems in scalability with relational databases become a point of failure. In particular, replication is not trivial. To understand why, consider the problem of having two database servers that need to have identical data. Having both servers for reading and writing data makes it difficult to synchronize changes. Having one master server and another slave is bad too, because the master has to take all the heat when users are writing information [11]. This statement leads to wish that there were a replication system that would come with an automatic conflict detection and resolution, by making it easy to synchronize data in both directions whenever wanted after being able to work offline independently [12].

REST (REpresentational State Transfer) technology [13] is considered to have great potential for solving scalability problems [14–17]. Fielding [13] states "The REST architectural style has been defined to describe the web architecture and to guide its future evolution, preserving the fundamental characteristics of scalability". It promotes software evolvability, efficiency, performance and reliability [12] using the protocols already available for the web.

Liu and Xu [18] reviewing some traditional technologies and systems for management of product information and discuss the need for their integration on a web environment as a means of having a more adaptive and flexible infrastructure. They also mention bandwidth and security concerns with data availability over the Internet. Unfortunately they do not say how the reviewed systems handle changes made to predefined data management models, leaving therefore, important questions unanswered.

Frutos and Borenstein [19] talk about mass customization and the role that customers should have in new product development. They report a web-based information system for flats' customization through interactions between building company web site and costumers. Customers can generate a variety of customized solutions, i.e. new products, with control on costs, despite flat variants' limitations.

One approach to solve these limitations can be done with the NoSQL concept, as in Not Only SQL [8]. Mateljan et al. [20] describe the several approaches to this concept and propose an ROI analysis. Leavitt [21] describes RDBMSs limitations and enumerates advantages and challenges of NoSQL databases, quoting several people's thoughts about them. One concept that has been used in production areas in several fields is the NoSQL concept. One of these types of databases is the Apache CouchDB [9] can be of great utility in managing NPD information since it is document-oriented and also provides easy implementation of RESTful web services due to its internal REST functionalities [13, 22–24].

The next sections will describe this work: the motivation behind it and our proposed approach followed by describing the system's architecture.

17.3 Motivation

Where do we quickly store our information on a daily basis? Perhaps taking a quick note on a piece of paper, put it on a post it, open a text editor and type it or whatever suits us best. These documents usually end piled up and we just browse through them and still have trouble finding what we want, if we ever do. Now, when this involves project documentation, documents are referencing other documents and back again. This is often seen as paper chaos. Even if it is well organized we still need to spend some time browsing them to find any information, going back and forth and eventually we end up picking up another piece of paper to sanitize collected information. In practice, this is how relational database systems work. Tables are referencing other tables and to retrieve little information, extreme computational work needs to be done by selecting and joining data from this structure.

Also, many different approaches have been developed for addressing the problem of adding structure back to semi-structured data, namely through relational databases, which are designed to store and report on highly structured, interrelated data. Moreover, in this kind of approach, as needs evolve the schema and storage of the existing data must be updated. This often causes problems as new needs arise that simply were not anticipated in the initial database designs, i.e., each change in the initial database schema is a challenge for local and distributed upgrades. When dealing with paperwork, this is common and happens when we need to adapt predefined paper forms by adding or removing information, fitting its strict form to the needs of the moment.

Therefore, in this paper, we propose a new approach to manage new product development information. This new concept is based on Document-oriented Databases (DoD) with native distributed properties, instead of the usual Relational

Database Managements System (RDBMS). This means that instead of the strict schema provided by tables we get to work with the documents and free schemas. This new concept is based on NoSQL databases [8], namely one Open Source Apache Project named Apache CouchDB or commonly referred as just CouchDB. CouchDB is an open source "distributed, fault-tolerant and schema-free document-oriented database, accessible via a RESTful HTTP/JSON API" [9].

The next sections will focus on explaining a new approach to the needs described on the previous sections followed by a description of the system architecture capable of fulfilling the need of organizing and sharing documentation on new product development.

17.4 Proposed Approach

In this work we propose the CouchDB for managing NPD information. CouchDB is an open source "distributed, fault-tolerant and schema-free document-oriented database, accessible via a RESTful HTTP/JSON API" [9]. Data is stored in documents, presented in key-value maps using the data types from JavaScript Oriented Notation (JSON) [25].

CouchDB is designed to store and report on large amounts of semi-structured data, simplifying the development of document-oriented applications, which is the bulk of collaborative web applications.

A CouchDB database can be described as a flat collection of JSON documents, identified by a unique ID (UUID). Each document is an object that consists of named fields. Field values can be strings, numbers and dates, ordered lists and associative maps. Figure 17.1 illustrates a JSON document including some of these substructures before.

Moreover, CouchDB should also not be seen as an object-oriented database, despite relationships between documents are possible.

With CouchDB, no schema is enforced, so new document types with new meaning can be safely added alongside the old ones.

CouchDB is a fully Atomic Consistent Isolated Durable (ACID) [26] storage engine, never overwriting committed data or associated structures, ensuring the database file to be always in a consistent state. Document updates are serialized and there are no locks. Instead it uses a Multi-Version Concurrency Control (MVCC) model, meaning any number of clients can be reading documents without being locked out or interrupted by concurrent updates, even on the same document [27]. Transactional commits provide consistent state to the database, i.e., in case of failure during a commit transaction, the commit is discarded and maintenance operations are run. From time to time, some jobs are run in order to compact data, freeing space on disk. Hence, CouchdB Documents are indexed in B-Trees and each update to the database generates a new sequential number to identify the document state in its history represented by the _rev field shown in Fig. 17.1.

```
{
    "_id": "1287072474",
    "_rev": "54-1867031578",
    "product_number": "1287072474",
    "name": "Round Table Anderatti Rosso",
    "description": "This is a round table for the Anderatti collection",
    "category": "Table",
    "version": "Rosso",
    "kind": "Round",
    "color": "Red",
    "leg_material": "Steel",
    "leg_height": "1.20",
    "leg_diameter": "0.800",
    "number_of_legs": "1",
    "top_diameter": 1.279,
    "top_material": "Wood",
    "created_at": {
        "json_class": "Time",
        "data": "2011/04/24 10:50:58 +0100"
    },
    "revisions": [
        {
            "accepted": "Yes",
             "doc_revision": 1,
            "description": "Created product number 1287072474"
        }
    ],
}
```


Fig. 17.1 JSON document in CouchDB

CouchDB is a peer based distributed database system. Here any number of CouchDB hosts (servers and offline-clients) can have independent "replica copies" of the same database, where applications have full database interactivity (query, add, edit, delete). When back online or on a schedule, database changes are replicated bi-directionally.

CouchDB also is characterized by having a built-in conflict detection and management and the replication process is incremental and fast, copying only documents and individual fields changed since the previous replication. Therefore, an updated document being inserted on a database should be replicated through the entire distributed database backend and easily retrieved. Hence, most applications require no special planning to take advantage of distributed updates and replication. With very little database work, it is possible to build a distributed document management application with granular security and full revision histories.

CouchDB also integrates a view model to add structure back to unstructured data. Views are the method of aggregating and reporting on the documents. CouchDB is based on Google's Map/Reduce programming model [28] for views, but the results of a Map/Reduce task always reflect the current state of the documents in the database by only reindexing the documents that have changed since the last index update [29]. Figure 17.2a shows an example of a JavaScript map function counting the elements of the array revisions seen on Fig. 17.1. It emits a view using emit (keys, values).

```
a)
function(doc) {
  if ((doc['category'] == 'Table') && doc['product_number']) {
    emit(doc['product_number'], doc['revisions'].length);
  }
}

b)
{
  "total_rows":2,
  "offset":0,
  "rows":[
    {"id":"1287072474","key":"1287072474","value":7},
    {"id":"1287072475","key":"1287072475","value":7}]
}
```

Fig. 17.2 JavaScript map function

Figure 17.2b is the JSON result document to the map function early described. The field total_rows holds a counter to the number of result documents mapped by the function. The offset field is where the document list has started. The rows array holds document objects used on emit function. Id is added automatically. A reduce function can now be applied using the values field and i.e. built-in reduce function is _sum. It returns the sum of all values in the view. _stats implements various numerical statistics on emitted data.

Any of these documents can be easily retrieved using an HTTP client, i.e. a browser, URL [30] since the communication protocol used by CouchDB is the HTTP [31].

Using these techniques, characteristics, functionalities and processes provided by CouchDB for producing and managing documents it should be easy to produce services that can generate and use information that needs to be shared on NPD. In the next section we will be proposing a system architecture that should support this.

17.5 System Architecture

New product development can be highly enhanced if new product ideas from all stakeholders involved and mainly from consumers are taken and the development process supported via Internet collaboration. The system based on CouchDB, here proposed can provide such support and the effective implementation of the product development process. Thus, it takes the advantage provided by couchDB Therefore, by using CouchDB development information, including all external attachments, including design information, reports, specifications, meaning all iterations should be stored.

There are several technologies being considered for implementation. CouchDB manages, with little maintenance efforts, the distributed environment keeping all peers synchronized on demand and/or on schedule. Since REST is used, operations to the database are simple.

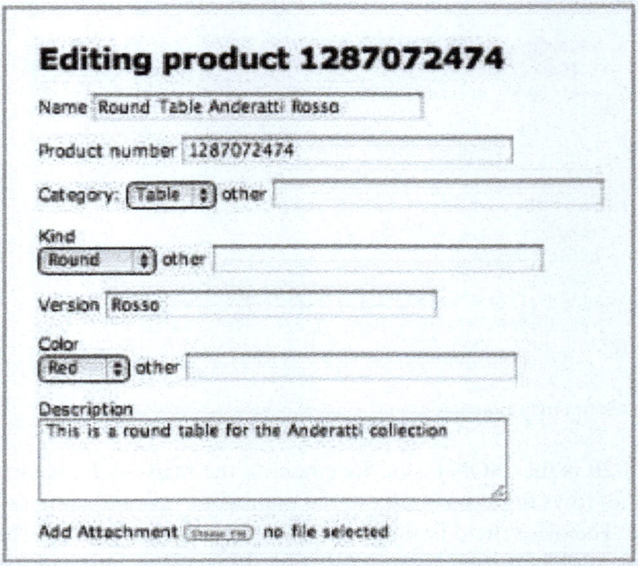

Fig. 17.3 Excerpt of a page from the web interface

To manage product information a web interface (see Fig. 17.3) is created using Ruby [32] and Ruby on Rails (or just Rails) [33].

Because Rails and CouchDB speak REST/HTTP homogeny is still assured with a small help from a Ruby wrapper. All of these technologies are open source and available for free.

If it is considered that all stakeholders speak in the same language, e.g., suppliers and NPD teams, they can dynamically supply resources for a product. In order to preserve suppliers' confidential information data would be filtered by web services, becoming a subpart of the product information. Figure 17.4 is a graphical representation of this idea.

Considering the example being used in this paper, Fig. 17.5 could be a document produced for sharing. This document would now be consumed by a web service to provide the information needed.

Moreover, there are also several use cases that could include those involved in distributed e-commerce websites where multiple partners are able to share their catalogs, patient medical records could be shared between different institutions, personal finances. Everyday needs involving different paper documents could be candidate since they are not immutable and keeping their history would be a prerequisite.

Shared catalogs, provided by distributed properties could be used in this context, since NPD teams would have (online and offline) access to suppliers' catalogs. Document revisions and attachments would be available through revision properties provided by Document-oriented Database Management Systems (DoDBMS).

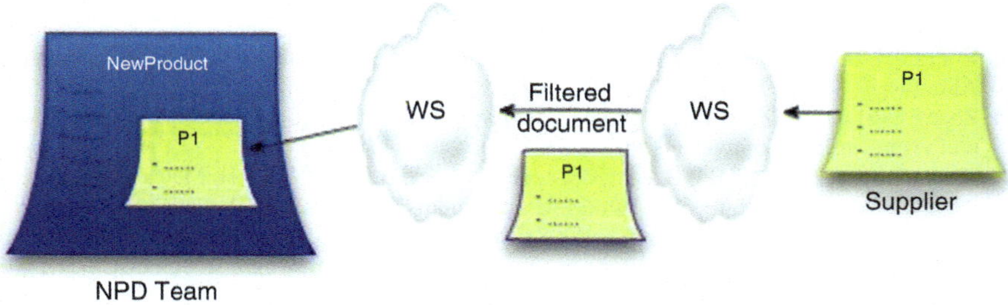

Fig. 17.4 Use of web services to filter/provide confidential information


```
{"total_rows":2,"offset":0,"rows":[
  {"id":"1287072474",
  "key":{"pn":"1287072474","name":"Round Table Anderatti Rosso"
  "value":null},
  {"id":"1287072475",
  "key":{"pn":"1287072475","name":"Round Table Anderatti Latte"
  "value":null}
]}
```


Fig. 17.5 Shared basic product information to be consumed by a web service

17.6 Conclusion

In this paper we make a proposal for a collaboration system for new product development (NPD), based on CouchDB emphasizing the importance and role of managing information and documentation during and after the development process.

Although we are still in an early stage of investigation and we could not carry out any practical implementation of the approach to NPD process and documentation management here proposed, similar concepts and approaches have been used in production on several fields [34, 35] and we have already performed some work, in order to put forward the proposed system soon.

In short, this work tries to manage information more naturally, integrating its several parts and using new technologies, based on CouchDB and REST web services.

Acknowledgments This work was developed under the scope of project PEst-OE/EME/UI0252/2011, financed by The Foundation for Science and Technology – FCT.

References

1. Yan W, Chen C, Huang Y, Mi W (2009) A data-mining approach for product conceptualization in a web-based architecture. Comput Ind 60:21–34
2. Dong B, Qi G, Gu X, Wei X (2008) Web service-oriented manufacturing resource applications for networked product development. Adv Eng Inform 22:282–295

3. Sharma A (2005) Collaborative product innovation: integrating elements of CPI via PLM framework. Comput Aided Des 37:1425–1434
4. Wang L (2002) Collaborative conceptual design—state of the art and future trends. Comput Aided Des 34:981–996
5. Zhang S, Shen W, Ghenniwa H (2004) A review of internet-based product information sharing and visualization. Comput Ind 54:1–15
6. Zhang L-J, Jeckle M (2004) Business & economics. In: Booth D, Haas H, McCabe F et al (eds) Web services architecture, W3C working group note, World Wde Web Consortium, 299 p
7. Wang H, Liu G, Han B, Zhang J (2008) Collaborative simulation environment framework based on SOA. In: Computer supported cooperative work in design, 2008. CSCWD 2008. In: Proceedings of the 12th international conference on CSCWin design, SC WD 2008, April 16–18, 2008, Nanyang Hotel, Xi'an Jiaotong University, Xi'an, China, pp 416–419
8. Http://nosql-database.org/ NOSQL Databases
9. Apache Software Foundation Apache CouchDB: the apache couchDB project, http://couchdb.apache.org/
10. Aziz H, Gao J, Maropoulos P, Cheung W (2005) Open standard, open source and peer-to-peer tools and methods for collaborative product development. Comput Ind 56:260–271
11. Shen W, Hao Q, Li W (2008) Computer supported collaborative design: retrospective and perspective. Comput Ind 59:855–862
12. Rodriguez K, Al-Ashaab A (2005) Knowledge web-based system architecture for collaborative product development. Comput Ind 56:125–140
13. Fielding RT (2000) Architectural styles and the design of network-based software architectures. Citeseer. R.T. Fielding University of California, Irvine, PhD Thesis, Irvine
14. Jing Dong, Sheng Yang, Dushyant S. Lad, Yongtao Sun (2006) Service oriented evolutions and analyses of design patterns, In: Proceedings of the second IEEE international symposium on Service-Oriented System Engineering (SOSE), pp 11–18
15. Xie S, Huang H, Tu Y (2002) A www-based information management system for rapid and integrated mould product development. Int J Adv Manuf Technol 20:50–57
16. Decandia G, Hastorun D, Jampani M, Kakulapati G, Lakshman A, Pilchin A, Sivasubramanian S, Vosshall P, Vogels W (2007) Dynamo: amazons highly available key-value store. In: SOSP '07 Proceedings of twenty-first ACM SIGOPS symposium on Operating Systems principles. ACM Press, New York, NY, USA. pp 205–220
17. Tseng K, Abdalla H (2008) A web-based integrated design system: its applications on conceptual design stage. Int J Adv Manuf Technol 35:1028–1040
18. Liu DT, Xu XW (2001) A review of web-based product data management systems. Comput Ind 44:251–262
19. Frutos JD, Borenstein D (2004) A framework to support customer-company interaction in mass customization environments. Comput Ind 54:115–135
20. International Conference on Computer Design and Applications ICCDA (2010) Qinhuangdao, China, vol 5, 25–27 June 2010
21. Leavitt N (2010) Will NoSQL databases live up to their promise? Computer 43:12–14
22. Anderson JC, Lehnardt J, Slater N (2009) CouchDB: the definitive guide. O'Reilly Media, Inc., CA, USA
23. Richardson L, Ruby S (2007) RESTful web services – Webservices for the real world. O'Reilly Media, USA
24. Allamaraju S (2010) Restful web services cookbook. O'Reilly Media/Yahoo Press, Beijing
25. Crockford D (2006) RFC 4627: the application/json media type for JavaScript Object Notation (JSON). In: Proceedings of the XML http://tools.ietf.org/html/draft-crockford-jsonorg-json-00
26. Gray J, Reuter A (1992) Transaction processing: concepts and techniques. Morgan Kaufmann Publishers, Inc., San Francisco
27. Reed DP (1978) Naming and synchronization in a decentralized computer system. Massachusetts Institute of Technology

28. Dean J, Ghemawat S (2008) MapReduce: simplified data processing on large clusters. Commun ACM 51:1–13
29. Orend K (2010) Analysis and classification of NoSQL databases and evaluation of their ability to replace an object-relational Persistence Layer, Master's thesis, Technische Universität München, Germany
30. Http://curl.haxx.se/ cURL and libcurl
31. Http://www.ietf.org/rfc/rfc2616.txt Hypertext Transfer Protocol – HTTP/1.1
32. Matsumoto Y (2001) Ruby programming language. O'Reilly Media, Inc., Beijing
33. Http://www.rubyonrails.com Ruby on rails
34. (2009) Http://wiki.apache.org/couchdb/CouchDB_in_the_wild CouchDB_in_the_wild – Couchdb Wiki
35. Http://www.couchbase.com/customers/case-studies Couchbase and CouchDB case studies | Couchbase

Chapter 18
Fuzzy Web Platform for Electrical Energy Losses Management

Gaspar Gonçalves Vieira, Maria Leonilde R. Varela, and Rita A. Ribeiro

Abstract Recently, electricity distribution companies have already done many investments to identify and calculate energy losses along the distribution network. Although these big efforts, most approaches for clearly identifying and solving the related problems still remain rather inefficient. The accurate identification and the precise calculation of electricity losses enables the clear specification of the critical points and segments in the networks and, consequently, the effective prioritization of actions and interventions in order to reduce those electricity losses and problems. Moreover, the work already performed on this issue, the existing approaches focus mainly on empirical and probabilistic data. Hence, there is still a clear gap between real information and the considered one, which tends to be poor and imprecise. Due to this reality and the lack of appropriate software applications, in this paper we propose a web platform for the management of the whole network of electricity distribution, from medium voltage (MV) down to low voltage (LV), including billing on the transformation centers (TCs), electricity losses calculation and proposals for solving actions, by means of a fuzzy decision-making model.

Keywords Electricity • Distribution network • Fuzzy • Based decision • Support system • Information and losses management

G.G. Vieira • M.L.R. Varela (✉)
Department of Production and Systems, School of Engineering, University of Minho, Azurém Campus, 4800-058 Guimarães, Portugal
e-mail: gaspar_vieira@hotmail.com; leonilde@dps.uminho.pt

R.A. Ribeiro
Department of Electronic Engineering, Faculty of Engineering, FCT/UNL, Monte de Caparica, Portugal
e-mail: rar@uninova.pt

A. Madureira et al. (eds.), *Computational Intelligence and Decision Making: Trends and Applications*, Intelligent Systems, Control and Automation: Science and Engineering 61, DOI 10.1007/978-94-007-4722-7_18, © Springer Science+Business Media Dordrecht 2013

18.1 Introduction

During the last decades electricity distribution companies have developed several improvements to their power distribution systems. Due to the high costs involved in the electricity generation, transmission and distribution, the efficient usage of the available energy along with the efficient management of the network are fundamental issues to be tackled. Furthermore, concerns with environmental issues must also be considered during the management and decisions processes. Companies are aware of the strong competition they are currently facing, largely driven by the liberalization of the electricity markets. Energy efficiency measures can represent the difference between having power quality at affordable prices, or having energy shortages and/or with poor quality.

Another big concern is how to minimize network energy losses. This leads not only to higher revenues, but also to improvement on the quality of the product offered to consumers. In addition, companies can expect to gain additional markets and to afford expanding network capacity. For those reasons, the loss of power and energy, which may occur in electrical systems, is still a major concern, both for electricity distribution companies and to Regulators. In fact, Energy Regulators are responsible for: proposing measures concerning the quality and maintenance of the assets of the distributor; to ensure the promotion of power quality; and to regulate tariffs for the market.

Electricity losses are divided in two main categories: technical and non-technical ones, which are precisely identified and calculated [1–3].

Therefore, in this paper we put forward a methodology and computational implementation based on real data, obtained through telemetry instead of empirical and probabilistic data, which enables to obtain precise results, namely in terms of losses determination. Moreover, the proposed system enables to clearly identify the losses in the distribution network points and activate a mechanism for analyzing and accurately solve the identified problems, by proposing corresponding repairing actions. The system architecture and its main functionalities are presented and briefly described in this paper. The system architecture follows a Peer-to-Peer (P2P) structure and includes a Data Base, where all relevant information, including parameters and data about diverse electricity losses types are stored, in order to enable an effective and efficient electricity information and losses management. The main system functionalities also include modules for billing and payment processing. The system is still under development although it is already at an advanced stage of implementation; hence some of the main functionalities are illustrated on this paper. A brief literature review is also included, in order to enable a comparative analysis between the proposed system and existing ones.

18.2 General Considerations

The growth of world population and economic conditions in developing countries necessarily implies an increase in energy consumption. However, energy production increase should be bounded by sustainable development and environmental

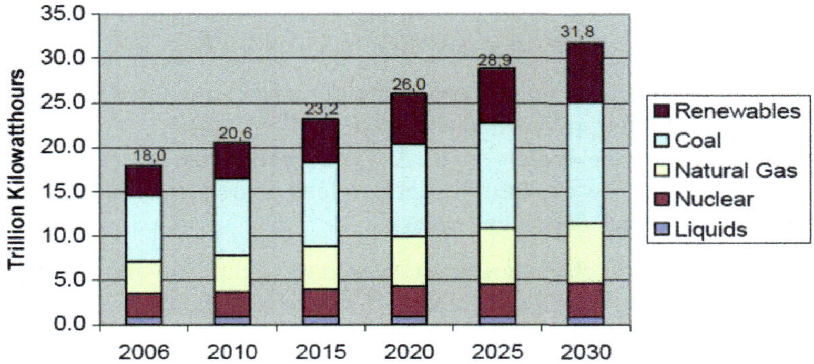

Fig. 18.1 Generation of electricity worldwide

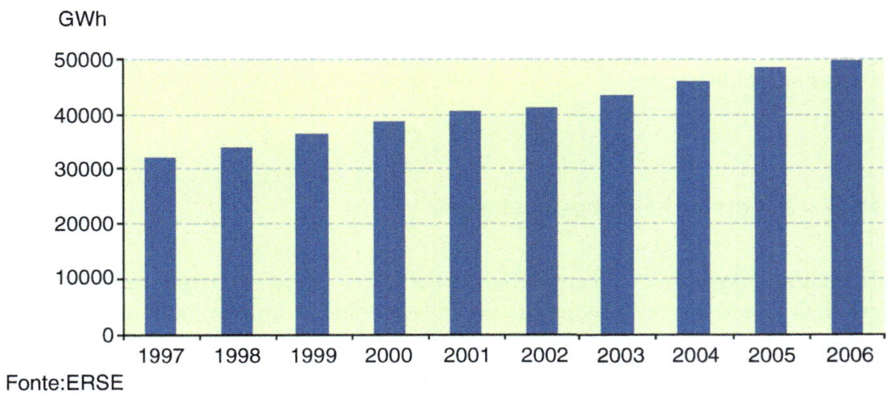

Fig. 18.2 Evolution of energy consumption in Portugal

responsibility. The graph in Fig. 18.1 [2] shows the growth in global electricity generation by fuel. According to this data, electricity generation is expected to grow by more than 50% during the next 20 years [2].

For example, in Portugal, although increasing concerns with energy efficiency have taken place, we can observe an increasing trend for energy consumption (see Fig. 18.2) [2].

The increasing energy consumption over time may partially explain the rising awareness with the quality of electricity. Namely, in Portugal, considerable improvements on the power systems quality and supply were achieved. The promotion of adequate levels of service quality in the electricity sector is a precondition for the welfare and needs of populations as well as increases on economic activities to become globally competitive [4].

The quality of service in the electricity sector can be analyzed in its two components:

- Quality of service, regarding its technical nature.
- Quality of service, regarding its commercial nature.

The quality of technical service in the electricity sector, also known as power quality, is associated with the analysis of [5, 6, 8, 9, 13–17]:

- Reliability of electricity supply (continuity of service) measured by the number and duration of interruptions in the supply.
- Characteristics of the waveform of AC voltage (voltage waveform quality), measured through the evolution of their values of frequency, amplitude, harmonic distortion, unbalance, and others.

To proceed with the characterization of voltage in the networks, exploited by operators of transmission and distribution, the system operators must perform the following tension measurements [7]:

- Frequency.
- Value of effective tension.
- Cavas voltage.
- Jitter (flicker).
- Imbalance of three-phase system of voltages.
- Harmonic distortion.

18.3 Electrical Energy Network

The current system of electricity distribution is based on large central power generation, which transmits power through systems of high voltage transmission, which is then distributed to distribution systems in medium and low voltage. In general the energy flow is unidirectional and the power is checked and controlled by local centers, in an order based on pre-defined requirements. Figure 18.3 shows a general outline of an electricity distribution network [10].

The share of electricity distribution is a segment of the electrical system, consisting on the primary electrical networks (distribution networks of medium voltage) and secondary networks (distribution networks for low voltage) [10] Electricity distribution companies are in charge of the construction, operation and maintenance of these networks.

The primary distribution networks [10] are electrical circuits with three-phase three-wire (three phase), connected at distribution substations and are usually built in the classes of voltage of 15, 23 or 34.5 KV.

In these classes of tension, the nominal operating voltage may be 11, 12.6, 13.2, 13.8, 21, 23, 33 and 34.5 KV. The levels of voltage: 13.8 and 34.5 KV are standardized by law [10]. The other levels exist and continue to operate normally. Primary distribution networks, are installed with distribution transformers, fixed on poles, whose function is to lower the voltage level to the primary side voltage level (e.g., for download from 13.8 KV to 220 V). The secondary distribution networks [10] are electrical circuits with three-phase four wires (three phases and neutral), typically operate at voltages (phase – phase/phase – neutral) 230/115, 220/127 and 380/220 V.

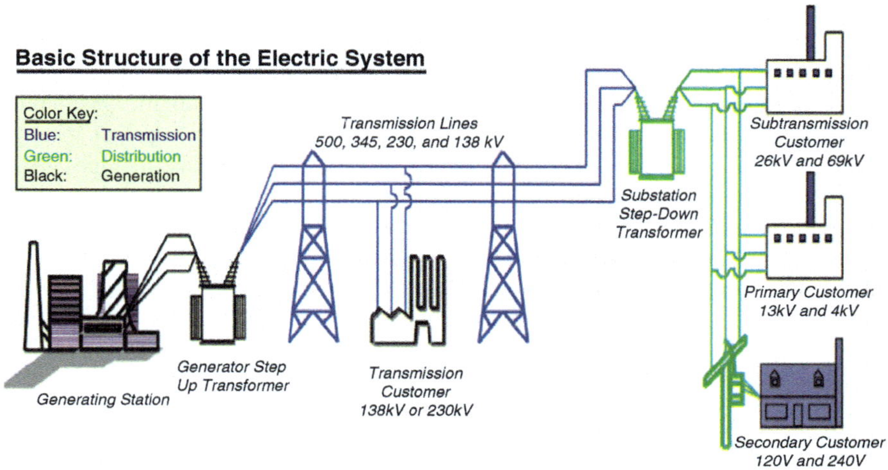

Fig. 18.3 Electricity distribution network

These networks are connected consumers, including residences, bakeries, shops, and so on, and also the fixtures for street lighting. These networks serve the large consumption centers (namely, population and large industry, among others). Establishments such as large buildings, shops and markets consume more power, and require individual processors of 75, 112.5 and 150 kva. In some cases, the tension between supplies varies from 380/220 to 440/254 V.

The entire distribution system is protected by a system composed by circuit breakers at the substations where the primary networks are connected, and with key fuse in distribution transformers, which in case of short circuit switch off the power grid [10].

18.4 System Architecture

The proposed system for managing an electricity distribution network includes a very comprehensive software architecture. One of the main purposes of this system consists on collecting information into the common server, about the primary needs of the system's users, and then transmits it to the supporting database. In this database information about the whole distribution network will be stored, from the transformer to the final consumer. This information will comprise the secondary network, the connection extensions, micro-generation billing of TCs, the compensating reactive power and the various types of acquisitions regarding parameters readings on TCs and end consumers, in particular, estimated readings and company lectures, including manuals, telemetry readings and lectures by PDA.

The system's underlying database follows a relational-based model and was developed through SQL Server. SQL Server is a database management system

(DBMS) for relational databases management, created by Microsoft, and consists on a platform for enabling reliable data manipulation. Moreover, it is a very robust platform and is used by corporate systems of different dimensions [11].

The database developed under the relational modeling paradigm, was "split" into various diagrams in the SQL server that will work linked together through their identity relations. Some important diagrams descriptions include information for the perception of the database schema, the transformer diagram, the network diagram about secondary liaison branch diagram, a diagram about microgeneration, an energy measurement diagram, a diagram for batteries capacitors, and also public lighting diagram, among others and three of the most important ones are going to be described on the next section.

18.5 System Functionalities

The proposed system, includes several functionalities to provide essential responses to questions and problems arising under the scope of electricity distribution networks. Some requests are about electricity power and quality and others are about parameters measurement or even with losses identification. The latter conduces to decision-making actions and intervention to solve the problem. <the whole system falls under the scope of global concerns including sustainability ones [12].

In general, the aims are: (a) the identification of all TC's; (b) their geographical location (map with TC's in Google Maps with photograph of the site); (c) the control of leakages and losses on TC; (d) the identification of types of clients at any given TC; (e) the emission of several different kind of reports, namely related with electricity consumption per day and respective tariff. This last feature is expected to support the analysis of the consumption time pattern. Other system features include the possibility of sending early warnings via SMS, GPRS or email in case of failure or other problems arise.

The system is also of great importance for collecting readings between previously defined time intervals, e.g., every 15 min from all counters, or to control the time of technical interventions for repairing and/or maintaining the TC's, including the index points of failures, consumption, losses, and necessary interventions management.

Summarizing, the main capabilities of the system are:

- Identification of all TC's. Geographical location (map with TC's in Google Maps with photograph of the site, used "snapshot" and contracted power), with information on the type of TC urban distribution (in cash), rural distribution, air (suspended pole) with cooling dry or oil, public or private.
- Control of electricity leakages and losses, namely on TCs (global, technical and nontechnical ones).
- Identification of the types of clients at each TC (micro ordinary consumers or producers) to detect whether there is injecting electricity network, or just usage.
- Reports with consumption per day and respective tariff to allow analyzing the times of highest consumption, compared to the same period (e.g. month) last year.

This capability allows appropriate consumption management and analysis of the evolution over time through graphical and/or table results for future prediction.

- Reports to the average monthly consumption for information and verification of the months of greatest consumption, and thus know if there is a power surge at a given time interval and prevent any failure of distribution and/or marketing of energy.
- Reports to compare the number of customers in the previous year, for information and verification of developments over time.
- View existing records to request a new network connection, which shows the power that the consumer requests and the voltage level of the request. Based on the information in the completed form online, the network manager can determine the need to conduct a study (which integrates the data contained in the request) in order to support the implementation of the budget to present to the requester.
- Early warning system via SMS, GPRS or email in case of failure, inconsistency of consumption, fraud, power and voltage exceeded the maximum permitted by the TC, among others.
- Collection of readings every 15 min from all counters, at a certain time automatically or by user request in real time and instantaneous.
- Control the time of intervention of technicians to repair and/or for maintaining each TC.
- Conducting comparative analysis about all TC's installed, including the index points of failures, consumption, losses, and necessary interventions, among others.
- Management of phases in case of failure or technical assistance for improved performance in cutting time.

18.6 Fuzzy Energy Losses Management

As mentioned before, energy losses management is one of the major problems of electricity distribution networks. Therefore, it is of paramount importance to identify which portion of the losses may be charged to consumers. Technical losses are inherent to the process, and therefore assumed to be liable to pay. Contrary, commercial losses, although technical, are related with the need of necessary investments in the electricity network (maintenance, expansion of capacity, reconfiguration, upgrading, etc.) and therefore, not liable to pay by consumers.

If commercial losses exist the problem of lack of maintenance of the electrical distribution network must be assessed and properly tackled. In order to properly address the energy losses reduction objective, there is a need for identifying and accessing the losses throughout the whole distribution system. Energy losses can be classified as to their origin and location [1]. Regarding their origin, a loss can be categorized into technical and non-technical one. A technical loss is the energy that is lost during distribution. It is inherent to the process and occurs on the materials used during the distribution, such as transformers, cables, and network connections, before the point of delivery. A non-technical loss is the energy actually delivered to the consumer, through their own or another distributor, but for some reason it was not billed, which is commonly associated with fraud energy consume, by some consumers in the network.

Moreover, the losses can be classified into global losses, losses in transmission and distribution losses. The global losses include losses in the generation, transmission and distribution. The transmission losses are associated with the systems for generation and transmission, distribution losses occur only in the distribution system.

The proposed system aims to take into account all these losses. Some important features are already implemented, namely the losses identification, along the electricity distribution network and energy losses determination. Regarding this last operation, a user is able either to introduce precise data, obtained from several distinct sources, such as by direct measuring along the distribution network or from other imprecise specification, for instance by defining some distribution function or through fuzzy parameters specification. In this last scenario, triangular functions are used to specify a set of different kind of parameters for losses determination, namely related with the estimation of losses associated to cable lengths and other parameters related to each kind of cable losses determination, since these are not always known precisely and they exist along the whole distribution network, from medium to low voltage. Moreover, other uncertain parameters, for losses determination, may occur, as for example the ones related with end users' electricity meters losses estimations. All these will also be able to be treated under the devised fuzzy sets approach.

18.7 System Illustration

The software was developed in Visual Basic 6.0, which will serve as a connection to the relational database developed in SQL Server. Considering the distance from consumers it was necessary to develop a friendly and intuitive graphic interface, enabling the reading of data from a geo-referenced database.

The system's main menu interface for billing purposes is illustrated on Fig. 18.4. Moreover, the proposed system can be used to intervene, from other TC characterization purposes down to all the electricity grid, enabling energy losses determination, micro-generation data processing and controlling, as well as the whole electrical energy data management along the distribution network, which is going to be illustrated next, among several kind of reports about various relevant information gathering about several parameters that are useful regarding the use requirements of the system.

The electricity network management menu enables to specify the main general and specific characteristics of the electricity network. These features include the identification of the electricity distribution centers, namely transformation centers.

Moreover, it is possible to identify the project type, in terms of ownership. In the example given the electricity distribution entity is identified as being the "Cooperativa Roriz". The Type of the installation can also be specified as urban or rural as well as its construction model, in the area "Modelo do Posto".

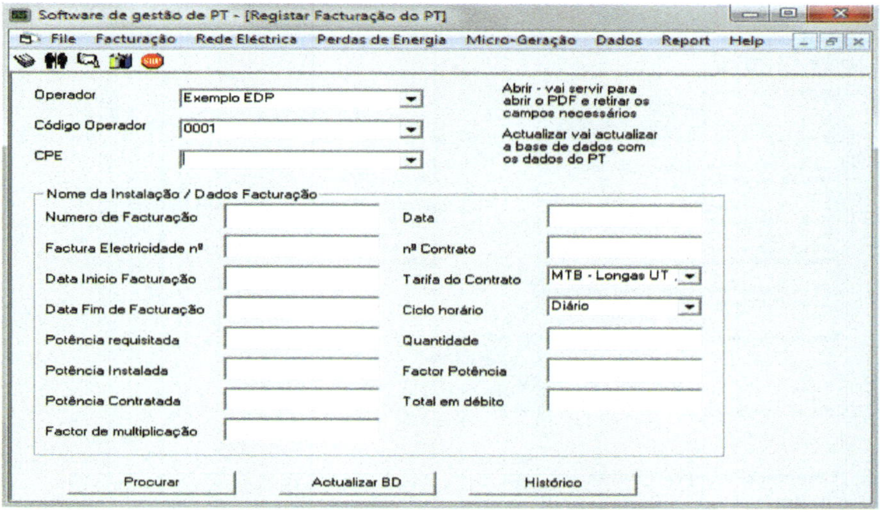

Fig. 18.4 Billing main menu

In terms of client type, the system also enables to identify it, for example, as "micro-generation" client.

The system can be used to define MV or LV electricity distribution, as well as detailed identification and determination of electrical energy losses, along a distribution network, as shown illustrated on Fig. 18.5. In case o MV it is possible to specify the service tension level, the level of isolation and the number of TCs, among other important information for managing the whole energy distribution network.

The system enables to identify, for each TC the output type, in LV, among other relevant data and store it on the corresponding database component. Figure 18.5 below illustrates the most important data input areas, in terms of electricity network losses specification and management. The system also allows specifying more details for each particular case, through subsequent menus.

18.8 Conclusion

This work discussed an ongoing project aimed to develop a system for monitoring a distribution network of electricity. The system features allow the definition of an interface for each user, customizing and configuring the system according to his interests and needs. The system is also easily adapted and new features can be implemented, according to future specific requirements. The work conducted so far allowed for the characterization of the distribution network, including the geodesic survey, the physical description in terms of infrastructure, the technical and behavioral characteristics of the network, and the assessment of connection points and ways of connecting networks and customers.

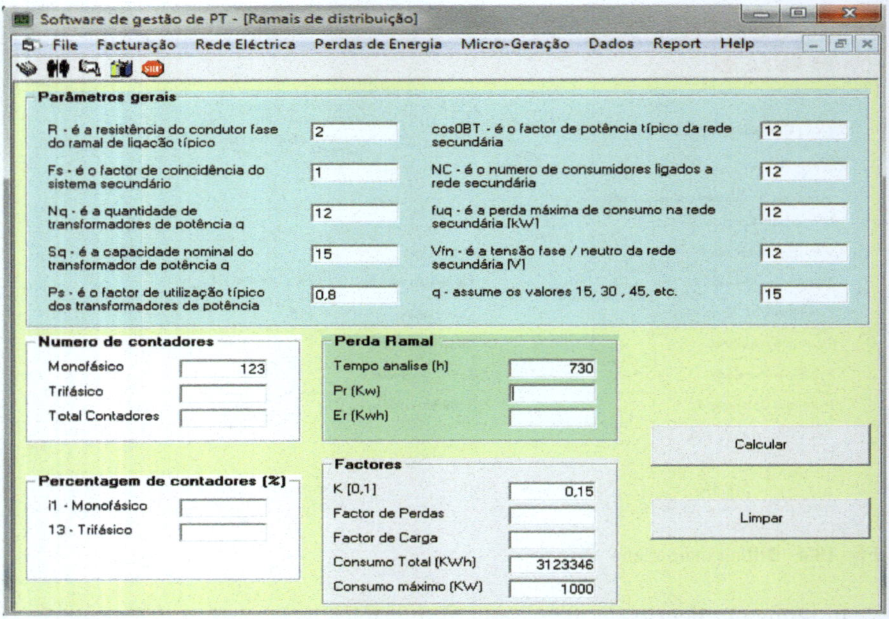

Fig. 18.5 Energy losses determination main menu

The main expected benefits for the complete system are the enhancement of the overall improvement in quality of service, the significant reduction in response time to failures of network management and better customer service line. The system being implemented will allow a continuous acquisition and recording on file of events that result in the detection of anomalies related to power quality and its losses among other important electricity distribution control parameters, combined with the application of fuzzy sets specification, in order to be able to more accurately deal with imprecision, regarding parameters determination. The information gained can be viewed in tables and/or graphics, and thereafter generate reports in order to support decision-making. The reports can be sent directly to the printer or inserted into other software applications to be accessed via the Internet by using a web browser.

A future goal of the system's operation is related to the possibility of being integrated with other systems or sub-systems, taking advantage of the potential it has for communicating with any Internet application based on web technology, for example, by implementing remote access to systems for data acquisition or even use wireless technology.

Acknowledgments This work was developed under the scope of project PEst-OE/EME/UI0252/2011, financed by The Foundation for Science and Technology – FCT.

References

1. Méffe A (2001) Metodologia para cálculo de perdas técnicas por segmento do sistema de distribuição. Msc thesis, Escola Politécnica da Universidade de São Paulo
2. Leão R (2009) GTD – Geração, Transmissão e Distribuição de Energia Eléctrica. Universidade Federal do Ceará
3. Perdas na rede de Distribuição (2009) Available at: http://www.erse.pt/pt/electricidade/factosenumeros/Paginas/Perdasnarededeedistribuicao_PC.aspx
4. Qualidade de Serviço (2010) Available at: http://www.erse.pt/pt/electricidade/qualidade-deservico/Paginas/default.aspx
5. Qualidade de serviço técnica (2009) Available at: http://www.erse.pt/pt/electricidade/qualidadedeservico/qualidadedeservicotecnica/Paginas/default.aspx
6. Qualidade Serviço nas redes de distribuição (2010) Available at: http://www.ariae.org/pdf/xi_reunion_espana/Sesion2_Mesa1/Jorge_Esteves_JE.pdf
7. Qualidade da onda de Tensão (2009) Available at: http://www.erse.pt/pt/electricidade/qualidadedeservico/qualidadedeservicotecnica/Paginas/Qualidadedaondadetensao.aspx
8. Qualidade de Serviço Comercial (2010) Available at: http://www.erse.pt/pt/electricidade/qualidadedeservico/qualidadedeservicocomercial/Paginas/default.aspx
9. Evolução das perdas nas redes de distribuição (2010) Available at: http://www.ariae.org/download/reuniones/xi_reunion_espana/Sesion2_Mesa1/Jorge_Esteves_JEpdf
10. Energia Eléctrica. Available at: http://pt.wikipedia.org/wiki/Energia_elétrica
11. European Commission (2006) European technology platform smartGrids. Vision and strategy for Europe's electricity networks of the future. Directorate-General for Research, Sustainable Energy Systems, (http://www.smartgrids.eu)
12. Strauch Mariana Torres (2007) Desenvolvimento de metodologia para cálculo de perdas eléctricas em redes de distribuição de baixa tensão. Msc thesis, Universidade Salvador
13. Todesco JL, Morales ABT, Veloso S (2005) Gestão da distribuição Secundária de energia eléctrica utilizando um sistema especialista. CELESC, Brasil
14. Manhães LR (2008) Cálculo de perdas com aplicação de curvas de cargas típicas de consumidores integradas a sistema gis. CPFL, Brasil
15. Aranha Neto EAC, Coelho J (2008) Combate às perdas não-técnicas no Brasil, Florianópolis, SC
16. Méffe A, Oliveira CCB, Campos MB, Fei SP (2005) Metodologia para definição dos níveis ótimos de perdas técnicas em sistemas de distribuição e priorização de obras para sua redução, Universidade de São Paulo. Metodologias de Investigação 58
17. Netto AV, Gouveia JD, Cateriano PSH Interface 3D para manipulação de dados em redes de distribuição de energia eléctrica. Available at: http://www.dcc.ufla.br/infocomp/artigos/v4.4/art09.pdf

Chapter 19
Web System for Supporting Project Management

Cátia Filipa Veiga Alves, André Filipe Nogueira da Silva, and Maria Leonilde R. Varela

Abstract This paper provides an alternative approach for project management through a web-based decision support system. This contribution integrates XML-based web services and decision support tools through a proposed platform, which is being developed for supporting large scale projects management, in order to enable to more easily work on projects, integrating many and complex tasks to be planned and controlled.

Keyword Web-based • Decision Support System • XML Web Service • Project Management • Gantt Chart

19.1 Introduction

Project management is necessary to ensure that the projects are completed with success in the time and budget establish initially. In order to enable a better decision making the project managers usually need to use several distinct software tools, which sometimes are very expensive and sometimes lack the possibility or being easily integrated with other applications.

In this paper we propose and described a web system for supporting project management (WSSPM). In order to enable a clear exposition of the proposed system, this paper is organized as follows. In Sect. 19.2 we include a brief contextualization about decision support systems. Section 19.3 summarizes a literature review about decision support systems for project management. Section 19.4 briefly describes the

C.F.V. Alves • A.F.N. da Silva • M.L.R. Varela (✉)
Department of Production and Systems, School of Engineering, University of Minho, Azurém Campus, 4800-058 Guimarães, Portugal
e-mail: a48853@alunos.uminho.pt; a43127@alunos.uminho.pt; leonilde@dps.uminho.pt

A. Madureira et al. (eds.), *Computational Intelligence and Decision Making: Trends and Applications*, Intelligent Systems, Control and Automation: Science and Engineering 61,
DOI 10.1007/978-94-007-4722-7_19,

proposed WSSPM architecture and its main functionalities and also includes an illustration about its potential use, for a project's management processes support. Finally, on Sect. 19.5, a conclusion is presented.

19.2 Decision Support Systems

Since the early development of Decision Support Systems (DSS), in the 1970s, the evolution of these kind of systems have not been homogeneous, which made possible the emergence of distinct sub-fields, and according to Arnott, seven can be distinguished: Personal Decision Support Systems (PDSS), Group Support Systems (GSS), Negotiation Support Systems (NSS), Intelligent Decision Support Systems (IDSS), Knowledge Management-Based DSS (KMDSS), Data Warehousing (DW), Enterprise Reporting and Analysis Systems [1]. These fields can be associated with the orientation of the DSS and its purpose.

While promoting a relationship with the user, Haettenschwiler [2] distinguished three different DSS categories: passive, active, and cooperative DSS. Another taxonomy can be found in the literature by Daniel Power, [3, 4] which characterized this kind or systems based on their mode of assistance, resulting five main categories, as follows: communication-driven, data-driven; document-driven; knowledge-driven and model-driven.

The DSS put forward in this paper can be included in the document-driven category, according to Daniel Power, since we can describe this as a category in which we can manage, retrieve and manipulate unstructured information in a variety of electronic format [3] combined with the use of web-based technologies for supporting documents about projects management sharing and processing, over the Internet.

Regarding the taxonomy put forward by Arnott [1], as described above, our proposed system for project data and knowledge management can also fill up the characteristics mentioned by this author about decision support systems for knowledge management (KMDSS), as the knowledge management feature is one of the most important ones of our proposed system, where developed web services aim at supporting decision-making, by enabling to easily aid project's data knowledge and to store it, in order to enable further data and knowledge retrieval, transference and processing, by supporting individual and organizational memory and inter-group knowledge access.

Moreover, as we are proposing a web-based system it also falls within the general category of web-based decision-support systems [4].

19.3 DSS for Project Management

Due the faster market changing, it becomes increasingly more important and necessary to ensure that projects developed by enterprises are completed with success in the right time and accomplishing tasks, by satisfying budget requisites, initially established. In order to achieve these goals, an accurate project management process plays a fundamental role. Munns and Bjeirmi refer that the project management could be defined as "the process of controlling the achievement of the project objectives" and it is "orientated towards planning and control activities" [5].

A project management system assists managers in order to schedule and visualize tasks and milestones within a certain project duration [6]. The most experienced project management practitioners recognize that usually there is more than one way to manage the project [7]. According to the Project Management Institute (PMI), some important concerns are about a Plan-Do-Check-Act cycle (PDCA), which is consists on a general concept for describing interactions among project management processes. The author in [6] refers that Process Groups are more complex than the cycle above mentioned and they are characterized by five processes: (1) Initiation, (2) Planning or Development, (3) Execution or Production, (4) Monitoring and Controlling, and (5) Closing. The Planning or Development process represents the "plan" and the Execution or Production process represents the "doing" components of a PDCA cycle. The Monitoring and Controlling process corresponds to the "checking" and "acting" components. Due to the integrative nature of the project management activity, it is necessary to handle interactions about Monitoring and Controlling processes in all other processes. The Initiation and Closing processes start and finish the cycle, respectively [7].

To achieve an accurate level of project management, many researchers have been investigating and developing new project management models.

Heindel and Kasten [8] have developed a prototype PC-based enterprise management system using off-the shelf, high-level, object-oriented tools for expanding the functionality of project management systems.

Latter, Petrie, Goldmann and Raquet [9] have recognized that traditional project management techniques were not sufficient for managing the many tasks in the design and development process. The most significant concern of the author was that these systems did not provide the proper mechanisms for notifying the correct people about the effects of changes at the correct time in the process.

Moreover, Shi and Deng [10] have developed a construction scheduling and planning method, named the object-oriented resource-based planning method (ORPM).

In the last years, contributions based on new technologies, based on the world wide web (www), or simply web, were created, through different researches, namely about decision support systems for supporting project management [1]. As web based DSS are very useful to help users in decision making activities [2] and, in general, DSS are interactive computer-based systems, which can be considered as tools that an organization may use to support and enhance decision-making activities [11], this kind of tools appear to be promising for supporting project managers.

The first steps, on the evolution of this kind of tools, were made in the 1970s, and since then a quick evolution has been noticed. Due the advance in the computers technology, data warehouses and the introduction of the web in the 1990s made it even easier to develop web platforms for using DSS [12, 13].

Since this tools, as referred above, are very helpful in the decision-making processes, they can be used in a huge variety of situations, for example to determine the cost of products, for establishing milestones for projects tasks, for resources management, and risk management among others.

Nowadays various kinds of web-based DSS already support the project management decision-making. Some models for project management were already developed using techniques like CPM/PERT and Gantt chart [14].

In the last few years some developments have been made to overtake unsolved problems behind some approaches and some new proposals and upgrades were made to help in better decision-making processes for project management, like the contribution presented in [15], where an architecture about a system for providing project management support through the web, by using open source solutions is described. This proposal includes tools like Work Breakdown Structure, Gantt chart, PERT diagram and Critical Path Method, but it is still subject to further developments.

Furthermore, there are some other approaches [16–18] that provide a contribution in terms of project management, also including data presentation and project planning and control based on graphical representation models, including Gantt charts, and other kind of data representation and processing models, for instance based on XML and related technologies, but these approaches are more directed to production scheduling purposes.

Marquez and C. Blanchar [19], proposed a decision support system for evaluating operations investments in high-technology business. Moreover, some other studies are being carried out, namely in order to properly evaluate the user's needs, in terms of project management processes. One interesting contribution was presented by Bharati and Chaudhury [20], which described an empirical investigation about levels of decision-making satisfaction, by end-users. Regarding this study as well as considering our daily-based perception about the availability of software and tools put forward for supporting projects management, based on the review carried out through this work, we may be able to say that although there are some important project management tools available, there is no historical data available yet about a full decision support system for project managers support, which is entirely free for being used by any end-user. Therefore, in this paper we propose a web-based system for supporting project planning, programming and controlling, which can be used by any user without of charge.

19.4 Web-Based Decision Support System

The web system proposed in this paper consists on a freely available tool for project managers that can be used by anyone, because it is an open source web platform and very user friendly, which intends to enable an easy and interactive use, based on XML data representation and processing, which can be used by any non-specialized user, through the web, instead of requiring special knowledge about complex data formats and its manipulation.

Next, we will briefly refer to the considered project description, as well as to the project management purpose in this work, followed by a summarized description of the proposed web system, including the presentation of its architecture, functionalities and also including an illustration about its use, for the project presented.

19.4.1 Project Description

In this study the project considered was developed in a call from the Seventh Framework Programme (FP7). According to this program it was necessary to build tasks that would enable to accomplish a project in the future years, if it was approved.

To a better task organization it is necessary to appeal to different tools, such as the Gantt chart and PERT analysis.

In the next section the main reason underlying the proposed decision support system will be described.

19.4.2 Project Management Purpose

Each project task includes a certain amount of variables and data in a structural way, such as the number of resources, each resource's cost, a set of precedences, and the starting and finishing times of each task, as well as milestones and other important project information.

Due to the large amount of tasks in a project it turns necessary to streamline the decision-making, by applying tools with functional planning capabilities, like the precious tool of Gantt chart.

To enable a project management success, the platform developed can help manager in:

- Planning with a commitment to complete the project.
- Correctly planning the activities in the project.
- Ensuring correct and adequate information flows.
- Change activities to adapt to frequent changes on a dynamic basis.
- Re-planning the plan of activities when mistakes have been identified.

Therefore, a web-based system enables to conciliate data complexity while assisting the project's management processes to obtain a quick response during project management activity.

Next the proposed system's architecture of the web-based decision support system for project management, as well as its main functionalities and illustrations, through an example of us, for the FP7 project tasks, will be provided.

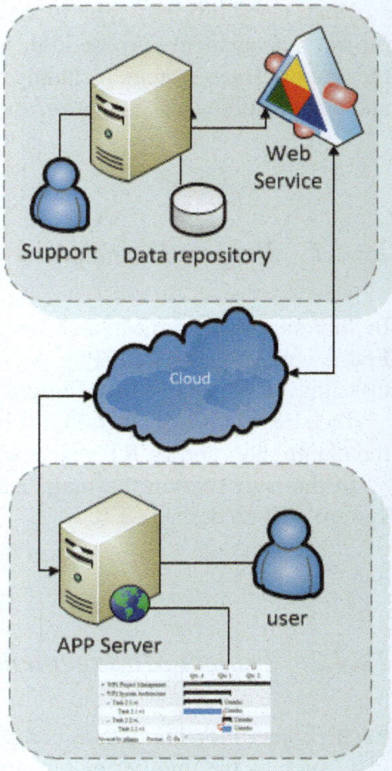

Fig. 19.1 Web-based decision support system architecture

19.4.3 Architecture of Web-Based Decision Support System

The WSSPM, as mentioned before, enables complex data structures, while assisting the whole project management activity. In Fig. 19.1 we can see the core system architecture. The data for the Gantt chart is saved into a data repository, based on a combination of a relational data based repository with other kind of data formats (txt, csv, xml, pdf, and others), where the most important one are based on XML (extensible markup language). This data is further processed through XML based web services (http://www.w3.org).

The main system's component is a core module for enabling users to be connected to an ASP based Server for enabling the project manager to clearly analyze the project data and to process it. The manager (user) can also easily load and upload data in its database, by using the previously different data formats put available through the system, as well as easily (re)edit data for project actualization.

The architecture shown in Fig. 19.1 is a simplified presentation about the whole system's structure, which is integrated in a more complex "cloud", including further components and functions, related to other closely related XML based web service for integrating the system with other sub-systems, including legacy systems and this integration process is still under development. Other functions to be considered in the future are related to real-time management capabilities, which are also under consideration and specification to be included in the proposed system.

19.4.4 System Functionalities

The main functionality of the WSSPM consists on assisting the manager in the project management activity. Therefore, with this work we aim at contributing to enable a better project planning and controlling, in order to be able to any end-user to perform project insertion, storing and data processing, in a simple and easy way, based on a temporal graphical visualization, which allows to easily identify the delayed tasks, the dependent tasks, completion percentage, the task resources, and the start and end date of each task, among other specific task's information, within the project.

Moreover, the WSSPM includes other more specific functionalities, for enabling a better project management support, like tasks addition, elimination and new task data saving on data repository for further utilization, among other functionalities, which are expressed through Fig. 19.2, represented by a flowchart about the main process the manager may follow by using the WSSPM.

A positive aspect is the WSSPM that may be referred is related to not being needed any software installation, i.e. it is only necessary that the user has available a personal computer linked on a network, through a server, for Internet access. The user has the possibility to export data through different support, such as PDA (Personal Digital Assistant).

19.4.5 System Illustrations

The WSSPM illustration of use is presented next, and Fig. 19.3 shows the system interface for users' registration for accessing the proposed web system.

The manager must be linked to a network server in order to be able to login into the system and in case of having forgotten its username and/or password a data retrieval process is provided by the system.

After the manager correctly performed his/her login in the system, a default table as shown in Fig. 19.4 will appear, where the manager can load his/her data files, if they already exist into a data repository, otherwise the manager can add new ones and/or remover some existing one.

The data insert process by manager must follow some standard format, in order to enable, for instance, Gantt charts presentation, and a set of formats, including xml are available, as described before. Moreover, it is important to refer that the Gantt chart component used in the system was developed by Shlomy Gantz and Brian Twidt (http://www.jsgantt.com/) which follows the standard format

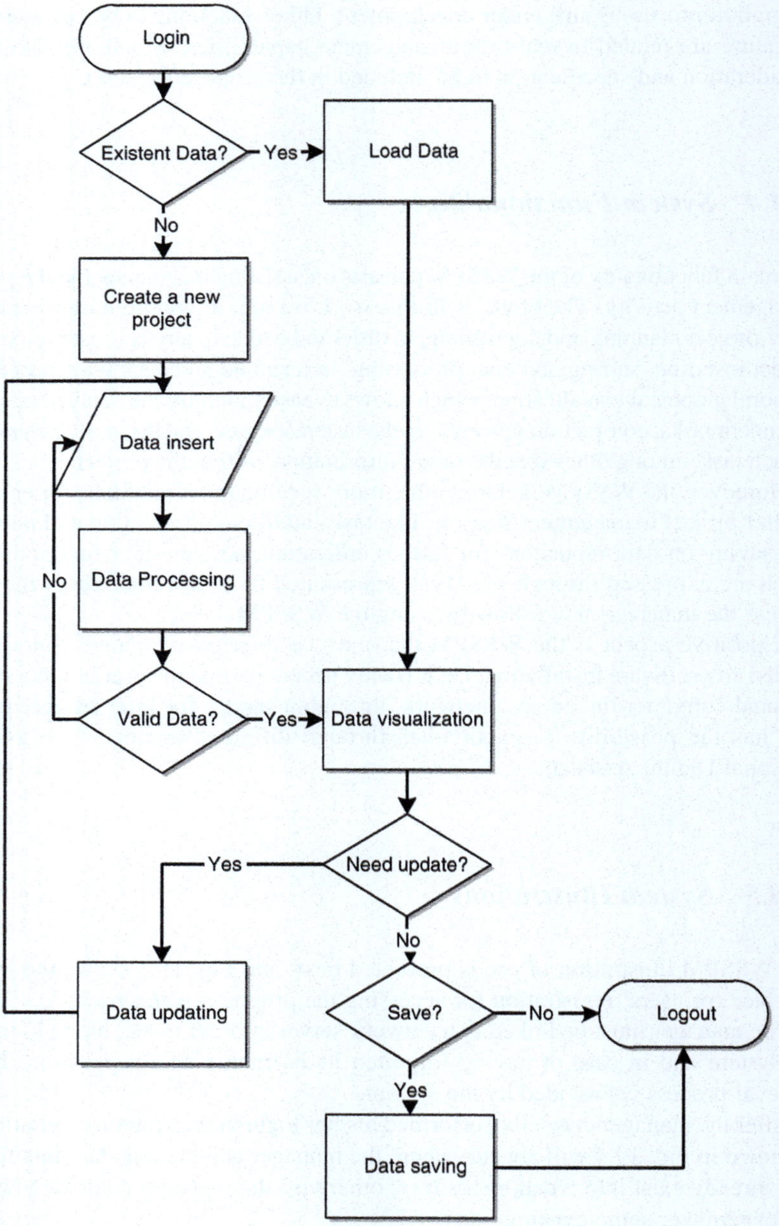

Fig. 19.2 System flowchart

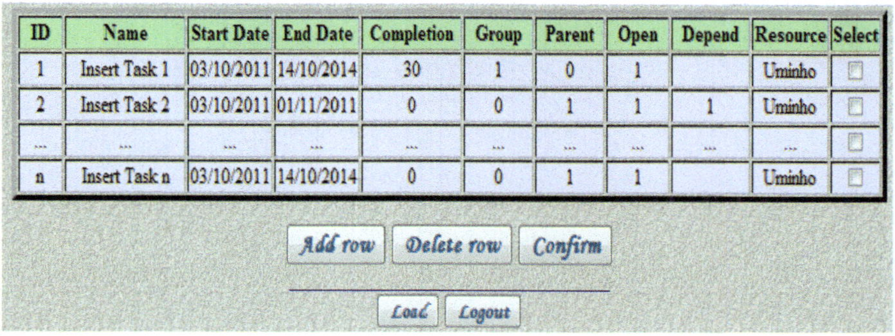

ID	Name	Start Date	End Date	Completion	Group	Parent	Open	Depend	Resource	Select
1	Insert Task 1	03/10/2011	14/10/2014	30	1	0	1		Uminho	
2	Insert Task 2	03/10/2011	01/11/2011	0	0	1	1	1	Uminho	
...	...	...	...	...	...	...	...	...	...	
n	Insert Task n	03/10/2011	14/10/2014	0	0	1	1		Uminho	

Fig. 19.3 Extract of a data table in the web system

a

ID	Name	Start Date	End Date	Completion	Group	Parent	Open	Depend	Resource	Select
10	WP1 Project Management	03/10/2011	14/10/2014	0	1	0	0			
20	Task 1.1	03/10/2011	01/11/2011	30	0	10	1		Uminho	
30	Task 1.2	03/10/2011	14/10/2012	0	0	10	1		Uminho	
120	WP2 System Architecture	03/10/2011	22/09/2014	0	1	0	1			
130	Task 2.1.vi	03/10/2011	21/05/2014	0	1	120	1		Uminho	
140	Task 2.1.vl	03/10/2011	01/02/2012	0	0	130	1		Uminho	
170	Task 2.2.vi	02/02/2012	20/06/2014	0	1	120	1		Uminho	
180	Task 2.2.vl	02/02/2012	04/03/2012	0	0	170	1	140	Uminho	

b

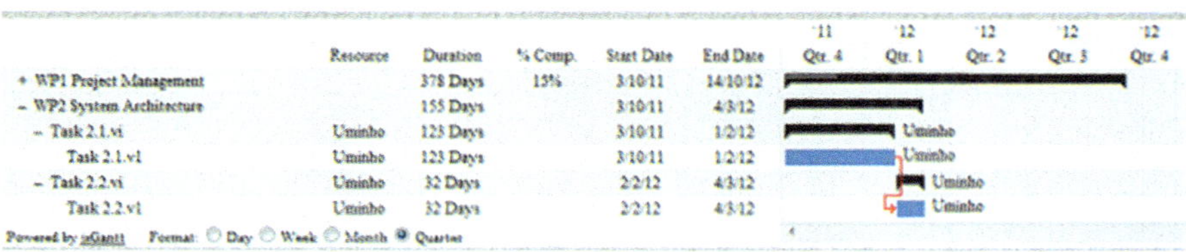

Fig. 19.4 Example tasks' table (**a**) and corresponding Gantt chart (**b**)

mentioned above. The description for data insertion on each tables column is as follows (Fig. 19.3):

- ID – is an unique ID used to identify each task.
- Name – is the task's name Label.
- Start Date – is the task's starting date/time.
- End Date – is the task's ending date/time.
- Completion – is the completion percentage of each task; by inserting only a given number.
- Group – indicates whether a task is a group (parent) or not: insert 0 if it is not a parent and 1 otherwise.
- Parent – identifies a parent ID. So, the task will be a child of identified task.
- Open – can be initially set to a closed folder when the project's chart is firstly drawn.

Fig. 19.5 General web system user interface

- Dependent – is the ID of the dependent task.
- Resource – is some resource name for performing a given task.

In the Fig. 19.4a is shown an example of tasks for the considered FP7 project example, to better illustrate the proposed system's functionalities. The Gantt chart presented through the web system also includes the capability of collapsing and expanding subtasks, as is illustrated through Fig. 19.4b.

A general user interface of the proposed WSSPM is presented in Fig. 19.5.

The proposed web system for supporting decision making enables to insert, transform and process information about large projects, as well as its visualization in different output formats, namely through simple tables or some other more complex format, like Gantt charts or other kind of charts, for instance obtained through excel transformations.

19.5 Conclusion

In a growing world enterprises must be alert due the projects complexity and it is becomes increasingly more necessary to be able to appeal to decision support systems in order to accurately manage projects for achieving planned goals, regarding projects constraints.

The web system proposed in this paper is still in progress and aims at assisting a project manager in a free and enriched way in order to be able to better support him/her in the project management activity, varying from simple tasks data insertion up to processing and visualizing complex project structures, for further accurate decision-making.

Therefore, the proposed Web Systems intends to provided a contribution not only in terms of project planning but also to support users in terms of projects' control activities, through a set of functionalities, namely related with projects tasks progress control and one of the main support tool is based on Gantt charts, which constitute and easy and clear way to enable comparing planned and on-going tasks' development.

In future developments of the proposed web system some additional functionalities are still planned to be incorporated, namely regarding an easy identification of critical resources, including different workloads and associated costs. Moreover, different kind of alerts to the end user, namely regarding delayed tasks or tasks that are close to a certain delay value will also be considered for further system's improvements.

Other emergent system's improvements being specified and implemented are related to real-time management capabilities, which are also being studied.

Acknowledgments The authors wish to acknowledge the support of: (1) The Foundation for Science and Technology – FCT, under the scope of the financed Project on "Ubiquitous oriented embedded systems for globally distributed factories of manufacturing enterprises" – PTDC/EME-GIN/102143/2008.

References

1. Arnott D, Pervan G (2008) Eight key issues for the decision support systems discipline. Decis Support Syst 44:657–672
2. Haettenschwiler P (1999) Neues anwenderfreundliches Konzept der Entscheidungsunterstützung. In: Gutes Entscheiden in Wirtschaft, Politik und Gesellschaft. vdf Hochschulverlag AG, Zurich, pp 189–208
3. Power DJ (2002) Decision support systems: concepts and resources for managers. Quorum Books, Westport
4. Power DJ, Kaparthi S (2002) Building web-based decision support systems. Stud Inform Control 11:291–303
5. Munns A, Bjeirmi B (1996) The role of project management in achieving project success. Int J Proj Manag 14:81–87

6. Massey AP (2008) Collaborative technologies. In: Frada Burstein, Holsapple CW (eds) Handbook on decision support systems, vol I. Springer, Berlin/Heidelberg, pp 341–354
7. PMI (2010) PMBOOk. A guide to the project management body of knowledge, PMI 4a edn. PMI Publishing, Project Management Institute, Newton Square, 2008, pp 27–35
8. Heindel LE, Kasten VA (1997) P++: a prototype PC-based enterprise management system. Int J Proj Manag 15:1–4
9. Petrie C, Goldmann S, Raquet A (1999) Agent-based project management. In: Wooldridge MJ, Veloso MM (eds) Artificial intelligence today: lecture notes in computer science, vol 1600. Springer, Berlin/Heidelberg, pp 339–363
10. Shi J, Deng Z (2000) Object-oriented resource-based planning method (ORPM) for construction. Int J Proj Manag 18:179–188
11. Sol H, Takkenberg CAT, de Vries Robbe PF (eds) (1987) Expert systems and artificial intelligence in decision support systems. In: Proceedings of the second mini Euroconference. Kluwer Academic Publishers, Lunteren, The Netherlands, pp 1–2
12. Power DJ (2000) Web-based and model-driven decision support systems: concepts and issues, In: AMCIS 2000 proceedings, Long Beach, California, pp 352–355
13. Shim JP, Warkentin M, Courtney JF, Power DJ, Sharda R, Carlsson C (2002) Past, present, and future of decision support technology. Decis Support Syst 33:111–126
14. Deepak Kumar C, Sarma V (1996) Towards building an intelligent decision support system for project management. Sadhana 21:327–343
15. Wojtera M, Sakowicz B (2006) Web application for project management based on open source solutions. In: Proceedings of the 13th international conference mixed design of integrated circuits and systems, Gdynia, Poland, pp 797–800
16. Varela MLR, Carmo-Silva S (2010) A P2P web decision support system for manufacturing scheduling. In: Putnik GD, Ávila P (eds) Business sustainability I, management technology and learning for individuals, Organisations and society in turbulent environments, University of Minho & ISEP, Portugal, pp 147–152. (ISBN: 147-152, 978-972-8692-48-3, 978-989-95907-1-7)
17. Carmo-Silva S, Varela L, Lemos A, Garcia A, Ribeiro C, Carvalho J (2007) Collaborative production scheduling. The Innovative Production Machines and Systems (IPROMS'07), Engelska, Sweden
18. Vieira G, Varela MLR, Putnik GD (2011) Technologies integration for distributed manufacturing scheduling in a virtual enterprise. In: First international conference on Virtual and Networked Organizations, Emergent Technologies and Tools (ViNOrg'11), Ofir, Portugal
19. Marquez AC, Blanchar C (2006) A decision support system for evaluating operations investiments in high-technology business. Decis Support Syst 41:472–487
20. Bharati P, Chaudhury A (2004) An empirical investigation of decision-making satisfaction. Web Based Decis Support Syst 37:187–197

Chapter 20
Generation Capacity Expansion Planning in Restructured Electricity Markets Using Genetic Algorithms

Adelino J.C. Pereira and João Tomé Saraiva

Abstract This paper describes an approach to model and to solve the Generation Expansion Planning Problem, GEP, using Genetic Algorithms. This approach was developed in order to help investors in new generation capacity to take decisions regarding new investments. This approach was developed in the scope of the implementation of electricity markets given that they eliminated the traditional centralized planning activities leading to the creation of several generation companies competing to supply the demand. As a result, the generation activity is more risky than in the past and so it becomes important to develop new tools to help decision makers to analyze the investment alternatives, having in mind the possible behavior of the competitors. The developed model aims at maximizing the expected profits that will be obtained by an investor, while it evaluates the reliability and the security of supply and it incorporates uncertainties related with the volatility of electricity prices, with the reliability of generation groups, with the evolution of the demand, and with the operation and investment costs The developed model and the implemented solution algorithm will be applied to a Case Study to illustrate the use of the developed approach to build the expansion plans.

A.J.C. Pereira (✉)
Departamento de Engenharia Eletrotécnica,
Instituto Superior de Engenharia de Coimbra Instituto Politécnico de Coimbra Rua Pedro Nunes, 3030-199 Coimbra, Portugal
e-mail: ajcp@isec.pt

J.T. Saraiva
Departamento de Engenharia Eletrotécnica e de Computadores;
FEUP/DEEC and INESC Porto Campus da FEUP, Rua Dr. Roberto Frias, 4200-465 Porto, Portugal
e-mail: jsaraiva@fe.up.pt

A. Madureira et al. (eds.), *Computational Intelligence and Decision Making: Trends and Applications*, Intelligent Systems, Control and Automation: Science and Engineering 61, DOI 10.1007/978-94-007-4722-7_20,

20.1 Introduction

Motivated by the success of deregulation in industries such as telecommunications, airlines, and transportation, the power industry went through a restructuring process that was introduced in many countries around the world. The restructuring process had several consequences, namely in the operation planning and in the expansion planning, both regarding generation capacity and transmission capacity. Among these new aspects, it is important to mention that these activities are now much more contaminated by uncertainties leading to a large level of risk. Reference [1] details some of the reasons explaining the increased complexity of the generation expansion planning exercises. The most relevant objectives of a generation expansion problem are the identification of the technologies on which to invest, the new capacity to install of each candidate technology and the timing these investments should be in place. The result of these problems corresponds to a set of investments distributed along a multiyear planning horizon that as whole constitute an expansion plan. Having in mind these ideas this paper is organized as follows. Section 20.2 details a number of models and solution algorithms in the literature for the GEP problem. Section 20.3 describes the mathematical formulation of the GEP problem and the application of genetic algorithm, and Sect. 20.4 presents a Case Study using a generation system with three generation agents and three candidate technologies. Finally, Sect. 20.5 presents the most relevant conclusions.

20.2 Generation Expansion Approach

The expansion planning of the generation system includes more and more uncertainties that result from the existence of different agents acting in the sector and from other aspects that as a whole contribute to increase the complexity of the problem. These uncertainties are very important to the preparation of the generation system expansion plans. In a competitive environment the available information is more reduced and, in some cases, it is simply not available or very much affected by uncertainties. This means that investors should obviously take this aspect in consideration. On the other hand, new models and approaches to this problem should also accommodate the nature of data above mentioned. From a traditional point of view the generation expansion planning problem was for instance addressed in [1]. References [1, 2] describe mixed integer problems and [2] consider transmission networks and uncertainties. The model in [2] use Benders decomposition profiting from the structure of the GEP problem. The model in [3] uses a multiobjective formulation considering investment, operation and transmission costs, environmental impact and the risk of getting a plan too much exposed to the volatility of the energy price. In the first place, the solution algorithm builds a set of non-dominated solutions that are afterwards ranked using Analytical Hierarchy Processes. Finally, [4–8] describe approaches using metaheuristics. The use of Genetic Algorithms is described in [4–8], Simulated Annealing in [4–7], ant colonies and PSO algorithms [5], expert systems and fuzzy logic [7] and combinations of Genetic Algorithms and Simulated Annealing in [7].

The introduction of competition determined several changes in the industry as briefly mentioned in the Introduction. Under the competitive paradigm, reference [9] describes a two-step approach in which individual generation companies prepare their investment plans that are then submitted to a regulatory agency for global check and approval. Reference [10] follows this idea and describes the use of the Cournot model to get the evolution of the electricity price along the planning horizon. On the other hand, reference [11] formulates the optimisation investment problem to maximize the profit of individual generation companies and the resulting global problem is then solved using Dynamic Stochastic Programming.

Given the long-term aspects of the GEP, recent approaches use System Dynamics. This approach is documented in [10] and Sect. 20.3 gives some insight on it. References [11–13] illustrate its application to electricity markets identifying the relations between several variables as the demand, the investment and operation costs, the electricity prices and formulate models to represent the long-term evolution of these variables enabling taking more sounded decisions regarding the GEP problem.

20.3 The Developed GEP Model

20.3.1 Structure of the Problem

Figure 20.1 details the general structure of the developed model. It includes the long-term model of the electricity market using System Dynamics, the coordination analysis and the investment problems solved by each agent. This approach is designed to help generation agents to prepare their own plans while considering the impact of the decisions of other agents and the evolution of fuel costs. As a result, a generation agent obtains its own plan and can eventually conduct sensitivity studies to evaluate the impact of changing input parameters. This will con-tribute to gain new insight to the problem and ultimately to take more sounded decisions. Regulatory agents can also use this approach to study the possible evolution of a generation system, or to design specific payments (for instance capacity payments) if that becomes necessary to turn generation investments more attractive.

20.3.2 Individual Optimization Problems

In the developed approach a particular generation agent solves the mixed integer problem (20.1), (20.2), (20.3), (20.4), (20.5), and (20.6) to identify the most

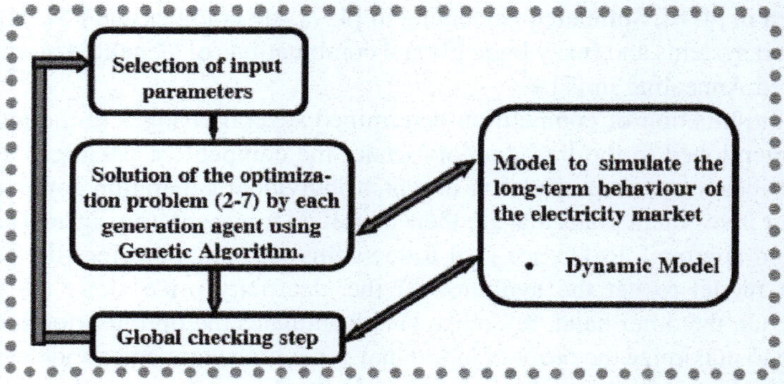

Fig. 20.1 Global structure of the developed application

adequate expansion plan considering a planning horizon with *T* years and *M* candidate technologies. The decision variables are and they represent the capacity to commission by GENCO *i* in year *t* of technology *j*. Each candidate technology is characterized by its investment cost, lifetime, fixed and variable operation costs and capacities available to be built of this technology.

$$\max z_i = \sum_{t=1}^{T}\left\{\sum_{j=1}^{M}\left[\sum_{n=1}^{tcj}\left(-Ccap_j X_t^{ij}\right)+\sum_{k=t+tcj}^{Tut+tcj}\left(\begin{array}{c}\left(\pi^t.X_k^{ij}.\alpha_k^{ij}+\left(1-\alpha_k^{ij}\right).X_k^{ij}.Pcap_k\right)-\\-Ccap_k.X_k^{ij}-\left(Cop_k^{ij}.X_k^{ij}.\alpha_k^{ij}+Cfix_k^{ij}.X_k^{ij}\right)\end{array}\right)\right]\right\} \tag{20.1}$$

$$\text{subj.} \sum_{t=1}^{Tn} X_t^{ij} \le MCapI_{Tn}^{ij} \tag{20.2}$$

$$\sum_{t=1}^{Tn}\sum_{j=1}^{M} X_t^{ij} \le MCapI_{Tn}^{i} \tag{20.3}$$

$$CAcum_{t=0}^{i} + \sum_{t=1}^{T}\sum_{j=1}^{M} X_t^{ij} \le CAcum_{Total}^{i} \tag{20.4}$$

$$\sum_{t=1}^{Tn}\sum_{j=1}^{M} X_t^{ij}.Cinv_t^{j} \le CapDisp_{Tn}^{i} \tag{20.5}$$

$$t = 1,\ldots T;\ j = 1,\ldots M;\ Tn \in \{1,\ldots,T\} \tag{20.6}$$

In this formulation:

T	Number of stages in the planning horizon;
t	Stage in the planning horizon (year);
Tut	Lifetime of a generation station (year);
tc_j	Construction time of a station of technology j (year);
T_n	Extension of the planning horizon (year);
i	GENCO index;
M	Number of candidate technologies;
j	Index for candidate expansion technology;
π^t	Electricity price in year t (€/MWh);
$\alpha_t^{i,j}$	Capacity factor of technology j in year t, GENCO *i*;
$Cinv_t^j$	Investment cost of technology j in year t (€/MW);
Cop_k^j	Variable operation and maintenance cost for technology j in year k (€/MWh);
$Pcap_k$	Capacity payment (€/MW) in year k;
$Ccap_j$	Capital cost due to the loan (€/MW.year);
$Cfix_k^{ij}$	Fix operation cost of technology j, in year k (€/MW.year);
X_t^{ij}, X_k^{ij}	New capacity of technology j, in year t or k for GENCO *i* (MW)*;*
$CAcum_{t=0}^i$	Installed capacity owned by GENCO *i* before the planning period (MW);
$CAcum_{total}^i$	Maximum value admitted for the installed capacity owned by GENCO *i* when the planning horizon ends (MW);
$CapDisp_{Tn}^i$	Financial resources available to invest in year t or in all $T_n$periods, by GENCO *i* (€);
$MCapI_{Tn}^i$	Maximum capacity that GENCO *i* can install in all M technologies in year t or in all T_n periods (MW);
$MCapI_{Tn}^{ij}$	Maximum capacity that GENCO *i* can install of technology j in year t or in all T_n periods (MW).

The objective function (20.1) corresponds to the maximization of the profit of GENCO *i* along the planning horizon. This profit is computed as the difference between the revenues from selling electricity at price π^t and from receiving the payment associated with the capacity not used to produce electricity. The first revenue is multi-plied by the capacity factor associated with each technology, $\alpha_t^{i,j}$, while the capacity payment term is multiplied by $1 - \alpha_t^{i,j}$. Regarding the costs, these include financial costs assumed before commissioning an unit due to the loan as well as financial, operation and fixed after starting the operation of an unit. This objective function is constrained by (20.2), (20.3), (20.4), (20.5), and (20.6). Constraints (20.2) represent limita-tions on the capacity to build of technology *j*. This limit can be specified by the GENCO itself in terms of its investment policy or can result from a regulatory decision or from an indication of global energy policy to contribute to diversify the technologies in use. Constraints (20.3) limit the total installed new capacity of GENCO *i*. This limit can be set for specific periods in the horizon or can be modelled in terms of a single constraint for the entire planning horizon. On the other hand, GENCO *i* already owns capacity prior to the period under analysis. Accordingly, constraint (20.4) limits the global capacity owned by GENCO in terms of the old and new capacity. Once again, this limit can be specified by a GENCO itself or can result from a regulatory decision as a way to reduce market power. Finally, constraints (20.5) enforce limitations on the available financial resources of GENCO i in each stage t or along sets of stages or along the entire horizon. These limits can be specified by each investor having in mind the financial available resources.

20.3.3 Application of GA to Solve GEP Problem

The problem (20.1), (20.2), (20.3), (20.4), (20.5), and (20.6) has a combinatorial nature because the available capacities are typically restricted to a small number of integer options. This characteristic suggested the use of Genetic Algorithms [14]. The application of GA's requires the definition of the number of populations to be used and the respective size. The populations are initialized in a random way using a random number generator. In each iteration of the algorithm, that is in each generation, each individual in each population is evaluated in order to check for its ability to solve the problem in a satisfactory way. The Fitness Function, FF, requires the computation of the value of the objective function (20.1) minus penalties over the violated constraints for the values of the decision variables codified in the chromosome.

Afterwards, the selection mechanism is used to select the individuals that pass to the next generation. The stochastic tournament was adopted in this work as the selection mechanism. This mechanism requires performing a competition between pairs of individuals selected in a random way in the population. The individual in the pair under analysis that has the best FF wins, that is passes to the next generation, with a probability that was set at 0.8. Apart from this technique, it was also used an elitist selection that involves selecting the individual in the population that has the best FF. This individual passes to the next generation without being changed by any of the genetic operators. This ensures that the best so far identified solution is not eliminated or modified and, in some way, this contributes to speed up the convergence of the algorithm. At the end of each iteration, the FF value of the mentioned copied individual is compared with the FF of the worse individual in the current population. If its FF is larger, the worse individual is eliminated. The crossover and mutation operators act over each of the populations that being used and transform them in order to introduce changes in some individuals contributing to enlarge the search space and to prevent a premature convergence until a satisfactory result is obtained. The developed application requires the user to set the cross over and the mutation rates.

As mentioned above, the developed approach uses more than one population. All these populations are changed according to the ideas detailed in the previous paragraph so that the efficiency of the AG is improved and better solutions are identified compared with the results that could be obtained if just one population was used. The most relevant advantage of this technique is related with the larger coverage of the search space so that the probability of identifying good quality solutions is enlarged. In order to further improve the AG it was also implemented the possibility of transferring individuals between the populations. In order to do this, we adopted a ring strategy in which we used a transfer rate of 15%. This means that at the end of each iteration 15% of the individuals in population 1 are transferred to the population 2 and so on till the ring is closed. The selection of the individuals to transfer is done randomly and when this process is finished we have built the new populations that will be used as inputs to the next generation.

The application of the AG to the generation expansion problem is done in two steps. In the first one, the algorithm runs *Nintmed* iterations in which the best individual identified in each iteration in all the populations is stored in a vector of the best so far identified solutions. Once we reach the iteration *Nintmed*, it is computed the mean value of the FF of the stored individuals, *MedAdpNintmed* and this value is compared with the value of the FF of the best individual identified in the current iteration, *AdpUlt_Ind*. Using these two values we check condition (20.7). If this condition is not valid, it is run a new iteration and the new best individual is stored in the mentioned vector from which it is eliminated the first element. The average of the FF values is recalculated and condition (20.7) is checked again. This process is repeated until condition (20.7) holds.

$$\left|MedAdt_{Nintmed} - Adp_{Ult\_Indi}\right| = 0 \tag{20.7}$$

In order to avoid a premature convergence of the algorithm, it is then activated a second cycle that requires running extra iterations. At the end of each iteration in this second cycle, the individual with the best value of the fitness function is stored in a new vector. When the *NintExt* iterations is reached, the algorithm checks weather it was stored an individual having better adaptation than the ones identified at the end of the first cycle. If that individual was not identified, the AG is stopped and it then presents the solution associated with the best identified FF. On the contrary, if the AG identified an individual with better FF, then convergence was not yet reached and the iterative process is resumed. In this case, the individual having the best FF identified in the second cycle is stored and it is run a new set of *NiterExt* iterations. This cycle is repeated until no individuals having better FF values are found, each time the second cycle is run and finished.

It is also important to mention that several input parameters can be subjected to uncertainty as for instance the operation and investment costs. The FF of an element of the population of the GA was already defined as the value given by (20.1) minus penalties on the violated constraints. To consider uncertainties, we use the Monte Carlo method to sample a large number of values of the uncertain data from their probabilistic functions to characterize each element of the population by the expected value of FF. The number of samples to take was controlled along the sampling procedure using an uncertainty coefficient that indicates if the current estimate of the expected value of FF is already of good quality in terms of being sufficiently stable to enable stopping the sampling procedure.

20.3.4 The Coordination Analysis and the Dynamic Model

The individual expansion plans prepared by each investor are then evaluated at the coordination level in order to check the ability of the global plan to supply the demand with the required security and reliability levels. This evaluation requires the computation of the reserve margin along the planning horizon comparing the installed capacity with the yearly peak demand, checking if the capacity to install in each candidate technology is not excessive given general defined policy objectives, and computing reliability indices (LOLP or LOLE) quantifying the ability of the generation system to supply the demand given the FOR values of each generator. This Coordination analysis was already detailed in [8].

Finally, it is important to mention that the presence of several agents in the planning process means that the decisions taken by one of them will impact in the output of the power stations owned by the other agents and this ultimately means that there will be an impact n the profits of other agents. The Dynamic Model already mention in Sect. 20.3. This long term model gives estimates of the evolution of the electricity market price along the planning horizon, of the reserve margin, of the capacity factors associated to each technology, of the impact of new capacities in the generation mix, of the decommission of old power stations and of the impact of new regulatory schemes adopted for instance to induce new investments.

The information obtained from the Dynamic Model is then used by the investors to update their investment plans solving new generation expansion problems given by (20.1), (20.2), (20.3), (20.4), (20.5), and (20.6) and submitting them again to the coordination level. This iterative process is repeated until the investment plans are not changed in two successive iterations. This approach enables generation agents to internalize in their decisions the information associated with the decisions of the other agents as well as the interdependency between the demand and the electricity price. This model using System Dynamics Model was already detailed in [8].

20.4 Case Study

20.4.1 Data

In this Case Study we considered a planning horizon of 15 years, three GENCO's and three candidate technologies. At the year previous to the planning horizon we admitted there were power stations using coal, natural gas, fuel and cogeneration. We also admitted there were power stations using renewable resources, namely run of river and reservoir hydro stations and wind parks. We decided to group coal stations in two sets in order to better model the characteristics of these stations in terms of costs and enabling computing different capacity factors for each group. A similar reasoning justifies the separation of hydro stations in run of river and in reservoirs. This allows a more refined and realistic representation of the hydro

Table 20.1 Initial installed capacities and respective ownership

Type of technology	GENCO_1	GENCO_2	GENCO_3	Other GENCO's
Coal_1	500	300	200	–
Coal_2	–	300	300	–
Gas turbine	250	250	–	–
CCGT	400	400	400	300
Oil	–	–	100	100
Wind parks	200	–	300	250
Hydro reservoirs	400	300	200	600
Hydro run-of-river	300	200	200	300
Cogeneration	–	–	–	500
Total (MW)	2,050	1,750	1,700	2,050

system. Regarding the wind parks, the installed capacity was modeled in an aggregated way. When running the Dynamic Model, the capacity factor is the same for all wind parks and it reflects the average use of this technology.

Table 20.1 includes the installed capacities of each technology before running the expansion exercise. These stations are owned by three GENCO's (GENCO_1 owns 27.15%, GENCO_2 owns 23.18%, GENCO_3 owns 22.52%) and by other companies 27.15% that we admit-ted are interested in participating in the planning exercise. In this planning exercise we considered the following decommissioning plan. In the 5th year 300 MW of coal type 1 and 200 MW of CCGT are decommissioned. In the 10th year 500 MW of gas turbines, 300 MW of coal type 1, 300 MW of coal type 2 and 200 MW of wind parks are decommissioned. In the 15th year are decommissioned 400 MW of CCGT and 200 MW of wind parks. The remaining thermal stations and wind parks in Table 20.1 are decommissioned in the 20th year. The life-time of hydro stations was set equal to the period considered in the simulation.

Regarding the candidate technologies, Table 20.2 contains information about the available capacities, the investment cost, the fix operation and maintenance cost, the cost of capital and the FOR. The life-time of these stations is 30 years and we admitted that the construction period is 2 years for the three technologies although different values can be easily set for each technology.

The value of the peak power in the starting year is 4,750 MW and the annual demand is 30.64 TWh. We admitted that the annual load duration curve was discretized in six steps as follows: 100% of the peak power during 5% of the year; 90% for 20% of the year; 80% for 45% of the year; 70% for 65% of the year; 60% for 85% of the year; 50% of the peak power during 100% of the year. Along the planning horizon we admitted that the annual demand increases as determined by the Dynamic Model but the above discretization rule remains unchanged.

On the other hand, GENCO_1 is interested in the three technologies. In the first 5 years he has 1,000 million € to invest and 1,000 million € more in the last 10 years. GENCO_2 is interested in Tech_1 and Tech_2. In the first 5 years he has 450 million € to invest, 450 million € more to invest from the 6th to the 10th year and 450 million € more in the last 5 years. GENCO_3 is interested in the three

Table 20.2 Characterization of the three candidate technologies

Type of technology	Available capacities (MW)	Investment cost (€/MW)	Fix operation and maint. cost (€/MW.year)	Cost of capital with the loan (€/MW. year)	FOR
Tech_1	100 or 150 or 200	450,000	6,000	42230.40	0.02
Tech_2	100 or 200	550,000	6,137	44133.40	0.02
Tech_3	100 or 200 or 250	800,000	7,655	49666.72	0.01

technologies. In the first 5 years he has 600 million € to invest, 600 million € more to invest from the 6th to the 10th year and 600 million € more in the last 5 years. In each year of the horizon each agent cannot install more than 300 MW in Tech_1, 300 MW in Tech_2 and 400 MW in Tech_3 and shall not own more than 40% of the installed capacity. The total new capacity shall not exceed 600 MW in Tech_1, 600 MW in Tech_2 and 800 MW in Tech_3. The reserve margin should range 20–40% and the maximum value admitted for LOLE is 2 h/year.

The capacity factor of wind parks was modelled by a stochastic process using a Normal FDP with mean 25% and standard deviation of 5%. Along the planning horizon we admitted that 3,000 MW of wind parks will be installed. The hydro stations are modelled in two groups: run of river and reservoirs. Their outputs are modelled by stochastic processes that use Normal FDP's. For run of river units the mean value is 30% and the standard deviation is 5% and for the reservoirs the mean value is 20% and the standard deviation is 5%. In order to use the approach described in Sect. 20.3 to the GEP problem it is necessary to specify a number of parameters associated with the GA. For this study, we considered 4 populations, the size of population was 100 individuals each, the migration rate was 4%, the crossover rate 35% and the mutation rate 4%.

20.4.2 Results of the Expansion Planning Exercise

After running the approach detailed in Sect. 20.3, we obtained the results to the expansion plans for GENCO's. The Table 20.3 detail the expansion plans for the GENCO_1.

At the end of the horizon, the distribution of the new capacity is as follows: 21.31% for Tech_1, 49.18% for Tech_2 and 29.51% for Tech_3. The distribution of the new capacity per GENCO is as follows: 37.70% for GENCO_1, 31.15% for GENCO_2 and 31.15% for GENCO_3. These results indicate that there is a large concentration of investments in the first half of the horizon due to the available financial resources and also due to the larger values of LOLE in the initial years (maximum value of 1.67 h/year in year 2 and minimum of 0.97 h/year in year 9).

The impact of new additions can also be seen in the evolution of the annual average electricity price. In the first 2 years the price increases since the

Table 20.3 Generation expansion plan obtained for GENCO_1

Stage	Tech_1 (MW)	Tech_2 (MW)	Tech_3 (MW)
1	100	200	200
2	100	200	200
3	–	100	–
6	100	200	200
7	100	200	200
8	–	–	200

construction period of the new stations is 2 years. The investments decided in year 1 will then only be in operation in year 3, contributing to reduce the price and LOLE from year 3 onwards. The evolution of the capacity factors of the three candidate technologies tend to decline along the horizon due to the increasing share of wind parks in the generation mix. Since we admitted that wind parks are paid by feed in tariffs, a larger share of wind generation contributes to reduce the electricity price in the market. In any case, Tech_3 has the largest capacity factors because it has the lowest marginal operation costs.

20.5 Conclusions

In this paper we described the mathematical formulation developed for the Generation Expansion Planning problem, GEP, considering that the power system now operates in terms of competition. The objective function aims at maximizing the profit of each investor during the planning horizon considering that each agent gets revenues from selling electricity at the market price. Considering the characteristics of the GEP problem to be solved individually by each GENCO, we adopted Genetic Algorithms together with the Monte Carlo method to sample sets of input data in order to reflect uncertainties in the results of the GEP problem.

The results that were obtained also enable us to conclude that the expansion plans are influenced by the characteristics of the candidate technologies in terms of technical aspects and costs. As a final comment, we consider that this approach has a number of merits that will turn it interesting for GENCO's and for state or regulatory agencies in order to get more insight on the possible evolution of the power systems in view of possible expansion generation plans.

References

1. Hobbs BF (1995) Optimization methods for electric utility resource planning. Eur J Oper Res 83:1–20
2. Sirikum J, Techanitisawad A, Kachitvichyanukul V (2007) A new efficient GA-Benders' decomposition method: for power generation expansion planning with emission con-trols. IEEE Trans Power Syst 22(3):1092–1100

3. Meza JLC, Yildrim MB, Masud ASM (2007) A model for the multiperiod multiobjective power generation expansion planning. IEEE Trans Power Syst 22(2):871–878
4. Park J-B, Park Y-M, Won J-R, Lee KY (2000) An improved genetic algorithm for generation expansion planning. IEEE Trans Power Syst 15(3):916–922
5. Zhu J, Chow M (1977) A review of emerging techniques on generation expansion planning. IEEE Trans Power Syst 12(4):1722–1728
6. Yildirim M, Erkan K, Ozturk S (2006) Power generation expansion planning with adaptive simulated annealing genetic algorithm. Int J Energy Res 30(14):1188–1199
7. Park J-B, Kim J-H, Lee KY (2002) Generation expansion planning in a competitive environment using a genetic algorithm. In: Proceedings of the 2002 I.E. PES summer meeting, vol 3, Chicago, July 2002, pp 1169–1172
8. Pereira AC, Saraiva JT (2010) A decision support system for generation expansion planning in competitive electricity markets. Electr Power Syst Res 80(7):778–787
9. Botterud A, Ilic M, Wangensteen I (2005) Optimal investment in power generation under centralised and decentralised decision making. IEEE Trans Power Syst 20(1):254–263
10. Forrester J (1991) System dynamics and the lessons of 35 years. In: De Greene KB (ed) The systemic basics of policy making in the 1990s. Sloan School of Management/MIT, Cambridge
11. Kadoya T, Sasaki T, Ihara S, Larose E, Sanford M, Graham AK, Stephens CA, Eubanks CK (2005) Utilizing system dynamics modelling to examine impact of deregulation on generation capacity growth. Proc IEEE 93(11):2060–2069
12. Olsina F, Garces F, Haubrich H-J (2006) Modelling long-term dynamics of electricity markets. Energy Policy 34(12):1411–1433
13. Pereira AC, Saraiva JT (2009) A decision support tool for generation expansion planning in competitive markets using system dynamics models. In: Proceedings of the PowerTech 2009, international conference, Bucharest, Romania, 28 June–2 July 2009
14. Michalewicz Z (1996) Genetic algorithms+data struc-tures=evolution programs, 3rd Rev ed. Springer, New York

Chapter 21
Decision Making in Maintainability of High Risk Industrial Equipment

José Sobral and Luis Ferreira

Abstract Considerations about availability and its dependence on reliability and maintainability have an increased importance when dealing with productivity or safety. This is particularly of outstanding importance if one is dealing with high risk industrial equipment, where maintainability plays an important and fundamental role in risk management when safety or huge economic values are in discussion. As availability is a function of reliability and maintainability, the objective is to improve one of these factors or both. This paper intends to show how maintainability influences availability, how to select to most important attribute for maintainability using the Analytic Hierarchy Process (AHP) and how improvements can be analysed assuming maintainability as a probability related with a restore probability density function [g(t)], and in particular using the Weibull distribution.

Keywords Maintainability • Multi-criteria Decision Making • Analytic Hierarchy Process • Weibull distribution

21.1 Introduction

As it is well perceived, in an industrial system there are repairable and non-repairable items. The repairable ones are maintained to operate and accomplish their functions after each failure occurrence. That can be done with more or less difficulty. Maintainability is "*a characteristic of design and installation, expressed*

J. Sobral (✉)
Mechanical Engineering Department, ISEL, Rua Conselheiro Emídio Navarro, 1, 1959-007 Lisbon, Portugal
e-mail: jsobral@dem.isel.ipl.pt

L. Ferreira
Mechanical Engineering Department, FEUP, University of Oporto, Rua Dr. Roberto Frias, s/n, 4200-465 Porto, Portugal
e-mail: lferreir@fe.up.pt

A. Madureira et al. (eds.), *Computational Intelligence and Decision Making: Trends and Applications*, Intelligent Systems, Control and Automation: Science and Engineering 61, DOI 10.1007/978-94-007-4722-7_21, © Springer Science+Business Media Dordrecht 2013

as the probability that an item will be retained in or restored to a specified condition within a given period of time, when the maintenance is performed in accordance with prescribed procedures and resources. It can also be characterized by the ease with which maintenance of a functional unit can be performed in accordance with prescribed requirements, when a failure occurs or a preventive action is performed" [1]. Probability of restore depends upon several factors as number of workers (workmanship), diagnostic tools and equipment, spare availability, workmanship qualification and maintenance planning.

Restore is a wider concept than repair, related to the action performed after a failure occurrence, including all necessary steps. Restore an item consists of confirming the malfunction, locating the fault responsible for malfunction, removing the faulty item or component (including time to gain access to it), installing the new item or the same if repaired (including time to test it), performing a final check to verify if the malfunction is cleared and documenting the action.

When dealing with high risk industrial equipment the time spent to repair and restore the normal operation plays an important role on equipment availability and in the inherent risk [2, 3].

While reliability measures an item ability to operate for a period of time, maintainability attempts to measure how quickly the item can be restored to its appropriate operating mode, or 'stated condition' once it experiences a fault or failure [4, 5]. The outcome of a reliable and maintainable system is really the primary goal. This goal, typically referred to as availability, is essentially a measurement of the ability to operate as a fully functional system. Availability is expressed as a ratio, comparing the amount of time that a system is 'up' over the total time measured. A typical expression for availability is:

$$A = \frac{uptime}{(uptime + downtime)} \tag{21.1}$$

Downtime includes all sources of delay or degraded status, including scheduled and unscheduled maintenance, waiting for spare parts, technicians or equipment. Availability is thus a measure of the combined effects of reliability and maintainability. Typically, in the design process issues related to items performance are primarily considered and reliability and maintainability usually are secondary targets. However, companies are increasingly viewing reliability and maintainability issues as responsible for equipment' mission availability. These characteristics together with operational readiness and design adequacy influence system effectiveness [6]. This paper intends to analyse maintainability as an important factor to be taken into account. It depends on all criteria that may affect maintenance when actions are carried out to bring the product back to its functional status [7]. In maintainability studies dealing with its influencing attributes, some factors shall be considered. These attributes or alternatives to success can be grouped into two distinct categories: Internal and External attributes. Table 21.1 shows a summary of these influencing factors.

Table 21.1 Maintainability attributes

Internal attributes	Accessibility	External attributes	Human competencies
	Reparability		Tools
	Interchangeability		Spares
	Standardization		Environment
	Redundancy		Detectability
	Simplicity		Logistics
	Modularization		Auto diagnostic
	Testability		

Obviously, each repairable unit has its own influencing attributes with high or low degree of importance. For the present study were chosen some of the above attributes or alternatives, considered the most significant at the industrial point of view. The selected attributes (alternatives) are:

- Accessibility (AC); Interchangeability (IC); Modularization (MD); Standardization (ST); Human Competencies (HC); Tools (TL); Spare Availability (SA); Logistics (LG) and Detectability (DT).

But, which one is the most important? To answer this question we can use the Analytic Hierarchy Process (AHP) as a tool to determine the weight of each maintainability attribute for a particular repairable unit assuming some criteria. The AHP is widely used for complex problems of decision and evaluation where the weight of each attribute is determined by a pair-wise comparison matrix [8]. In this paper we established some criteria, mostly considered in high risk industrial equipment concerns, such as:

- Logistics Costs (LC); Production Loss (PL); Safety Impact (SI); Environmental Impact (EI) and Fluid Loss (FL).

21.2 Multi-criteria Decision Making (MCDM) and Analytic Hierarchy Process (AHP)

Multi-criteria decision making (MCDM) is a decision tool for many scientific, financial, political and engineering challenges [9]. Among all MCDM methodologies the AHP is widely applied due to its consistency.

When dealing with maintenance issues, the AHP is often used to select the best maintenance strategy [10] or select a facility problem [11]. It can also be applied as a decision support system for inspection and maintenance [12], to identify important parameters in a reliability assessment [13] or to select the maintenance activity that best fits in specific equipment [14]. It is also common to use the AHP process with a fuzzy approach to face these decision problems [15–18].

Analytic Hierarchy Process deals with a matrix using the relative importance of the alternatives in terms of each criterion. According to this process the purpose is to select the alternative with highest preference value of M alternatives and N criteria. For each criterion its importance or weight (W) is determined.

Table 21.2 Scale of relative importance

Intensity	Definition
1	Equal importance
3	Weak importance
5	Essential or strong importance
7	Demonstrated importance
9	Absolute importance
2, 4, 6, 8	Intermediate values between the two adjacent judgements

Table 21.3 Random consistency index

N	3	4	5	6	7	8	9
RCI	0.58	0.90	1.12	1.24	1.32	1.41	1.45

The pair-wise comparison is a crucial step of this process because it is necessary to have real knowledge and information about all elements to determine their relative importance. This is the keystone of every decision making process. An approach proposed by Saaty [19] for pair-wise comparison points out a single pair-wise comparison at a time. Table 21.2 shows the scale of relative importance, according to Saaty.

After pair-wise comparison the next step is to determine the relative importance of those comparisons. So, one has to estimate the eigenvalue of the matrix using the geometric mean of each row, followed by the normalization of these numbers dividing them individually by their sum. The consistency of comparisons is done by the Consistency Ratio (CR) which must be less than 0.1. Closer the values of CR to zero better the judgement consistency.

Prior to determine CR, we must estimate the Consistency Index (CI) by:

$$CI = \frac{\lambda \max - n}{n - 1} \tag{21.2}$$

where: λ_{max} = maximum eigenvalue and n = number of items to be compared

The Consistency Ratio (CR) is then determined dividing the Consistency Index (CI) by the Random Consistency Index (RCI), of Table 21.3 introduced by Saaty as an upper limit on how inconsistency may be tolerated in a decision process [9].

$$CR = \frac{CI}{RCI} \tag{21.3}$$

If CR is higher than the referred limit value a new pair-wise comparison must be done reformulating the used judgement.

After alternatives comparison for each criteria the priority vectors became the columns of the decision matrix where the final priorities of the alternatives for all criteria are determined by:

$$Pi = \sum_{j=1}^{N} ai, j.Wj \tag{21.4}$$

for i = 1, 2, 3, ..., M

21.3 Maintainability Background

Maintainability, as a probability, can be mathematically analysed and times to repair statistically allocated to a specific distribution. Mathematical models applied to reliability can also be used to quantify maintainability [20]. Statistical distributions used in maintainability studies are characterized by a restore probability density function. It is common to fit the available data to one of the following statistical distributions:

(a) Weibull;
(b) Exponential;
(c) Lognormal;
(d) Normal.

In maintainability the Weibull distribution is the widest applied and the most accurate and practical among all distributions and covers almost practical case studies. This is possible due to the shape parameter influence. This parameter will define the shape of the restore probability density function [g(t)]. It also defines maintainability deviation during component or system life. Exponential distribution can be considered a particular case of Weibull distribution when shape parameter assumes the value 1 ($\varepsilon = 1$). It is applied when restore rate is constant. Lognormal distribution is mainly applied when a significant restore value occurs at the beginning of repairs and few at their end (asymmetric to a medium value). Normal distribution is applied when restores occur around a medium value. This distribution is characterized by a symmetric shape and is quite similar to Weibull distribution when shape parameter assumes a value of 3.4 ($\varepsilon \approx 3.4$).

As Lognormal, Normal or Exponential distributions can be considered specific situations of the Weibull distribution one must assume and justify Weibull distribution as the starting point for maintainability studies.

21.3.1 Maintainability

Starting from the definition of Weibull restore probability density tri-parametric function:

$$g(t) = \varepsilon.{\tau_0}^{\varepsilon}.(t-\gamma)^{\varepsilon-1}.e^{-[\tau 0.(t-\gamma)]^{\varepsilon}} \tag{21.5}$$

Weibull maintainability function can be presented as:

$$M(t) = \int_{-\infty}^{t} g(t).dt \tag{21.6}$$

$$M(t) = 1 - e^{-[(\frac{t-\gamma}{\delta})]^{\varepsilon}} \tag{21.7}$$

where:

t = Variable Time (hours, days, weeks, ...);
γ = Position parameter;
ε = Shape parameter;
δ = Scale parameter

21.3.2 Maintainability Improvement

Maintainability of some components (units or systems) can be improved. This can be done modifying the design, the process or any other conditions that change their restore operational characteristics. Maintainability improvement can be expressed by [21]:

$$[\mathrm{M}] = M_B - M_A \tag{21.8}$$

If:

[M] > 0 better maintainability
[M] = 0 no change in maintainability
[M] < 0 worse maintainability

21.4 Proposed Methodology

Based on the above theory, a methodology is proposed to analyse maintainability improvements in a mathematically way, using the well-known Weibull distribution. This methodology has the following steps:

1. Equipment/System definition and description;
2. Selection of the most important repairable unit;
3. Restore times analysis (Weibull);
4. Selection of the most important attribute to improve maintainability;
5. Maintainability improvement (actions);
6. Maintainability improvement analysis (Weibull).

Table 21.4 Analytic Hierarchy Process structure

Objective	Select the most important attribute for maintainability	
Criteria	Logistics costs	LC
	Production loss	PL
	Safety impact	SI
	Environmental impact	EI
	Fluid loss	FL
Alternatives (attributes)	Accessibility	AC
	Interchangeability	IC
	Modularization	MD
	Standardization	ST
	Human competencies	HC
	Tools	TL
	Spare availability	SA
	Logistics	LG
	Detectability	DT

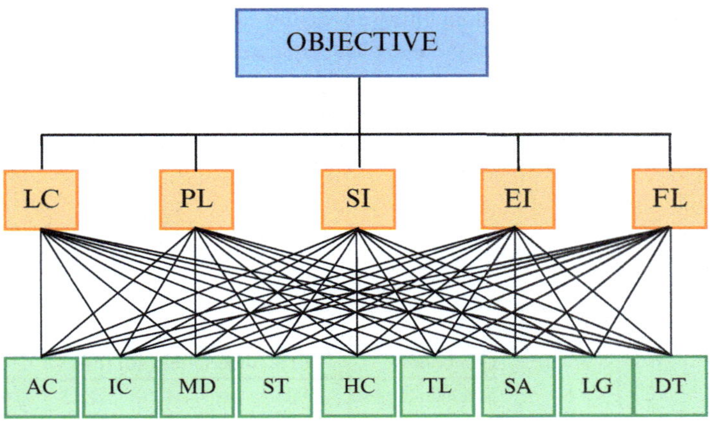

Fig. 21.1 Schematic AHP

In the first step a description of the equipment or system is done. Then, the most important repairable unit in terms of maintainability belonging to the system previously selected is determined and data concerning historic values related to times to restore is managed using Weibull distribution. Upon this analysis the maintainability function is known. The next step of the proposed methodology is to apply the AHP to this particular unit. In this methodology the AHP has the structure proposed in Table 21.4, and the corresponding Fig. 21.1.

Now it is possible to identify the best alternative to improve maintainability and their relative importance in improvement, regarding all other alternatives. The following step is to suggest modifications in design or process in a way to improve the discussed item maintainability. Finally, a new maintainability analysis is done to compare maintainability values before and after improvements and some conclusions are made.

21.5 Demonstrative Example

Data used in this work was collected in a petrochemical industry and is related with centrifugal pumps used for injection of chemical products. Due to its importance this kind of unit has a high level of maintenance care. The failure of this equipment drives to a serious failure of the system and therefore of all process. After the failure modes and inherent repairs were analysed and after having interviewed the maintenance crew it was concluded that the main maintenance problem for these centrifugal pumps are the mechanical seals. So, our study was focused on mechanical seals due its importance. Next step of the methodology concerns the data analysis related with times to repair (TTR) and other significant information (Table 21.5).

Assuming these values and using commercial software [22], the adherence test shows the bi-parametric Weibull distribution as the most adequate with $\varepsilon = 2.3439$ (shape parameter) and $\delta = 13.2101$ (scale parameter). As it was pointed out in paragraph 1 availability is related with probability of survive (reliability) and restore probability (maintainability). Availability can then be enhanced if better values of reliability and maintainability are achieved. Reliability improvements can be done acting in a better design, using sensors and/or redundancies. Maintainability improvements can be done changing design and acting in maintenance. Step four of the proposed methodology concerns the selection of the most important attribute (or alternative) to improve maintainability. For that purpose we use the AHP method previously explained. Based on the first pair-wise comparison between the criteria established and presented in Table 21.4, the hierarchy for those elements under study was achieved, as it can be seen in Table 21.6.

Table 21.5 Historic data

Component	TTR (time to restore) [h]					
Mechanical seal	2.0	6.0	12.0	13.5	14.5	16.5
	3.0	9.0	12.0	14.0	15.0	16.5
	3.0	10.0	12.0	14.0	15.0	17.0
	3.0	10.0	12.0	14.0	15.0	17.0
	3.0	10.0	12.5	14.0	15.0	18.0
	4.0	10.5	13.0	14.0	15.5	
	4.0	10.5	13.0	14.0	15.5	
	5.0	11.0	13.0	14.5	16.0	
	5.0	11.0	13.0	14.5	16.0	
	5.0	11.5	13.0	14.5	16.0	

Table 21.6 Criteria hierarchy

LC	=	Logistics costs	=	0.0498
PL	=	Production loss	=	0.4915
SI	=	Safety impact	=	0.2296
EI	=	Environmental impact	=	0.0786
FL	=	Fluid loss	=	0.1506

So, the first most important criterion for maintainability of mechanical seals of these centrifugal pumps is the production loss, followed by the safety impact. The least important criterion is the logistics costs. To each criterion was performed a pair-wise comparison between alternatives under study. This is the main part of this process that can be crucial for the results to obtain and recommendations to be taken in the future concerning maintainability improvement. From the hard work of making these pair-wise comparisons some results were obtained. Table 21.7 states these results, showing the four most important attributes (and respective values covering approx. 75% of importance) for each criteria.

All the values achieved allow to establish the desired hierarchy, taking into account all criteria and alternatives to select the most important attribute in order to act and improve maintainability. From all process calculations the results are shown in Table 21.8.

Table 21.7 Ranking attribute

	First attribute	Value
Logistics costs	Logistics	0.2644
	Standardization	0.2597
	Interchangeability	0.1476
	Modularization	0.1181
Production loss	Spares availability	0.3085
	Accessibility	0.1859
	Human competencies	0.1497
	Modularization	0.1325
Safety impact	Detectability	0.2934
	Human competencies	0.2355
	Accessibility	0.1358
	Tools	0.1280
Environmental impact	Human competencies	0.2431
	Detectability	0.2291
	Tools	0.1330
	Accessibility	0.1316
Fluid loss	Detectability	0.3206
	Accessibility	0.2141
	Standardization	0.1043
	Modularization	0.1043

Table 21.8 Attributes hierarchy

Spares availability	=	0.1826
Accessibility	=	0.1665
Human competencies	=	0.1556
Detectability	=	0.1490
Modularization	=	0.1053
Tools	=	0.0806
Standardization	=	0.0677
Interchangeability	=	0.0556
Logistics	=	0.0370

In order to finish the fourth step it remains only to verify if the judgements consistency for each matrix. According to formula (21.4) we reached in every situations a Consistency Ratio (CR) less than 0.1, validating the process.

Next step is a consequence of the results previously obtained. In this particular case study, although not detected a very strong attribute comparing the ranking obtained, were suggested some improvements in maintainability of mechanical seals and thus assuring higher availability for the centrifugal pumps.

Analysing this situation it was established a compromise to act in three areas, namely: (1) Change the type of mechanical seal (not so complex as the previous) and provide a minimum stock to face possible failures avoiding the time lost to contact supplier and wait for equipment delivery, (2) give technical formation to workers to improve personnel technical skills and (3) promote better access to mechanical seal, mainly to assure a clean area to operate if necessary.

Assuming these improvements, and after a data collecting period with new restore times, a new maintainability analysis was performed (similar to the previous analysis). Assuming the bi-parametric Weibull distribution as the most adequate, the new parameters are $\varepsilon = 2.9587$ (shape parameter) and $\delta = 4.5533$ (scale parameter). This shows a great progress in maintainability, which can be observed by the decrease of time restore dispersion (higher shape parameter) and huge reduction in characteristic restore time (lower scale parameter).

Now, instead of a probability of 0.5941 to restore a failed mechanical seal in 10 min we have now the same probability but to restore in 3.65 min. In the same way the time related to a restore probability of 0.90, with a confidence level of 95% (bilateral is now 2.13 min with 2.83 min in upper bound and 1.60 min in lower bound).

It needs to be referred that an economical study must always be done and this positive maintainability and availability results must be analysed comparing the cost observed to achieve such maintainability improvements. This is important because roughly 38% of the life cycle cost is directed toward maintainability issues [23].

21.6 Conclusions

In availability analysis the main contributing factor is usually considered to be the reliability factor. The effect of maintainability on availability is not so well studied although in some real cases it can have a significant weight. In this paper it was shown that maintainability can be a very important issue, especially in high risk industrial equipment. It was presented a methodology to study the maintainability of an item using a multi-criteria decision making process (AHP), just to determine which factor or attribute is the most important to a specific situation, and then act to improve maintainability. As it was also exposed by a practical application, maintainability improvements can have high effect in system maintainability and subsequent availability. When dealing with high risk industrial equipment related with safety (people, equipment and environment) or huge economic losses due to unavailability these kinds of studies may be very useful.

References

1. Department of Defense (DoD) (1997) MIL-HDBK-470A Designing and Developing maintainable Products and Systems – vol 1, Department of Defense, USA. http://www.weibull.com/mil_std/mil_hdbk_470a.pdf. Accessed 12 Jul 2011
2. Endrenyi J et al (2001) The present status of maintenance strategies and the impact of maintenance on reliability. IEEE Trans Power Syst 16(4):638–646
3. Eisinger F (2008) Risk based maintenance prioritisation for safety functions. In: Proceedings of the 34th ESReDA seminar, Lecture 15, San Sebastian
4. ReliaSoft (2003) Relationship between availability and reliability, reliability HotWire. Issue 26 (April), ReliaSoft. http://www.weibull.com/hotwire/issue26/relbasics26.htm. Accessed 22 Aug 2011
5. O'Neill G (2005) Maintainability: Theory and Practice, Paper presented at ISHEM - First International Forum on Integrated System Health Engineering and Management in Aerospace, California, USA, 7–10 Nov 2005
6. Sharma R, Kumar S (2008) Performance modelling in critical engineering systems using RAM analysis. Reliab Eng Syst Saf 93:891–897
7. Coulibaly A, Houssin R, Mutel B (2008) Maintainability and safety indicators at design stage for mechanical products. Comput Ind 59:438–449
8. Zhong L, Youchao S (2007) Research on maintainability evaluation model based on fuzzy theory. Chin J Aeronaut 20:402–407
9. Triantaphyllou E et al (1997) Determining the most important criteria in maintenance decision making. J Qual Maint Eng 3(1)
10. Bevilacqua M, Braglia M (2000) The analytic hierarchy process applied to maintenance strategy selection. Reliab Eng Syst Saf 70:71–83
11. Bertolini M, Bevilacqua M (2006) A combined goal programming – AHP approach to maintenance selection problem. Reliab Eng Syst Saf 91:839–848
12. Dey P (2004) Decision support system for inspection and maintenance: a case study of Oil pipelines. IEEE Trans Eng Manag 51(1):47–56
13. Zio E, Cantarella M, Cammi A (2003) The analytic hierarchy process as a systematic approach to the identification of important parameters for the reliability assessment of passive systems. Nucl Eng Des 226:311–336
14. Hijes F, Cartagena J (2006) Maintenance strategy based on a multicriterion classification of equipments. Reliab Eng Syst Saf 91:444–451
15. Wang L, Chu J, Wu J (2007) Selection of optimum maintenance strategies based on a fuzzy analytic hierarchy process. Int J Prod Econ 107:151–163
16. Al-Najjar B, Alsyouf I (2003) Selecting the most efficient maintenance approach using fuzzy multiple criteria decision making. Int J Prod Econ 84:85–100
17. Dagdeviren M, Yuksel I (2008) Developing a fuzzy analytic hierarchy process (AHP) model for behaviour-based safety management. Inform Sci 178:1717–1733
18. Vaidya O, Kumar S (2006) Analytic hierarchy process: an overview of applications. Eur J Oper Res 169:1–29
19. Saaty T (1980) The analytic hierarchy process. McGraw-Hill, New York
20. Monchy F (1996) La Fonction Maintenance – Formation à la gestion de la maintenance industrielle, Masson
21. Pallerosi C (2007) Mantenabilidade e Disponibilidade, vol 5. ReliaSoft Brasil, Brasil
22. Software Weibull ++7 (2007) ReliaSoft Corporation, Version 7.5.1
23. Barringer (2001) Problem of the month – Maintainability. http://www.barringer1.com/jul01prb_files/jul01prb.pdf. Accessed 18 May 2011

Chapter 22
The Classification Platform Applied to Mammographic Images

P.J.S. Gonçalves

Abstract In this paper is tackled the mammographic image classification problem. From the previous developed CLAP – *CLAssification Platform for use with Matlab*, several computational paradigms emphasizing neural networks, support vector machines and fuzzy systems, were used to classify mammographic images in two classes, i.e., with or without tumour. To perform the classification task, features must be extracted from the mammographic images. Amongst the methods implemented in CLAP, features obtained from the co-occurrence matrix and wavelets were used, to describe the texture of the region of interest in the image. Results obtained while training and validating the mentioned computational paradigms, show that support vector machines outperform the other two types of classifiers, independently of the features selected.

22.1 Introduction

The CLAP – CLAssification Platform [1], is a software tool designed to classify images. CLAP can extract features from images in order to organize the images, e.g., mammographic images with or without tumour. CLAP is suitable to obtain data for classification purposes using images of any type, and then classify them.

This paper applies the CLAP to breast cancer diagnosis to help medical doctors in diagnostic tasks, using mammographic images. In this paper, mammograms were classified in two classes: with tumour and without tumour.

P.J.S. Gonçalves (✉)
School of Technology, Polytechnic Institute of Castelo Branco, Av. Empresário, Castelo Branco 6000-767, Portugal

IDMEC/LAETA, Technical University of Lisbon (TU Lisbon), Av. Rovisco Pais, Lisbon 1049-001, Portugal
e-mail: paulo.goncalves@ipcb.pt

A. Madureira et al. (eds.), *Computational Intelligence and Decision Making: Trends and Applications*, Intelligent Systems, Control and Automation: Science and Engineering 61, DOI 10.1007/978-94-007-4722-7_22,

From the diagnostic data, mammograms, the software tool can propose to the medical doctor a possible assessment. The basis of the tool is twofold. The first part is related to image processing algorithms that outputs the descriptors used in the second part, which is related to classification.

The first part of the CLAP extracts the image features from the captured image, a mammogram. The algorithms used in this paper are the wavelet decomposition for texture and the gray-level co-occurrence matrix (GLCM). These algorithms give a large number of features that are used in the second part of the system. The features to be used can be chosen by the user.

The second part of the system uses the image features obtained in the image processing part of the system. The soft computing systems implemented in the platform are based in fuzzy logic, neural networks and support vector machines. From fuzzy logic theory, the system model was built using takagi-sugeno fuzzy models, where clustering is performed using Fuzzy C-Means and Gustafson-Kessel. The neural networks implemented in CLAP are: cascade-forward back-propagation network, Elman back-propagation network, feed-forward back-propagation network, probabilistic neural network and learning vector quantization network. The support vector machines based classification can be performed using linear and quadratic penalization of misclassified examples, and classification with nearest point algorithm.

The database used was the Mammographic Image Analysis Society Database, MIAS [2], to validate both, the image processing and the classification parts. The database is freely available in the web. After the validation of the two subsystems, the overall system will work with the MIAS database to classify the images with or without tumour.

The CLAP is developed in MATLAB© and can be used by medical doctors in clinical practice, by teachers and students. The major benefit of CLAP is that has implemented a large number of algorithms that the user can choose from, to perform a valid diagnose.

The paper is organized as follows. Section 22.2 describes briefly image features extracted from the mammographic images. Section 22.3 presents very briefly the computational paradigms used in this paper: neural networks, support vector machines and fuzzy systems. Section 22.4 presents the mammographic images and the performance indexes, used to evaluate the quality of the classifiers estimated, and also the results obtained are discussed. Finally, Sect. 22.5 presents the conclusions and the future work.

22.2 Image Features

Visual feature extraction applied to content based image retrieval has been thoroughly studied for the last years. Most work concentrates on low level visual features such as colour, shape, texture, etc. Classical segmentation algorithms use often edges and/or texture as useful and discriminating features or attributes. This is still an open research field. In this work, we implemented, wavelet decomposition for both texture [3], gray-level co-occurrence matrix [4].

22.2.1 Co-occurrence Matrix

Grey-Level Co-occurrence Matrix (GLCM) texture measurements have been the workhorse of image texture since they were proposed by Haralick in the 1970s [4]. Whether considering the intensity or grayscale values of the image or various dimensions of colour, the co-occurrence matrix can measure the texture of the image. Because co-occurrence matrices are typically large and sparse, various metrics of the matrix are often taken to get a more useful set of features. Features generated using this technique are usually called co-occurrence (Haralick) features. The GLCM estimates the properties of the image related with second order statistics, i.e., the measures consider the relationship between groups of two neighbouring pixels in the original image. First order texture measures are statistics calculated from the original image values, like variance, and do not consider pixel neighbour relationships. Thus, for each spacial relation considered in the image, one GLCM is obtained. In this paper, are presented the results obtained from four spacial relations, i.e., the vectors $[4, 0]$, $[4, 4]$, $[0, 4]$, $[-4, 4]$. These directions were chosen based on the results of dozens of tests performed with the available mammographic image data. Increasing the size of the vectors and even choosing a lower direction step, e.g., 22.5°, will increase the complexity of the GLCM and its computational cost, without showing better classification results.

22.2.2 Wavelet Filters

The use of wavelet transform has been widely used for texture analysis [3]. This transform provides a robust methodology for texture analysis in different scales. The wavelet transform allows for the decomposition of a signal using a series of elemental functions called wavelets and scaling. In practice, wavelets are applied as high-pass filters, while scalings are equal to low-pass filters. As a result of this, the wavelet transform decomposes the original image into a series of images with different scales. In this paper, are presented the results obtained with the Spline BiOrtoghonal 3.5 wavelet, e.g. 'bior3.5'. Other wavelets available in CLAP, i.e., the ones available in the Fuzzy Logic Toolbox for MatLab, were tested during the dozens of tests performed with the available mammographic image data. Bior 3.5 presented the best results.

22.3 Classification Paradigms

Classification is an important subject for a variety of fields, including pattern recognition, artificial intelligence, and vision analysis. In a classification problem the aim is to learn the behaviour between the input and output of the training data to classify any given new input data of the same type. CLAP – CLAssification Platform, currently, have three different types of classifiers, which are based in fuzzy systems, neural networks and support vector machines. The classifiers based in Fuzzy Logic and Neural Networks were based in the correspondent Matlab toolboxes. Thus it is very likely that this toolbox will continue to evolve, making our toolbox very interesting from that point of view.

22.3.1 Neural Networks

Neural networks have found profound success in the area of pattern recognition, function approximation, optimization, pattern matching and associative memories. By repeatedly showing a neural network inputs classified into groups, the network can be trained to discern the criteria used to classify, and it can do so in a generalized manner allowing successful classification of new inputs not used during training [5]. In CLAP are implemented the following types of neural networks (NN): cascade-forward backpropagation network (newcf), Elman backpropagation network (elm), feedforward backpropagation network (ff), probabilistic neural network (pnn), learning vector quantization network (lvq) and self-organizing map network (som). The later algorithm showed no theoretical improvement to the task at hand. This result is based on the dozens of tests performed.

22.3.2 Fuzzy Models

From the modelling techniques based on soft computing, fuzzy modelling is one of the most appealing. If no a priori knowledge is available, the rules and membership functions can be directly extracted from process measurement. Fuzzy models provide a transparent description of the system, which can reflect a possible nonlinearity of the system. The fuzzy models implemented in the CLAP are Takagi-Sugeno fuzzy models [6] where the consequents are crisp functions of the antecedent variables and linguistic or Mandani [7, 8] fuzzy models where both the antecedent and consequent are fuzzy propositions. Different classes of fuzzy clustering algorithms can be used to approximate a set of data by local models. Each of these models is represented by a fuzzy subset in the data set available for identification. In this paper, two fuzzy clustering methods were used, e.g., Fuzzy C-Means (FCM) [9] and Gustafson-Kessel (GK) [10]. Other clustering algorithms are presented in CLAP, like Possibilistic

C-Means (PCM) [11], Fuzzy Possibilistic C-Means (FPCM) [12], Possibilistic Fuzzy C-Means (PFCM) [13] and subtractive clustering [14]. The later algorithms show no theoretical improvement to the task at hand. The antecedent membership functions can be obtained by projecting the fuzzy partition onto the antecedent variables, or by computing the membership degrees directly in the product space of the antecedent variables. Optimal consequent parameters are estimated by the least-squares method. This method was used, as it aims at the minimization of the global prediction error and yields an optimal predictor.

22.3.3 Support Vector Machines

The support vector machine (SVM) maps an input vector x into a high-dimensional feature space z through some nonlinear mapping, chosen a priori. In this space, an optimal separating hyperplane is constructed. In the case of pattern recognition, SVM perform classification between several classes by finding a decision hyper surface determined by certain points of the training set, termed Support Vectors [15]. This part of the CLAP, is based in the toolbox presented in [16]. In this paper were used the *oneaone* – one against one, *oneaall* – one against all, $\{2d\}$ and *ls* – large scale SVM based approaches implemented in CLAP, that are theoretical candidates for the classification task at hand.

22.4 Experimental Results

The CLAP – Classification Platform, was applied to mammographic images using the MIAS database [2]. This image database was chosen because it is nowadays an accepted benchmark for image based classification tools, e.g., as confirmed by the 400 citations in google scholar.

The MIAS database, from the Mammographic Image Analysis Society [2], is a free database available on the web for research purposes. The database has 322 images, divided in normal or with abnormalities. The abnormalities present are: calcification, well-defined/circumscribed masses, speculated masses, other or ill-defined masses, architectural distortion, and asymmetry. The abnormal mammograms, i.e., with a tumour, are further divided in malignant or benign. In this paper, the authors applied the CLAP – classification platform to classify mammograms in Normal or Abnormal, i.e., the classes Negative and Positive, respectively.

The remainder of the section presents the training and validation results for the different classification paradigms used, based on the features extracted using the co-occurrence matrix, and wavelets. The classification is performed in a small subset of the MIAS database, with 60 images, that can be easily extended to the all database. In Fig. 22.1a, b are presented two images of each class, i.e., without or with tumour. In Fig. 22.2a, b are presented the region of interest (ROI) of both

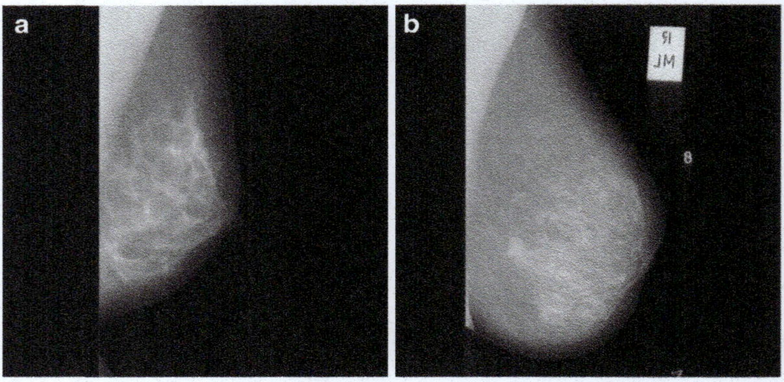

Fig. 22.1 Mammographic Images. (**a**) without tumour (*left*). (**b**) with tumour (*right*)

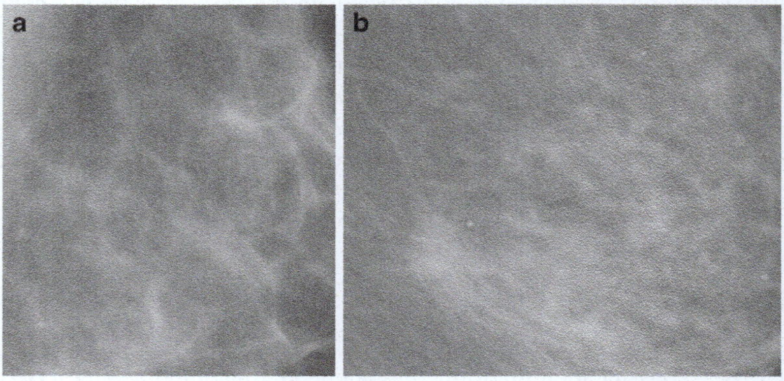

Fig. 22.2 ROI of Mammographic Images. (**a**) without tumour (*left*). (**b**) with tumour (*right*)

images used to extract the features. These images belong to the validation dataset, composed by five images.

22.4.1 Performance Indexes

After obtaining the features, using wavelets and the co-occurrence matrix, this data must be prepared to train the classifiers. Classification algorithms often use distance norms, which are sensitive to variations in numerical ranges of the different features. The Euclidean distance, for example, assigns more weighting to features with wide ranges than to those with narrow ranges. The result of clustering can thus be negatively influenced when choosing different measurement units. A linear transformation that moves the origin and gives a uniform change of scale, i.e., not changing the information enclosed is used in CLAP. This transformation guarantees that all the features are between zero and the unity.

Fig. 22.3 Confusion matrix

		Classifier Output		
		Positive	Negative	Total
True diagnosis	Positive	a	b	a + b
	Negative	c	d	c + d
	Total	a + c	b + d	N

Having trained the classifier, the next step is to assure the quality of the developed model, i.e., to evaluate its performance. For this purpose, the indexes true positive (*TP*) and true negative (*TN*), obtained from the confusion matrix, Fig. 22.3, were used. These indexes are defined in (22.1) and (22.2).

$$TP = \frac{a}{a+b} \times 100\% \tag{22.1}$$

$$TN = \frac{d}{c+d} \times 100\% \tag{22.2}$$

22.4.2 *Training the Models*

The training dataset, is a subset of the 60 images used in this paper of the MIAS database. The training data, i.e., the co-occurrence or the wavelet features, was obtained from the first 55 images of the MIAS database [2].

Tables 22.1, 22.2, and 22.3, present the performance indexes obtained using the classification paradigms presented in Sect. 22.3, i.e., Fuzzy Modelling (Sugeno and Mamdani), Support Vector Machines (SVM) and Neural Networks (NN). These tables also present the results using the co-occurrence features or the wavelets.

Several conclusions can be drawn form these tables. Fuzzy modelling presented the worst results for the classification of the training data, being wavelets the best features in this case. For the SVM model, the co-occurrence features are best suited for the training test, having 100% for (*TP*) and (*TN*). The Probabilistic Neural Network (NN-pnn) is the best classifier of its type having also 100% for (*TP*) and (*TN*). Thus, several classifiers are candidates to be used in classification task for the validation data. The candidates are all the classifiers that obtained 100% for (*TP*) and (*TN*) in Tables 22.1, 22.2, and 22.3, e.g., SVM-oneaone using co-occurrence features.

22.4.3 *Validating the Models*

The validation dataset, is a subset of the 60 images used in this paper of the MIAS database. The training data, i.e., the co-occurrence or the wavelet features, was obtained from the images number 56–60 of the MIAS database [2].

Table 22.1 Results obtained for the Fuzzy model training

	Sugeno-GK		Mamdani-GK		Sugeno-FCM		Mamdani-FCM	
Training	Wavelet	Co-ocurrence	Wavelet	Co-ocurrence	Wavelet	Co-ocurrence	Wavelet	Co-ocurrence
TP %	18.0	36.7	31.3	26.0	13.3	20.0	28.7	6.0
TN %	78.0	75.8	83.8	81.5	82.3	70.0	78.8	86.5

Table 22.2 Results obtained for the SVM model training

	SVM-oneaone		SVM-oneall		SVM-2d		SVM-ls	
Training	Wavelet	Co-ocurrence	Wavelet	Co-ocurrence	Wavelet	Co-ocurrence	Wavelet	Co-ocurrence
TP %	91.7	100	100	100	91.7	91.7	100	100
TN %	100	100	96.9	100	96.8	100	96.9	100

Table 22.3 Results obtained for the Neural Networks model training

	NN-elm		NN-ff		NN-lvq		NN-pnn	
Training	Wavelet	Co-ocurrence	Wavelet	Co-ocurrence	Wavelet	Co-ocurrence	Wavelet	Co-ocurrence
TP %	10.8	8.3	8.3	7.5	69.2	23.3	100	100
TN %	87.5	98.1	95.3	97.8	97.5	100	100	100

Table 22.4 Results obtained for the Fuzzy model validation

	Sugeno-GK		Mamdani-GK		Sugeno-FCM		Mamdani-FCM	
Training	Wavelet	Co-ocurrence	Wavelet	Co-ocurrence	Wavelet	Co-ocurrence	Wavelet	Co-ocurrence
TP %	50.0	–	20.0	20.0	50.0	20.0	0.0	0.0
TN %	100.0	–	73.3	86.7	33.3	66.7	100	83.3

Table 22.5 Results obtained for the SVM model validation

	SVM-oneaone		SVM-oneall		SVM-2d		SVM-ls	
Training	Wavelet	Co-ocurrence	Wavelet	Co-ocurrence	Wavelet	Co-ocurrence	Wavelet	Co-ocurrence
TP %	50.0	100	50.0	50.0	50.0	50.0	50.0	50.0
TN %	100	66.7	100	100	66.7	66.7	100	100

Tables 22.4, 22.5, and 22.6, present the performance indexes obtained using the models trained in Sect. 22.3, i.e., Fuzzy Modeling (Sugeno and Mamdani), Support Vector Machines (SVM) and Neural Networks (NN).

Several conclusions can be drawn form these tables. Fuzzy modeling and Neural Networks failed to identify the tumours, i.e., the *TP* rate is very low, being the neural networks the worst classifier. The SVM model using the co-occurrence features (SVM-oneaone | co-occurrence) showed the best results for the validation

Table 22.6 Results obtained for the neural networks model validation

	NN-elm		NN-ff		NN-lvq		NN-pnn	
Training	Wavelet	Co-ocurrence	Wavelet	Co-ocurrence	Wavelet	Co-ocurrence	Wavelet	Co-ocurrence
TP %	10.0	0.0	5.0	0.0	30.0	0.0	0.0	0.0
TN %	86.7	100	93.3	93.3	73.3	100	66.7	66.7

images, having 100% for *TP* and 66.7% for *TN*. The SVM models showed the best results when applying the trained models to new mammogram images.

22.5 Conclusions and Future Work

This paper proposes the application of the CLAP – CLAssification Platform to classify mammographic images without or with tumours. Several computational paradigms emphasizing neural networks, support vector machines and fuzzy systems, were used. To perform the classification task, features were extracted from the mammographic images amongst the methods implemented in the CLAP, e.g., the co-occurrence matrix and wavelets.

The obtained experimental results showed that the CLAP – CLAssification Platform can be used to classify mammographic images. It was also demonstrated that the Support Vector Machine based classifier, using the co-occurrence matrix features is the best approach for the mammographic images tested during the training and validation procedures.

As future work, the CLAP – CLAssification Platform will be used in all of its capabilities to the mammographic image classification, e.g., using the already implemented feature extraction and classification algorithms. New feature extraction and classification algorithms are being incorporated to the CLAP – CLAssification Platform.

Acknowledgments To the developers of CLAP – CLAssification Platform at the Center of Intelligent Systems of IDMEC/ LAETA. Also a special thanks for the students at Polytechnic Institute of Castelo Branco, Health Equipment Technology degree, that are using the platform for the past 2 years in their classes. This work was partly supported by the Strategic Project, PEst-OE/EME/LA0022/2011, through FCT (under IDMEC-IST, Research Group: IDMEC/LAETA/CSI).

References

1. Almeida RJ, Vieira S, Sousa JMC, Pinto JRC, Gonçalves PJS (2007) Clap – classification package for use with matlab. In: Proceedings of RecPad 2007 – 13th Conferência Portuguesa de Reconhecimento de Padrões, Lisbon, Portugal, Oct 2007
2. Suckling J, Parker J, Dance D et al (1994) The mammographic image analysis society digital mammogram database. Exerpta Med Int Congr Ser 1069:375–378

3. Alexandre H, Pinto JRC (2006) New wavelets based features for natural surface indexing, VISAPP 2006. In: Proceedings of the first international conference on computer vision theory and applications, vol 1, Setúbal, Portugal, Feb 2006, pp 311–316
4. Haralick RM, Shanmugam K, Dinstein I (1973) Textural features for image classification. IEEE Trans Syst Man Cybern 3(6):610–621
5. Zhang G (2000) Neural networks for classification: a survey. IEEE Trans Syst Man Cybern C Appl Rev 30(4):451–462
6. Takagi T, Sugeno M (1985) Fuzzy identification of systems and its application to modeling and control. IEEE Trans Syst Man Cybern 15(1):116–132
7. Zadeh L (1973) Outline of a new approach to the analysis of complex systems and decision processes. IEEE Trans Syst Man Cybern 3(1):28–44
8. Mamdani E (1977) Application of fuzzy logic to approximate reasoning using linguistic systems. IEEE Trans Comput 26(12):1182–1191
9. Bezdek JC (1981) Pattern recognition with fuzzy objective function algorithms. Plenum Press, New York
10. Gustfson D, Kessel W (1979) Fuzzy clustering with a fuzzy covariance matrix. In: Proceedings of IEEE conference on decision and control. CDC, San Diego, CA, USA, pp 761–766
11. Krishnapuram R, Keller J (1993) A possibilistic approach to clustering. IEEE Trans Fuzzy Syst 1(2):98–110
12. Pal N, Pal K, Bezdek J (1997) A mixed c-means clustering model. In: Proceedings of the sixth IEEE international conference on fuzzy systems, vol 1, Barcelona, Catalonia, July 1997, pp 11–21
13. Pal NR, Pal K, Bezdek JC (2005) A possibilistic fuzzy c-means clustering algorithm. IEEE Trans Fuzzy Syst 13(4):517–530
14. Chiu S (1994) Fuzzy model identification based on cluster estimation. J Intell Fuzzy Syst 2(3):267–278
15. Vapnik VN (1998) Statistical learning theory. A Wiley- Interscience Publication, New York
16. Canu S, Grandvalet Y, Guigue V, Rakotomamonjy A (2005) Svm and kernel methods matlab toolbox. Perception Systmes et Information. INSA de Rouen, Rouen, France. http://asi.insa-rouen.fr/enseignants/~arakotom/toolbox/index.html

Chapter 23
On an Optimization Model for Approximate Nonnegative Matrix Factorization

Ana Maria de Almeida

Abstract Within image and video signal processing several tasks must be performed in order to obtain an efficient and low-dimensional encoding for transmission. Studies have shown that Nonnegative Factorization of the data Matrix (NFM) is a suitable tool for achieving ranking reduction and sparse coding. Thus, developing efficient optimization algorithmic methods for video/image encoding presents a key technical challenge in the effort to minimize distortion and obtain good quality reduced encodings. In this paper, we propose a discrete optimization framework where NMF is formulated as a nonlinear programming optimization problem, and apply an Spectral Projected Gradient (SPG) algorithm to this problem. The empirical experience indicates that the proposed approach can, in fact, produce good quality local minima.

Keywords Nonnegative Matrix Factorization • Constrained Large Scale Optimization • Image Signal Processing

23.1 Introduction

The focus of this work lies in an optimization problem that can be applied to several and potentially different applications. Nevertheless, our main motivation and examples will come from the area of Video and Image Signal Processing. In some models for physical phenomena, the data to be analyzed is non–negative in nature. Consequently there's a need to apply processing tools that assure that the non–negativity of the data is preserved in order to maintain an accurate interpretation of the captured phenomena.

This processing involves dimensionality reduction and subsequent reconstruction of the data, i.e., the object that it originally represented. The mathematical

A.M. de Almeida (✉)
CISUC – Center for Informatics and Systems of the University of Coimbra, Coimbra, Portugal

A. Madureira et al. (eds.), *Computational Intelligence and Decision Making: Trends and Applications*, Intelligent Systems, Control and Automation: Science and Engineering 61, DOI 10.1007/978-94-007-4722-7_23,

model that allows for the necessary preservation of non–negativity while achieving low–rank factorization that will be the basis for this work can be simply described as: Given a matrix $V \geq 0 \in R^{m \times n}$, find a decomposition $V = WH$, where $W \in R^{m \times r}$, $H \in R^{r \times n}$ and such that $W \geq 0$ and $H \geq 0$. Therefore, this problem is commonly termed as Nonnegative Matrix Factorization, or NMF for simplicity. The value r is the decomposition parameter called factorization rank.

Nonnegative Matrix Factorization has been intensively applied in diverse fields within data mining and machine learning, noise reduction and spectral data analysis. In fact, in all of these applications, V is no more than a data matrix and this type of factorization becomes necessary, not only because of hardware limitations for data analysis that imply some reduction of dimensionality but also because the required low–rank representation must still be non–negative in order to retain the interpretability of the physical realities that the data is supposed to model. This is the case within video and image signal processing, where each data matrix V is made up of several images, showing a composite object in many articulations or poses, or a sequence of video frames. In a simplified way this means that W represents the basis elements for the image space and H denotes the coefficient sequences representing n images in the basis elements. NMF as a decomposition technique is unique in the fact that completely obeys the nonnegative constraints, which with images is especially important since a pixel presenting a negative value has no interpretation whatsoever.

Browsing the related literature we find that Alternating Constrained Least Squares methods were used right at the introduction of NMF, in a work dating from 1994 by Paatero and Tapper [17]. The authors attempted to perform analysis on measurements taken from environmental data. Their analysis was merely empirical and their work did not present any theoretical foundations for the use of NMF or the algorithm that they presented. A few years later Lee and Seung [15] introduced the concept of NMF in a machine learning environment, using a multiplicative update rules algorithm as a Fixed Point approach to extract semantic features from images like basic patterns for faces (eyes, lips, eyebrows, etc.). The authors also provided a proof of convergence for their algorithm, however C. Lin later corrected this proof. Since the previous demonstration only worked for the feasible interior points, Lin extended the proof to the frontier of the feasible region [16]. Since then a surge of works has been published with extensions and new applications of this algorithm or technique. For instance, Plemmons and its group consistently used unconstrained Newton-type methods to find stationary points [1, 4]. Finally, Chichoki, Zudnek and others have been recently exploring the use of fast projected gradient descent methods with considerable success in their experiments [19]. This is especially true for the area of blind image separation, where, over the last 3 years, they have contributed with important results, namely for approximating Tensor Factorizations [7].

The paper is organized as follows: the next section presents the definition of an NMF optimization problem using the most common formulation as well as highlights some of the issues concerning exact and approximate matrices decompositions. In Sect. 23.2.3 we introduce the new approach: the Spectral

Projected Gradient Algorithm for NMF. Next we present some computational experiments, ending with Sect. 23.3 were conclusions are drawn and trends for future work are discussed.

23.2 NMF as an Optimization Problem

In order to obtain an exact factorization, the parameter r must obey some necessary conditions, namely it must be, at least, equal to the positive rank of V (see Sect. 23.2.1). Due to the fact that, in general, the determination of the value of the positive rank is still an open problem, there exists no known exact algorithmic factorization method for NMF. Usually the approach used is that of formulating NMF as an approximation optimization problem for which r is a parameter. The most common formulation is the minimization of the Fröbenius norm of the decomposition, that is, given a nonnegative matrix $V \in R^{m \times n}$,

$$\begin{aligned} \phi(W,H) &= \min \frac{1}{2} \|V - WH\|_F^2 \\ s.t.\ W_{n \times r} &\geq 0,\ H_{r \times m} \geq 0. \end{aligned} \tag{23.1}$$

The solution W H is an approximate factorization of rank r for V. The choice of the value for the parameter r is usually problem dependent, being generally chosen so that $1 < r \ll \min\{m, n\}$ and therefore WH can be thought of as a compressed form of V.

There are $(n + m)r$ variables in formulation (23.1), which implies that the bigger the dimension of the data matrix the larger the scale of the problem resolution. Other important challenge affecting the numerical optimization of the formulation (23.1) comprises the existence of local minima due to the non convexity of the objective function in both W and H. The fact that φ is not a convex function implies that a stationary point is not necessarily a minimal solution for (23.1). Nevertheless, (23.1) constitutes a global optimization problem for which the minimum possible value is known since zero is the global optimum[1] for ϕ.

23.2.1 Preliminaries

Let's begin by noting that whenever there is one exact factorization there are an infinity of possible pairs for the decomposition since if (W^*, H^*) is a solution for

[1] Again, a good estimation for the value of the positive rank is especially important within this context. When we know that exact factorizations exist, then we know an optimal lower bound to use for algorithmic performance evaluation.

NMF so are all the pairs of the form $(W^*D, D^{-1}H^*)$, for any nonnegative invertible matrix D. Of course that the important question is not about how many decompositions are there but when does a decomposition exist. The fact that the existence of exact factorization for a given matrix V is dependent on the factorization parameter r is a classical result (see reference [8]).

Definition 1. Given $V^{m\times n} \geq 0$, the minimum r such that there exists $W^{m\times r}$, $H^{r\times n}$ for the exact factorization $V = WH$, is called the positive rank for V and denoted as *posrank*(V).

Theorem 1. Given $V^{m\times n} \geq 0$, $\mathrm{rank}(V) \leq \mathrm{posrank}(V) \leq \min\{m, n\}$.

The optimum or the exact factorization, can be achieved fixing $r \geq \mathrm{posrank}(V)$. In fact, the lowest rank exact factorization can only be obtained by using $r = \mathrm{posrank}(V)$. For values below this limit, only approximate solutions exist.

For some applications, the question of exactness is not an issue. In fact, when noise reduction is intended the approximation may constitute a better solution than an exact factorization, which would probably reproduce the extra signals that should have been eliminated! For image recognition, since our eyes cannot detect but a limited degree of distortion, the dimensionality reduction achieved with a smaller value for r may overcame the loss produced by the approximate factorization. Nevertheless, for accurate feature extraction, an exact factorization implies that all the basic features were, in fact, extracted, which can constitute a major tool when, for instance, security is at stake. However, the determination of the value posrank is still unknown and, presently, considered a very difficult problem on its own [9].

Notice that another important (and common) issue raised by optimization techniques is the generation of the iterative initial point. From the related literature, it can be empirically concluded that, in general, the random generation of initial points tends to perform with final good quality measures. However, this approach does not give any guarantee for obtaining a stationary point for any merit function that can be used for the optimization.

23.2.2 Narrowing the Optimization Model for Approximate NMF: Projected Gradient and Active-Set methods

The use of constrained optimization methods for solving problem (23.1) for large scale problems may be most suited than unconstrained optimization. However, Lee-Seung type approaches for large-scale matrices are not efficient (namely for general image processing). Multiplicative algorithms are known to be very slowly convergent for large-scale problems. Moreover, they easily get stuck in local minima. Since many of the applications for which NMF is being proposed as a processing tool must handle large data matrices there is a pressing need to overcome the complexity issues encountered. Furthermore, when pursuing low– rank solutions the extracted features need to produce recovered (reconstructed) images of good quality.

In recent years, there has been a considerable growth of interest in Projected Gradient (PG) methods: they are highly efficient in solving large-scale minimization problems subject to linear constraints. Nonnegative Matrix Factorization (NMF) of large matrices is a perfect model for this class of problems. Since Lin's work [16] many other works were published attempting to present faster and better Projected Gradient (PG) methods or Projected Alternating Least-Squares (ALS) (among other we can refer [5, 6, 13, 16]). Although all the mentioned methods use frameworks were some projection is required, especially to insure the non-negativity of the constraints, this is still yet a major issue to be tackled. Projected gradient methods and projected quasi-Newton methods [14] apply traditional techniques for unconstrained optimization with modifications for accommodating the non-negativity constraints. Active set methods [12] search for the optimal active and passive sets by exchanging a variable at each of the iterations. Either way, the question of which is the best method for obtaining a projection and insuring that the projected direction has a good convergence rate has not yet been settled. In particular, the main drawback for the application of PG methods to arbitrary problems is the expensiveness of computing projections in general cases.

Next we will discuss projected gradient methods that are generally expressed by iterative updates when applied to NMF approximate resolution. These iterative updates can be expressed by the following formulas:

$$\begin{aligned} W_{k+1} &= P_\Omega(W_k - \eta W_k D W_k) \\ H_{k-1} &= P_\Omega(H_k - \eta H_k D H_k) \end{aligned}$$

where P_Ω (z) is the projection of z onto a feasible set Ω where the nonnegative constraints are enforced. DW_k and DH_k are descent directions for W and H, and ηW_k and ηH_k are learning rules.

One of the most used gradient projection techniques in SPG methods is the Barzilai-Borwein gradient projection. This method is motivated by the quasi-Newton approach, i.e. the inverse of the Hessian is replaced with an identity matrix $(H)_k = \alpha_k$ I, The scalar α_k is of major importance for the spectral method and is generally selected so that the inverse Hessian behaves similarly to the true Hessian in each iteration. Under this particular conjunction, the required projection onto Ω is the spectral projected gradient direction P_Ω $(X_k - \alpha_k\ \phi'_X(W_k, H_k))$ with X standing for W or H and φ' for the function gradient. Thus, α_k is the spectral choice of step length. If we consider $u = x_k - \alpha_k D^{-1} g_k$ then, to obtain the projection onto Ω, we need to solve the following optimization problem:

$$\text{minmize } \left\{ ||u - y||^2_{D_k} : x \in \Omega \right\}. \tag{23.2}$$

This is then subjected to several alternative methods to alleviate the computational burden imposed by yet another optimization problem. Nevertheless, no technique has yet been found that overcomes all the others. In fact, the described approaches and several others have been compared over and over again to find that no algorithm strictly dominates the other. The selection of the more suited algorithm depends mainly on the size of the problem to be solved. However, in the numerical experiments that we designed, projected gradient algorithms seem to behave much better than multiplicative algorithms.

23.2.3 A Particular Optimization Model for Approximate NMF: The Spectral Projected Gradient Method with BPP Projection

An interesting approach to the projection solution and that is especially important for active-set methods is the Block Principal Pivoting (BPP) method [11].

An active set method searches for the optimal active and passive sets by exchanging a variable at each iteration. The BPP tries to overcome this limitation by exchanging several variables per iteration, so that it may find the optimal passive set of variables faster. On the other hand, in contrast to the multiplicative algorithms Projected Gradient methods use additive updates. One particular subclass of PG methods is the Spectral Projected Gradient (SPG) method [2, 3] a most suitable scheme for solving large-scale (convex) optimization problems. The most important characteristic of SPG algorithms is that they only require extremely low memory access, which turns them into very attractive schemes for large scale problems.

In this paper, we present an on-going case-study to adapt the Block Principal Pivoting method into a Spectral Projected Gradient extension for the NMF computation. This method was originally designed for optimization over convex sets like in

$$\begin{aligned} &\min \quad f(x) \\ &s.t. \quad x \in K \end{aligned} \tag{23.3}$$

where $f : \mathrm{R} \rightarrow \mathrm{R}$ and K is a closed convex set. The general algorithm for the SPG method can be simply described as:

Algorithm 1 General SPG Algorithm

For x

1. Determine the project gradient direction

$$d = (P\ x - \eta\ \nabla f(x)) - x$$

2. If d $\approx$ 0 Then x is a stationary point: Stop
3. Find a value α by Exact Line Search
4. x = x + αd.

For each iterative step k = 0, 1, 2, ..., the parameter ηk can change its value accordingly with

$$y_k^T s_k > 0 \Rightarrow \eta_k = \frac{s_k^T s_k}{y_k^T s_k}$$

where $s_k = x_k - x_{k+1}$ and $y_k = \nabla f(x_k) - \nabla f(x_k + 1)$.

This has been proving itself as a robust and efficient approach for a variety of large-scale problems, comparing very positively with commercial codes such as LOQO/AMPL or MINOS/GAMS. We then propose a new framework for NMF by applying the Spectral Projected Gradient method together with a BPP projection to the approximate resolution of problem (23.1).

Algorithm 2 SPGANMF

Given $k = 0$, $W_0 \geq 0$, $H_0 \geq 0$, and $\zeta \in [0, 1]$,

1. Evaluate

$$\nabla_W \phi_k \leftarrow \nabla_W \phi(W_k, H_k)$$
$$\nabla_H \phi_k \leftarrow \nabla_H \phi(W_k, H_k)$$

2. Determine the project gradient direction:

$$(z_W, z_H) \leftarrow P_{BPP}((W_k, H_k) - \eta_k(\nabla_W \phi_k, \nabla_H \phi_k))$$
$$(d_W, d_H) \leftarrow (z_W, z_H) - (W_k, H_k)$$

3. If $\|(d_W, d_H)\| \approx 0$
   Then (W_k, H_k) is a stationary point: *STOP*
4. If $\phi(z_W, z_H) \leq \phi(W_k, H_k) - \zeta(d_W, d_H)^T(\nabla_W \phi_k, \nabla_H \phi_k)$
   Then $\delta_k \leftarrow 1$
   Otherwise determine $\delta_k \in [0, 1]$ by Exact Line Search
5. $(W_{k+1}, H_{k+1}) \leftarrow (W_k, H_k) + \delta_k (d_W, d_H)$
6. $k \leftarrow k + 1$

To achieve this goal we adapted the general SPG algorithm for this particular function and solution set obtaining the Spectral Projected Gradient Algorithm for NMF (SPGANMF).

The BPP has been proved to be a robust and extremely fast projection method and we are now attempting to finer tune the spectral choice with the final goal of obtaining some important results within NMF projected gradient approaches.

23.3 Conclusions

This study proposes a new optimization model for approximate Nonnegative Factorization of nonnegative matrices. The approach here presented uses a particularization of the Spectral Projected Gradient algorithm for the optimization problem (23.1) integrated with a Block Principal Pivoting method. Although we have no assurance that a stationary point will be determined, the empirical experiences already performed indicate that this approach ends up with fair enough good results when random starting points are used. There are, however, a large amount of iterative steps spent under this type of initialization, which seems to indicate that some more work can be done, namely on the determination of the gradient projection and on exact line search technique for computing the step length δk.

The issue of setting or estimating the decomposition parameter r to values $\text{rank}(V) \leq r \leq \min\{m, n\}$ is an important step, either to assure that good quality images can be achieved as well as to determine as few basis as possible in order to obtain good compression of data and exact feature extraction. The values used in our test beds for the decomposition parameter r always assured that an exact decomposition existed.

Nevertheless, good initializations or seeding is an important open problem, even when the exact decompositions are guaranteed. The need of heuristics for setting this question remains independently of the optimization method used.

Sparsity in W and H is quite important, not only for algorithms to efficiently compute the decomposition pair of matrices, but also for subsequent image processing (coding) so that it can decrease the amount of loss of information. It should therefore be further exploited. Some work related to this issue can be found in references like [10, 18].

Acknowledgments The author would like to thank J. J. Júdice for highlighting the possibility of an SPG algorithmic approach for NMF and also for introducing her to the previous works on BPP methods.

References

1. Berry MW, Browne M, Langville A, Pauca VP, Plemmons R (2007) Algorithms and applications for approximate nonnegative matrix factorization. Comput Stat Data Anal 52(1):155–173
2. Birgin EG, Martinez JM, Raydan M (2000) Nonmonnote spectral projected gradient methods on convex sets. SIAM J Optim 10:1196–1211
3. Birgin EG, Martinez JM, Raydan M (2001) Algorithm 813: software for convex-constrained optimization. ACM Trans Math Softw 27:340–349
4. Chu MT, Diele F, Plemmons R, Ragni S (2004) Optimality, computation, and interpretation of nonnegative matrix factorizations, preprint, 2004. http://www.wfu.edu/plemmons
5. Cichocki A, Zdunek R (2006) Multilayer nonnegative matrix factorization. Electron Lett 42(16):947–948
6. Cichocki A, Zdunek R (2007) Multilayer nonnegative matrix factorization using projected gradient approaches. Int J Neural Syst 17:431–446
7. Cichocki A, Zdunek R, Phan AH, Amari S (2009) Nonnegative matrix and tensor factorizations: applications to exploratory multi-way data analysis and blind source separation. John Wiley & Sons, Chichester
8. Cohen JE, Rothblum UG (1993) Nonnegative ranks, decompositions, and factorizations of nonnegative matrices. Linear Algebra Appl 190:149–168

9. de Almeida AM (2011) About nonnegative matrix factorization: on the posrank approximation. In: Dobnikar A, Lotrič U, Šter B (eds) ICANNGA 2011, Part II, LNCS 6594. Springer, Berlin/Heidelberg, pp 295–304
10. Hoyer PO (2004) Nonnegative matrix factorization with sparseness constraits. J Mach Learn Res 5:1457–1469
11. Júdice JJ, Pires FM (1996) A block principal pivoting algorithm for large- scale strictly monotone linear complementarity problems. Comput Optim Appl 5(51):44–77
12. Kim J, Park H (2008) Nonnegative matrix factorization based on alternating nonnegativity constrained least squares and active set method. SIAM J Matrix Anal Appl 30(2):713–730
13. Kim J, Park H (2008) Toward faster nonnegative matrix factorization: a new algorithm and comparisons. In: Proceedings of ICDM'08, eighth IEEE international conference on data mining
14. Kim D, Sra S, Dillon I (2007) Fast newton-type methods for the least squares nonnegative matrix approximation problem. In: Proceedings of the 2007 SIAM international conference on data mining
15. Lee DD, Seung HS (1999) Learning the parts of objects by non–negative matrix factorization. Nature 401:788–791
16. Lin C (2007) Projected gradient methods for non-negative matrix factorization. Neural Comput 19:2756–2779
17. Paatero P, Tapper U (1994) Positive matrix factorization: a nonnegative factor model with optimal utilization of error estimates of data values. Environmetrics 5:111–126
18. Pascual-Montano A, Carabo JM, Kochi K, Lehmann D, Pascual-Marqui R (2006) Nonsmooth nonnegative matrix factorization (nsNMF). IEEE Trans Pattern Anal Mach Intell 28(3): 403–415
19. Zdunek R, Chichoki A (2008) Fast nonnegative matrix factorization algorithms using projected gradient approaches for large-scale problems. Comput Intell Neurosci 2008:939567

Chapter 24
Random Walks in Electric Networks

D.M.L.D. Rasteiro

Abstract The problem considered is the optimization of the value of a certain utility function defined over a general resistor random network. A probabilistic interpretation to the qualities in electric network will be given. To the network arcs there are associated parameters which represent the time/length/probability that an object takes to travel from a node into another. Starting from a source node, one wants to determine the optimal path to a sink node that optimizes the value of the utility function. Using network optimization it will be determine the network path which as the smallest resistance (highest conductance).

Keywords Optimal paths • Probabilities • Probable location of an object inside a network

24.1 Introduction

In this work it will be given a probabilistic interpretation of voltage and current in an electric network as well as description of what is considered a random walk in this type of network.

Given an undirected network of resistors other network will be constructed using the resistance and/or the conductance. We will show that the network path with de smallest resistance/highest conductance may be determined using the knowledge of optimization network problems namely shortest path algorithms.

In the optimal path problem, a real function is considered which assigns a value to each path that can be defined between a given pair of nodes in a given network; a path with the *best* value in a subset of paths between that pair of nodes is what has to be determined. For a general study of the optimal path problem considering

D.M.L.D. Rasteiro (✉)
Mathematics and Physics Department, Coimbra Superior Engineering Institute, Coimbra, Portugal
e-mail: dml@isec.pt

A. Madureira et al. (eds.), *Computational Intelligence and Decision Making: Trends and Applications*, Intelligent Systems, Control and Automation: Science and Engineering 61,
DOI 10.1007/978-94-007-4722-7_24,

a general objective function the reader is invited to look through and in [7] and [6] a general study of the stochastic optimal path problem where the arc parameters are real random variables is presented.

24.2 Optimal/Shortest Path Problem

In the shortest path problem a directed probabilistic network (N, A) is given, where to each arc $(i,\ j) \in A$ is associated to the value C_{ij} which is called the parameter of the arc $(i,\ j) \in A$. The values C_{ij} are sometimes referred as cost, time or distance.

In the electric networks case the parameters may represent the resistance, the conductance or the probability of walking from node i to node j. We will assume that the network dimension is always a finite value. If an appropriate utility measure is assigned to each possible consequence, then the best action is to consider the alternative with the highest utility (e.g. the smallest utility value). Different axioms that imply the existence of utilities with the property that utility is an appropriate guide to consistent decision making are presented in [5, 8, 9]. The choice of an adequate utility function for a specific type of problem can be taken using direct methods presented in Keeney and Raiffa's book.

The utility of the arc $(i,\ j) \in A$ in the optimal path is measured taking into account the utility function previously considered. It is worth mention that from node i emerges a set of nodes called the forward star[1] i and, in the optimal path only one emerging arc is chosen.

Associated to the path p, we define the real value $C_p = \sum_{(i,j)\in p} C_{ij}$ representing the utility value of the loopless path $p \in P$.

With the objective of determining the optimal path, we consider a real function $U : P \rightarrow \Re$, called *utility function*, such that, for each loopless path p, $U(p)$ depends on the parameters associated to the arcs of p.

In the shortest path problem, we want to determine the loopless path $p^* \in P$ that minimizes the value of the utility function. The loopless path $p^* \in P$ is called *optimal solution* of the referred problem. The problem can then be mathematically defined as

$$\min_{p\in P} U(p) = \min \sum_{(i,j)\in A} C_{ij}X_{ij}$$

$$s.t.$$

$$\sum_{(i,j)\in A} X_{ij} - X_{ji} = \begin{cases} 1 \ , & i = s \\ 0 \ , & i \notin \{s,t\} \\ -1 \ , & i = t \end{cases}$$

$$X_{ij} \in \{0,1\}$$

[1] The forward star of node i is the set formed by the terminal nodes of its outgoing arcs.

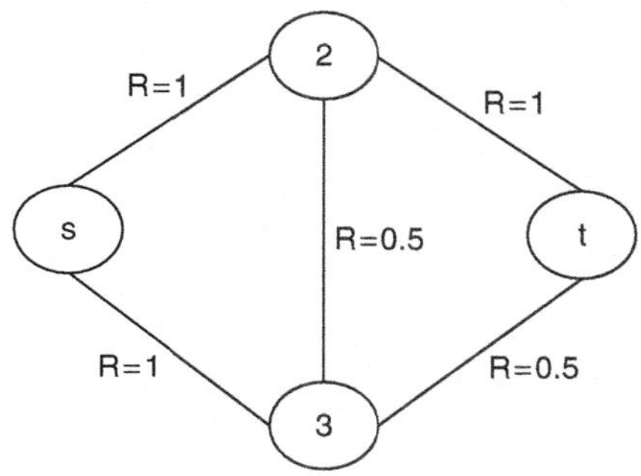

Fig. 24.1 Resistances network

Since the constraint matrix is totally unimodular, by the integrality property, the solution of the previous problem is equal to the solution of its linear relaxation.

24.3 Probabilistic Interpretation of a General Resistor Network

Given a network of resistors, to each undirected arc we associate the value of a resistance R_{ij} and, between i and j, two directed arcs (i,j) and (j,i) are considered. The conductance of the undirected arc will be denoted by $C_{ij} = \frac{1}{R_{ij}}$.

A resistance network example is and in this case we have $C_s = 2, C_2 = 4, C_3 = 5$ and $C_t = 3$.

Let $G = (N, A)$ be the resulting network. We define the probability of walking from node i to node j by $\mathrm{P_{ij}} = \frac{\mathrm{C_{ij}}}{\mathrm{C_i}}$, where $C_i = \sum_{(i,j)} C_{ij}$. For the previous example we will have

In the following it will be assumed that the network is connected, i.e., there exits always a path between any two nodes.

It is possible to define a random walk on G to be a Markov chain with transition matrix P given by $P = (P_{ij})_{n \times n}$, where n is the network number of nodes. Since the network is connected, it is possible to walk between any two stages (nodes) and thus we may say that this is an ergodic Markov chain. For an ergodic chain, there exists a unique probability vector (called fixed vector) f such that $P = f$. The component f_i represents the proportion of times, in long run, that the state (node) i will be visited. For random walks determined by electric networks, the fixed vector is given by $f_i = \frac{C_i}{C}$, where $C = \sum_{i \in N} C_i$. In addition to being ergodic, Markov chains associated with networks have also a property called reversibility. An ergodic chain is said to be reversible if $f_i P_{ij} = f_j P_{ji}, \quad \forall i, j \in N$. For the present network chains we have

$$C_i P_{ij} = C_i \frac{C_{ij}}{C_i} = C_{ij} = C_{ji} = C_j \frac{C_{ji}}{C_j} P_{ji}.$$

Consider two points of the networks of resistors s and t and put a one-volt battery across these points establishing a voltage $v_s = 1$ and $v_t = 0$.

Fig. 24.2 Changed network

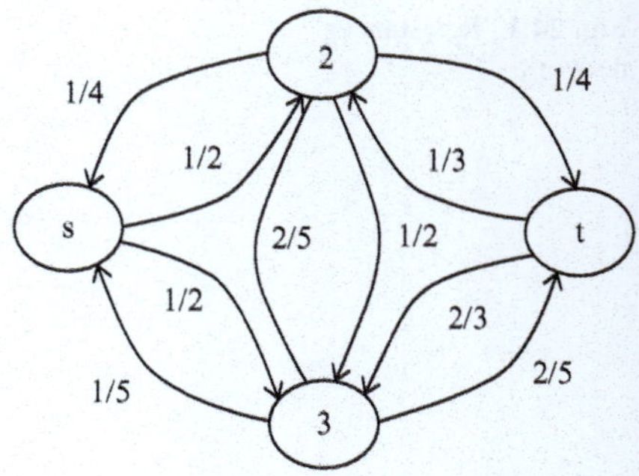

We are interested in finding the voltages v_i and the currents IC_{ij} in the circuit and in giving a probabilistic interpretation to these quantities. The voltage will be interpreted as a hitting probability, observing that both functions are harmonic and they have the same boundary values. By Ohm's law, the currents through the resistors are determined by the voltages by $IC_{ij} = \frac{v_i - v_j}{R_{ij}} = (v_i - v_j)C_{ij}$. Note that $IC_{ij} = -IC_{ji}$. Kirchhoff's current law requires that the total current flowing into any point other than s and t is 0. Thus for $i \neq s, t$ we have $\sum_{(i,j) \in A} IC_{ij} = 0$ which implies that the voltages values have the following property $v_i = \sum_{(i,j) \in A} \frac{C_{ij}}{C_i} v_j = \sum_{(i,j) \in A} P_{ij} v_j$, for all nodes i different from s and t, i.e., the voltage v_i is harmonic all points $i \neq s, t$.

Let p_i be the probability, starting at i, that node s is reached before node t. Then, p_i is also harmonic at all points $i \neq s, t$. Furthermore $v_i = p_i = 1$ and $v_t = p_t = 0$. If we modify the transition matrix P by making s and t absorbing states, we obtain an absorbing Markov chain Q and v and p are both solutions of the Dirichlet problem for the Markov chain with the same boundary values, hence $v = p$.

To summarize we may say that when a unit voltage is applied between s and t, making $v_s = 1$ and $v_t = 1$, the voltage v_i at any point represents the probability that a walker starting from i will return to s before reaching t.

In this probabilistic interpretation of voltage, we have assumed a unit voltage, but we could have assumed an arbitrary voltage v_s between s and t. Then the hitting probability p_i would be replaced by an expected value in a game where the player starts at s and is paid v_s if s is reached before t and 0 otherwise.

The probabilistic interpretation of the current is made supposing that positively charged particles enter the network at node s and walk around from node to node until they finally arrive at node t, where they leave the network (being more realistic it should the other way around, i.e., negative particles entering at t and leaving at s but in theory it works the same way).

To determine the current IC_{ij} along the edge between i and j, we consider that the walker passes once or several times along the edge from i to j and in the opposite direction. We can assume, by hypothesis, that the current IC_{ij} is proportional to the expected net number of movements along the edge from i to j, where movements from j to i are counted as negative. The walker begins at s and walks until he reaches t

(if he returns to s before reaching t, he keeps on going). Let u_i be the number of visits to node i before reaching t. Then $u_t = 0$ and, for $i \neq s, t$, $u_i = \sum_{(i,j)\in A} u_j P_{ji}$. Since, $C_i P_{ij} = C_j P_{ji}$ we have $u_i = \sum_{(i,j)\in A} u_j \frac{P_{ij}C_i}{C_j} \Leftrightarrow \frac{u_i}{C_i} = \sum_{(i,j)\in A} P_{ij}\frac{u_j}{C_j}$. This means that $v_i = \frac{u_i}{C_i}$ is harmonic for $i \neq s, t$. Since we also have $v_t = 0$ and $v_s = \frac{u_s}{C_s}$ we may conclude that v_i is the voltage at s when we put a battery from s to t that establishes a voltage $\frac{u_s}{C_s}$ at s and voltage 0 at t. The current that flows from i to j is $IC_{ij} = (v_i - v_j)C_{ij} = \left(\frac{u_i}{C_i} - \frac{u_j}{C_j}\right)C_{ij} = u_i P_{ij} - u_j P_{ji}$.

Thus $u_i P_{ij}$ is the expected number of times the walker will go from i to j and $u_j P_{ji}$ is the expected number of times the walker will go from j to i and the current IC_{ij} is the expected value for the net number of times the walker passes along the edge from i to j. Note that the currents here are not the original ones where a one-volt battery was applied but they are proportional to them. To determine the constant of proportionality we note that $\sum_{(s,j)\in A} IC_{sj} = 1$ once this sum represents the expected value of the difference between the number of times the walker leaves s and enters s and this number must necessarily be 1. This unit current flow can be obtained from the currents in the original circuit, corresponding to a one-volt battery, by dividing through by the total amount of current $\sum_{(s,j)\in A} IC_{sj}$ flowing into s. This shows that the constant of proportionality is the reciprocal of the amount of current that flows through the circuit when a one-volt battery is applied between s and t. This quantity is called the effective resistance between s and t. In fact if we impose a voltage v between nodes s and t, a voltage $v_s = v$ is established and $v_t = 0$ and a current $IC_s = \sum_{(s,j)\in A} IC_{sj}$ will flow into the circuit from the outside source. The amount of current that flows depends upon the overall resistance in the circuit which will be define as effective resistance and denoted by R_{eff} between s and t. Thus, $R_{eff} = \frac{v_s}{IC_s}$. Similarly the effective conductance may be defined as $C_{eff} = \frac{1}{R_{eff}}$. If the voltage between s and t is multiplied by a constant the currents are multiplied by the same constant, so R_{eff} depends only on the ratio of v_s and IC_s.

The effective conductance can be interpreted as an escape probability. When $v_s = 1$, the effective conductance equals the total current IC_s flowing into s. This current is

$$IC_s = \sum_{(s,j)\in A} (v_s - v_j)C_{sj} = \sum_{(s,j)\in A} (v_s - v_j)\frac{C_{sj}}{C_s}C_s$$
$$= C_s\left(1 - \sum_{(s,j)\in A} P_{sj}v\right) = C_s P_{esc},$$

where P_{esc} is the probability of, starting at s, that the walker reaches t before returning to s. Thus, $C_{eff} = C_s P_{esc} \Leftrightarrow P_{esc} = \frac{C_{eff}}{C_s}$.

24.4 Shortest Resistance Path/Highest Probability Path

Given a network of resistors, to each undirected arc we associate the value of a resistance R_{ij} and, between i and j, two directed arcs (i,j) and (j,i) are considered.

The conductance of the undirected arc will be denoted by $C_{ij} = \frac{1}{R_{ij}}$.

Let $G = (N,A)$ be the resulting network. We define the probability of walking from node i to node j by $P_{ij} = \frac{C_{ij}}{C_i}$, where $C_i = \sum_{(i,j)\in A} C_{ij}$. The shortest resistance path can be found by solving the problem mathematically defined as

$$\min_{p\in P} U(p) = \min \sum_{(i,j)\in A} R_{ij}X_{ij}$$

$$s.t.$$

$$\sum_{(i,j)\in A} X_{ij} - X_{ji} = \begin{cases} 1, & i = s \\ 0, & i \notin \{s,t\} \\ -1, & i = t \end{cases}$$

$$X_{ij} \in \{0,1\}.$$

To obtain the optimal solution one should use Bellman's equations and the resulting algorithm is

Algorithm – Shortest Resistance Path

Initialize $l[ending\ node] = 0$;
Initialize all label nodes (except the ending node) to "∞";
Initialize the set L with the ending node;
While $L \neq \phi$ do
    Remove node i from L;
    For each successor j of i do
    If $l[j] > C_{ij} + l[i]$
        $l[j] = C_{ij} + l[i]$
            If $j \notin L$
            $L = L \cup \{j\}$

Analogously we may obtain the highest probability path using the above algorithm as shown in the following reasoning. The highest probability path optimization problem is defined as:

$$\max_{p\in P} U(p) = \max \prod_{(i,j)\in A} P_{ij}X_{ij}$$

$$s.t.$$

$$\sum_{(i,j)\in A} X_{ij} - X_{ji} = \begin{cases} 1, & i = s \\ 0, & i \notin \{s,t\} \\ -1, & i = t \end{cases}$$

$$X_{ij} \in \{0,1\}.$$

Since $0<P_{ij}\leqslant 1$ the utility function with the product is not well behaved during the algorithm (the products will tend to zero and no solution is obtained). In order to escape from this problem the arc parameters must be changed to $-\ln P_{ij}$ and by doing that we obtain

$$\begin{aligned}\min_{p\in P} U(p) &= \min \sum_{(i,j)\in A} -\ln P_{ij}X_{ij} = \max \sum_{(i,j)\in A} \ln P_{ij}X_{ij}\\ &= \max \ln \prod_{(i,j)\in A} P_{ij}X_{ij} = \max \prod_{(i,j)\in A} P_{ij}X_{ij}.\end{aligned}$$

Observe that the highest probability optimal path, the shortest resistance optimal path and the highest conductance optimal path have the same optimal solution which is, for the network example considered above, $s \rightarrow 3 \rightarrow t$.

24.5 Conclusion

To summarize if a probabilistic interpretation of a resistor network is given, when applying network optimization we are able to determine the resistance/conductance of the network and the corresponding paths and also the highest probability path which are all the same although with different utility function values. The voltage and the current are also possible to determine.

References

1. Ahuja RK, Magnanti TL, Orlin JB (1993) Network flows theory, algorithms, and applications. Prentice-Hall, Upper Saddle River
2. Bard JF, Bennett JE (1991) Arc reduction and path preference in stochastic acyclic networks. Manag Sci 31(7):198–215
3. Bellman R (1958) On a routing problem. Q Appl Math 16:88–90
4. Cheung RK, Muralidharan B (1995) Dynamic routing of priority shipments on a less-than truckload service network. Technical report. Department of Industrial and Manufacturing Systems Engineering, Lowa State University
5. Luce RD, Raiffa H (1957) Games and decisions. Wiley, New York
6. Rasteiro DMLD (2008) Finding optimal paths in networks where arc parameters are not deterministic. In: Putnik GD, Cunha MM (eds) Encyclopedia of networked and virtual organizations. Information Science Reference, Hershey, pp 1151–1163. doi: 10.4018/978-1-59904-885-7.ch151
7. Rasteiro DDML, Anjo AJB (2004) Optimal paths in probabilistic networks. J Math Sci 120(1): 974–987. Springer, New York. ISSN: 1072–3374
8. Savage LJ (1954) The foundations of statistics. Wiley, New York
9. von Neumann J, Mogenstern O (1947) Theory of games and economic behavior. Princeton University Press, Princeton

Chapter 25
Business Intelligence Tools

Jorge Bernardino and Marco Tereso

Abstract Business intelligence platforms allows managers and business analysts monitor data and statistics, make decisions, analyze market trends, study the insertion of new products and maintain profitable and long-lasting relationships with customers in order to stay competitive in a global world. Consequently, the importance of using business intelligence tools is crucial for the success of actual organizations. Yet, many organizations are learning that this no easy task to decide which business intelligence tool to choose to expand their business. In this work we evaluate eight major business intelligence tools: four commercial (IBM Cognos, Microsoft BI, MicroStrategy and Oracle BI) and four open source solutions (JasperSoft, Pentaho, SpagoBI and Vanilla). The main objective is to present the differences in functionality between the various platforms, estimating overall usability, helping decision managers to choose the best tool to their business.

25.1 Introduction

Business Intelligence (BI) is a broad category of applications and technologies for gathering, storing, analyzing, and providing access to data to help managers make better business decisions. Business Intelligence applications include the activities of decision support systems, query and reporting, online analytical processing (OLAP), statistical analysis, text mining, data mining and visualization [1].

The benefit of business intelligence platforms for organizations is extremely important. Business intelligence platforms allows managers and business analysts monitor data and statistics, make decisions, analyze market trends, study the

J. Bernardino (✉) • M. Tereso
Departamento Engenharia Informática e de Sistemas, IPC – Instituto Superior de Engenharia de Coimbra, Coimbra, Portugal
e-mail: jorge@isec.pt; a21190968@alunos.isec.pt

A. Madureira et al. (eds.), *Computational Intelligence and Decision Making: Trends and Applications*, Intelligent Systems, Control and Automation: Science and Engineering 61, DOI 10.1007/978-94-007-4722-7_25, © Springer Science+Business Media Dordrecht 2013

insertion of new products and maintain profitable and long-lasting relationships with customers in order to stay competitive in a global world.

Currently, there are several known cases of success from implementing business intelligence tools in various business segments. A strategy which ensures business intelligence platforms is extremely important in order to monitor market trends, technological innovation and support both management and decision making. The business intelligence tools are intended to assist business managers in making decisions, drawing on data analysis and applying them to metrics that enable them to extract knowledge. The processes of business intelligence help make business more profitable and dynamic. When is time to choose one business intelligence platform is important that companies have some knowledge about existing platforms on the market. There are several tools in the business intelligence market with different types of licenses, which we can break up in two main groups: commercial and open source platforms. Before making a decision for the business intelligence platform to acquire, business managers must take into account the features of each tool to satisfy the needs of the organization, and check whether the minimum requirements of the platform are supported by the company computer system.

The implementation of business intelligence platforms in an organization requires some additional efforts. Some of the problems regarding the adoption of business intelligence for organizations is directly related to: acquisition of skilled human resources in the area of databases, changes in organizational structure forcing the inclusion of new employees, implementation of new standards of regulation and planning of new work methods. This study compares the features of four commercial business intelligence platforms: IBM Cognos [2], Microsoft BI [3], MicroStrategy [4] and Oracle BI [5], with four open source platforms: JasperSoft [6], Pentaho [7], SpagoBI [8] and Vanilla [9].

This paper is structured as follows: Sect. 25.2 analyzes the advantages and drawbacks of open source and commercial platforms. In Sect. 25.3 we describe the four major commercial tools, and Sect. 25.4 the four major open source tools. In Sect. 25.5 is presented a comparison between all tools. Finally, Sect. 25.6 concludes the paper by summarizing the main conclusions of our research and pointing out directions for future work.

25.2 Open Source vs Commercial Tools

In this section we describe some of the benefits and disadvantages of using open source and commercial tools. The most relevant aspect of open source tools is that they are free and allow access to the code, with the possible modification of the various modules. In contrast to the open source tools, commercial tools are paid and often represent an extra expense, unbearable for the vast majority of companies. One of the advantages of open source platforms are if they do not serve the needs of the organization, can be replaced by other platforms without a cost; this does not happen with commercial platforms. The open source tools generally require lower system requirements than commercial applications. This factor is justified in assuming that those who can not invest in software may not invest in hardware. As open source has advantages compared to commercial tools, it also has disadvantages. The open source projects are in constant development, which means that some projects have not yet reached the desired maturity. The fact that open source tools are developed by a community of contributors raises some concerns about product quality. A lack of development methods and quality standards can be a reality in a community of open source development. The open source tools are usually seen as less reliable, given that the testing process is typically narrow in scope, although in recent years has improved. Absence of documentation is one of the other critical factors of open source tools; only a minority of successful open source projects provides some documentation to support the use of the solution. The continuity of the open source projects is not always guaranteed, this being more a factor to take into account.

The idea that the acquisition of open source software does not involve costs is sometimes wrong. Often the open source tools do not satisfy all the pretensions of the customer, costing to adapt them to meet their needs, involving costs. Such adjustments in open source platforms can be expensive. Upon the acquisition of open source software should be checked if it meets the requirements of the organization, to decide for a more viable solution.

The commercial business intelligence platforms have some advantages, which we can summarize: the maturity of the platforms, the level of confidence that they offer, the amount of documentation that usually provide, the customer service, the continuity of the platform on the market and the usability with interfaces more user friendly.

Some of the disadvantages of commercial platforms are: high acquisition costs, the requirement to be connected to the sellers, typically require more powerful hardware and the difficulty of transition to other platforms, taking into account the costs and terms of the contract.

When we have to decide which platform to acquire is important to take in consideration the points above and if the tool fulfills the needs of the organization. In the following sections we will review four major commercial and four open source platforms.

25.3 Commercial Business Intelligence Tools

In this section we will analyze the four major commercial business intelligence platforms, according to the study of Gartner: "Magic Quadrant for Business Intelligence Platforms Report"[10]. The four major commercial business intelligence platforms are IBM Cognos [2], Microsoft BI [3], MicroStrategy [4] and Oracle BI [5].

Usually commercial tools are more attractive and have a simpler user interface, more user-friendly. Apart from a greater confidence on commercial platforms, these platforms evolve to facilitate adaptation of users, also providing a good deal of documentation. The commercial business intelligence platforms are more confidence due to its rigid testing procedures. The commercial tools present new features and more attractive techniques. On the other hand the acquisition costs are expensive.

We analyze the features of each of the four major commercial business intelligence tools independently.

25.3.1 *IBM Cognos*

Cognos was a company acquired by IBM and is its business intelligence platform, currently in version 10.1 [2]. IBM Cognos allows: query and reporting (gives the performance of ad-hoc queries, issuing and structuring of reports, enabling adaptation of the models' own reports), OLAP analysis, scorecarding, dashboarding, real-time monitoring, statistics, planning and budgeting, extending BI (convenient methods for delivering business intelligence to a wider audience), collaborative BI (modern tools for collaborating, building stronger teams and stimulating better decisions) and importing/exporting data.

Cognos has also a mobile version, where we can use its benefits anywhere at any time online or offline. The mobile version was supported by: Apple iPhone and iPad; RIM BlackBerry smart phones and PlayBook. The mobile version has support for devices using the Android, Symbian and Windows Mobile operating systems [11].

25.3.2 *Microsoft BI*

According to Gartner [10], Microsoft emerged as the leading tool of business intelligence tools market.

In the business intelligence processes Microsoft [3] uses two of its solutions packages: Microsoft Office and Microsoft SQL Server. The features of Data Warehouse, Data Marts and Operational Data are supported by SQL Server; data integration by SQL Integration Services; data analysis by SQL Analysis Services and operations with reports served by SQL Reporting Services.

Excel supports the features of end-user analysis tool and the business intelligence portal uses SharePoint. The business scorecards, analytics and planning tools are supported by PerformancePoint.

This platform allows: querying, analysis, reporting, data integration, synchronization, searches, cloud storage and importing/exporting data.

Webalo Inc, a software company, developed a application that support the Microsoft BI, Mobile Webalo Dashboard. This application functions as a hosted service. This application allows to transform the SQL Server, Excel, Performance-Point and SharePoint contents in Dashboards. This service has the benefit not be developed for a specific device but for a range of devices that support operating systems. The mobile devices compatible are: RIM BlackBerry, Windows Mobile 2003 for Pocket PC, Windows Mobile 5, Windows Mobile 6 and Java MIDP 2.0 smartphones [12].

25.3.3 MicroStrategy

According to Gartner [10], Microstrategy [4] holds the third podium position of commercial business intelligence platforms market leaders.

Currently at version 9, it is available in optimized solutions for mobile environments, including iPad, iPhone and BlackBerry.

The main features of the MicroStrategy are: querying, reporting (allows the development and structuring of reports) OLAP analysis (allows the analysis and construction of an intelligent data cubes, MDX connection and incorporates a ROLAP server), dashboarding, data mining and importing/exporting data.

The MicroStrategy mobile version is developed around the Apple applications (iPhone and iPad) but is supported by other devices as BlackBerry smartphones. MicroStrategy provides a free mobile suite [13].

25.3.4 Oracle BI

According to Gartner [10], Oracle BI [5] holds the second podium position of commercial business intelligence platforms market leaders.

Oracle BI is available in the market as package, called Oracle Business Intelligence Enterprise Edition (OBIEE). The OBIEE offers a range of solutions necessary to create and view reports. OBIEE allows the development of several tests for the data.

Oracle Business Intelligence Enterprise Edition allows: to create interactive dashboards, reporting and publishing, answers (ad-hoc analysis), delivers (proactive detection and alerts), disconnected analytics, Microsoft Office Plug-in and Web Services.

The OBIEE provides a mobile version. The OBIEE mobile version is supported by: BlackBerry, iPhone, HTC, Android, Windows Mobile, Symbian and Palm [14].

25.4 Open Source Business Intelligence Tools

The communities of developing open source business intelligence solutions are required along with their employees to optimize their platforms, contributing to a steady development, making them more functional solutions.

In this section we present a brief description of the potential of JasperSoft [6], Pentaho [7], SpagoBI [8] and Vanilla [9] platforms. The analysis platforms have one feature in common, the fact that all have a GNU General Public License (GNU GPL).

We analyze the features of each of the four major open source business intelligence platforms independently.

The open source business intelligence tools have followed the evolution of the commercial tools. Developed by non-profit open source communities, these tools present good business intelligence features.

25.4.1 JasperSoft

JasperSoft [6] is an open source business intelligence platform developed in Java and Perl languages. JasperSoft has two proprietary versions, Enterprise and Professional in addition to the Community version. The Community is a free version that is available in three individual modules: Jasper Reports Server, JasperSoft OLAP and ETL JasperSoft. The Jasper Reports Server module enables the creation, structuring and displaying BI reports; JasperSoft OLAP module is the tool that allows linking of data and parameterization of the different views of the data cube, ETL JasperSoft is the module that allows the tool creating the structure of production, processing and loading data.

The set of modules that make up the community version of the tool features available for creating and viewing reports, various graphics, dashboards, OLAP and ETL processes, analysis of geo-referencing, export data and ad-hoc queries. This platform was developed for Linux and Windows operating systems.

The last JasperSoft open source version is 4.2. This release includes the iOS software development kit, which will allows developers to embed BI capabilities into native mobile apps [15].

25.4.2 Pentaho

Pentaho [7] is one of the most popular open source business intelligence platforms. Developed in Java, the tool is available in two versions, Community and Enterprise. Like other platforms, the Community version is free and the Enterprise version is a paid solution with more features. Pentaho is a platform consisting of a set of integrated tools. Pentaho incorporates the Kettle application for ETL processes, Mondrian and Ramsetcube in OLAP processes, and Weka for properties of data mining. Pentaho allows creating reports, creating charts, dashboards, data manipulation processes through ETL and OLAP, data mining processes, KPI's, exports of data and ad-hoc queries. Pentaho uses MDX to access and manipulate data from databases. The tool is available for Linux, Windows and Unix operating systems.

Pentaho provides the ad hoc analytical reporting for iPad. The mobile version allows creating the reports directly from the iPad. The touch-enabled technology provides more information about reports from user [16].

25.4.3 SpagoBI

SpagoBI [8] is an open source business intelligence platform developed in Java. SpagoBI is available in a single Community version. This is the most comprehensive free platform, because has all the main features of business intelligence and provides them as a single release covering its users with a good solution that offers all its capabilities in a package completely free.

SpagoBI contains features for creating and exporting reports, charting several, development of dashboards, data integration using ETL processes, analyzes, processes using OLAP, data mining processes, implementation of filters in queries, export data, ad-hoc queries and features GEO/GIS. The tool is available for Linux, Windows and Unix operating systems.

The SpagoBI provide a mobile version for platforms Apple and Android support. This version allows remote connection with SpagoBI Server. This mobile version include: user authentication; documents selection, according to end user's role, and consequent controlled data download; data update on request; and periodical and automatic check to identify possible alarms and notifications [17].

25.4.4 Vanilla

Vanilla [9] is an open source business intelligence platform developed in PHP language. Vanilla is available in a single version, community. Vanilla is a complete platform taking into account that integrates all main functionalities of business intelligence. The platform allows to create reports, perform analysis, generate and analyze data tables, charting several, dashboards, ad-hoc queries, integrate OLAP and ETL processes, define KPIs, data export and use of procedures of Data Mining.

Vanilla uses the tools BIRT and iReports for the purpose of metadata integration. Vanilla allows filing several separate projects, and an historical book.

The platform is available in a version for mobile operating system Android, highlighting the monitoring effort of innovation on the part of the development group. Vanilla is available for Linux, Windows and Unix operating systems.

Vanilla was the first open source tool to provide a mobile version, in version 3.4, where is possible to browse reports and run dynamic reports from smartphones. It is also possible to browse OLAP cubes. This version is compatible with iPhone, iPod Touch and iPad [18].

25.5 Comparison Between Commercial and Open Source Business Intelligence Tools

In this section we intend to illustrate a comparative table of the different features offered by each platform. Reports are information elements such as information tables and several charts. Dashboards are graphical elements that allow view the data in graphical mode. The On-Line Analytical Processing (OLAP) is the process that allows visualize the information in data cubes. ETL is the process in charge of data extract, data transformation and data load. Data mining is a complex process of extract information by data sequences. The KPIs are the key performance indicators that usually arise associated with the data represented. The data export allows export data to Excel, CSV files etc. GEO/GIS describes any information system that displays geographic information for informing decision-making. Ad-hoc queries are the typical queries made by decision makers where any field can be queried at any time. The results of this work on the eight BI platforms are shown in Table 25.1.

In open source platforms, Pentaho and SpagoBI are identical platforms, having all the main features and are available for major operating systems. JasperSoft has neither Data Mining features nor KPI's. JasperSoft does not support UNIX

Table 25.1 Comparison between open source and commercial business intelligence

	Business intelligence platforms							
	Open source				Commercial			
Features	JasperSoft	Pentaho	SpagoBI	Vanilla	IBM	Microsoft	MicroStrategy	Oracle
Reports	✓	✓	✓	✓	✓	✓	✓	✓
Dashboards	✓	✓	✓	✓	✓	✓	✓	✓
OLAP	✓	✓	✓	✓	✓	✓	✓	✓
ETL	✓	✓	✓	✓	✓	✓	✓	✓
Data mining		✓	✓	✓	✓	✓	✓	✓
KPIs		✓	✓	✓	✓	✓	✓	✓
Data export	✓	✓	✓	✓	✓	✓	✓	✓
GEO/GIS	✓	✓	✓		✓	✓	✓	✓
Ad-hoc queries	✓	✓	✓	✓	✓	✓	✓	✓
Linux	✓	✓	✓	✓	✓		✓	✓
Windows	✓	✓	✓	✓	✓	✓	✓	✓
Unix		✓	✓	✓	✓		✓	
Mobile	✓	✓	✓	✓	✓	✓	✓	✓

operating system. Vanilla just does not have GEO/GIS features. Microsoft Business Intelligence platform is not available to Linux and Unix, but if resort to Wine tool it is possible.

The commercial platforms are identical, processing all the features analyzed. Relatively to operating system, Oracle does not support Unix.

Pentaho and SpagoBI are equivalent to commercial platforms, being an alternative with lower acquisition costs.

All the analyzed tools have a mobile version for several devices. In general the devices supported are the Apple mobile devices and devices with Android operating system.

25.6 Conclusions and Future Work

The importance of implementing business intelligence tools in an organization is unquestionable, given the recognized importance of use of such platforms, it is important when seeking to implement one that meets the needs, from a standpoint of cost/quality.

It is important to realize the capabilities of the organizational structure in terms of basic knowledge of business intelligence tools. Aspects such as knowledge of data modeling languages are sometimes essential and organizations must have qualified people with such knowledge.

The choice between open source or commercial tools should take into account the financial situation of the company or organization and their capacity for investment and maintenance.

The commercial adoption of evaluation tools have the advantage of providing interfaces easier to use and which do not require much knowledge of informatics by users, this is the main advantage that sometimes makes the difference in business without people with knowledge. As for open source tools, already have very large features common to commercial tools but in return are more difficult to use.

Given the current state of BI software we have two choices: purchasing commercially available BI tools or using open source tools. Each approach has advantages and disadvantages. The tradeoffs are development time and customization vs. money and standardized products.

Looking at available commercial BI applications, we have robust platforms with a plenty of features. The leading business intelligence vendors are waking up to the potential Linux market and are increasingly supporting the platform. In addition to Linux compliant commercial tools, the market continues to pressure vendors to provide more open source or free options as well.

Looking at what is available from the open source community, clearly the open source software development community has a long way to go. There are many Linux open source database projects, only a few have analytical tools or report generation capabilities. Mondrian is often mentioned as an application with enough OLAP (online analytical processing) utilities to become a springboard for some serious business intelligence tools – if developers are willing to take the time to build them. It implements the Multi-Dimensional Expressions (MDX) language, and the XML for Analysis and Java OLAP specifications. It reads from SQL and other data sources, and aggregates data in a memory cache. For the immediate future, more companies are looking for increased value for their BI spending and the open source community increasingly devotes energy to projects beyond the systems and back office markets.

In our analysis, the open source BI tools with the greatest potential are clearly Pentaho and SpagoBI. The evolutions of these tools to mobile versions allows more users and make easy the work of decision-makers that can decide anywhere and anytime.

For the commercial platforms, we highlight the two platforms on top of market leadership, Microsoft BI and Oracle BI. This conclusion is partly motivated by the publication of Gartner [10].

As future work we intend to determine the average cost of maintaining two types of solutions and evaluate the need to have skilled labor in the organization to implement business intelligence processes.

References

1. Anandarajan M, Anandarajan A, Srinivasan CA (2004) Business intelligence techniques: A Perspective from Accounting and Finance. Springer, Berlin
2. http://www-01.ibm.com/software/data/businessintelligence/. Accessed 08 June 2011
3. http://www.microsoft.com/BI/en-us/pages/home.aspx. Accessed 08 June 2011
4. http://www.microstrategy.com/Software/businessintelligence/index.asp. Accessed 08 June 2011
5. http://www.oracle.com/technetwork/middleware/bi-enterprise-edition/overview/index.html. Accessed 08 June 2011
6. http://www.jaspersoft.com/. Accessed 08 June 2011
7. http://www.pentaho.com/. Accessed 08 June 2011
8. http://www.spagoworld.org/xwiki/bin/view/SpagoBI/. Accessed 08 June 2011
9. http://www.bpm-conseil.com/. Accessed 08 June 2011
10. http://www.fdsystems.co.uk/files/Gartner%20BI%20Magic%20Quadrant%20QlikView.pdf. Accessed 08June 2011
11. http://www-01.ibm.com/software/analytics/cognos/mobile/features-and-benefits.html. Acessed 03 Jan 2012
12. http://news.softpedia.com/news/Microsoft-Business-Intelligence-Available-For-Most-Mobile-Phones-via-Webalo-95013.shtml. Accessed in 03 Jan 2012
13. http://www.microstrategy.com/freemobilebi/. Accessed 04 Jan 2012
14. http://www.fdbnetwork.com/OBIEE-blackberry-PDA-BI-app. Accessed in 04 Jan 2012
15. http://www.microscope.co.uk/news/vendor-news/jaspersoft-claims-world-first-open-source-mobile-bi-package/. Accessed 04 Jan 2012
16. http://www.pentaho.com/mobile-bi/. Accessed 04 Jan 2012
17. http://www.spagoworld.org/xwiki/bin/view/SpagoBI/MobileBI. Accessed 04 Jan 2012
18. http://downloads.zdnet.com/abstract.aspx?docid=3131407. Accessed 04 Jan 2012

Chapter 26
Food Service Management Web Platform Based on XML Specification and Web Services

Pedro Sabioni, Vinícius Carneiro, and Maria Leonilde R. Varela

Abstract Company's importance and competitiveness is no longer just to enable automatic processes but also to fulfill needs of humans as we nowadays also live concerned with enjoyment and pleasure, in order to complement people's lives. Built from the decision to bring to market a fully customized product to the customer comes the Fragus Company. This company enables services of Personal Chef and Bartender aiming to bring convenience in one of life's greatest pleasures: eating. This publication aims at briefly describing a web platform of this company and its usefulness, which is to some extend provided by the use of structured data based on XML and related technologies, for better supporting the Company's services and to enable an easy way for information storing and processing over the web, namely through web services.

Keywords Mass customization • Minimum standardization • Food service management • Web platform • Xml specification and web services

26.1 Introduction

The minimum standardization is present in several areas of study now widely investigated, varying from projects and systems developments or process mapping up to simulation tasks, this standard consists on a way for achieving expected goals, regarding minimum standardization [1–3]. The base line segment of this issue follows the same reasoning of a macro-activity, aiming at, among other things, to properly be able to progressively subdivide a concept down to a minimum level of significance. This, in turn, mostly includes more than one pattern. Therefore, this

P. Sabioni • V. Carneiro • M.L.R. Varela (✉)
Department of Production and Systems, School of Engineering, University of Minho, Azurém Campus, 4800-058 Guimarães, Portugal
e-mail: pedrosabioni@gmail.com; vinicius@fragus.com.br; leonilde@dps.uminho.pt

A. Madureira et al. (eds.), *Computational Intelligence and Decision Making: Trends and Applications*, Intelligent Systems, Control and Automation: Science and Engineering 61, DOI 10.1007/978-94-007-4722-7_26,

patterns are something like an indivisible task, resource, or an attribute that is defined for a given application context, which in this work is related to food services provided to end-users, in a daily basis.

The purpose of this work arose from a very simple idea, in order to enable a better service providing to each internet user that seeks for some high level of service customization. In order to make a contribution in this direction, this paper intends to use and express this concept within an adjustment for DTD (Document Type Definition) that was applied to this food service company, for developing a management system, in order to enable the company's main activity to be enriched and progressively more personalized, through the creation and configuration and processing of customized menus to perform the service of a personal chef's and bartender services, which are characterized by realization of customized meals and drinks directly to the customer's home or in some required space.

This paper is divided into five main sections, the next being one to introduce a brief explanation and illustration of this personal chef's and bartender processes, in different areas, regarding the minimum standardization and mass customization principles supporting this work and the third section shows the context on which the minimum standardization was applied in the context of the Company, including a brief description and illustration of the Company's web platform. In the fourth section an explanation an illustration of the underlying Company's web platform XML and DTD supporting documents is presented, for data and processes information storing and processing in an integrated way. Finally, the fifth section shows the conclusions drawn from this work, regarding the implementation of XML and related technologies.

26.2 Minimum Standardization

The minimum standardization, in the context of this work, is intended to designate a minimum set of elements for enabling from a macro analysis down to a detailed analysis of food services. Other concepts, which are also being applied to this Company are the ones related to the use of three issues in force over the recent years, which are: Work Breakdown Structure, Simulation and Mass Customization, which are going to be briefly described next.

26.2.1 Work Breakdown Structure

The Work Breakdown Structure (WBS) is a concept that is being widely used in the area of project management, which is also used as a concept of Project Breakdown Structure (PBS) that is known as being development in order to be used intrinsic to the minimum standardization concept.

The concept of WBS implies to be in constant search for some derivable or endpoints also called deliverables. In order to reach such a point of concepts integration, regarding WBS combined with the MS one, an effort is required so that project scope and goals are able to be progressively subdivided for reaching lower levels of deliverables that must be met.

Therefore, in this context the minimum standardization about facts it enables to progressively extend concepts from the first level down to lower levels. Obviously this does not mean that it is a standard itself, but it surely enables to achieve a growing level of standardization, as it enables and aims at a detailed and structuring of sub-concepts at progressively lower levels. The application of this concept enables to constitute a line of reasoning to assemble the tree ones, also regarding WBS and a third one, regarding Mass Customization (MC), by following a streamline of minimum standardization. For example, let us consider the activity of "painting a room", which may require that each room requires different type of developments, not resulting in a pattern known in advance, but instead to consider the construction of a new one, regarding some new combination of existing patterns for "painting rooms". Thus each element of a second level has a pattern of elements which in this case represents the third level. So for example it would greatly facilitate the work of selection of various types of materials to purchase in a second level, because there would be available various combinations of patterns, for performing out the third level, for finally being able to answer the questions: "how to buy ink in blue or yellow, ladder medium or high, roll small, medium or large and so on.

But the third level does not have a minimum standardization yet, since this is not yet the basic one, so information about this particular situation, at this level, has to be inserted into the decision support system, in a simple, straightforward and usual way.

Therefore, at the last level we will have the minimum standardization, regarding all levels above them, enabling to create flexibility in the introduction of variations of the standard for further selection situations occurring, regarding the activity of "painting rooms".

26.2.2 Simulation

The simulation consists on a science but also on an art to represent events and behaviors of a system over a period in which it is subject to interferences provided. So, it can be seen like a mimic of the evolution of a system.

The simulation tool is used in this work to help to model possible scenarios for delivering services to a set of clients, during a certain planning horizon. This kind of tool combined with the concepts of Mass Customization and Minimum Standardization objectives enables to better reach appropriate responses for planned actions, in terms of services programming. Therefore, some parameters are dynamically specified for each concrete context in order to maximize Mass Customization based services, which are briefly described below.

26.2.3 Mass Customization

The concept of Mass Customization (MC) was originated to serve the consumer market enabling more personalized and affordable products. To meet these requisites, the MC concept has the advantage of promoting a kind of the Mass Production concept coupled with customized product specification.

Compared to Mass Production, Mass Customization includes a much larger level of information exchange [4–9] because the specific needs of each client must be transmitted through some appropriate way throughout the supply chain. This process requires some co-creation, within the Company and across outside ones, in order to generate a flow of information, which requires a robust management and coordination of it, so that the level of service is according to each client's preference.

According to Zipkin [3], in "The Limits of Mass Customization", (2001), mass customization systems have three key capabilities: elicitation (a mechanism for interacting with the customer and obtaining specific information); process flexibility (production technology that processes the product according to the information); and logistics (subsequent processing stages and distribution that are able to maintain the identity of each item and to deliver the right one to the right customer at the right time). Those elements are connected by powerful communication links and thereby integrated into a seamless whole.

This concept is being used in several distinct contexts, namely by the Dell computer company, which is considered a pioneer in customized mass production. The company has a structure of customer service working 24 h a day, 7 days a week, to answer each customer's questions, asking for personal computer components, among other related activities. The Dell Company configures computers according to the specifications of each customer making the request directly to the company, which has is known by its excellent level of service provided (http://www.dell.com/).

26.3 Fragus' Bartenders & Personal Chef

Fragus is a Brazilian Company, located in the city of Belo Horizonte which has its primary focus on the Personal Chef Service. This service is highly personalized and consists of making gourmet meals in the clients' homes, which in turn choose every detail of the service on which he/ she is interested on.

The Company's development and expansion was greatly due to a growing number of clients promoted through an extremely noticeable and visible trade-off out performed by applying Mass Customization services.

To better exemplify the concept behind this company we can cite an historical persons regarding the development of industrial management that revolutionized the way of thinking of managers at that time, namely Taylor and Ford. These two persons aimed at extending and expanding the production output and did revolutionize the standardization of forms of production. Moreover, they realized that a

Fig. 26.1 Fragus platform home page

certain specific final product should be delivered to the corresponding customers. Thus the famous Model T Ford at that time came to be a success by the amount of scales of production achieved and we can even consider it as the beginning of the popularization of cars.

But over the years, technological developments and especially the increased competition and the increasingly growing need for improved products for fulfilling customer requirements lead firms to try to customize their products so that customers can feel more unique and special. This is also the situation that this Company faces today and, therefore, it wants to fulfill the customer needs, by using available technological developments to meet these needs and to meet demand for its core services.

From the reasoning presented throughout the previous sections it is now time to put all concepts together and show inside its main services provided, through a summarized description. Therefore, in the next section an illustration of the application of those concepts combinations will be presented, part of which was used to perform the Minimum Standardization activities at the Company.

26.3.1 Fragus Web Platform

The address of the web platform of the FRAGUS Company is http://www.FRAGUS.com.br, and Fig. 26.1 illustrates its main page.

In Fig. 26.1 we can see which directions can be taken at the site, which is divided into five areas addressed by the links on the top, which are: "who we are", which provides information about the Company, integrating a brief presentation of it;

Fig. 26.2 Fragus main services

"services", which describes the services provided by the Company, which are "personal chef" and "bartender". Moreover, "customers", presents information regarding historical customers with the highest expression with which the Company works or did work at the past. The "budgets" link enables customers to request an estimate and detailed information about some kind of service, which can be even better fulfilled through the "contact" option, so that the person can communicate with someone in the Company to accomplish an order.

26.3.2 Fragus Main Services

Figure 26.2 shows the two links to the main pages for the two types of services that FRAGUS provides to its customers, which are the bartender and the personal chef services. Through this links further sites related to the Company are possible to be reached, including for obtaining detailed information about dishes, drinks and how to personalize events and dishes and drinks ingredients, among other useful information and processes, namely for being able to actively participate on some event's organization and budgeting activities.

Figure 26.3 represents the form to connect to the application of budgeting, which enables to provide basic information, about a proposed event, in the Company. Once this form is submitted it generates an XML file that is sent directly to one of the chefs.

Once an ordering service reaches the Company, the corresponding chef will be responsible, along with a manager to outperform it, being informed simultaneously for contacting the client, in order to arrange a meeting between himself and the team of the Company to define details about customer requirements for the event. According to Zipkin [3] this is a real expression of the Mass Customization concept, by enabling a real perception and extraction of customer needs through a highly personalized service providing.

Fig. 26.3 Fragus's ordering service

26.4 XML, DTD and Implementation

The main tool used to perform the standardization of processes within the Company was the Document Type Definition (DTD).

A DTD consists on a kind of grammar, which combined with appropriate XML documents and related technologies, including web services, enables to express domain-knowledge in a simple a compact way, even for complex data structures [10–12] and this was one of the main motivations for using this type of data semantic specification. On the other hand, this is also a standard that has the advantage of being possible to provide families of XML documents, for different document categories, which also fits well for the purpose of this work, in terms of defining all information related to the Company's services. Through XML and corresponding DTD documents, all kind of data used in the Company can be accurately stored and processed, as these documents also can be processed and transmitted within and across companies through web services, by using the Fragus platform. For more information about XML and DTD see, for example, (http://www.w3schools.com).

Therefore, with DTDs and related technologies all data structures and business procedures were developed for mapping the whole Company's activity, as we can find in some other application examples, namely the ones described in the following papers [13–16].

Next an example will be provided, through Fig. 26.4 In this figure a fragment of the DTD that has been formulated to standardize a dish is presented. In DTDs bookmarks are used for a technical presentation of dishes, promoting to consider all

Fig. 26.4 DTD code example

```
<!ELEMENT menu (Sheet)>
<!ELEMENT Sheet (id, name, type, ingredient+)+>
<!ELEMENT id (#PCDATA)>
<!ELEMENT name (#PCDATA)>
<!ELEMENT type (#PCDATA)>
<!ELEMENT ingredient (namIng, quantity, measure)>
<!ELEMENT nameIng (#PCDATA)>
<!ELEMENT quantity (#PCDATA)>
<!ELEMENT measure (#PCDATA)>
```

information required that can be used in certain events preparation, considering all explicit and implicit needs. Thus, for instance, for accomplishing an immediate response to a client about all kind of events, regarding the request of some kind of dish and the underlying ingredients needed to prepare it is necessary to promote a quick selection of information regarding the considered dish and related information and conditions, which in turn will generate all the values embedded in the corresponding available options.

Thus, for each event a different and personalized menu item is created, according to the corresponding DTD, which consists of one or more alternatives, representing the options chosen by the customer for the required dish for some kind of event. Each sheet, regarding some specific peace of information, includes an identifier (id), a name with the designation of the dish, a type that characterizes the type of meal, in order to be able to define it, as well as the ingredients for making the dish. For each element on the associated XML document, the information contained can be presented through free text, which is expressed through the #PCDATA data type. Therefore, for instance, an ingredient may be constituted by several distinct other elements, each one being composed by three elements types: "namIng", the ingredient's name, "quantity", for the measurement of the ingredient, and the ingredient measurement type, expressed through the element "measure", which may also further vary, assuming different possible sub-types of measurement, like: kg, L, etc. Therefore, considerably different menus can be defined, once they are validated through a corresponding DTD and this happens for each family of documents, which can also be grouped together regarding "families of dishes", and so on.

Within the DTD fragment presented in Fig. 26.4, standardization is not only present and expressed through the textual elements but also through the composition of each one. In the example provided, the composition of each ingredient, based on three elements enables to better clarify the Minimum Standardization concept previously referred, where a standardization of ingredients is performed at the bottom level of each ingredient's structure. And this is important not only in terms of each element definitions and sub-elements of data, and so on, as well as for defining the way they are inter-related and can be integrated in the whole system and "communicate" with other documents and applications, within the entire platform structure and also across platforms when required. For example when querying the system for the generation of a budget for an given event that will take place at a certain moment, on a given place, by using a set of resources, of human and materials

```
<menu>
    <Sheet>
      <id>000321</id>
      <name>Bromua de Bacalhau</name>
      <type>principal</type>
      <ingredient>
        <nameIng>Bacalhau Site</nameIng)
        <quantity>0.25</quantity>
        <measure>Kg</measure>
      </ingredient>
    </Sheet>
</menu>
```

Fig. 26.5 XML code example

kinds, the system, through the Company's platform, will enable to perform all the calculations needed and retrieve all the information necessary to accurately enable the realization of the given event. Moreover, the processing required to accomplish the event are also possible, through an integrated and distributed interconnection of applications behind the platform, based on corresponding web services. Once again, regarding the important information carried out through the minimum standardization principles, for instance, for each ingredient, the correct amount used, expressed in the appropriate measure, among other important information regarding a given ingredient type can be provided by the system, as well as outside information retrieval, namely regarding connections to other sites, for obtaining, for example, information about the best ingredient's price and set the budget.

In order to better illustrate a concrete example of an XML document instantiation that can result from the validation of a corresponding DTD, as previously described, Fig. 26.5 shows a simple extract of XML code.

Thus this fragment gives an illustration of how the DTD could be used for obtaining the above described definition of the data of a dish. According the corresponding DTD, this is a valid XML document excerpt, as it includes a valid document structure, composed by a "menu" element, which includes one or more "sheet" elements, which, in turn, includes an "ingredient", and so on. This document allows the incorporation of a huge range of combinations of information that characterizes the ability to exchange different kind of documents, based on the standardization of the concept "menu", being possible to obtain quite different combinations of MS elements.

Regarding the whole platform functionalities, a fundamental issue overcomes through a process that enables to verify the Minimum Standardization for turning possible a completely integrated communication system, which is created to undergo a procedure of search or even to allow the generation of a consistent database to promote an appropriate decision-making process through an integrated team, working on events preparation and processing within the Company and this is one of the greatest advantages of the Company's platform. Moreover, it enables a good communication system characterized by important features regarding the integration of external business partners and customers, when preparing services for planned events.

A truly integrated communication system enables a truly implemented Mass Customization service, which is the main purpose of this work. Therefore, the three major skills necessary for Mass Customization cited by Zipkin, 2001 [3] turns possible, by extracting the customers' wishes, which occurs when a representative of the Company meets a client to define what are the dishes, beverages, and other events requisites he/she wants to specify. Then this information is entered into the Company's system through XML documents.

Moreover, the Company, working with custom-oriented services, also includes flexible approaches regarding a high quality relation and service in terms of personal relations with their clients and business partners. This is performed in different ways, namely by enabling to combine dishes through several distinct menus, but also by enabling training support for clients that may want to directly participate on the event's organization. Therefore, let us reminder that the third skill of Mass Customization principles, referred by Zipkin is logistics, and this aspect can also be facilitated by using information registered and processed through XML documents. Moreover, this is a quite important aspect, since the time between the auscultation of clients specifications about events and dishes, and so on, and the event starting data is usually very narrow, which makes it even more important and necessary to be able to provide responsibility sharing with clients and associated business partners in the preparation of the whole event's scenario. For this purpose, the XML documents and related technologies perform well as they enable to also perform fast, through and interconnected and integrated way, for instance, in terms of gathering signed documents, which is another important step that can be carried out precisely and fast, in a real-time basis, through the Company's integrated system infrastructure based on its platform.

26.5 Conclusion

Given the complexity in terms of number of variables and possible combinations of dishes, drinks and desserts, and, moreover, the complex and amount of data that turns out from events organization and accomplishing, the Fragus Company needed it realized that a truly integrated system was needed, for enabling to easily use and allow a set of activities, including a quickly way for creation of new dishes or new combination on existing baseline dishes in an accurately and integrated process, through a really customer-oriented and personalized way, and this was promoted by the use of XML web service oriented procedures and related web technologies.

Therefore, the use of XML and related technologies presented an appropriate alternative because they permitted an easy and simple handling code through which complex and diverse data structures turned out to be able to be obtained. For these reasons, the XML specification language was used and allowed to easily integrate sub-systems through the Company's web platform. Moreover, this kind of data specification also enabled to properly integrate principles of Mass Customization and Minimum Standardization combined with the use of simulation procedures.

From this integration processes directly outcome fundamental XML documents with a hierarchical and precise structure, enabled through documents validation by using corresponding DTDs, which act as a kind of grammar for specifying valid data structures.

Moreover, the Mass Customization principles were also directly and indirectly turned possible, through appropriate integrated processes, not only with customers, but also with business partners, which also play a fundamental role in a daily basis in the whole scenario of the Company's events organization processes. Therefore, XML and related technologies suit well the Company's needs and are recommended for systems that use large databases, besides being a specification language that allows the availability of data structures for several programs even cross platforms and enterprises collaborating on events organization. So, by using XML and related web technologies it was possible to greatly enrich the FRAGUS' platform, designed and enhanced to make possible easy information and processes interactions, turning documents truly integrated and accessible to its employees, customers and business partners, which turns out to be extremely important due to the need for an agile response for events organization and accomplishing. Moreover, under the scope of the Mass Customization concept, it was easier for the Company to consider and incorporate customer needs and new specifications proposals for events preparation, while being able to remain itself competitive in the scope of the nowadays extended global market context, which is currently occurring in all kind of areas, including industrial and service oriented ones.

Acknowledgments The authors wish to acknowledge the support of: (1) The Foundation for Science and Technology – FCT, under the scope of the financed Project on "Ubiquitous oriented embedded systems for globally distributed factories of manufacturing enterprises" – PTDC/EME-GIN/102143/2008.

References

1. Eason G, Noble B, Sneddon IN (1955) On certain integrals of Lipschitz-Hankel type involving products of Bessel functions. Phil Trans Roy Soc Lond A247:529–551
2. Castro AA, Varela MLR, Carmo-Silva S (2011) An architecture for a web service based product configuration information system. Conference on Enterprise Information Systems (CENTERIS'10), Viana do Castelo, Portugal
3. Zipkin Pau (2001) The limits of mass customization. MIT Sloan Manage Rev 42(3):81, 7p, 1 cartoon
4. Piller FT, Moeslin Katharin, Stotko CM (2004) Does mass customizaton pay? An economic approach to evaluate customer integration. Prod plan Control 15(4):435–444
5. Brabazon PG, Maccarthy B (2003) Virtual-build-toorder as an order fulfilment model for mass customization. MCPC 2003 Proceedings of the 2003 world congress on mass customization and personalization, Technische Universitaet Muenchen
6. Chamberlin EH (1962) The theory of monopolistic competition. Harvard University Press, Cambridge, MA
7. Cox MW, Alm R (1998) The right stuff. America's move to mass customization. Federal Reserve Bank of Dallas, Dallas

8. Dellaert B, Syam N (2002) Consumer-product interaction: a strategic analysis of the market for customized products. Review of marketing science working paper #1, Berkeley Electronic Press, Berkeley
9. Helander MG, Jiao JX (2002) Research on E-product development (ePD) for mass customization. Technovation 22:717–724
10. Ceponkus A, Hoodbhoy F (1999) Applied XML. Wiley Computer Publishing, New York
11. Clark J (1999) XSL Transformations (XSLT), version 1.0. W3C, at: http://www.w3c.org/TR/xslt
12. Chinnici R et al (2006) Web services description language (WSDL) version 2.0. http://www.w3.org/TR/wsdl20
13. Varela L, Aparício J, Silva C (2005) A scheduling web service. In: Kendall G, Burke E, Petrovic S, Gendreau M (eds) Multidisciplinary scheduling – theory and applications, ISBN: 0-387-25266-5, Springer, 15 p
14. Xiang A, Liu L, Luo Q (2008) VPeers: a peer-to-peer service discovery framework for Virtual Manufacturing Organizations. Comput Ind 59(5):411–419
15. Papazoglou MP, Krämer BJ, Yang J (2003) Leveraging web-services and peer-to-peer networks. In: Proceedings of the Advanced Information Systems Engineering – 15th international conference, CaiSE, Klagenfurt, Austria, pp 485–501, 16–18 June 2003
16. Rocha F, Varela MLR, Carmo-Silva S (2009) A peer based distributed database for supporting collaborative product design, iDEMi'09 – international conference on Integration of Design, Engineering and Management for innovation, Porto, Portugal

Chapter 27
Differential Evolution on the Minimization of Thermal Residual Stresses in Functionally Graded Structures

T.A.N. Silva and M.A.R. Loja

Abstract Global optimization techniques present considerable advantages when applied to non-linear and/or non-convex design spaces, where local search techniques can easily be trapped in local minima. In the present work, it is considered the application of Differential Evolution to the optimization of thermal residual stresses distribution in a sandwich panel, which is composed by an aluminium core and functionally graded outer layers. With this aim, numerical examples were carried out in order to evaluate the influence of different design parameters on the thermal residual stresses distribution. From those results, it is possible to conclude from the adequacy of the Differential Evolution strategy to minimize thermal residual stresses values, under different scenarios. It is worth to note the obtained increasing smoothness of residual stresses distribution, specially on the material transition interface.

Keywords Functionally graded material • Residual stresses optimization • Differential evolution

T.A.N. Silva (✉)
ADEM/ISEL– Instituto Superior de Engenharia de Lisboa, Av. Conselheiro Emídio Navarro, 1, 1959-007 Lisboa, Portugal
e-mail: tasilva@dem.isel.ipl.pt

M.A.R. Loja
IDMEC/LAETA – Instituto de Engenharia Mecânica, Av. Rovisco Pais, 1, 1049-001 Lisboa, Portugal
e-mail: amelialoja@dem.isel.ipl.pt

A. Madureira et al. (eds.), *Computational Intelligence and Decision Making: Trends and Applications*, Intelligent Systems, Control and Automation: Science and Engineering 61, DOI 10.1007/978-94-007-4722-7_27, © Springer Science+Business Media Dordrecht 2013

27.1 Introduction

Dual-phase functionally graded materials (FGM) are a particular type of composite materials, whose properties are tailored to vary continuously, depending on the composition distribution of its two constituents, and which use is increasing on the most diverse application fields. These materials are known to provide superior thermal and mechanical performances when compared to the traditional laminated composites, because of its continuous properties variation characteristic, which enables among other advantages smoother stresses distribution profiles. Moreover, it is well known that abrupt transitions in material properties within a composite structure can cause stress concentrations, which can be mitigated by gradually varying the microstructure and/or composition of materials in a gradient architecture [1].

The concept of FGM emerged in the 1980s [2, 3]. Since then, one can conclude of a great research involvement on the study of these materials. In fact one of the significant capabilities of these materials are related to their behaviour in high temperature conditions, thus being initially thought as thermal barriers [4].

Birman and Byrd [5] presented a work where the principal developments in FGMs are focused. Diverse areas relevant to various aspects of theory and applications of FGM, such as homogenization of particulate FGM, heat transfer issues, stress, stability and dynamic analyses, testing, manufacturing and design, applications, and fracture. The critical areas where further research is needed for a successful implementation of FGM in design are outlined in the conclusions.

Among several other published works, one can refer the one due to Reddy [6], which proposed theoretical formulation, Navier's solutions of rectangular plates, and finite element models based on the third-order shear deformation plate theory for the analysis of through-thickness functionally graded plates. The plates are assumed to have isotropic, two-constituent material distribution through the thickness, and the modulus of elasticity of the plate is assumed to vary according to a power-law distribution in terms of the volume fractions of the constituents.

Often the behaviour of complex systems, which exhibit large oscillations and various local minima, results from the interaction of multiple factors or design variables. Optimization approaches must be able to deal with this non-linear and/or non-convex design spaces and still find the best global solutions.

Population-based evolutionary algorithms, such as genetic algorithm (GA) [7], particle swarm optimization (PSO) [8], differential evolution (DE) [9] and ant colony optimization (ACO) [10], are nature inspired computing paradigms able to predict complex emergent behaviours, being suitable to be used in structural optimization problems.

Kitayama et al. [11] submitted DE through benchmark test functions including some examples of structural optimization problems, such as topology optimization of truss structures and optimum design of elastic springs. In order to evaluate DE performance the authors compared its results with PSO and generalized random tunnelling algorithm (GRTA) and highlight similarities between those techniques. DE proved to be very robust and suitable to be applied with advantages as a global optimizer. Another applications of DE algorithm to topology optimization appear on [12, 13]. DE algorithm was also used to optimize the backface geometry of a centrifugal impeller, constructed using a commercial CAD software, with respect to the calculated maximum von Mises stress, in order to extend its overspeed limits [14]. The backface geometry is defined using a Bezier curve, which parameters are used as design variables. The optimization results showed a significant improvement in admissible maximum rotational speed, with respect to a reference geometry. The authors also compared DE with a simulated annealing (SA) algorithm, reporting a higher robustness of DE, which converges to a better solution for the same number of evaluations and computation time, due to the capability to avoid local minima presented by DE algorithm. A different approach can be found in [15], where DE was applied to the optimization of the continuous processing technology, based on twin screw extrusion (TSE), by the integration of a performance model of a FGM burning rate to identify the optimal gradient material microstructure for rocket motor applications. Babu and Munawar [16] applied DE to find the optimal design of heat exchangers, through the estimation of the minimum heat transfer area. During the optimization process, DE proved to be a potential source for accurate and faster optimization, being significantly faster when compared to GA.

The present work is focused on the optimization of thin and/or moderately thin sandwich functionally graded panels with metallic core. Its main goal is to minimize thermal residual stresses magnitudes and gradient profile. The achievement of this objective will be attainable by carrying out a study on the influence of optimization parameters on the optimization process and goals.

27.2 Modelling of an FGM Sandwich Structure

A FGM is typically a particulate composite material, on which we encounter spherical or nearly spherical particles embedded in an isotropic matrix. In this paper it is considered a dual-phase FGM constituted by metallic and ceramic particles.

This concept can also be found in typical laminated composites with fibre reinforcement, for bending stiffness improvement purposes. Therefore, in the outer layers a grading embedding scheme can be adopted without a significant weight addition. Concerning to the manufacturing of such solutions, it is easier to adjust the volume fraction of particles in these layers than to proceed to a variation of the fibre volume fraction or a variation of the fibre orientation.

Precise information about the size, the shape and the distribution of particles is not available, thus the effective properties of graded microstructures must be evaluated based on the volume fraction distribution and the approximate shape of the dispersed phase (the ceramic). This approach allows us to make a prediction on the overall material properties, bearing in mind that the validity of the underlying assumptions can be affected by the detailed graded microstructure.

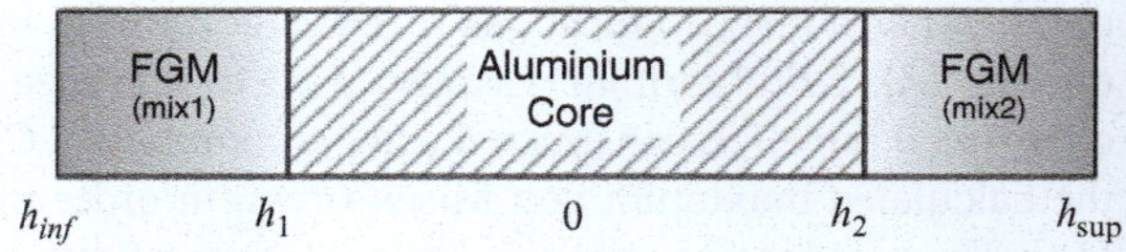

Fig. 27.1 Schematic representation of the FGM beam under study

Usually FGM structures are submitted to high temperature environments. Hence, it is necessary to take into account the effective properties of these materials, which depend not only on the position but also on the temperature [17]. In order to allow for a better prediction of the mechanical behaviour of these structures, we explicitly consider these parameters to define the effective material properties $P_{eff}(z,T)$ of a FGM, using the rule of mixtures of Voigt [18], where the material property P associated to a generic thickness coordinate z and a certain temperature T, depends on the corresponding metal and ceramic properties, P_m and P_c, weighted by the corresponding volume fractions, V_m and V_c.

In this study, it is considered a sandwich panel which core is metallic and its outer layers are made of a FGM. In these layers, the distribution of the metal volume fraction across the thickness $V_m(z,p)$, with $z \in [h_{\text{inf}}, h_{\text{sup}}]$, being h_{inf} and h_{sup} the coordinates of the lower and upper outer surfaces of the sandwich (Fig. 27.1) and the exponent of the metal volume fraction p [19].

27.3 Thermal Residual Stresses

Thermal residual stresses often arise when there is a cooling after high temperature processing, but also due to problems related to chemical compatibility. Regardless the motive, the fact is that the difference, sometimes significant, in the thermal expansion characteristics of the constituents is an inevitable drawback.

Ravichandran [20] presented calculations of thermal residual stresses limited to one-dimensional linear elastic cases, reporting the invariable increase of residual stresses when the inclusion of fully metal and/or ceramic regions is used. However, the lower magnitude of thermal residual stress is achieved if a linear gradation ($p = 1$) of the FGM constituents is considered. On the other hand, if $p>1$ a concave down stress profile can be obtained. Shaw [21] conducted studies on plates and coatings in order to define the effect of different materials combinations on the magnitude and distribution of thermal residual stresses in FGMs and multi-layered materials. The author states that it is possible to adjust the distribution and magnitude of thermal residual stresses through a proper combination of materials and compositional gradient. By this, it is possible to generate compressive residual stresses on the FGM surfaces, which can result in superior strength and fracture resistance. In order to estimate the residual stress state in FGM structures, Becker Jr. et al. [22] proposed an approximated method. The performance of this method was compared with analytical solutions and finite element method results for a series of problems with nonlinear spatial variations in the thermal expansion coefficient. The proposed method satisfactorily approximates the solution at a lower computational cost, allowing to avoid stress concentrations or to localise compressible residual stresses in a FGM free surface.

In this study, it is considered a sandwich FGM panel, infinitely long in the width direction and with the thickness characteristics already mentioned (Fig. 27.1), in order to minimize the thermal residual stresses defined according to Ravichandran [20]. Note that, in the present work, it was neglected the temperature dependent properties.

27.4 Optimal Design of FGM Sandwich Structures

27.4.1 Differential Evolution Fundamentals

Developed by Price and Storn [9] between 1994 and 1996, the DE algorithm is considered to be an improved version of GAs, which placed 3rd on the 1st International Contest on Evolutionary Computation, and 1st if one excludes the non-GA type algorithms [23]. DE basis was built upon four key requirements: (i) ability to handle non-differentiable, nonlinear and multimodal cost functions; (ii) ability to run in parallel computation to deal with computationally intensive cost functions; (iii) ability to use few control variables, being those robust and easy to choose; and, (iv) capability to have consistent convergence to the global minimum in consecutive independent trials. DE is a population-based approach to function optimization that generates trial individuals by calculating vector differences between other randomly selected members of the population. Given an objective function to be minimized, DE begins by randomly generating population vectors of n members per design variable. These vectors, also known as individuals, will evolve over the progression of the optimization procedure, simulating an evolving population. For each population or iteration, the objective function is evaluated n times, as much as the number of members per individual, trying to find the best population member until a termination criterion is achieved. An implementation of DE algorithm can be found in a pseudo code version on [9]. DE congregates characteristics from well-known global optimization and meta-heuristics algorithms. Although it is a global search technique, DE preserves a search direction vector, which gives an important descent property, and the randomness of populations, which improves the method's robustness [11]. The choice of DE's control variables is not a difficult task and reasonable values are easy to find in the literature. Several algorithms' schemes are available and differ according to the number of difference vectors computed and if the current individual or the global best one will be used or not as part of that computation. As we can find in the literature, DE/best/1/bin seems the most promising DE strategy of all [16]. Recent developments regarding DE optimization techniques and applications can be found on [24, 25].

In the present work, we use the DE/best/1/bin optimization strategy, tailored for small population sizes and fast convergence, where *best* specifies the vector of lowest cost from the current population to be mutated, *1* the number of difference vectors used for calculation, and *bin* a binomial crossover scheme.

27.4.2 Optimal Design Problem

A generic minimization problem can be stated as:

$$\min \Omega(b) \tag{27.1}$$

subject to:

$$b_i^{low} \leq b_i \leq b_i^{up},\ i = 1, \ldots, n_{dv}\ and\ \Psi_j(q, b) \leq 0,\ j = 1, \ldots, n_{bc}$$

where $\Omega(b)$ is the objective function, b is the design variables vector, which values are bounded by the lower and upper limits b_i^{low} and b_i^{up}, respectively, and $\Psi_j(q,b)$ are the n_{bc} inequality constraint equations. Note that the length of b is limited by the total number of design variables n_{dv}.

In this context, the design variables can be thicknesses and/or the exponents of the metal volume fraction law that dictate how the material combination will be characterized across the thickness coordinate of the FGM layers. These limits imposed to the design variables are used as boundary constraints of the optimization process. The objective function considered in this study is the maximum value of the thermal residual stress function associated to a thin or moderately thin FGM sandwich panel with metallic core.

27.5 Numerical Examples

A dual-phase functionally graded sandwich beam with a metallic core is considered in this study (Fig. 27.1). For this purpose the core of the sandwich is made of aluminium and the outer layers are a FGM obtained by mixing titanium alloy and zirconia according to the Voigt rule of mixtures. The properties of these materials are given in [17].

The authors establish the maximum value of the thermal residual stresses as common objective function to be minimized for the considered test cases. There are some constraints that are common to the analysed cases, namely the total thickness, $h_t = 0.04m$, and the thickness of both FGM layers are equal, preserving symmetry. The test cases under study, described on the following subsections, differ with respect to the considered design parameters, namely, the thickness of the FGM

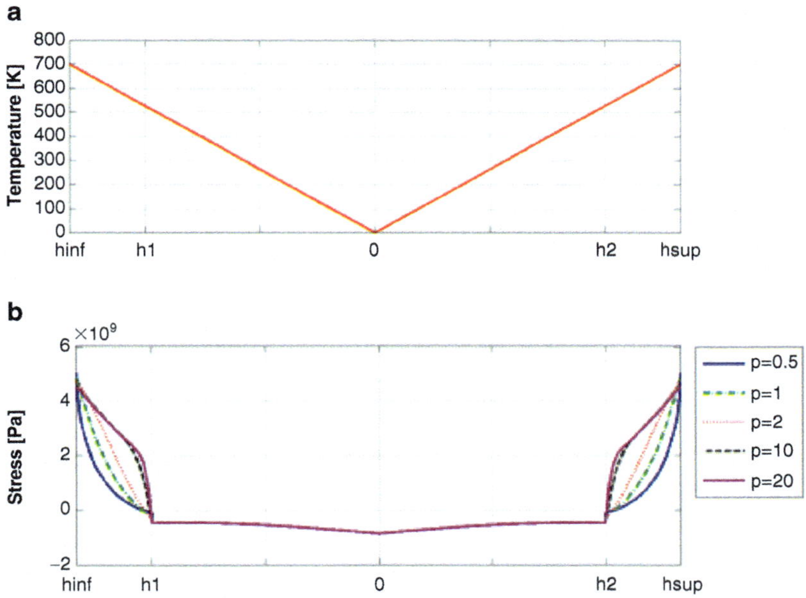

Fig. 27.2 (**a**) Temperature distribution along material thickness z; and (**b**) Comparison of stress profile distributions for different values of p, preliminary results

layer (h_{FGM}) and the exponent (p) of the metal volume fraction power law. In terms of DE parameters, it was considered for all cases DE/best/1/bin with a selection weighting factor $F = 0.5$ and a binomial crossover probability constant $CR = 1$.

27.5.1 *Test Case – Symmetric Temperature Distribution*

Considering a symmetric temperature distribution as shown on Fig. 27.2a a DE optimization procedure was run. From preliminary results, we found that the higher value of thermal residual stresses decrease with an increasing value of p (Fig. 27.3). However, a maximum value of $p = 10$ was defined. This value was established within the range of values published [17], after several trial values. Note that we assume the following reference values: $p = 10$ and $h_{FGM} = 0.005m$.

27.5.1.1 For Constrained p and Fixed h_{FGM}

In this case, it was considered $0 \leq p \leq 20$ and $h_{FGM} = 0.005m$. As expected the best solution was found for the value of p which matches the maximum limit value allowed for this design variable (Fig. 27.3a), however with no significant

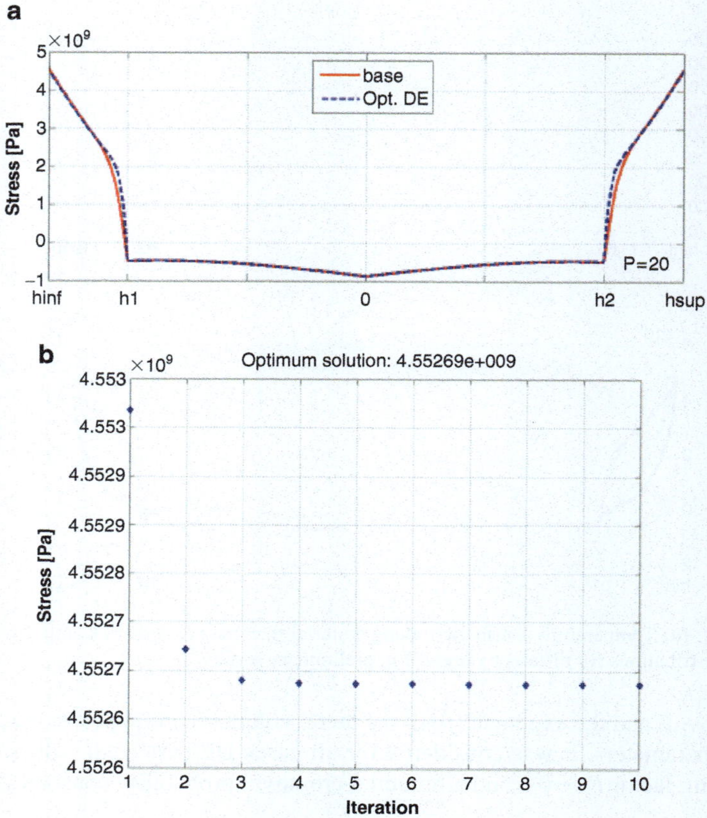

Fig. 27.3 (**a**) Comparison of stress profile distributions for different values of p, with fixed h_{FGM} Case I.1; and (**b**) Optimum value evolution during optimization process – Case I.1

improvement. From Fig. 27.3b it can be concluded from a fast convergence, corresponding to the optimum maximum stress value and concerning to the design variables.

27.5.1.2 For Side Constraints p and h_{FGM}

In this case, the parameters were bounded as follows: $0 \leq p \leq 10$ and $5.0 \times 10^{-5} \leq h_{FGM} \leq 1.5 \times 10^{-2}[m]$. The maximum number of iterations considered was 10. The results obtained in this process are presented in Fig. 27.4. The optimum solution is now 4.30805 GPa corresponding to the design variables combination: $p = 9.9998$ and $h_{FGM} = 0.01246$m.

In order to analyse the convergence a similar study was carried out considering now 100 populations (iterations), being the corresponding obtained results presented

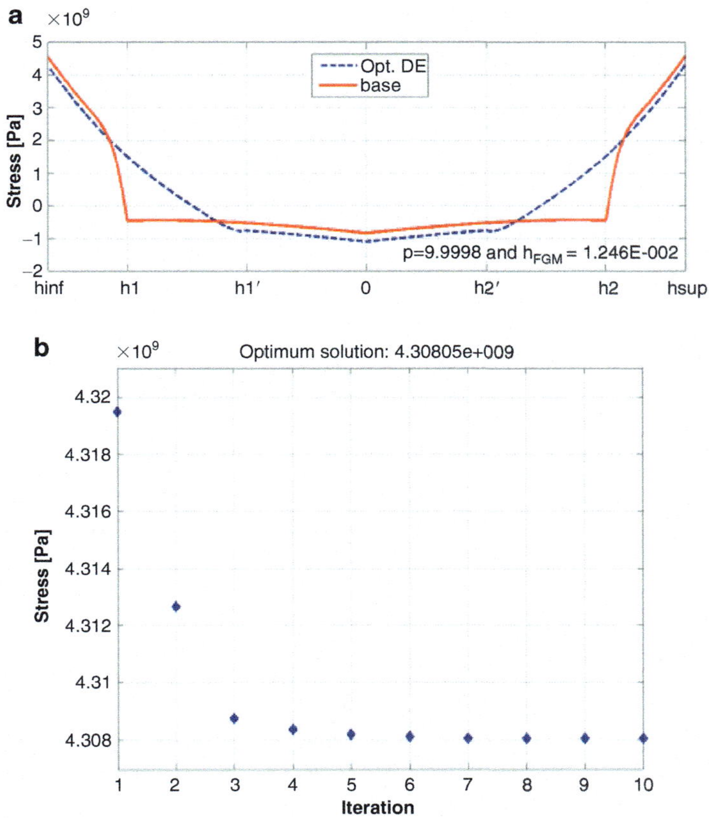

Fig. 27.4 (**a**) Comparison of stress profile distributions for different values of p and h_{FGM} Case I.2 (10 populations); and (**b**) Optimum value evolution during optimization process Case I.2 (10 populations)

in Fig. 27.5. With a higher number of iterations, no visible improvement occurs, being the optimum solution equal to 4.30805 GPa for the design variables combination: $p = 10.00$ and $h_{FGM} = 0.01246$m.

27.5.2 Conclusions

The present work deals with thermal residual stresses magnitudes and profile distributions optimization in FGM panels using Differential Evolution strategy. A very important conclusion to withdraw is related to the smoother stresses distribution profile obtained using DE. This is a fundamental matter because it is

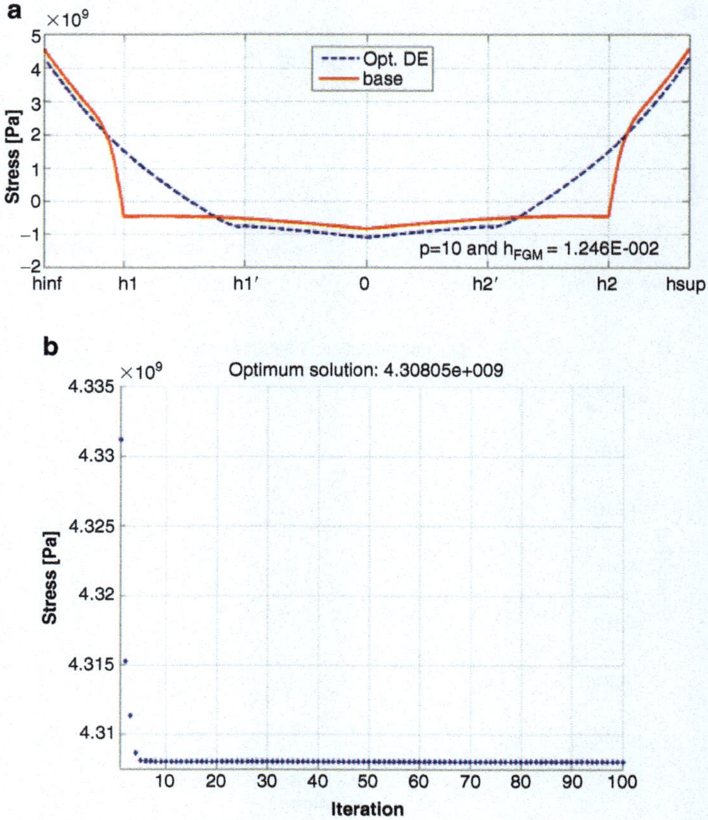

Fig. 27.5 (**a**) Comparison of stress profile distributions for different values of p and h_{FGM} Case I.2 (100 populations); and (**b**) Objective function value evolution during optimization Case I.2 (100 populations)

well known that abrupt transitions in material properties within a composite structure can cause stress concentrations, which can be mitigated by a gradual variation of the microstructure and/or composition of materials in a gradient architecture. In this sense, an advantageous concave up stress profile is obtained. Furthermore, the considered structure only has one of each FGM's surfaces constrained by the metallic core, which results in a decrement of the residual stresses magnitude both in the material transition area and in the free surfaces.

Acknowledgments This work was partially supported by the Portuguese Foundation for Science and Technology (FCT) through the Project PTDC/EME-PME/120830/2010 and the PhD grant SFRH/BD/44696/2008.

References

1. Surendranath H, Bruck HA, Gowrisankaran S (2003) Enhancing the optimization of material distributions in composite structures using gradient architectures. Int J Solids Struct 40 (12):2999–3020
2. Yamanoushi M, Koizumi M, Hiraii T, Shiota I (eds) (1990) In: Proceedings of the first international symposium on functionally gradient materials, Sendai
3. Koizumi M (1993) The concept of FGM. Ceram Trans 34:3–10
4. Uemura S (2003) The activities of FGM on new applications. Mater Sci Forum 423–425:1–10
5. Birman V, Byrd L (2007) Modeling and analysis of functionally graded materials and structures. Appl Mech Rev 60:195–216
6. Reddy J (2000) Analysis of functionally graded plates. Int J Numer Method Eng 47:663–684
7. Goldberg DE (1989) Genetic algorithms in search, optimization and machine learning, 1st edn. Addison-Wesley Longman Publishing Co, Reading
8. Kennedy J, Eberhart RC (1999) The particle swarm: The social adaptation in information-processing systems. In: Corne D, Dorigo M, Glover F (eds) New ideas in optimization. McGraw-Hill, London, pp 379–388
9. Storn R, Price K (1997) Differential evolution – a simple and efficient heuristic for global optimization over continuous spaces. J Global Optim 11:341–359
10. Deneubourg J, Aron S, Goss S, Pasteels J (1990) The self-organizing exploratory pattern of the argentine ant. J Insect Behav 3:159–168
11. Kitayama S, Arakawa M, Yamazaki K (2011) Differential evolution as the global optimization technique and its application to structural optimization. Appl Soft Comput 11(4):3792–3803
12. Wu C-Y, Tseng K-Y (2010) Topology optimization of structures using modified binary differential evolution. Struct Multidiscip Optim 42:939–953
13. Vu A-T, Werner F (2009) Optimization of steel frame structures based on differential evolution algorithm, In: 18th International Conference on the application of computer science and mathematics in architecture and civil engineering, Weimar
14. Valakos I, Ntipteni M, Nikolos I (2007) Structural optimization of a centrifugal impeller using differential evolution in Catia environment. Oper Res 7:185–211
15. Bruck H, Gallant F, Gowrisankaran S (2002) Development of a novel continuous processing technology for functionally graded composite energetic materials using an inverse design procedure In: 2002 SEM annual conference and exposition on experimental and applied mechanics, Milwaukee, pp 296–302
16. Babu B, Munawar S (2007) Differential evolution strategies for optimal design of shell-and-tube heat exchangers. Chem Eng Sci 62(14):3720–3739
17. Reddy JN, Chin CD (1998) Thermo-mechanical analysis of functionally graded cylinders and plates. J Therm Stresses 21(6):593–626
18. Li L, Wang T (2005) A unified approach to predict overall properties of composite materials. Mater Charact 54(1):49–62
19. Nguyen T-K, Sab K, Bonnet G (2008) First-order shear deformation plate models for functionally graded materials. Compos Struct 83(1):25–36
20. Ravichandran K (1995) Thermal residual stresses in a functionally graded material system. Mater Sci Eng A 201:269–276
21. Shaw LL (1998) Thermal residual stresses in plates and coatings composed of multi-layered and functionally graded materials. Composites B 29B:199–210
22. Becker T Jr, Cannon R, Ritchie R (2000) An approximate method for residual stress calculation in functionally graded materials. Mech Mater 32:85–97
23. Babu B, Jehan M (2003) Differential evolution for multi-objective optimization. In: Proceedings ofthe 2003 Congress on Evolutionary Computation (CEC-2003), Canberra, Australia, pp 2696–2703
24. Price K, Storn RM, Lampinen JA (2005) Differential Evolution: a practical approach to global optimization, ser. Natural computing series. Springer, New York
25. Chakraborty UK (2008) Advances in differential evolution, 1st edn. Springer, New York

Chapter 28
Magnetic Wheeled Climbing Robot: Design and Implementation

M.F. Silva, R.S. Barbosa, and A.L.C. Oliveira

Abstract This paper presents a climbing robot, with wheeled locomotion and adhesion through permanent magnets. This machine was developed to be used in the inspection of several types of man-made ferromagnetic structures, such as towers for wind turbines, fuel storage tanks, ship hulls, etc., in order to, for instance, detect weaknesses due to corrosion. In this article are presented the main aspects taken into consideration for its design, as well as several constructive aspects, among which are detailed its mechanical and electrical construction, the implemented control architecture and the Human-Machine Interface (HMI) for its control. The distinguishing characteristic of this robot is its dynamic adjustment system of the permanent magnets to assure the machine adhesion to the surfaces when crossing slightly irregular and curved surfaces with a large radius.

28.1 Introduction

The interest in the development of climbing robots has grown rapidly in recent years, since these devices can be adopted in varied applications such as maintenance, building, inspection and safety, mainly in the process and construction industries. They are mainly adopted in places where direct access by humans is expensive, because of the need for scaffolding or special structures, or dangerous, due to the presence of a hostile environment. The main motivations for its use are to increase the operation efficiency, by eliminating the costly assembly of scaffolding, or to protect human health and safety in hazardous tasks. Several climbing robots

M.F. Silva (✉) • R.S. Barbosa • A.L.C. Oliveira
Departamento de Engenharia Electrotécnica, GECAD – Grupo de Engenharia do Conhecimento e Apoio à Decisão, ISEP – Instituto Superior de Engenharia do Porto, Rua Dr. António Bernardino de Almeida, Porto 4200-072 Porto, Portugal
e-mail: mss@isep.ipp.pt; rsb@isep.ipp.pt; 1900166@isep.ipp.pt

A. Madureira et al. (eds.), *Computational Intelligence and Decision Making: Trends and Applications*, Intelligent Systems, Control and Automation: Science and Engineering 61,
DOI 10.1007/978-94-007-4722-7_28,

have been developed, and others are under development, for applications ranging from cleaning to inspection of difficult to reach constructions [1].

Magnetic adhesion is a principle adopted for the creation of an adhesion force, in cases where the surface allows it, and it can be highly desirable due to its inherent reliability. This method is fast but, depending on the weight of the robot, may involve the use of heavy actuators to obtain the required adhesion force [1]. It is possible to use electromagnets [2] or permanent magnets for assuring the robot adhesion to surfaces, combined with the use of wheels (as in this case) or tracks to move [3]. The main advantages of using permanent magnets is that there is no need to spend energy in the adhesion process, the robot does not experience any loss of adhesion in the case of a power failure and the fact that they are suitable for application in hazardous environments, such as the ATEX zones associated with flammable products stored in warehouses or tanks [4]. Among the drawbacks of this adhesion strategy, one can mention the fact that if the surface is very thin, it can deform and bend. In the limit situation, the surface can enter in contact with the magnets, this way making difficult the locomotion of the robot. Another problem that can arise with this technology is associated with the fact that the surface can present irregularities. These irregularities can make the magnets very close to the surface, even causing their contact, leading to the manifestation of an increased friction; the opposite situation can also occur, i.e. an exaggerated clearance, which might jeopardize the ability of the robot to adhere to the surface where it is moving, that could even lead to his downfall [5].

To overcome these situations, in this machine is implemented a device to vary the distance between the magnets and the surface where the robot moves, depending on the dimensions of the detected irregularities, in order to maintain this distance constant and controlled. Thus, we propose a robot with permanent magnets, which are adjusted in real time to the surface of displacement, through the adoption of a system to detect the magnets distance from the surface, using two inductive sensors combined with a support structure coupled to an actuated worm shaft.

Bearing these ideas in mind, the sequel of this paper is organized as follows. Section 28.2 introduces the robot main design considerations and Sect. 28.3 its mechanical structure. Section 28.4 describes its control architecture and, based on this, on Sect. 28.5 is presented the programming architecture for the vehicle. Finally, on Sect. 28.6 are presented the main conclusions of the work.

28.2 Robot Main Design Considerations

The robot described in this article was developed as a prototype of a vehicle intended to inspect man-made ferromagnetic structures, such as fuel storage tanks, towers for wind turbines, and ship hulls. These structures generally present a smooth surface, without big curvatures and are characterized by the existence of welding cords (protrusions), in the majority of the cases of small height, along the surface. Given these specifications it was decided to adopt a wheeled robot, with

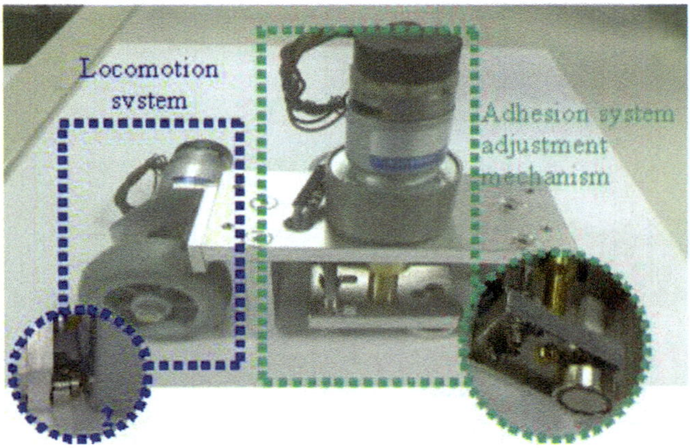

Fig. 28.1 Modular standard unit, where it is visible the locomotion system (*left*) and the adhesion system dynamic adjustment mechanism (*right*)

four wheels, and with some sort of "suspension system", to which we refer as the dynamic adjustment of the adhesion system, able to cope with the welding cords.

To choose the adequate permanent magnets for the robot under development some considerations regarding the weight of the final vehicle were first made.

The idea was to build a robot adopting a modular structure, being constituted by a rigid PVC base and four equal standard units, assembled on it [6]. Each of these units (Fig. 28.1) is composed of an actuated locomotion system (a set motor/wheels for locomotion) composed by a pair of wheels, one in rubber and the other magnetic (that was not considered in the initial phase of the project) (Fig. 28.1, lower left corner), and an actuated structure to adjust the permanent magnet distance to the displacement surface (a set motor/magnet for adhesion) (Fig. 28.1, lower right corner). The assembly of the four modules, to give rise to the robot, can be seen in Fig. 28.2, where it is possible to clearly identify the four equal standard units.

The weight of each motor, given by the manufacturer is 1.78 N. Since each module possesses two motors, their weight is 3.56 N. Each rubber wheel weights 1.36 N. For the modular metallic structure (in steel) we estimated a weight of approximately 5 N, for each unit. Summing up these values, it was considered a weight of 9.92 N for each module. On top of this structure was assumed that would be assembled an Allen Bradley MicroLogix 1100/1763 Programmable Logic Controller (PLC) weighting 8.3 N, without considering the I/O modules. These modules add about 4.9 N to the vehicle weight. Furthermore, the adoption of a few printed circuit boards (PCB) was predicted, for the implementation of the discrete control of the motors of the vehicle. For all the needed PCB we estimated a weight of 5 N. Summing all these values, we concluded that the total estimated weight of the vehicle would be around 57.88 N.

Based on these estimates for the total robot weight, the worst condition was assumed: the robot would need to be supported on the displacement surface upside down. This meant that the four permanent magnets had to develop a force, at least, equal to 57.88 N (or approximately 15 N for each magnet).

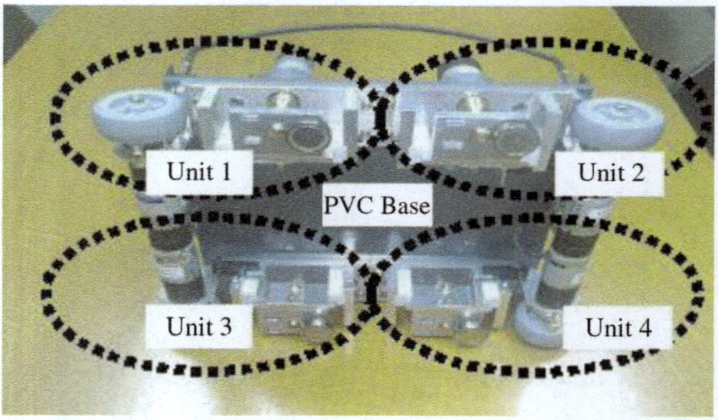

Fig. 28.2 Robot built using four equal sets, for locomotion and vertical position magnet adjustment

The magnetic force developed by permanent magnets depends heavily on the following factors: distance between magnet and object (the force decays rapidly with the distance), material of object of magnetic attraction (there is a loss of 30% of the magnetic force if the surface where the robot moves is built of construction steel, which is the usual for the applications under consideration), direction of the force, and surface area and thickness of the object of magnetic attraction.

Given these considerations, was assumed the rather unfavourable situation that on normal robot operating conditions the magnet would only develop approximately 15% of the maximum possible force. This implied that the maximum achievable force by the permanent magnet would be around 100 N. Looking at the SUPERMAGNETE manufacturer catalogue permanent magnets characteristics, were chosen SUPERMAGNETE FTN-20 permanent magnets. Each of these permanent magnets presents a maximum magnetic force of approximately 108 N, under "optimum conditions".

28.3 Mechanical Structure

28.3.1 Adhesion System

The adhesion module is responsible for supporting the robot when it is placed in a vertical ferromagnetic surface. This unit is composed by the permanent magnets and by two inductive sensors, responsible for detecting the distance to the locomotion surface, as can be seen in Fig. 28.3.

The actuated mechanism for the real time adjustment system is composed by the motor (Gearmotor 12 V DC, 33 rpm) coupled to the support structure, the permanent magnet (SUPERMAGNETE FTN-20) and the inductive sensors (IFM IY5049). Using the distance information obtained by the two inductive

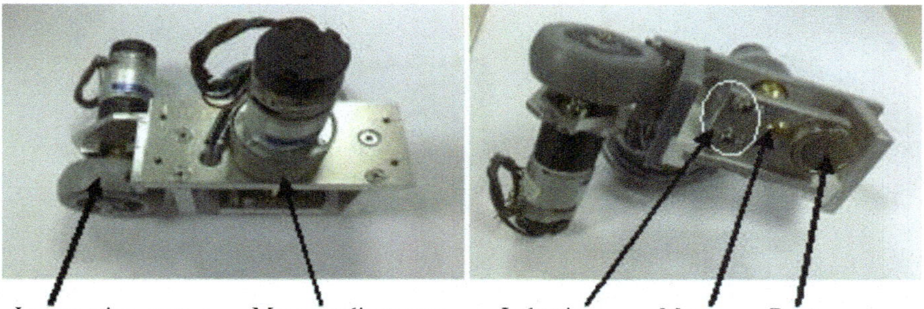

Fig. 28.3 The two inductive sensors, motor screw and magnet

sensors, it dynamically adjusts the magnet distance to the contact and locomotion surface using an actuated worm shaft [7], to keep a constant distance between the permanent magnets and the surface where the robot is moving, in the presence of irregularities.

28.3.2 Locomotion System

Regarding the locomotion system, each standard unit consists of a motor (Gearmotor 12 V DC, 62 rpm, with encoder) and a 50 mm diameter wheel, with rubber exterior to improve the adhesion to the surface of displacement. Each of these structures, which are mounted on a set that also contains the system of vertical adjustment of the magnet, presents in an almost parallel assembly to the outermost part of the wheel, a 10 mm diameter circular magnet with the possibility of rotation in synchronism with its wheel. Its purpose is to maximize the adherence to the contact surface, and allowing the robot to move from a horizontal to a vertical surface.

In order to improve the adherence of the four locomotion units to the contact and locomotion surface, even when it does not present perfectly flat and regular characteristics, an auto-levelling structure was implemented in the two rear systems (Fig. 28.4), which is composed of a mechanism with a sliding guide and compensated through a spring.

28.4 Control Architecture

28.4.1 PLC Inputs/Outputs

The control of this robot is based on an Allen Bradley MicroLogix 1100/1763 PLC. One characteristic of this PLC is that it presents a rather limited number of I/O, being necessary to attach one expansion board with 8 digital inputs (1762-IQ8) and

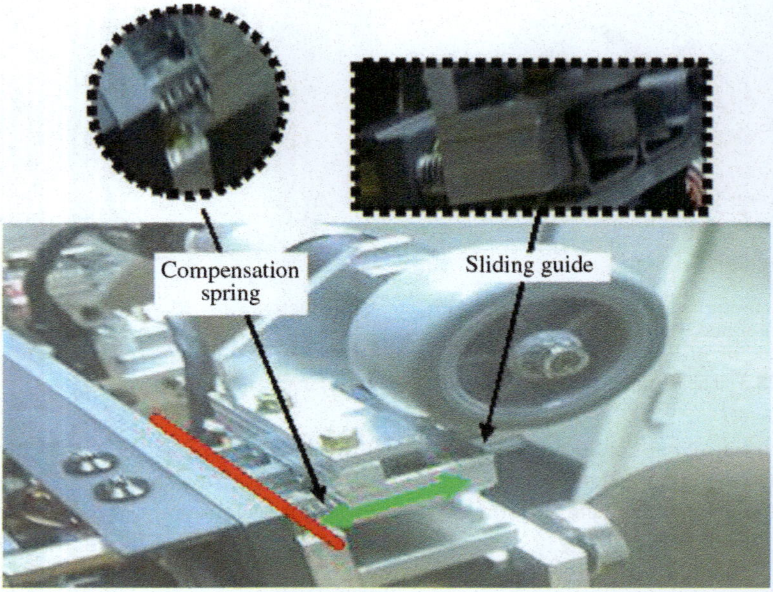

Fig. 28.4 Photo of the auto-levelling system, with the sliding guide and spring compensated structure

another with 8 digital relay outputs (1762-OW8). Since the PLC allows the use of 10 inputs, there are a total of 18 inputs available for use, of which only two are direct analog inputs. These can be configured to be connected to devices that have voltage output (in this case, the sonars in the front and rear of the robot, used to detect obstacles).

The used outputs are obtained directly from the PLC (two using Fast FET, two using FET and two using relays) and from the expansion module (8 outputs, all of them using relays), which is coupled by a dedicated bus to the PLC.

28.4.2 Logical and DC Power Circuits

There was also the need to develop several dedicated PCBs to support the adequate control of the system [7]. This electronics is devoted to power the entire system and to its configuration and logical control (the Adhesion and the Locomotion Units are associated with a logic circuit that allows its optimized control by the PLC and have each a DC Power Supply circuit associated [7]).

In the case of the Adhesion Unit control logic electronic circuit, its working principle is based on a binary up/down counter (74HCT4516 IC), aided by logical inverters and Quad Push-Pull Drivers. Since the control over this unit has to be fast, are used PLC outputs of the type FET and Fast FET [7].

The Quad Push-Pull Drivers are of the L293D type and serve to drive the four motors, with a voltage of 12 V DC. In each L293D are combined the four Half-H Drivers to form two H bridges, which drive two motors (on the same side of the climbing robot). This system allows setting the rotation direction and the start of each motor in an individual and independent way.

Regarding the Locomotion Unit control logic electronic circuit, the operating principle is similar to what was described for the control of the Adhesion Unit, with minor differences associated with the particularities of this Unit. Since this type of control does not need to be as fast as the one in the Adhesion Unit, the Reset and Parallel Load signals are commanded by PLC relay outputs, while the Clock signal is commanded by a PLC FET output.

Regarding the Adhesion and the Locomotion Units DC Power Supply circuits these are implemented based on monolithic type regulators. The climbing vehicle possesses two of these circuits. One of them is used to provide the 5 V DC that powers all the digital circuits and 2 × 12 V DC for each pair of motors of the Locomotion Unit. The other circuit is used to provide the 5 V DC that powers the Access Point (AP) Wi-Fi 802.11 g (the system for the remote communication with the robot), and 2 × 12 V DC for each pair of motors of the Adhesion Unit.

The remaining electronic circuitry, such as the PLC and the whole set of sensors, are powered directly from the 24 V DC supply (except for the sonars, that are 5 V DC powered by the digital part). Therefore, there is a system of power distribution at the 24 V DC voltage level.

The Main Power Control Panel allows powering the entire mobile robot, and presents the future possibility of integration with batteries and their electrical charging circuit.

There is also integrated into the power supply system a DC-to-DC converter to dissipate as heat, through a controlled fan, the excess of power delivered from the 24 V DC supply, given the lower voltages also needed, such as 5 V DC and 12 V DC. This converter is powered from the 24 V DC (Main Power) and provides two distinct voltage levels at its output: 9 V DC, used to supply the monolithic type circuit that provides 5 V DC / 700 mA to the AP Wi-Fi communication system, and 12 V DC, supplying the monolithic circuit that provides 5 V DC / 200 mA to all logic / digital circuits complementary to the PLC logical operation.

28.5 Programming of the Control System

28.5.1 PLC Program

The control program adopted for the vehicle operation is implemented on the PLC with the aid of the RSLOGIX 500 programming software (Rockwell Automation).

In this program was established a main routine and four subroutines. The main program (main routine) is associated with LAD 2, and the remaining LAD (LAD 3, LAD 4, LAD 5, and LAD 6) are called by the main routine.

Besides calling the other routines, the main routine (LAD 2) is responsible for some initialization logic of the robot and for the main control over the locomotion and adhesion systems. It is also responsible for handling the logic required to connect the external logic circuits to the PLC, allowing an interface with the motors through the binary counters and the H bridges. The LAD 3 routine implements the logic required to configure and manually control (by a robot operator) the locomotion and adhesion systems. Routine LAD 4 has all the logic needed to automatically control the adhesion and locomotion systems of the robot, without requiring any operator intervention. It is this routine that controls the magnets position regarding to the robot locomotion surface. This is also the routine that handles the possibility of autonomous surface exploitation by the robot. The LAD 5 routine processes the signals sent by the sensors that control the movement of the robot, in particular as regards the sonars for obstacle detection, and the motors rotational asymmetry detected by the encoders. Finally, LAD 6 is associated with the logic control of a local manual control physical interface, which is coupled to two digital inputs in the PLC. This interface allows acting on the adhesion and locomotion systems, enabling some local features, such as:

1. Activation of the adhesion system.
2. Displacement of the four adhesion units towards the locomotion surface.
3. Stop the adhesion and locomotion systems.
4. Move the four adhesion units to the security zone, i.e. in a direction opposite to the locomotion surface.

The local manual command interface is also used for the connection of the robot sonars to the +5 V DC power supply and to the IV1 and IV2 PLC analog inputs.

28.5.2 Human-Machine Interface

Although it can be manually controlled, the vehicle is designed to have a semi-autonomous behaviour, allowing a remote inspection process controlled by a technician, this way reducing the risks associated with the human inspection of tall structures and ATEX places.

In order to have a friendlier graphical environment for those who remotely operate the climbing robot, a software application was developed and is used as a HMI. This HMI application is of the Supervision Control and Data Acquisition (SCADA) type. The used software is the RSView32 (Rockwell Automation), which provides greater ease of manipulation of the variables defined in the program that runs on the MicroLogix 1100 PLC.

It is possible to access all the configuration parameters of the adopted adhesion and locomotion process, and have feedback from the physical variables measured by the climbing robot sensors, from any computer running this SCADA, and the dedicated programming associated, with its graphical interface directed towards the remote operation of the robot (HMI).

In developing the SCADA/HMI program, which runs on the RSView32 software, it was essential to create Tags in an organized way, of the digital, analog or character (Strings) types, which are stored in the Tag Database, with the aim of obtaining a structured and synchronized programming with the PLC program and its labels.

In this project were created, and programmed with the labels previously created, seven graphs, whose general function is associated with the remote control of the climbing robot. The characteristic of each chart and its particular function is described below.

1. The graphic/display "General", which is the HMI display screen.
2. The graphic/display "Choice", which is the "gateway" to the other graphics/displays.
3. The graphic/display "Alarms Register" allows the operator to view the list of active alarms and the occurrences log file, besides allowing to act on its recognition, activation or deactivation.
4. The graphic/display "Auto" is responsible for monitoring the remote operation capability, over the surrounding environment, on automatic mode.
5. The graphic/display "CFG General" is used to set some parameters related to the functioning of the robot.
6. The graphic/display "Manual Adhesion" allows to manually operate the various features of the climbing robot that constitute the adhesion system.
7. The graphic/display "Manual Locomotion" allows to manually operate the various features of the climbing robot integrating the locomotion system.

To assist the manual remote control of the robot locomotion, it is shown in the graphic/display "Manual Locomotion" the distance travelled for each motor, allowing to identify asymmetries in the movement. There is also an indication of the distance to any obstacles that are on the front or rear of the robot. When an obstacle is within the preset distance to the robot, it stops, unless this option is disabled.

With all these graphics/displays, provided by the remote monitoring system of the robot (its HMI), it is possible for an operator to have "access" to all equipment placed on board the autonomous climbing robot. This allows the internal and external monitoring of the mobile system and the operation, in a controlled manner, of the various features available.

28.6 Conclusions

This paper presented a climbing robot with wheeled locomotion and adhesion through permanent magnets (Fig. 28.5), to be used in the inspection of various types of ferromagnetic structures. The distinguishing characteristic of this machine is its real time system for the adjustment of the magnets in order to assure the machine adhesion to the surfaces, even when crossing irregular and curved surfaces.

Fig. 28.5 Photo of the final prototype of the developed climbing robot with locomotion through wheels and adhesion with permanent magnets

The robot development is now complete and several tests have confirmed the adequacy of the vehicle to the intended application, while manually and remotely controlled, verifying its locomotion in ferromagnetic vertical surfaces with irregularities and curvatures, while in autonomous mode.

Acknowledgments The authors would like to acknowledge FCT, FEDER, POCTI, POSI, POCI, POSC, and COMPETE for their support to R&D Projects and GECAD.

References

1. Silva MF, Tenreiro Machado JA (2008) A survey of technologies for climbing robots adhesion to surfaces. In: Proceedings of the ICCC 2008 – IEEE International Conference on Computational Cybernetics, Stara Lesná, Slovakia, 27–29 Nov 2008, pp 127–132
2. Armada MA, González de Santos P, García E, Prieto M, Nabulsi S (2005) Design of mobile robots. In: Proceedings of the 2005 CLAWAR: introductory mobile robotics workshop, London, UK, 12 Sep 2005, pp 2890–2895
3. Sánchez J, Vázquez F, Paz E (2006) Machine vision guidance system for a modular climbing robot used in shipbuilding. In: Tokhi MO, Virk GS, Hossain MA (eds) Climbing and walking robots, Springer, London, UK, Feb 2006, ISBN 3-540-26413-2, pp 893–900
4. Berns K, Braun T, Hillenbrand C, Luksch T (2005) Developing climbing robots for education. In: Armada MA, Gonzaléz de Santos P (eds) Climbing and walking robots, Springer, Madrid, Spain, Sept 2005, ISBN 3-540-22992-2, pp 981–988
5. Akinfiev T, Armada MA (2001) On the optimal location of a magnet gripper for a climbing robot. In: Berns K, Dillmann R (eds) Climbing and walking robots – 4th international conference on climbing and walking robots and the support technologies for mobile machines, Professional Engineering Publishing, Karlsruhe, Germany, Sept 2001, pp 877–882

6. Oliveira ALC, Silva MF, Barbosa RS (2010) Development of an wheeled climbing robot for metallic surfaces with permanent magnetic system dynamic adjustment. In: Fujimoto H, Tokhi MO, Mochiyama H, Virk GS (eds) Emerging trends in mobile robotics, World Scientific, Nagoya, Japan, Aug 2010, pp 1340–1346
7. Oliveira ALC, Silva MF, Barbosa RS (2010) Architecture of an wheeled climbing robot with dynamic adjustment of the adhesion system. In: Proceedings of the SISY 2010 – 8th IEEE International Symposium on Intelligent Systems and Informatics, Subotica, Serbia, 10–11 Sep 2010, pp 127–132

Chapter 29
Development of an AGV Controlled by Fuzzy Logic

Ramiro S. Barbosa, Manuel F. Silva, and Dário J. Osório

Abstract This paper presents the development of a behavior-based AGV using fuzzy logic. A robot platform and a fuzzy logic controller (FLC) are developed for the embodiment of different behaviors. Experimental results are given to assess the performance of the AGV and to validate the proposed design schemes for its construction and control.

29.1 Introduction

Fuzzy logic has emerged in the 1960s, more precisely in 1965, when the first article was published by Zadeh. However, only in the following decade were developed the first applications in the field of automatic control by Mamdani. The control by fuzzy logic allows a different approach to the problem, in which intuitive knowledge on how to best control the process should be acquired, and this information will be part of the FLC. FLCs have been successfully applied in the control of many physical systems, particularly those with uncertainty, unmodelled, disturbed and/or with nonlinear dynamics [1, 2]. Nowadays, one of the main applications of FLC is in the area of autonomous robotics, where several works on behavior-based fuzzy control have been developed [3–5].

Bearing these ideas in mind, this paper presents the development of a mobile robot, with an open and distributed architecture, controlled by fuzzy logic, and capable for the embodiment of different behaviors. The rest of the paper is organized as follows. Section 29.2 addresses the design of the AGV. Section 29.3

R.S. Barbosa (✉) • M.F. Silva • D.J. Osório
GECAD – Knowledge Engineering and Decision Support Research Center, Institute of Engineering – Polytechnic of Porto (ISEP/IPP), Porto, Portugal
e-mail: rsb@isep.ipp.pt; mss@isep.ipp.pt; darioosorio@msn.com

A. Madureira et al. (eds.), *Computational Intelligence and Decision Making: Trends and Applications*, Intelligent Systems, Control and Automation: Science and Engineering 61, DOI 10.1007/978-94-007-4722-7_29,

is concerned with the design of the fuzzy controller and the implementation of the several behaviors. Section 29.4 presents experimental results showing the effectiveness of the proposed design schemes for the construction and control of the robot. Finally, Sect. 29.5 draws the main conclusions.

29.2 Design of the AGV

29.2.1 AGV Structure

The robot is built adopting a modular structure (Fig. 29.1, left), being constituted by a rigid PVC base structure, two DC motors with encoder in the front, four ultrasonic sensors (also called sonars) and a free-wheel in the rear, to support part of the robot weight. The motors can develop a maximum torque of 6.12 kg/cm and a maximum velocity of 66 rpm. The sonars are of the type SRF05, with range from 3 cm to 4 m. Three microcontrollers of the PIC 18F family are used, namely the 18F4585, with a CAN interface for the communication between the several modules of the system. Among other characteristics, the 18F4585 possesses timers and A/D converters, can

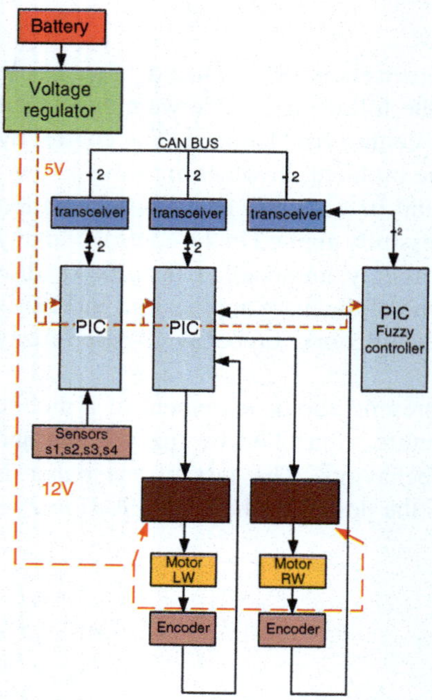

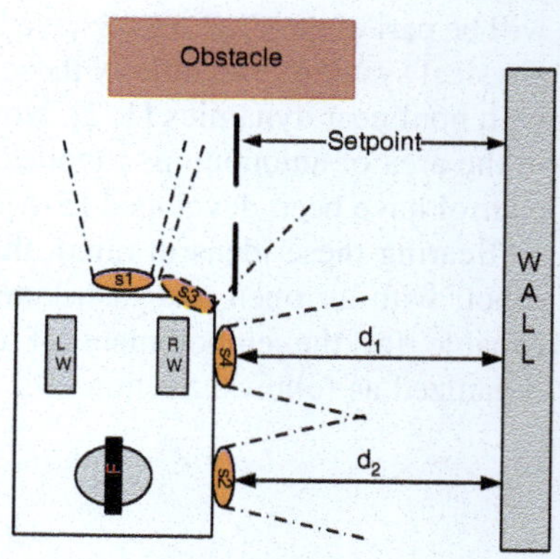

Fig. 29.1 AGV architecture (*left figure*) and schematic representation of the AGV physical structure (*right figure*)

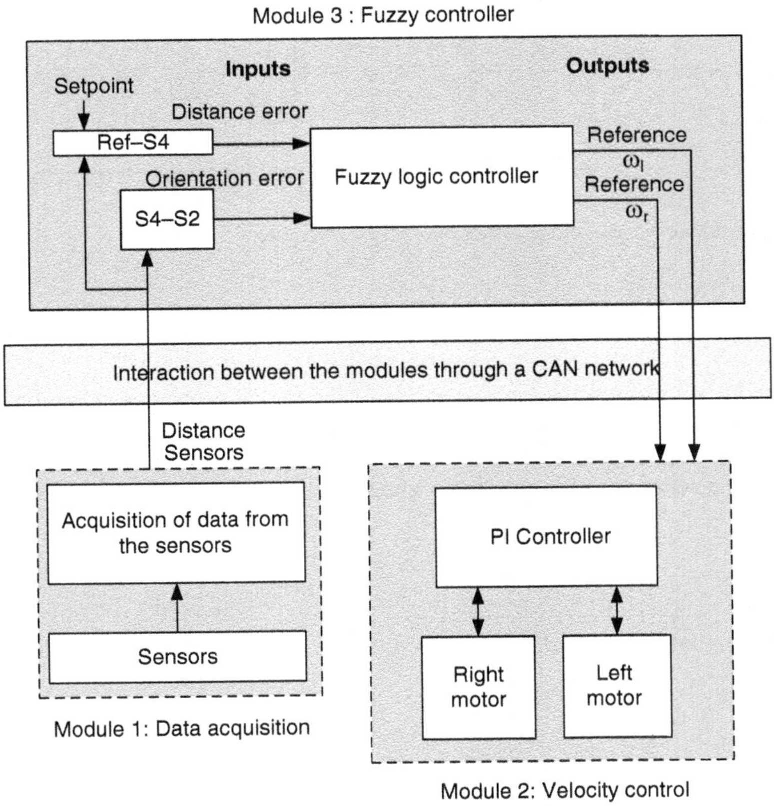

Fig. 29.2 Features of the modules of the AGV and interaction between them

operate at frequencies up to 40 Mhz and presents 48 kbytes of program memory, 3,328 bytes of data memory and 1,024 bytes of EEPROM.

The schematic representation of the AGV physical structure is presented in Fig. 29.1 (right). The two right side sonars (S2 and S4) are used to determine the orientation and the distance of the robot from the wall. This configuration of the sonars is used for the implementation of the right wall-following behaviour. The front sonars (S1 and S3) allow the detection of obstacles and the implementation of the emergency behaviour.

Figure 29.2 illustrates a block diagram showing the features implemented in the three modules that constitute the AGV and the interaction between them. Module 1 performs the data acquisition from the sonar sensors. Module 2 is responsible for the velocity control of the left wheel (LW) and right wheel (RW) differential traction allowing orienting the robot in space. Finally, Module 3 implements the FLC. Each of these modules has an interface for communication according to the CAN protocol. The use of the CAN bus in this work aimed not only to distribute the several tasks across the multiple nodes, but was also chosen due to the flexibility that this network offers in the incorporation of future modules into the system.

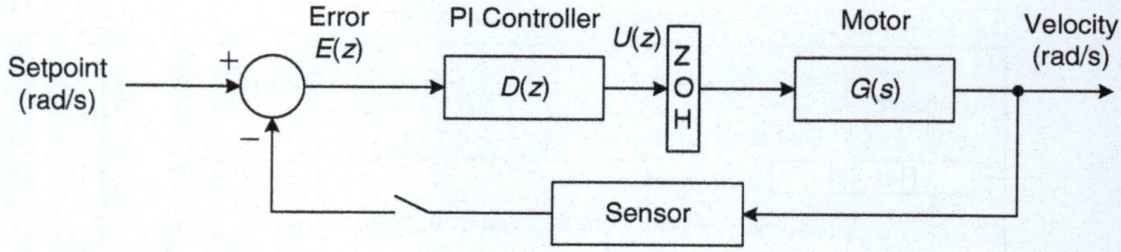

Fig. 29.3 Block diagram of the system with the PI controller

29.2.2 Velocity Control

Figure 29.3 illustrates the block diagram of the digital control system with the PI controller used for the velocity control of both motors (RW and LW) of the AGV.

The transfer function between the output velocity $\omega(s)$ and the applied voltage $U(s)$ of the set motor plus actuator is given by the first-order model [6]:

$$G(s) = \frac{\omega(s)}{U(s)} = \frac{6.72}{0.0454s + 1} \tag{29.1}$$

The digital PI controller was designed through the discrete root-locus [7], yielding:

$$D(z) = \frac{U(z)}{E(z)} = K_c \frac{z - \alpha}{z - 1} \tag{29.2}$$

with $K_c = 0.0472$, $\alpha = 0.2869$, and a sampling period of $T = 0.01$ s.

29.2.3 AGV Kinematics

Figure 29.4 shows a scheme of the kinematic model of the robot, where $v_r(t)$ is the linear velocity of the right wheel, $v_l(t)$ is the linear velocity of the left wheel, $\omega_{system}(t)$ and $V_{system}(t)$ are respectively the angular and linear velocities of the robot, θ is the orientation angle of the system, and L is the distance between the two wheels.

The linear velocity for each motor is calculated using the relation $v = \omega r$ where r is the radius of the wheel. The linear and angular velocities of the robot system, V_{system} and ω_{system}, respectively, can be obtained through the equations:

$$V_{system} = \frac{v_r + v_l}{2}, \quad \omega_{system} = \frac{v_r - v_l}{L} \tag{29.3}$$

where r and L are 3.5 and 21.7 cm, respectively.

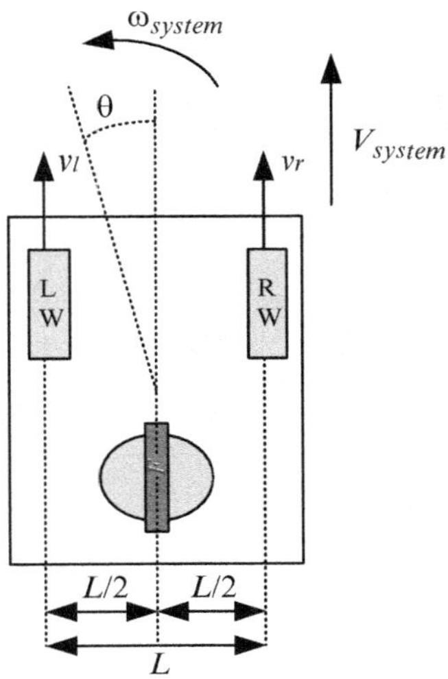

Fig. 29.4 AGV kinematics

29.3 Design of the Fuzzy Controller

29.3.1 Behaviors

The robot platform is used to incorporate right wall-following, avoid-obstacles and emergency behaviors, as illustrated in Fig. 29.5. For the wall-following behavior, the input variables are the distance error from the wall (*ErrorDistance*) and the orientation error of the robot (*Orientation*), determined from readings of the side sonars S2 and S4 (Fig. 29.1, right). For the avoid-obstacles and emergency behaviors are used the readings of the robot front sonars (S1 and S3) (Fig. 29.1, right). On the other hand, the outputs are the linear velocity (*LinearVelocity*) and the angular velocity (*AngularVelocity*), which will determine the velocities to be applied to each motor of the AGV.

29.3.2 Wall-Following Behavior

Figure 29.6 shows the membership functions associated with the input and output variables for the wall-following behavior. For each fuzzy input five membership functions are constructed: *BigRight*, *Right*, *Zero*, *Left* and *BigLeft* (Fig. 29.6, left). The output *LinearVelocity* is characterized by four membership functions (*Zero*, *Low*, *Midle* and *High*) while the output *AngularVelocity* is defined through five membership functions: *BigRight*, *Right*, *Zero*, *Left* and *BigLeft*, as illustrated in Fig. 29.6 (right).

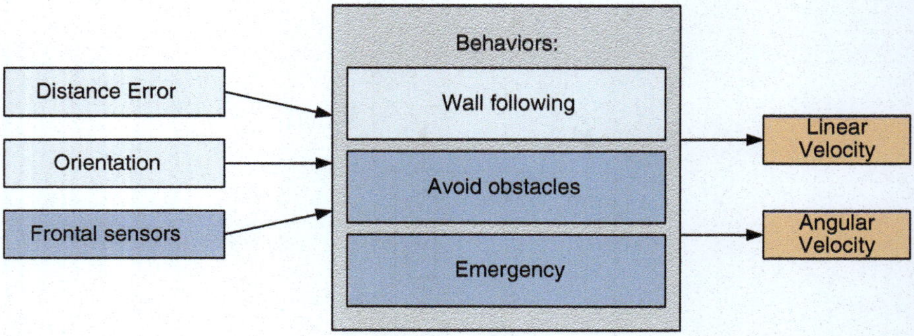

Fig. 29.5 Designed and implemented behaviors

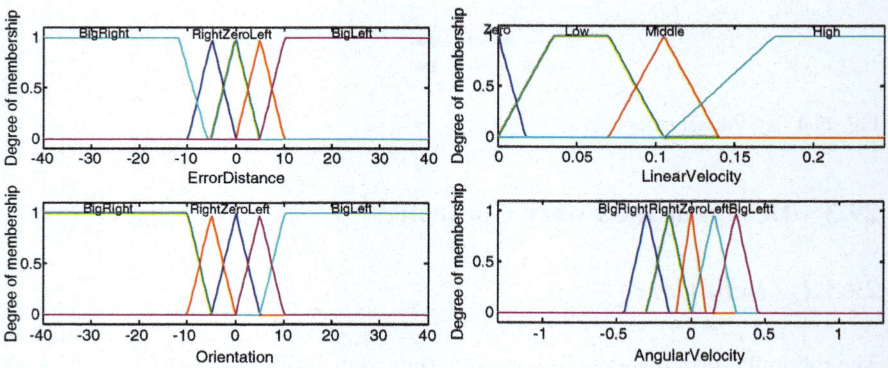

Fig. 29.6 Membership functions of input and output variables for the wall-following behavior

The rule bases containing the control fuzzy rules for the proposed FLC are listed in Tables 29.1 and 29.2, for the outputs *LinearVelocity* and *AngularVelocity*, respectively.

29.3.3 Implementation

The fuzzy implication and the aggregation of the rules were implemented by using the *min* and *max* operators, respectively. The centroid method was used for the defuzzification of the output fuzzy sets. Since, ω_{system} is calculated as in (29.3), the difference of velocities between the motors is $\Delta v = \omega_{system} L$. Since the desired linear velocity is already known, the velocities to be applied in each motor are given by:

$$v_r = V_{system} - \frac{\Delta v}{2}, \quad v_l = V_{system} + \frac{\Delta v}{2} \tag{29.4}$$

Table 29.1 Rule base of output variable *LinearVelocity* for the wall-following behavior

Linear velocity		ErrorDistance				
		Big Left	Left	Zero	Right	Big Right
Orientation	Big Right	Middle	Middle	Middle	Middle	Middle
	Right	Middle	Middle	Middle	Middle	Middle
	Zero	Middle	Middle	Middle	Middle	Middle
	Left	Middle	Middle	Middle	Middle	Middle
	Big Left	Middle	Middle	Middle	Middle	Middle

Table 29.2 Rule base of output variable *AngularVelocity* for the wall-following behavior

Angular velocity		ErrorDistance				
		Big Left	Left	Zero	Right	Big Right
Orientation	Big Right	Big Left	Left	Left	Zero	Zero
	Right	Left	Left	Zero	Left	Zero
	Zero	Left	Zero	Zero	Zero	Right
	Left	Zero	Right	Zero	Right	Right
	Big Left	Zero	Zero	Right	Right	Big Right

The resultant linear velocities are divided by the radius of the corresponding wheel and their values are sent, through the CAN bus, to Module 2 (Velocity control).

29.3.4 Avoid-Obstacles Behavior

The avoid-obstacles behavior helps the robot move freely without crashing against objects. This behavior uses the readings of the two front sonars (S1 and S3, see Fig. 29.1, right). Figure 29.7 shows the membership functions associated with the input and output variables. In order to improve the obstacles avoidance capability of the AGV, two new terms were added in the definition of the output variable *AngularVelocity* (*BigBigRight* and *BigBigLeft*). The corresponding rule bases are presented in Tables 29.3 and 29.4.

29.3.5 Emergency Behavior

The emergency behavior is fundamental for the operation of the robot, since it guarantees a safe distance between the robot and the objects. This behavior results from a particular set of rules that are not considered in the avoid-obstacles behavior. Table 29.5 presents the actions that should be considered in the emergency behavior.

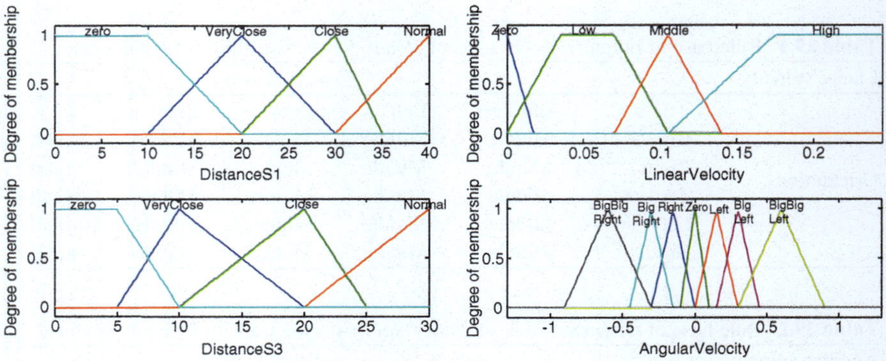

Fig. 29.7 Membership functions of input and output variables for the avoid-obstacles behavior

Table 29.3 Set of rules of output variable *LinearVelocity* for the avoid-obstacles behavior

Linearvelocity		Sensor S1 – DistanceS1			
		Zero	VeryClose	Close	Normal
Sensor S3 – DistanceS3	Zero	Emergency behavior	Emergency behavior	Emergency behavior	Emergency behavior
	VeryClose	Emergency behavior	Middle	Middle	Middle
	Close	Emergency behavior	Middle	Middle	Middle
	Normal	Emergency behavior	Middle	Middle	Wall following behavior

Table 29.4 Set of rules of output variable *AngularVelocity* for the avoid-obstacles behavior

Angular velocity		Sensor S1 – DistanceS1			
		Zero	VeryClose	Close	Normal
Sensor S3 – DistanceS3	Zero	Emergency behavior	Emergency behavior	Emergency behavior	Emergency behavior
	VeryClose	Emergency behavior	BigBigLeft	BigBigLeft	BigBigLeft
	Close	Emergency behavior	BigBigLeft	BigBigLeft	BigLeft
	Normal	Emergency behavior	BigBigLeft	BigLeft	Wall following behavior

Table 29.5 Set of rules of output variables *LinearVelocity* and *AngularVelocity* for the emergency behavior

		Sensor S1 – DistanceS1			
		Zero	VeryClose	Close	Normal
Sensor S3 – DistanceS3	Zero	Zero	Zero	Zero	Zero
	VeryClose	Zero	Avoid obstacles behavior	Avoid obstacles behavior	Avoid obstacles behavior
	Close	Zero	Avoid obstacles behavior	Avoid obstacles behavior	Avoid obstacles behavior
	Normal	Zero	Avoid obstacles behavior	Avoid obstacles behavior	Wall following behavior

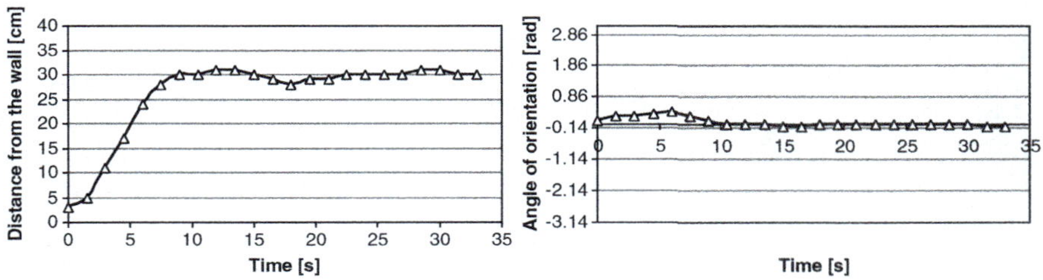

Fig. 29.8 Evolution of the distance from the wall (*left figure*) and of the angle of orientation of the robot (right figure)

Since different behaviors are implemented in the same FLC, it is necessary to define an arbitration method to coordinate the activation of the different behaviors. The arbitration was done by a set of rules that define different contexts. These context rules make the management of the active behaviors in order to guarantee that no conflicts occur in the output actions. The management of the behaviors is performed by a set of rules with the format **IF** *DistanceS1* is *Zero* **Or** *DistanceS3* is *Zero* **Then** *Emergency Behavior*.

If no conflicts exist in the output fuzzy sets, the defuzzification process can be performed in order to get the control values to the system. Note that these context rules, that make the coordination of the different behaviors, guarantee that only the rules of the active behaviors are processed, so some processing time is saved.

29.4 Experimental Results

In order to verify the performance of the robot system, several experimental tests were performed [6]. Figure 29.8 shows the evolution of the distance from the wall and the angle of orientation of the robot for a setpoint (desired distance from the wall) of 30 cm. As can be seen, the robot starts at a distance from the wall of approximately zero (near to the wall) and tends to follow the wall keeping the desired distance.

After the implementation of all behaviors in the FLC, a test was realized that allowed analyzing the behavior of the robot in different situations. Fig. 29.9 shows the environment where the robot was tested (left) and presents the experimental results obtained from the experiment (right). In the first instants the robot tries to reach the setpoint of 30 cm from the wall; then (at around 10 s) an obstacle is detected and the robot turns left, avoiding it. When the obstacle is no longer at the right side of the robot, it turns right until detecting the wall. Once the robot detects the wall, it considers that the wall is an obstacle and turns left again until staying parallel to it. This test allowed validating the system, since the robot was capable to traverse the path, in the desired way, without colliding with the obstacle.

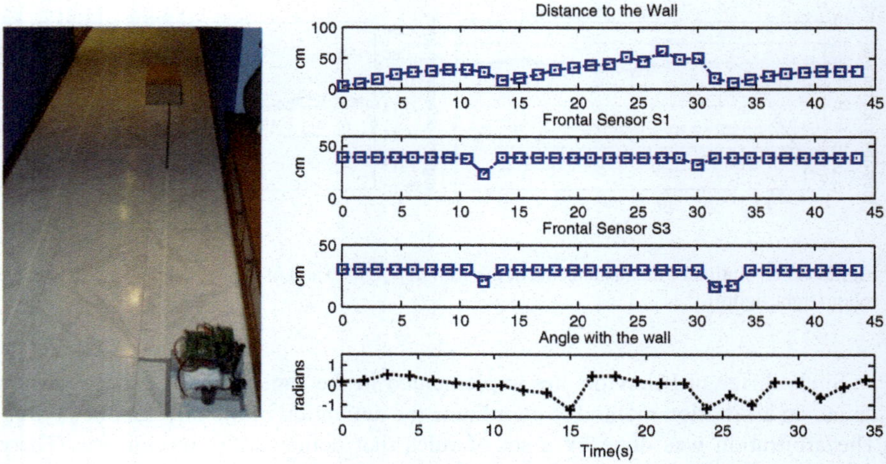

Fig. 29.9 Environmental conditions (*left figure*) and experimental results (*right figure*)

29.5 Conclusions

This paper presented the development of an AGV, with an open and distributed architecture, controlled by fuzzy logic, and capable for the embodiment of different behaviors. It was verified that the robot performs as expected, assessing the design options adopted for its construction and control. The vehicle was designed to have mainly reactive features, in order to follow walls and avoid obstacles. It was verified that the robot is able to perform these tasks with success.

References

1. Passino KM (1998) Fuzzy control. Addison-Wesley, Menlo Park
2. Reznik L (1997) Fuzzy controllers. Newnes, Oxford
3. Saffiotti A (1997) The uses of fuzzy logic in autonomous robot navigation. Soft comput 1:180–197
4. Thongchai S, Suksakulchai S, Wilkes DM, Sarkar N (2000) Sonar behaviour-based fuzzy control for a mobile robot. In: Proceedings of the IEEE international conference on systems, man, and cybernetics, Brisbane, Australia, 8–11 Oct 2000, p 16
5. Zein-Sabatto S, Sekmen A, Koseeyaporn P (2003) Fuzzy behaviors for control of mobile robots. J Syst Cybern Inform 1:68–74
6. Osório DJ, Barbosa RS, Silva MF (2011) Fuzzy control architecture for a mobile robot system. In: Proceedings of the 31st IASTED international conference on modelling, identification, and control, Innsbruck, Austria, 14–16 Feb 2011, pp 286–293
7. Franklin GF, Powell JD, Workman M (1998) Digital control of dynamic systems. Addison-Wesley, Menlo Park

Chapter 30
Affect Recognition

Raquel Faria and Ana Almeida

Abstract This survey describes the advances in affect recognition from the first emotion theories and models to affective computer techniques used in affect detection. These techniques allow the recognition of human body physical features, such as the facial expression, voice intonation, gestures or movements and physiological aspects such as respiration, skin color, temperature, heartbeat, blood pressure, pupillary dilation. Emotions can then, be inferred from the referred features analysis performed according to any chosen emotion theory.

30.1 Introduction

Human to human communication depends on the interpretation of a mix of audio-visual and sensorial signals. To simulate the same behavior in a human-machine interface the computer has to be able to detect alterations in the affective state and behavior and modify its interaction accordingly. The field of Affective Computing (AC) develops systems and mechanisms that are able to recognize, interpret and simulate human emotions [1] so closing the gap between man and machine. The AC was developed by Rosalind Picard in 1995 as a tool to improve the interface human-machine by including affective connotations. One of the problems of understanding

R. Faria (✉)
Computer Science Department, Institute of Engineering – Polytechnic of Porto (ISEP/IPP), Porto, Portugal
e-mail: arf@isep.ipp.pt

A. Almeida (✉)
Computer Science Department, Institute of Engineering – Polytechnic of Porto (ISEP/IPP), Porto, Portugal

GECAD – Knowledge Engineering and Decision Support Research Center, Institute of Engineering – Polytechnic of Porto (ISEP/IPP), Porto, Portugal
e-mail: amn@isep.ipp.pt

A. Madureira et al. (eds.), *Computational Intelligence and Decision Making: Trends and Applications*, Intelligent Systems, Control and Automation: Science and Engineering 61, DOI 10.1007/978-94-007-4722-7_30,

affects and its effect on cognition is that affective state information is hard to measure. To sense without interfering implies the development of techniques that are nonintrusive, but with the highest percentage of accuracy.

30.2 Affective Theories

The study of affect is included in different fields of science such as psychology, cognitive science, neuroscience, engineering, computer science, sociology, philosophy, and medicine. This has contributed to different understanding of basic terms related to affect such as emotions, feelings, moods, attitudes, affective styles, temperament, motivation, attention, reward, and so many others [1]. One of the problems of studying affect, is the definition of what it is emotion. Moreover, the definition and the terms associated to it. Nearly hundreds of definitions of emotion have been registered since 1981 [2]. Emotion plays an important role in the decision process and knowledge acquisition of an individual. Therefore, they directly influence our perception, learning process, the way we communicate, and the way we make rational decisions. Emotions have two components: the mental component (cognitive) and the physical component (body) and can be classified in two categories: primary emotions and secondary emotions. The primary emotion occurs as a response to some kind of event, which can cause a detectable physical response and trigger emotions such fear, anger, happiness, sadness, shame or desire. The secondary emotions can be the result of a primary emotion or the result of a cognitive process such as embarrassment.

To analyze emotion, there are several theories which attempted to specify the interrelationships of all the components involving an emotion and also the causes, the reasons and the function of an emotional response. Although there are several works summarizing these approaches, some of these theories are very controversial among the intellectual community. Nevertheless they were the starting points for most of the research works done today in Affect Recognition (AR).

Darwin was the first to develop a theory about emotions [3]. According to Darwin observations human face and body expressions were alike to other animals. He found out that the behavior corresponds to emotional experience therefore it was a consequence of the evolutionary processes. He also concluded that emotion expressions evolved through a natural selection process and they have a corresponding universal cross-culturally emotional expression.

Darwin proposed three principles: the serviceable associated habits, antithesis, and expressive habits, or nervous system discharge. The first reinforced the idea that expression could be used to discover a state of mind. This principle stated that useful habits reinforced previously by the course of an experience were genetically inherited by offspring. One of the examples he used was the example of contracting the eyebrows. Darwin noticed that eyebrows were used to prevent too much light from entering the eyes. He also noticed that the raise of the eyebrows increased de field of vision. To link this facial expression to a human behavior, he mentioned examples of people trying to remember something and raising their brows, as though they could see what they were trying to remember [4]. The second principle proposed that some actions or habits were carried out simply because they were opposite in nature to a serviceable habit, but were not serviceable themselves. The example used was the shoulder shrugging passive action to an aggressive expression [3]. The third principle suggested that some habits were created because of a build-up to the nervous system, which caused a discharge of the excitement. The examples of this principle included foot and finger tapping, as well as vocal expressions [3].

Although Darwin failed to explain some emotional expression and behaviors, there are a number of expressions that are still considered as universal although several studies have been carried out to challenge this view like the ones we can find in [5, 6]. Other researchers tried to expand Darwin theories like Paul Ekman and assumed that the emotion expressions are universal and share basic emotions like fear, sadness, happiness, anger, and disgust. His research was focused on the physical displays of emotion including body language. In fact, most of Facial Expression Recognition (FER) research is based on his work [7]. FER is a nonintrusive technique that it enables the discovery of the emotional state without having sensors and wires attached to the human body

James-Lange Theory of Emotion was one of the first theories to attempt to describe the process of emotional reactions. James-Lange Theory states that emotions are the embodiments of certain physiological changes (basically physiological changes causes an emotion). The physiological changes are triggered by the autonomic nervous system and include alterations in the heart rate, muscular tendencies, and skin conductance. This theory did not explain which physiological changes caused an, but paved the way to the gathering of physiological data for emotion analyze. With the advances in physiological sensors it is possible to determine the emotional state with accuracy. Nevertheless these methods are a little bit intrusive, but possible to mitigate with the development of wearable sensors. Wearable sensors can be used in a wide variety of areas from health care to social behavior motoring and gaming applications.

In emotion psychology, the term appraisal specifically refers to the cognitive evaluation antecedent of an emotional episode. The cognitive theory of emotions states that reason is a part of emotion, in essence this theory gives the notion that different individuals (with different reasons, objectives, rules) will appraise the same experience/circumstances in a different way and, consequently, present diverse emotional reactions [8, 9]. Magda Arnold was the first to introduce the "cognitive theory" in the 1960s, which indicated that the first step in emotion is an appraisal of the situation. This line of research was followed by several psychologists that influenced AR research with their theories of emotion. Like Lazarus's cognitive-motivational-relational theory of emotion or Roseman model where emotions are the result of a multidimensional appraisal process. Also the cognitive theory by Ortony, Clore, and Collins (OCC) consider emotions as reactions to situational appraisals of events, actors, and objects [10]. The OCC model has recognized itself as the standard model for emotion synthesis, and it is often used to model user emotional states. The model specifies 22 emotion categories based on the reactions of a subject to an event, the reactions of the subject to an action of a responsible agent, the reactions of the subject to approaches of attractive or unattractive objects. Roseman and Ortony models are alike in several points. Both are based in the universality of the appraisal process, and are used to predict user emotional state [11].

Scherer‘s Geneva Expert System on Emotion (GENSES) [12] is described as a system motivated by the need to: use computer modeling and research as a tool to further theoretical development and collect data on the emotion-antecedent appraisal process. Thus creating a computational model to evaluate and improve emotional theories. The system was composed by a computer application which invites the subject to think of an experience containing a strong emotion. This is performed with the purpose of diagnosing the emotion felt. The system was built upon a knowledge base that links appraises to different emotions predicted in Scherer theory. The system mapped over 200 emotional situations with an accuracy of 78% [13]. Despite the system significance as a tool to test emotional theories, it was too limited for most real life applications.

Averill [14] argued that emotions cannot be explained strictly on the basis of physiological or cognitive terms, suggesting that, emotions are socio-culturally constructed which determine the behavior and value patterns. The relationship between emotion language and consciousness is an important factor of human interaction as part of the emotion experience. Social constructionists sustain that emotions are influenced by a social awareness concerning when, where, and what to feel as well as when, where, and how to act. This led to the questioning of Ekman work universality by the social constructivists. Certain cultures might not have a correspondent word to an emotion or have labels that cannot be translated [15].

Psychology has focused over time on the importance of the social process to explain the emotional phenomenon. The review "Sociology of emotions" presents five theoretical perspectives on human emotions: Dramaturgical Theories, Symbolic Interactionist Theories, Interaction Ritual Theories, Power and Status Theories and Exchange Theories [16]. Although the progresses made by these theories, some questions still persist, like the nature of emotions, feelings, and affect; the point to which emotions are biologically based or socially constructed; the breach between social psychological theories on emotions and macro structural theorizing; and the relatively narrow array of emotions theorized, joined with the structural and cultural conditions creating these emotions [16].

Emotions are believed to be associated to brain activity. There are brain areas directed linked to our attention, influencing our behavior, and controlling the understanding of what is going on around us. Neuroscience has given several contributes to the study of emotion and its relation with the brain, mainly the field of Affective Neuroscience (AN) which aids in the understanding of the neural process behind an emotional experience [17, 18]. The techniques used in the AN include imaging (e.g. Functional Magnetic Resonance Imaging (fMRI)), lesion studies, genetic, and electrophysiology [12]. The last decades have provided evidence that the neural substrates of emotion and cognition significantly overlap [17].

Refuting the idea that emotion is strictly subcortical and limbic, whereas cognition is cortical. Cognitive processes take place always in the emotional experience. Methods like Electroencephalography (EEG) that record electrical activity along the scalp produced by the firing of neurons within the brain, provide and alternative method to self-reports and have been increasing appearing in AC literature [19, 20].

30.3 Recognizing Emotions

A facial expression is the result of the movements or positions of face muscles. Facial expression recognition plays an important role in natural human-machine communication [14]. Most of Facial Expression Recognition (FER) research was influenced by the theories described in section (A) based on the work of Ekman [7]. His work was based on the assumption that the emotions are universal across individuals as well as human ethnics and cultures [21]. With nowadays advances in facial recognition software, it is possible to recognize basic facial expressions like anger, disgust, fear, happiness, sadness and surprise.

A study carried out by Ekman and Friesen [22] developed the Facial Action Coding System (FACS), which is a manual technique for the measurement of facial movement. FACS can code practically any anatomically possible facial expression, decomposing it into the specific Action Units (AU) that identifies independent face motion in a temporal order. Manually coding a segment of video is a timely and expressive method performed by highly trained human experts. Each minute of video take nearly an hour to code [23]. Studies trying to automate this, see [24, 25].

Recognizing emotion through analysis of body language and posture appears to be an overlooked area in detriment of facial and speech in AC literature. There are some studies indicating the advantages of body language [26, 27]. One of the first systems created using automated posture analyses to determine a student affective state of in a learning environment is Tekscan's Body Posture Measurement System (BPMS) [28]. The BPMS system recognizes postures related with affective states linked to a child's interest level while carrying out a learning task on a computer. This system consists of two pressure sensors mounted on the chair, from those sensors the student postures are extrapolated in real time. There is a neural network to classify the postures with an accuracy of 87.6%. Later this model was improved by Mello et al. [29] creating a system that can detect learners' affect by monitoring their gross body language. This system was designed to detect boredom, confusion, delight, flow and frustration with accuracy between 73 and 83%.

Eye tracking is the process of measuring either the point of gaze (what are we looking at) or the motion of an eye in relation to the head position. The analyses of eyes properties can be used to measure the human response to visual, auditory or sensory stimuli [30, 31]. Emotion is measured through the eye tracking hardware and a statistical program which determines the excitement level to a visual image [30]. Parameters analyses include: pupil size, blink properties and gaze. The pupil size is related to an emotional reaction, a study performed by Partala et al. [31]

indicates there is an alteration in pupil size when a subject faces a positive or negative stimulus opposite when facing a neutral stimulus. Blink has also been associated with emotional responses, for example with defensive reactions, the eye is modulated blink startle [32]. Finally, gaze patterns have been linked to emotional reactions [33]. There are several areas where the eye tracking process can be applied: neuroscience, psychology, computer science, industrial engineering and marketing [34].

Speech recognition consists in the ability of a machine or program in identify words or phrases from the spoken language. The main application of speech recognition resides in assisted technology to help people with disabilities. The vocal aspect of a communication also carries information about speech emotional contents. So, speech can be divided in two parts: an explicit message, consisting of what was said; and an implicit emotional expression, entailing how the message was said. Speech Emotion Recognition (SER) aims to recognize the user emotional state his speech signal [35].

Most acoustic features which have been used for emotion recognition can be divided into two categories: prosodic and spectral. Prosodic features have been shown to provide key speaker emotional cues [35]. Prosodic features include itch, intensity, delta-pitch, delta-intensity and speaking rate. Literature in SER reached important conclusions that can be used in AC applications [6, 36]. The major conclusion relies in the possibility to recognize emotion from a speech. Although there is some difficulty in recognising certain kind of emotions like disgust, there are others that can be recognize with more accuracy like sadness and fear. Also, the pitch of a speech seems to be connecting to the level of arousal, in addition, affect discovering in a speech has accuracy rates lower than facial expression for recognizing basic expressions.

Affection recognition by identifying patterns in the physiological activity. Physiology areas like psychology and psychophysiology try to understand what physiological patterns are behind them and the implications in behavior. To monitor these physiological patterns, the electrical activity produced by brain, heart, eyes, skin and muscles is recorded

In Andreassi book [37] there is a review of some affect detection relevant concepts to the based in physiology. These concepts include Law of Initial Values (LIVs), Stimulus–response (SR) specificity, Individual-Response (IR) specificity, Cardiac-somatic features, habituation and rebound. The LIVs is a physiological and psychological concept which states that with a certain level of stimulation, the degree of alteration produced tends to be greater when the initial value of that variable is low. Cardiac-somatic features are related to the heart activity modification triggered by a body behavior response. Habituation and rebound are the result of a stimuli situation, for instance, if subject to a repeated stimulus the physiological response diminishes, which causes habituation to a give stimulus. On the other hand the presence of a stimulus can cause, after a time period the return to the level of pre-stimulus. Stimulus or elicitation techniques used in most studies submit the subjects to the use of personal and mental images to generate an emotion response [12].

The field of AN tries to map the neural circuitry that occurs during an emotional experience [17, 38], the techniques used by neuroscientists include fMRI, EEG, among others. fMRI is based on the MRI technology. It is a noninvasive test that uses a strong magnetic field and radio waves to create detailed body images. It monitors the blood flow in brain to detect activity areas. Therefore fMRI provides a map of which parts of the brain are active during an emotion or feeling. A study performed by Yang and Damasio [39] with brain damaged patients show that the emotional course was required for learning and it essential to the decision making process.

EEG records brain spontaneous electrical activity through mapping. Due to the absence of an emotion neural model this method is limited and frequently supported by other techniques like the fMRI. One of the measures taken with EEG is the Event-Related Potential (ERP). ERP is any measured brain response that is directly the result of a thought or a perception. It basically measures any electrophysiological response to an internal or external stimulus. Most AC researchers focus on using ERP [40].

Affective detection in text consists in determine the emotional or attitude context within the written language or transcripts of oral communication. The initial work on this matter aimed to understand how text could express an emotion or how text could generate different emotions. These studies began by find resemblance in how people of different cultures communicate [41]. Osgood used a Multidimensional scaling (MDS) procedure to create models of affective words based on words evaluations provide by diverse cultures. Other research includes lexical analysis of the text so the affective state can be inferred [42, 43]. Most of these studies use the Linguistic Inquiry and Word Count (LIWC) [44]. LIWC is text analysis software that determines if the text has positive or negative emotions, self-references, causal words, in over 70 language dimensions. Other approaches accept that people sharing the same language have comparable notions for different emotions. Based on this, a lexical database was built containing emotional terms. Databases like WordNet-Affect domain have a hierarchy of affective domain labels representing affective concepts annotated.

Several points of views state that an emotion response has multiple manifestations both physical and behavioral. Also the human computer paradigm advocates that in the future human centered interfaces should have the capacity to sense changes in user affective behavior. So the development of affective Multimodal Human-Computer Interaction (MHCI) is growing. MHCI relies on several research areas such as computer vision, artificial intelligence, psychology and others. The study of MHCI entails the knowledge of three variables: user, system and the interaction between them [45]. So we are approaching a future that is likely to include new ways of interaction between humans and computers, such as: visual interaction, auditory, tactile, brain signals and many others. Examples of affective MHCI include: The Gaze-X: Adaptive Affective Multimodal Interface [46] is able to learn and analyse the user's behavioural patterns and adapt the interface accordingly. The automated Learning Companion [47] assembles data from cameras, a sensing chair, and mouse, wireless skin sensor, and task state to detect frustration so that it can predict when the user requires assistance. The Multimodal

Computer-aided Learning System [48] creates a computer avatar based on the information of the user's facial expression, keywords, eye movement, and task state creates suitable tutoring plan.

30.4 Applications

The applications of AC can be divided in five categories: security, education, health and medicine, military and home. The development of AR based systems could play a vital role in the next generation of biometric surveillance systems. The application of this kind of systems in security industry helps prediction in real-time human behavior and intention. Examples of applications are the development of smart cameras, lie detectors and emotional speech processing. In education, the knowledge about the emotional state of learners can improve teacher styles, thus learning process itself. Examples of applications are electronic tutoring, e-learning, affective mirror and text-to-speech applications. In Health and Medicine accessing patient emotions can help in the treatment, for instance it could be used to help people with mental illness or create an affective mirror. Military applications include devices with emotional intelligence like training equipment and simulations. Home applications include video games and toys.

30.5 Concluding Remarks

In this document we highlighted some key approaches to affect detection. We focused techniques such as facial and speech recognition among others and their role in the discovery of emotion. Emotion shapes almost all modes of human communication form facial expression, posture, tone of voice, choice of words, respiration, skin temperature and many others. Today advances in recognition allow creation of new ways of humans-computer interaction, making this interaction more usable and intuitive.

Acknowledgments This work is supported by FEDER Funds through the "Programa Operacional Factores de Competitividade – COMPETE" program and by National Funds through FCT "Fundação para a Ciência e a Tecnologia" under the project: FCOMP-01-0124-FEDER-PEst-OE/EEI/UI0760/2011.

References

1. Picard RW et al (2004) Affective learning — a manifesto. BT Technol J 22(4):253–269
2. Kleinginna PR, Kleinginna AM (1981) A categorized list of emotion definitions, with suggestions for a consensual definition. Motiv Emot 5(4):345–379

3. Darwin C (1872) The expression of the emotions in man and animals. John Murray, London
4. Hess U, Thibault P (2009) Darwin and emotion expression. Am Psychol 64(2):120–128
5. Ortony A, Turner TJ (1990) What's basic about basic emotions? Psychol Rev 97(3):315–331
6. Russell JA, Bachorowski J-A, Fernandez-Dols J-M (2003) Facial and vocal expressions of emotion. Annu Rev Psychol 54:329–349
7. Ekman P, Friesen WV, Ellsworth P (1982) Emotion in the human face. In: Ekman P (ed) Emotion in the human face, vol 2. Cambridge University Press, New York, p 464
8. Manstead A, Tetlock PE (1989) Cognitive appraisals and emotional experience: further evidence. Cognit Emot 3:225–240
9. Smith CA, Ellsworth PC (1985) Patterns of cognitive appraisal in emotion. J Personal Soc Psychol 48(4):813–838
10. Ortony A, Clore GL, Collins A (1988) The cognitive structure of emotions, vol 18. Cambridge University Press, Cambridge, p 224
11. Roseman IJ, Spindel MS, Jose PE (1990) Appraisals of emotion-eliciting events: testing a theory of discrete emotions. J Personal Soc Psychol 59(5):899–915
12. Calvo RA, Mello SD (2010) Affect detection: an interdisciplinary review of models, methods, and their applications. Rev Lit Art Am 1(1):18–37
13. Wehrle T, Scherer KR (2001) Towards computational modeling of appraisal theories. In: Scherer KR, Schorr A, Johnstone T (eds) Appraisal theories of emotions: theories, methods, research. Oxford University Press, New York, pp 350–365
14. Chibelushi CC, Bourel F (2002) Facial expression recognition: a brief tutorial overview. School of Computing, Staffordshire University, Staffordshire
15. Russell JA (2003) Core affect and the psychological construction of emotion. Psychol Rev 110 (1):145–172
16. Turner JH, Stets JE (2006) Sociological theories of human emotions. Annu Rev Sociol 32 (1):25–52
17. Dalgleish T, Dunn B, Mobbs D (2009) Affective neuroscience: past, present, and future. Emot Rev 1(4):355–368
18. Davidson R (2003) Seven sins in the study of emotion: correctives from affective neuroscience. Brain Cognit 52(1):129–132
19. Liu Y, Sourina O, Nguyen MK (2010) Real-time EEG-based human emotion recognition and Visualization. In: 2010 international conference on cyberworlds, pp 262–269. doi:10.1109/CW.2010.37
20. Shen L, Wang M, Shen R (2009) Affective e-learning: using 'emotional' data to improve learning in pervasive learning environment related work and the pervasive e-learning platform. Learning 12:176–189
21. Fischer R (October 2004) Automatic facial expression analysis and emotional classification. S. M. thesis, Math & Natural Sciences, MIT/University of Applied Science Darmstadt, Germany
22. Ekman P, Friesen WV (1978) Facial Action Coding System: A Technique for the Measurement of Facial Movement. Consulting Psychologists Press 1978. Consulting Psychologists Press. Retrieved from http://www.citeulike.org/user/DocNero/article/3171485
23. Donato GL, Bartlett MS, Hager JC, Ekman P, Sejnowski TJ (1999) Classifying facial actions. IEEE Trans Pattern Anal Mach Intell 21(10):974–989
24. Gunes H, Piccardi M (2007) Bi-modal emotion recognition from expressive face and body gestures. J Netw Comput Appl 30(4):1334–1345
25. Bänziger T, Grandjean D, Scherer KR (2009) Emotion recognition from expressions in face, voice, and body: the Multimodal Emotion Recognition Test (MERT). Emotion 9(5):691–704, Washington, DC
26. Bull P (1987) Posture and gesture. Pergamon Press, New York
27. Meijer M (1989) The contribution of general features of body movement to the attribution of emotions. J Nonverbal Behav 13(4):247–268
28. Mota S, Picard RW (2003) Automated posture analysis for detecting learner's interest level. Media 5:49

29. Mello SD, Graesser A, Hall D (2009) Automatic detection of learner's affect from gross body language. Appl Artif Intell 23:123–150
30. Jakob de Lemos OJ, Sadeghnia Golam Reza, Ólafsdóttir Íris (2008) Measuring emotions using eye tracking. Int J 2008:2008–2008
31. Partala T (2003) Pupil size variation as an indication of affective processing. Int J Hum Comput Stud 59(1–2):185–198
32. Dawson ME, Schell AM, Bohmelt AH (1999) Startle modification: implications for neuroscience, cognitive science, and clinical science: implications for neuroscience, cognitive science, and clinical science, vol 14. Cambridge University Press, Cambridge, 383
33. Calvo MG, Lang PJ (2004) Gaze patterns when looking at emotional pictures: motivationally biased attention. Motiv Emot 28(3):221–243
34. Duchowski AT (2002) A breadth-first survey of eye-tracking applications. Behavior research methods, instruments, & computers. J Psychon Soc Inc 34(4):455–470
35. Wu S, Falk TH, Chan W-Y (2010) Automatic speech emotion recognition using modulation spectral features. Speech Commun 53(5):768–785
36. Scherer KR, Johnstone T, Klasmeyer G (2003) Vocal expression of emotion. In: Davidson RJ, Scherer KR, Barrett LF (eds) Handbook of affective sciences. Oxford University Press, New York, pp 433–456
37. Andreassi JL (2007) Human Behaviour and Physiological Response. Taylor & Francis e-Library 2009
38. Panksepp J (1998) Affective neuroscience: the foundations of human and animal emotions. Oxford University Press, New York, p 466
39. Immordino-Yang MH, Damasio A (2007) We feel, therefore we learn: the relevance of affective and social neuroscience to education. Mind Brain Educ 1(1):3–10
40. Olofsson JK, Nordin S, Sequeira H, Polich J (2008) Affective picture processing: an integrative review of ERP findings. Biol Psychol 77(3):247–265
41. Osgood CE, May WH, Miron MS (1975) Cross-cultural universals of affective meaning. University of Illinois Press, Urbana, p 486. Retrieved from http://books.google.com/books?id=Ax9CFVxXYD4C&printsec=frontcover&source=gbs_navlinks_s#v=onepage&q&f=false
42. Shields CG et al (2005) Emotion language in primary care encounters: reliability and validity of an emotion word count coding system. Patient Educ Counc 57(2):232–238
43. Hancock J, Landrigan C, Silver C (2007). Expressing emotion in text-based communication. In: Proceedings of the ACM conference on human factors in computing systems (CHI 2007), pp 929–932
44. Pennebaker JW, Francis ME, Booth RJ (2001) Linguistic inquiry and word count. Word J Int Linguist Assoc: 2–21
45. Jaimes A, Sebe N (2007) Multimodal human–computer interaction: a survey. Comput Vis Image Underst 108(1–2):116–134
46. Maat L, Pantic M (2006) Gaze-X: adaptive affective multimodal interface for single-user office scenarios. Human Factors. In: Proceedings of ACM international conference on multimodal interfaces, pp 171–178
47. Kapoor A, Burleson W, Picard R (2007) Automatic prediction of frustration. Int J Hum Comput Stud 65(8):724–736
48. Multimodal Human Computer Interaction Project (n.d.) http://itr.beckman.uiuc.edu/index.html. Retrieved 9 May 2011

Chapter 31
Web 2.0: Tagging Usefulness

Joaquim Filipe P. Santos and Ana Almeida

Abstract User tagging in social networks has emerged in the recent years like a very promising way of new knowledge extraction. Our work focus is to understand tags importance and the correlation between them and user emotions. Understanding the meaning of certain tags for a single and unique user will allow the creation of better profiles almost without asking anything directly to users. This knowledge generation will allow the creation of a Social Recommender System for Tourism only based in the knowledge previous referred.

Keywords Social networks • Tags • Tourism • Emotions • Recommendation

31.1 Introduction

This work has its main focus directed to the understanding and evolution/enhancement of several current Web 2.0 mechanisms, in which the most important are tags. Web 2.0 arrival a few years ago has started a new world of opportunities in social systems that appeared to respond to users' needs of network interaction.

The choice of these topics is due to current worldwide impact of the underlying technologies and theories and finally, and most important, the existence of a

J.F.P. Santos (✉)
Computer Science Department, Institute of Engineering – Polytechnic of Porto (ISEP/IPP), Porto, Portugal
e-mail: jpe@isep.ipp.pt

A. Almeida
Computer Science Department, Institute of Engineering – Polytechnic of Porto (ISEP/IPP), Porto, Portugal

GECAD – Knowledge Engineering and Decision Support Research Center, Institute of Engineering – Polytechnic of Porto (ISEP/IPP), Porto, Portugal
e-mail: amn@isep.ipp.pt

A. Madureira et al. (eds.), *Computational Intelligence and Decision Making: Trends and Applications*, Intelligent Systems, Control and Automation: Science and Engineering 61, DOI 10.1007/978-94-007-4722-7_31, © Springer Science+Business Media Dordrecht 2013

strong improvement space which can be exploited in order to enhance utility and usefulness of such technologies.

In order to fulfill our goals, in this work is essential to start by understanding how several Web 2.0 based technologies (tagging, blogging, twitting, etc.) are being currently applied and detect improvement techniques by which one or many of them can be further perfected, for instance, in order to get better and richer knowledge about users that can be used to create automated profiles, along with the development of a system in which those techniques are indeed applied.

Some knowledge based techniques like, Case-based reasoning, Neural networks, Bayesian networks seems to be promising approaches in achieving good results.

This research assumes a vital importance in areas like commercial, marketing, tourism, and any other area where individual user knowledge is a key factor but is not available.

This paper starts, in Sect. 31.2, with a state-of-art that is divided in the main themes regarding tagging and social recommender systems. Section 31.3 describes the proposed approach to create new knowledge about users and the way we are going to use it to make better tourism recommendations. Section 31.4 presents some conclusions and a analysis of the produced work.

31.2 State of Art

Web 2.0 is unconsciously making part of everyone's lives. Nowadays, almost everyone uses Twitter, Flickr and other related applications for their own benefit or for social purposes. Given the current average literacy of people who use such technologies, it's incredible to notice how these ideas rapidly grow. Twitter, for instance, grew 752% in 2008 [1]. The majority of current studies which analyze these technologies, tries, in a very theoretical approach, to explain these phenomena and to understand how people are influenced by them. However, despite the logical conclusions that are taken, the fact is that the assessed information is not used in order to evolve the studied systems [2–7]. Other studies just point out future guidelines, from the user point of view or the application [8, 9].

The exemplifying technology used in this state of the art is tags, which will be the great building block of the devised proposal. Regarding tags, there are three sub-categories that are generally referred, quoting [10]: (1) folksonomy itself (2) tagging and (3) social tagging systems. Each one of them can be further studied and perfected, namely option 2, since research denoted less work done in that subject. As it was already referred, the study performed in the cognitive aspect of tags is existent, but inconsequential [11–15]. Also, the use of emotional information, important in a social environment, seems absent. Plus, tag analysis considering negative aspects of such technologies are rarer and much more superficial [7].

Researching works regarding tags seem to be performed using the same datasets: Flickr and del.icio.us to name the mostly used [5, 6]. The data collection process is generally realized in a very manual and random fashion, resulting in conclusions poorly confident and solid, which is sometimes even admitted by authors. Only some of the analyzed works make use of a prototypical developed application, or a much less known application, such as [3, 14] and [16], which deals mainly with real-time mobile tagging. These technical endeavors are successful, but the sheer size and ambition of projects could be much more fruitful; none of those works indicates any real advantage for being chosen in place of the current major applications.

31.2.1 Tags – Relevance, Ranking and Ambiguities

The task of extracting the opinion/emotion expressed in tags is challenging due to different reasons like relevance, ranking and ambiguities solving but is a very promising research area because instead of taxonomies which uses a hierarchical approach to the categorize the content, folksonomies allow users to characterize content with freedom. Like we are going to see in the works described next this freedom can become a problem because produces a very uncontrolled and unpredictable world but it has some capabilities that can overcome these problems, like the extraction of.

In [17] an approach (OpAL) to solve tag ambiguities is proposed. This work is based on three different strategies: "(a) the evaluation of the polarity of the whole context using an opinion mining system; (b) the assessment of the polarity of the local context, given by the combinations between the closest nouns and the adjective to be classified; (c) rules aiming at refining the local semantics through the spotting of modifiers. The final decision for classification is taken according to the output of the majority of these three approaches".

Another important work in this area is [18] where the authors want to determine the impact of Ambiguity and Redundancy on the effectiveness of tag recommendation. They use a cluster-based approach to define ambiguity and redundancy measure where resources and tags are "clustered into highly cohesive partitions based on co-occurrence". In their opinion a tag is considered ambiguous if several resources from different clusters have been annotated with it. Tags from the same tag cluster are considered redundant. This paper claims that the cluster-based approach provides a more general and language independent method for defining ambiguity and redundancy over a variety of semantic and linguistic approaches. This approach defines metrics, based on the resulting clusters, for measuring the degree of ambiguity for a tag, and the level of redundancy for pairs of tags. To test this approach an extensive evaluation on three real world Folksonomies was made by the authors in order to determine the impact of ambiguity and redundancy in several common tag recommendation algorithms.

Another approach to determine the ambiguity of a set of (user-contributed) tags is proposed in [19]. This work suggests new tags to disambiguate the original tag introducing a probabilistic framework that allows us to "find two tags that appear in different contexts but are both likely to co-occur with the original tag set. If such tags can be found, the current description is considered ambiguous". These tags are recommended to the user for further clarification.

The work presented in [20] shows a personalization algorithm for recommendation in folksonomies based on hierarchical tag clusters that is independent of the clustering method. They use a context-dependent variant of hierarchical clusters which is related with user actual context. This work was tested against some real world data sets that gave very successful results, which suggest "that folksonomies encompassing only one topic domain, rather than many topics, present an easier target for recommendation, perhaps because they are more focused and often less sparse". The authors also conclude that "context dependent cluster selection, an integral step in our personalization algorithm, demonstrates more utility for recommendation in multi-topic folksonomies than in single-topic folksonomies". These results allow them to suggest that topic selection is an important strategy for recommendation in multi-topic folksonomies.

A possible solution to solve tagging problems like the one described above some works propose a solution like the approach presented in [21] that allows to "overcome the cold start problem for tagging new resources". To fulfill this work, they use Latent Dirichlet Allocation (LDA) to display important topics from resources with a stable and complete tag set. With it, they can recommend topics for new resources with only a small set of tags that allows other tags belonging to the recommended topics to be used. Based on authors opinions compared to an approach using association rules, suggested previously for tag recommendation, their approach achieves a significantly better precision which permits that recommended tags are more specific for a particular resource, and thus more useful for searching and recommending resources to other users.

TagEz [22] is a comprehensive way of solving tag related problems that consists of a "language, vision, and an aggregation component" applied to Flickr. The language component uses some initial tags inserted by user to characterize the image, to find commonly co-occurring tags throughout the rest of Flickr. At the same time, the vision component "learns correlations between the contents of the image and tags, but ignores any previously applied tags". This allows a completely pure vision based approach when the user has not supplied any tags yet and a very efficient and accurate language component when the user has already inputted some tags.

Another proposal to solve tag problems is presented in [23], which is a system for personalized tag suggestion for Flickr: While the user is entering/selecting new tags for a particular picture, the system is suggesting related tags to the user, combining tags that were already used with the tags entered at the time. The suggested tags are updated in a very dynamically manner with all other tags that were inserted along the time. They describe a set of algorithms which can be applied to solve this problem allowing, in their opinion, the achievement a better performance (10–15%) when compare to methods similar to currently used by Flickr.

A multi-modality recommendation based on both tag and visual correlation is proposed in [24] which consider the tag recommendation as a learning problem. Each modality is used to generate a ranking feature, and Rankboost algorithm is applied to learn an optimal combination of these ranking features from different modalities.

A Flickr Tag Recommendation tool based on Collective Knowledge is developed in [25] where authors analyses a representative snapshot of Flickr and present the results by "means of a tag characterization focusing on how users tags photos and what information is contained in the tagging". This analysis results allow them to "present and evaluate tag recommendation strategies to support the user in the photo annotation task by recommending a set of tags that can be added to the photo". The results demonstrated by the empirical evaluation made by authors show that relevant tags can effectively recommend avoiding most of the time the exhaustiveness tagging in a non-controlled manner.

In [26] the practical aspect of tag recommendation is discussed and is proposed a system that tries to address the problem of learning in tag recommendation, without losing efficiency. The authors say that "learning is realized in two aspects: adaptation to newly added posts and parameter tuning. The content of each added post is used to update the resource and user profiles as well as associations between tags. Parameter tuning allows the system to automatically adjust the way tag sources (e.g., content related tags or user profile tags) are combined to match the characteristics of a specific collaborative tagging system". They claim that they evaluated this learning method against a few collaborative tagging systems which confirmed it importance. They propose an architecture based on text indexing and claims that this makes the system efficient enough to serve in real time collaborative tagging systems with number of posts counted in millions, given limited computing resources.

In [27] a system that participated in the SemEval 2010 task of disambiguating sentiment ambiguous adjectives for Chinese is described. The system uses text messages extracted from Twitter, to create a dataset with emotional texts that allows the system to classify "the meaning of adjectives into positive or negative sentiment polarity according to the given context in a fully automatic manner".

One problem that most researchers in this area have is understanding why users use a specific tag in a specific context. In [28] this problem is analyzed recurring to non-subject related tags in social bookmarking tools like Del.icio.us, Connotea and Citeulike. Some older studies based on Del.icio.us and Citeulike have reached the conclusion that "many common tags are not directly subject related". A set of non-subject tags like time and task related, emotions tags, some prepositions and conjunctions from the previous studies was used in this work to collect posts with non-subject tags from the referred social tools. In this study these tags have been analyzed to examine their role in the tagging process. Conclusions reached in this study suggest that "users relate information to time related tasks, activities and their own emotional reactions".

Tag ranking is another important area of research in this field because most of the times is really fundamental to understand the importance of tags along contexts and users experiences. In [29] a Social Ranking method is proposed that "exploits recommender system techniques to increase the efficiency of searches within Web 2 .0". They measure users' similarity based on their previous inserted tags in order to extract tags relationships based on their association to content. These relations are then used to rank content based "on the inferred semantic distance of the query to the tags associated to such content, weighted by the similarity of the querying user to the users who created those tags".

An alternate tag ranking scheme, aiming to automatically rank the tags associated with a given image according to their relevance to the image content is proposed in [30]. Authors estimate initial relevance scores for the tags based "on probability density estimation, and then perform a random walk over a tag similarity graph to refine the relevance scores". They use the generated tag ranking to make tag-based image search, tag recommendation, and group recommendation.

Some researchers are trying to use other users work to define tag relevance. In [31] a neighbor voting algorithm is proposed which "accurately and efficiently learns tag relevance by accumulating votes from visual neighbors". Under a set of well-defined and realistic assumptions, authors claim that their algorithm is a "good tag relevance measurement for both image ranking and tag ranking".

Another common problem addressed by tag relevance is the time centered question. A tag or a set of tags can have a specific relevance between millions of others in a very specific time so is really important to center extracted knowledge in a period of time that improves accuracy and relevance. In [32] is described a system that presents a visualization of interesting tags over time, for any specified timescale. The backend must provide data to support this visualization.

Qtag (Qualitative Tag) [33] is hybrid tagging system that allows users tags to rate and express opinions. Qtag tries to minimize the potential danger of diversified vocabulary and redundant task imposition in order to create better user experiences studying how users understand the data proposed by the system.

HAMLET [34] is "a suite of principles, scoring models and algorithms to automatically propagate metadata along edges in a document neighborhood". This approach can be very important because it can boost and improve tag prediction in community-based Web 2.0 tagging applications without being always analyzing data that was already evaluated and scored. Experiments conducted by some authors using real-world data demonstrate the viability of this approach in large-scale environments because it promotes an efficient and precise reuse of shared metadata in highly dynamic systems.

31.2.2 Emotion Detection in Tags

In [35], authors focus their interests on the target-specific polarity determination of adjectives. "A domain-specific noun, the target noun, is modified by a qualifying adjective". They affirm that rather than having a prior polarity, adjectives are often bearing a target-specific polarity, which, for example, allows an adjective to switch polarity depending on the accompanying noun. This work proposes a method to make "sentiment disambiguation". This method is composed by "a two stage model: (a) Identification of domain-specific target and the construction of a target-specific polarity adjective lexicon; (b) a bootstrapping approach to determine the target-specific adjective polarity".

Sentiment Tagging of Adjectives is an important field to improve efficiency of these kind of systems and in [36] is described a system that is "based on multiple bootstrapping runs on WordNet synsets and glosses using different non-intersecting seed lists of manually annotated words". This system is enhanced with the addition of a "module for partial sense disambiguation of sentiment-bearing adjectives using combinatorial patterns". With this the system can enable sentiment annotation at the sense, rather than whole word level, and provide an effective tool for the automatic cleaning of the lists of sentiment-annotated words. The results allowed cleaning a set of sentiment-bearing adjectives with a performance similar to human annotation.

In summary, the majority of research seems to focus on studying and understanding major systems, rather than on evolving them and proposing improvements. The proposal at hand will try to go directly into the source and develop a type of application such as the major ones, such as Twitter, Blogger, Flickr, etc. [37, 38, 39], as far as tags are concerned. The truth is that all these technologies have found incredible adherence over the years and many studies seem to try to find out exactly why that happens. However, the proposed work will try to take a step further, by pinpointing where and how those technologies can be further evolved and by prototyping those exact environments, therefore proposing both theoretical and technical innovations.

31.3 Proposed Approach

To achieve these project goals we need to propose and develop several tools. In Fig. 31.1 we show the model of the entire system that starts with the development of a social tag network with a tag suggestion mechanism that will give support to users when they are tagging and a same time will solve some problems that focused above in start-of -art like ranking and ambiguities solving.

This tag suggestion mechanism will in our opinion improve the accuracy of knowledge acquired by the next module.

The tag analysis tools will be based in several mechanisms like Case-based, Neural networks and Bayesian networks that are usually used in systems where emotional feelings analysis is very important. Affective aspects of tags are in this case our main goal because they will allow the extraction of user characteristics and interests behind the simple word analysis.

These tools will provide a simple and comprehensive method of creating a user profile without asking several boring and repetitive questions.

The tourism recommender system will be able of using the profile based on the extracted knowledge give to user's recommendations of site seeing tours.

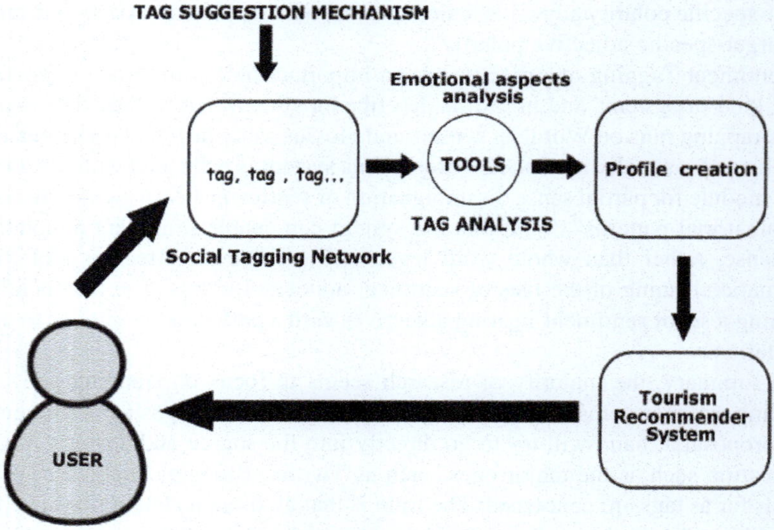

Fig. 31.1 Proposed System

The best way of measure the quality of the produced results is to compare the results of the tourism recommender system when it uses the profile created by our analysis system and other created using direct user answers

In our opinion considering unpredictability and constant changes that are occurring in this field, the Action-Research methodology [40] seems to be a good option to guide us thought this work. This methodology starts by identifying the problem so that a hypothesis can be formulated on which the development will be based. Subsequently, the information is recompiled, organized and analyzed, continuously building a proposal for solving the identified problem. Finally, one can make its conclusions based on the results obtained during the investigation.

31.4 Conclusions

The proposed approach will be entirely based in tags to extract new knowledge to use in a tourism recommender system, this consist in a first phase in the development of a social network system where the set of tags allowed must be controlled. The act of tagging will be driven by a specific context related set of tags that can allow a better performance in tag analysis and further profile creation. Even with driven tagging there is a great risk of users makes tagging more related with uncontrolled aspects like time and space then only with emotional aspects. Emotional aspects are the biggest clue of this project, because our main goal is to evaluate a tag in a context not a tag like a simple word so, must probably in future research task a deep analysis of Affective Computing research area must be done in order to discover if existing approaches in this area can be used to solve our emotional versus tag relation.

Recommendation systems are increasing today their influence in modern world because users seems to be trusting more in their results, but still remains the need of ask users a set most of the times very boring questions to create a profile and give recommendation algorithms the necessary data that will allow them to give a recommendation solution with great accuracy and efficiency. Our goal is to reach the recommendation phase almost without asking anything to users "know the users without knowing them" allowing a better user experience and, in our opinion, a more trustable solution, because most of the times when we ask some more personal data to a user, the answer tends to be more vague. If we analyze data that the user left on internet without great preoccupations we will be able of generate better conclusions.

This work based in web 2.0 mechanisms tries to draw a possible solution to answer to one of most problematic internet problems, the huge amount of data available without a visible usefulness. If newer systems can demonstrate to users that the time that they usually spent tagging can be useful to them in future, probably more and better data will be available.

Acknowledgments This work is supported by FEDER Funds through the "Programa Operacional Factores de Competitividade – COMPETE" program and by National Funds through FCT "Fundação para a Ciência e a Tecnologia" under the project: FCOMP-01-0124-FEDER-PEst-OE/EEI/UI0760/2011

References

1. Ostrow A (2009) Twitter's massive 2008: 752 percent growth. Mashable – the social media guide. http://mashable.com/2009/01/09/twitter-growth-2008/
2. Shreeharsh K, Ajita J, Doree Duncan S (2007) An activity-based perspective of collaborative tagging. In: Proceedings of the international conference on weblogs and social media (ICWSM), Boulder, 26–28 Mar 2007
3. Otis Gospodnetic (2005) Web 2.0 quick study: tag use patterns. Blog
4. Kipp MEI (2006) Exploring the context of user, creator and intermediary tagging. In: Proceedings of the Information Architecture Summit. Vancouver
5. Golder S, Huberman BA (2006) Usage patterns of collaborative tagging systems. J Info Sci 32(2):198–208
6. Marlow C, Naaman M, Boyd D, Davis M (2006) Tagging paper, taxonomy, Flickr, academic article, to read. In: Proceedings of the seventeenth conference on hypertext and hypermedia, Odense, 22–25 August 2006
7. Spalding T (2007) When tags work and when they don't: amazon and library thing. Thingology blog
8. Mejias U (2005) Tag literacy. Ulises mejias's Blog. http://blog.ulisesmejias.com/2005/04/26/tag-literacy/
9. Shilad S, Lam SK, Al Mamunur R, Cosley D, Frankowski D, Osterhouse J, Maxwell Harper F, Riedl J (2006) Tagging, communities, vocabulary, evolution. In: Proceedings of the 2006 20th anniversary conference on computer supported cooperative work (CSCW '06). ACM, New York, pp 181–190

10. Trant J (2009) Studying social tagging and folksonomy: a review and framework. J Digit Info 10:1 [Online]
11. Rashmi (2005) A cognitive analysis of tagging, Rashmi's blog. http://rashmisinha.com/2005/09/27/a-cognitive-analysis-of-tagging/
12. Zollers A (2007) Emerging motivations for tagging: expression, performance and activism. In: Proceedings of the 16th international conference on world wide web 2007. Banff, Alberta
13. Morrison PJ (2007) Why are they tagging, and why do we want them to? ASIS&T bulletin: special section: Folksonomies. http://www.asis.org/Bulletin/Oct-07/morrison.html. Oct/Nov 2007
14. Morgan A, Naaman M (2007) Why we tag: motivations for annotation in mobile and online media. In: Proceedings of the SIGCHI conference on human factors in computing systems (CHI '07). ACM, New York, pp 971–980
15. Lawley L (2005) Social consequences of social tagging. Many2ManyBlog, comment posted 20 Jan 2005
16. Maarek YS, Marmasse N, Navon Y, Soroka V (2006) Tagging the physical world. Collaborative web tagging workshop at WWW2006, Edinburgh
17. Balahur A, Montoyo A (2010) OpAL: applying opinion mining techniques for the disambiguation of sentiment ambiguous adjectives in SemEval-2 task 18. In: Proceedings of the 5th international workshop on semantic evaluation (SemEval '10). Association for Computational Linguistics, Stroudsburg, pp 444–447
18. Gemmell J, Ramezani M, Schimoler T, Christiansen L, Mobasher B (2009) The impact of ambiguity and redundancy on tag recommendation in folksonomies. In: Proceedings of the third ACM conference on recommender systems – RecSys '09, New York, p 45
19. Weinberger KQ, Slaney M, Van Zwol R (2008) Resolving tag ambiguity. In: Proceeding of the 16th ACM international conference on Multimedia – MM '08, New York, p 111
20. Shepitsen A, Gemmell J, Mobasher B, Burke R (2008) Personalized recommendation in social tagging systems using hierarchical clustering. In: Proceedings of the 2008 ACM conference on Recommender systems – RecSys '08, New York, p 259
21. Krestel R, Fankhauser P, Nejdl W (2009) Latent dirichlet allocation for tag recommendation. In: Proceedings of the third ACM conference on Recommender Systems – RecSys '09, New York, p 61
22. Anderson A, Raghunathan K, Vogel A (2008) TagEz: Flickr tag recommendation. Association for the advancement of artificial intelligence (www.aaai.org). http://cs.stanford.edu/people/acvogel/tagez/
23. Garg N, Ingmar I (2008) Personalized tag suggestion for flickr. In: Proceedings of the 17th international conference on World Wide Web (WWW '08). ACM, New York
24. Wu L, Yang L, Yu N, Hua XS (2009) Learning to tag. In: Proceedings of the 18th international conference on World Wide Web – WWW '09, New York, p 361
25. Sigurbjörnsson B, Van Zwol R (2008) Flickr tag recommendation based on collective knowledge. In: WWW '08: Proceeding of the 17th international conference on World Wide Web. ACM, New York
26. Lipczak M, Milios E (2010) Learning in efficient tag recommendation. In: Proceedings of the fourth ACM conference on recommender systems. ACM, New York, pp 167–174
27. Pak A, Paroubek P (2010) Twitter based system: using twitter for disambiguating sentiment ambiguous adjectives. In: Proceedings of the 5th international workshop on semantic evaluation. Association for Computational Linguistics, Stroudsburg, pp 436–439
28. Kipp MEI (2007) @toread and cool: tagging for time, task and emotion. In: Proceedings of the 8th information architecture summit, Las Vegas, 22–27 March 2007
29. Zanardi V, Capra L (2008) Social ranking. In: Proceedings of the 2008 ACM conference on Recommender Systems – RecSys '08, New York, p 51
30. Liu D, Hua XS, Yang L, Wang M, Zhang HJ (2009) Tag ranking. In: Proceedings of the 18th international conference on World Wide Web – WWW '09, New York, p 351

31. Li X, Snoek CGM, Worring M (2009) Learning social tag relevance by neighbor voting. IEEE Trans Multimed 11(7):1310–1322
32. Dubinko M, Kumar R, Magnani J, Novak J, Raghavan P, Tomkins A (2006) Visualizing tags over time. In: Proceedings of the 15th international conference on World Wide Web – WWW '06, New York, p 193
33. Lee SE, Son DK, Han SS (2007) Qtag. In: Proceedings of the 25th annual ACM international conference on design of communication – SIGDOC '07, New York, p 189
34. Budura A, Michel S, Cudré-Mauroux P, Aberer K (2008) To tag or not to tag: harvesting adjacent metadata in large-scale tagging systems. In: SIGIR '08: Proceedings of the 31st annual international on research and development in retrieval, Singapore
35. Fahrni A, Klenner M (2008) Old wine or warm beer: target-specific sentiment analysis of adjectives. In: Proceedings of the AISB 2008 symposium on affective language in human and machine. The Society for the Study of Artificial Intelligence and Simulation of Behaviour Press, Aberdeen, pp 60–63
36. Andreevskaia A, Sabine Bergler S (2006) Senses and sentiments: sentiment tagging of adjectives at the meaning level. In: Proceedings of the 19th canadian conference on artificial intelligence, AI'06, Quebec. Lecture Notes in Computer Science vol. 4013, pp 336–346, 7–9 June 2006
37. Twitter (2009) http://twitter.com
38. Blogger (2009) https://www.blogger.com/start
39. Flickr (2009) http://www.flickr.com
40. Somekh B (2005) Action research: a methodology for change and development. Open University Press, London

Chapter 32
Multidimensional Scaling Analysis of Electricity Market Prices

Filipe Azevedo and J. Tenreiro Machado

Abstract This paper studies the impact of energy and stock markets upon electricity markets using Multidimensional Scaling (MDS). Historical values from major energy, stock and electricity markets are adopted. To analyze the data several graphs produced by MDS are presented and discussed. This method is useful to have a deeper insight into the behavior and the correlation of the markets. The results may also guide the construction models, helping electricity markets agents hedging against Market Clearing Price (MCP) volatility and, simultaneously, to achieve better financial results.

32.1 Introduction

Electricity is a non-storable product that, associated with the fact that electric power systems must always be balanced, poses special financial challenges. In fact, wide fluctuations in spot prices associated with weather high or low temperatures, can lead to prices climbing up to 1,000% during short periods [1]. The overall behavior is of high volatility, even when compared with the prices that occur in oil, gas and stock markets. A second implication of the electricity non-storability is the impossibility of transferring a certain amount of energy without considering the electric

F. Azevedo (✉)
INESC TEC - INESC Technology and Science (formerly INESC Porto) and ISEP/IPP - School of Engineering, Polytechnic Institute of Porto, Rua Dr. António Bernardino de Almeida, 4200-072 Porto, Portugal

Portugal and INESC TEC (formerly INESC Porto), Porto, Portugal
e-mail: fta@isep.ipp.pt/filipe.azevedo@inescporto.pt

J.T. Machado
Institute of Engineering, Department of Electrical Engineering, Polytechnic of Porto (ISEP/IPP), Rua Dr. António Bernardino de Almeida, 4200-072 Porto, Portugal
e-mail: jtm@isep.ipp.pt

A. Madureira et al. (eds.), *Computational Intelligence and Decision Making: Trends and Applications*, Intelligent Systems, Control and Automation: Science and Engineering 61,
DOI 10.1007/978-94-007-4722-7_32, © Springer Science+Business Media Dordrecht 2013

transmission constraints. Besides the instantaneous nature of the electricity commodity, factors such as the uncertainty associated with fuel prices, energy demand, generation availability, or even technical restrictions, may have also considerable impact on electricity volatility [2–4]. Complementary, the structure and management rules of any specific electricity market may constitute sources of price volatility as well [5].

Due to these facts electricity market agents are required for understanding the factors underlying the market price evolution. A comprehensive knowledge allows agents to develop strategies for selling, or buying, electric energy both in the spot and futures market. Furthermore, those strategies are important to practice the hedge against electricity market volatility and to increase their profits.

Derivatives were introduced in electricity markets for allowing agents to eliminate the risk of credit and to increase market liquidity. The superior liquidity is mainly due to the appearance of new agents, usually operating in classical stock exchanges, which see in electricity markets an opportunity to withdraw dividends and to increase the efficiency in risk management. In addition, it should be noted that some of those agents are also active participants in energy markets, like the cases of oil and natural gas.

The first power plants were based on water power or coal, but today we rely mainly on coal, nuclear, natural gas, hydroelectric and petroleum, with a small amount from solar energy, wind generators and geothermal sources. Figure 32.1 illustrates the production of electricity in the U.S. by source for the year 2009.

From Fig. 32.1 it is clear that, in the U.S. and for the year 2009, the main sources for the production of electricity are coal, natural gas and nuclear.

For analyzing the electricity markets and their correlation with energy prices and stock markets indices, a Multidimensional Scaling (MDS) technique is adopted [6–10].

MDS is used in several scientific areas such as, for example, visualizing time-varying correlations across stock markets [11, 12], signal processing [13, 14], digital communications [15], adaptive controllers [16] and music [17]. However, to the authors' best knowledge MDS was not yet applied in the study of electricity markets and therefore, can constitute a good tool to analyze their correlation with energy prices and stock indices.

Monthly average historical data from July 2007 up to August 2010, for stock markets indices, energy and electricity markets prices is used. It is considered historical data from July 2007 because OMEL just set prices for Portugal and Spain separately due to market splitting from that date. Tables 32.1, 32.2 and 32.3 present the energy markets, stock markets indices, and electricity markets, respectively. For PJM Interconnection electricity market Locational Marginal Price (LMP) Load Weighted Mean Price was used.

Bearing these ideas in mind, the paper is organized as follows: Sect. 32.2 presents an overview of the Multidimensional Scaling (MDS), Sect. 32.3 formulates the problem and Sect. 32.4 develops a case study. Finally, Sect. 32.5 outlines the most relevant conclusions.

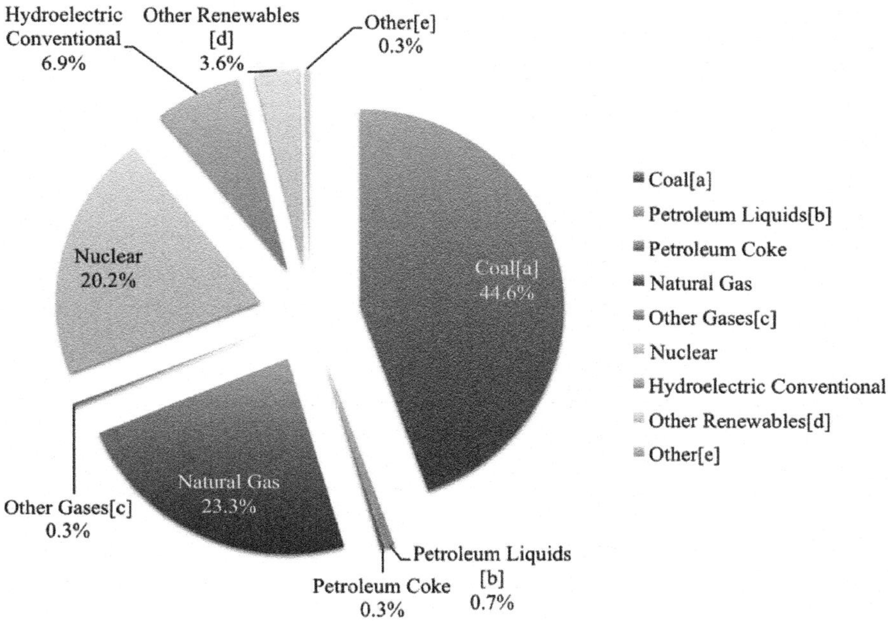

Fig. 32.1 Sources of electricity in the U.S. during 2009 ((**a**) Anthracite, bituminous, subbituminous, lignite, waste coal, and coal synfuel. (**b**) Distillate fuel oil, residual fuel oil, jet fuel, kerosene, and waste oil. (**c**) Blast furnace gas, propane gas, and other manufactured and waste gases derived from fossil fuels. (**d**) Wood, black liquor, other wood waste, biogenic municipal solid waste, landfill gas, sludge waste, agriculture byproducts, other biomass, geothermal, solar thermal, photovoltaic energy, and wind. (**e**) Non-biogenic municipal solid waste, batteries, chemicals, hydrogen, pitch, purchased steam, sulfur, tire-derived fuel, and miscellaneous technologies)

Table 32.1 Energy Markets

Energy market	Abbreviation	Country
West Texas Intermediate	WTI	USA
BRENT Crude	BRENT	UK
Natural Gas	NG	USA

Table 32.2 Stock Markets

Stock market index	Abbreviation	Country
Portuguese Stock Index	PSI-20	Portugal
Iberia Index	IBEX	Spain
NASDAQ −100	NDX	USA
Standard & Poor’s	S&P500	USA
Dow Jones Industrial	DJI	USA
Footsie	FTSE	UK

Table 32.3 Electricity Markets

Electricity market	Abbreviation	Country
OMEL Electricity Market	OMEL-PT	Portugal
OMEL Electricity Market	OMEL-ES	Spain
Energy Exchange Austria	EXAA	Austria
		Germany
Gestore Mercati Energetici	GME	Italy
PJM Interconnection	PJM	USA

32.2 Multidimensional Scaling

Multidimensional scaling (MDS) is a technique for the analysis of similarities, or of distances, of a set of objects [18]. Its main purpose is to find a configuration of the data points in a low-dimensional space SL, such that the proximity between objects in the full-dimensional space SF is represented with some degree of fidelity by the distances between points in the SL. This means that observations that are close together in a SF should be close in the SL. Many aspects of MDS were originally developed by researchers in the social science community, and the method is now available in several statistical packages [19].

32.2.1 Classical Multidimensional Scaling

Classical multidimensional scaling, or classical scaling, is also known under the names Torgerson scaling and Torgerson-Gower scaling, because the first practical method was a technique presented in [6, 8, 9], and also because it is based on the theorems formulated in [7, 10].

32.2.2 Classical Multidimensional Scaling

Nonclassical multidimensional scaling creates a configuration of points, such that the resulting distances approximate the given dissimilarities. This is sometimes an high strict requirement, and non-metric scaling is designed to relax it a bit. Instead of trying to approximate the dissimilarities themselves, non-metric scaling approximates a nonlinear, but monotonic, transformation of them. Because of the monotonicity, larger or smaller distances on a MDS plot will correspond to larger or smaller dissimilarities, respectively. However, the nonlinearity only makes an attempt to preserve the ordering of dissimilarities. Thus, there may be a contraction or expansion of distances at different scales.

There are two forms of nonclassical multidimensional scaling, namely metric scaling, and nonmetric scaling.

With metric MDS it is created a configuration of points such that their inter-point distances approximate the original dissimilarities. One measure of the goodness of fit of that approximation is known as the "stress" and to check the fit of the output configuration to the dissimilarities. To understand the disparities, it helps to make a Shepard plot that shows a configuration of points in two dimensions whose inter-point distances approximates the disparities, which, in turn, are a nonlinear transformation of the original dissimilarities. However, nonmetric MDS has a slightly less ambitious goal than metric scaling. Instead of attempting to create a configuration of points for which the pairwise distances approximate the original dissimilarities, it attempts only to approximate the ranks of the dissimilarities. Another way of saying this is that nonmetric MDS creates a configuration of points whose inter-point distances approximate a monotonic transformation of the original dissimilarities [8, 9, 18–21].

32.3 Case Study

This study aims to study the correlation of energy and stock markets on electricity markets using Nonmetric scaling form of Nonclassical Multidimensional Scaling. To achieve this goal, historical data from major energy, stock and electricity markets is adopted. Are only used natural gas monthly average prices for U.S. market due to difficulties finding those prices for Europe.

32.3.1 Stock, Energy and Electricity Markets used in This Study

Tables 32.1, 32.2 and 32.3 present the energy markets prices, stock markets indices and electricity markets prices used in this study, respectively.

32.3.2 Nonmetric Scaling Stress Function

The Nonmetric Scaling form of Nonclassical Multidimensional is adopted.

The *Minkowski* and *Chebychev* metric function presented in (32.1) and (32.2), respectively, are compared and used to measure the distance (d_{st}) between pairs of observations. The expressions are given by:

$$d_{st} = \left(\sum_{j=1}^{m} \left| x_{sj} - x_{tj} \right|^{p} \right)^{\frac{1}{p}} \tag{32.1}$$

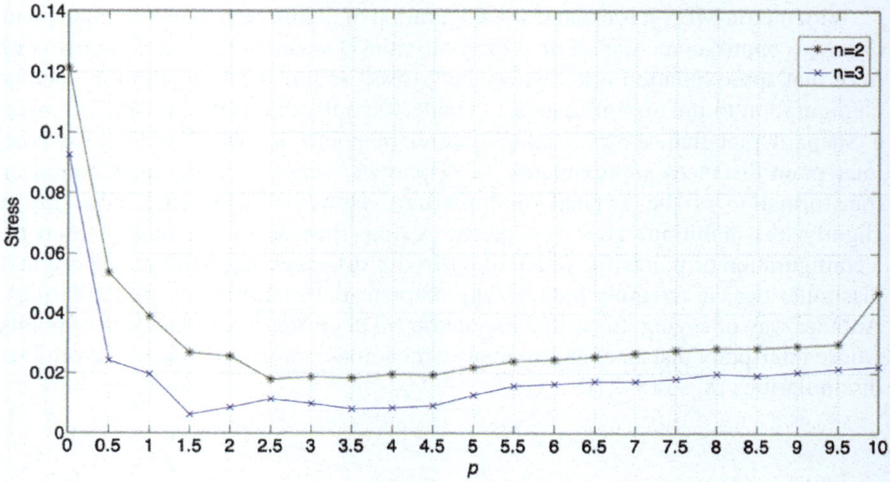

Fig. 32.2 Stress *vs* the parameter p of the *Minkowski* metric

$$d_{st} = \max_j \{ |x_{sj} - x_{tj}| \} \tag{32.2}$$

where p is a parameter, m represents the number of observations and x the variables.

To find the better value of parameter p the evaluation of the "stress" is represented in Fig. 32.2. We conclude that, for two dimensions ($n = 2$) the value of p that minimizes the "stress" is $p = 2.5$, while for three dimensions ($n = 3$) we get $p = 1.5$.

32.3.3 *Number of Dimensions*

Figure 32.3 presents the variation of the "stress" versus the number of dimensions n of the MDS map when considering the *Minkowski* function for $p = \{0.5, 1, 1.5, 2\}$. We conclude that the required number of dimensions for the MDS plot is $n = 4$. However, the difference between the "stress" for $n = 3$ and $n = 4$ is small and, therefore, is adopted a three-dimensional representative for the MDS chart.

The variation of the "stress" value with the number of dimensions n of the MDS map using the *Chebychev* and the *Minkowski* metrics, with parameter $p = 1.5$, is depicted in Fig. 32.4. We conclude that, for $n = 3$, the "stress" value for *Minkowski* metric, with parameter $p = 1.5$, is lower than for *Chebychev* metric. Therefore, is adopted the *Minkowski* metric with parameter $p = 1.5$.

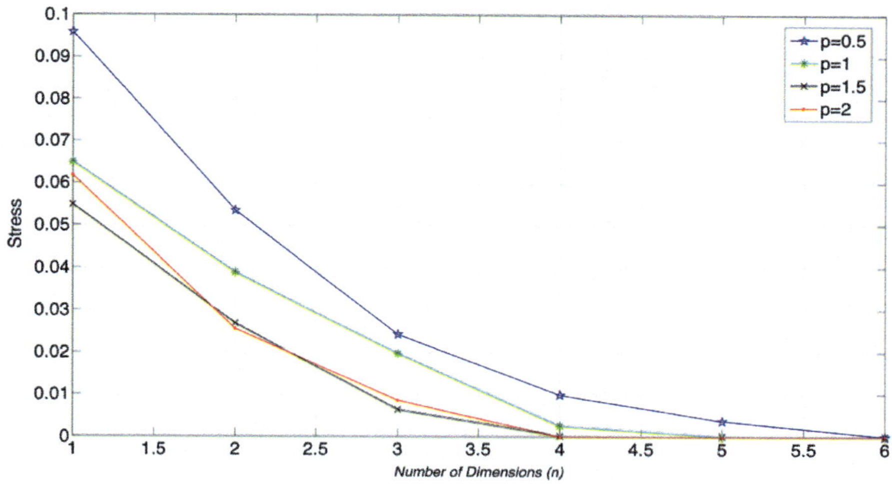

Fig. 32.3 Stress *vs* n the number of dimensions of the MDS plot for $p = \{0.5, 1, 1.5, 2\}$ of *Minkowski* metric

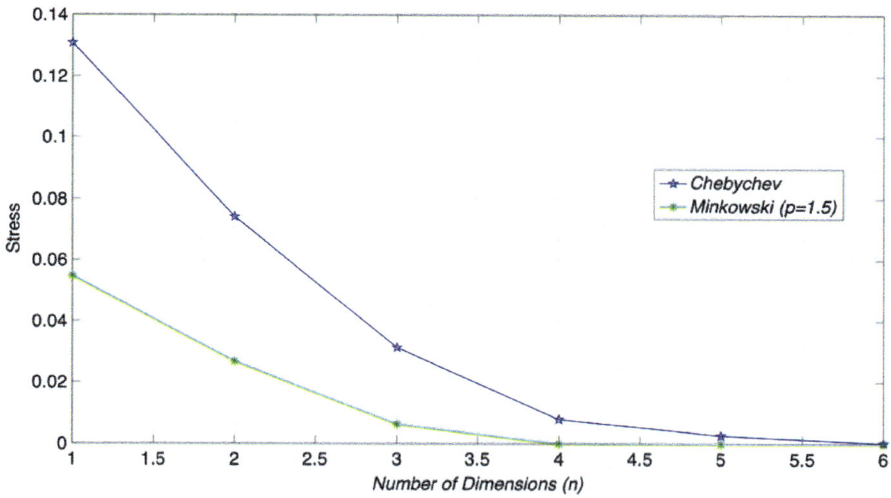

Fig. 32.4 Stress *vs* the number n of dimensions for the MDS plot for *Chebychev* and *Minkowski* ($p = 1.5$) metrics

32.4 Results

The nonmetric MDS solution plot of the configuration in three dimensions is depicted in Fig. 32.5. We verify that emerge three major groups: oil {WTI, BRENT}, stock markets {PSI20, DJI, SP500, FTSE, IBEX} and electricity markets {PJM, EXAA, GME, OMEL-PT, OMEL-ES}.

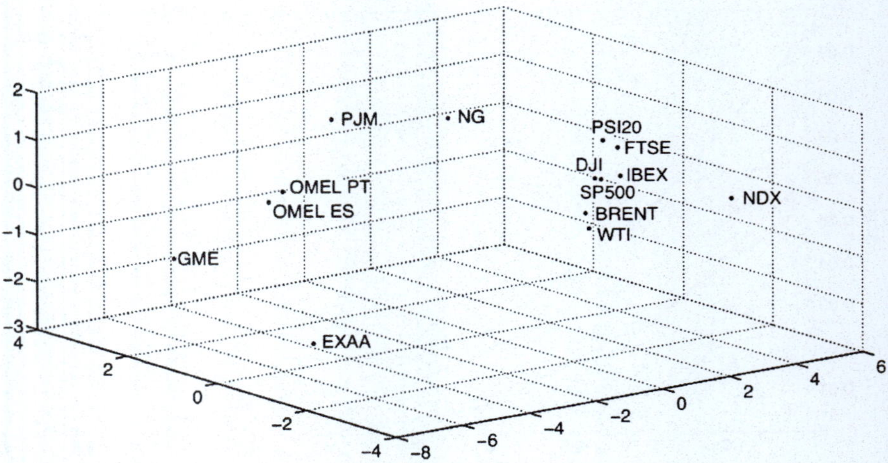

Fig. 32.5 Nonmetric MDS solution for $n = 3$ and $p = 1.5$

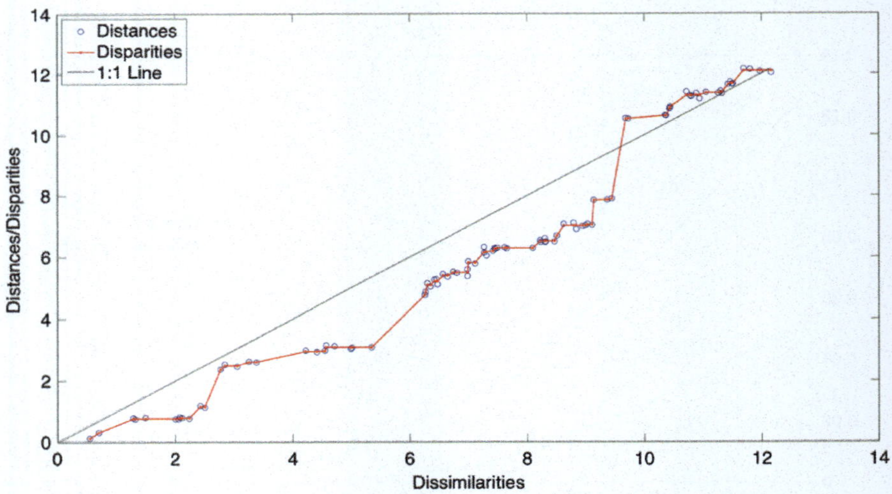

Fig. 32.6 Shepard plot, disparities *vs* dissimilarities, for the MDS with $n = 3$ and $p = 1.5$

From Fig. 32.1 we see that in U.S. the main sources for the production of electricity are coal, natural gas and nuclear. This is the reason why the PJM electricity market is closer to the natural gas energy market (NG), than to the oil energy markets (WTI and BRENT). Moreover, the natural gas market (NG) used in this case study is for U.S., which reinforces its proximity to the PJM electricity market.

Nonmetric scaling measures the how well the inter-point distances of the output configuration approximate the disparities. To check the fitting of the output configuration relatively to the dissimilarities is often used the Shepard plot presented in Fig. 32.6.

Figure 32.6 shows that MDS has found a configuration of points in three dimensions whose inter-point distances approximate the disparities which, in turn, are a nonlinear transformation of the original dissimilarities. The concave shape of the disparities as a function of the dissimilarities indicates that fit tends to contract small distances relative to the corresponding dissimilarities. This may be perfectly acceptable in practice.

In conclusion, we verified that MDS produced simple and assertive results for a set of financial data and that the plots are reliable.

32.5 Conclusions

This paper addressed a graphical and statistical tool to visualize time-varying correlations between several distinct financial indices arising in stock, energy and electricity markets. The proposed MDS-based method makes use of monthly price average of six stock markets, three energy markets and five electricity markets.

As results it is clear the emergence of three major clusters: oil group {WTI, BRENT}, stock markets group {PSI20, DJI, SP500, FTSE, IBEX} and electricity markets group {PJM, EXAA, GME, OMEL-PT, OMEL-ES}. From the described groups, oil group is closer to stock markets than natural gas, and natural gas is closer to the electricity markets than the oil group. This result is due to the increasing importance of combined cycle power plants on electricity production. Since it was used natural gas market prices for U.S., the PJM electricity market is closer to natural gas than European electricity markets, as expected.

Finally, there are several aspects relevant for future research. A first issue concerns applying MDS to alternative data sets to check how informative the method can be in other cases. A second issue concerns taking the MDS maps seriously and to incorporate them in decision models for improving empirical specification strategies.

Acknowledgments The author thanks the following organizations for allowing access to the data:

- U.S. Energy and Information, http://www.eia.doe.gov/
- Yahoo! Finance, http://finance.yahoo.com/
- OMEL Mercado de Electricidad, http://www.omel.es/inicio
- PJM: Interconnection, http://www.pjm.com/
- EXAA Energy Exchange Austria, http://www.exaa.at/
- GME Gestore Mercati Energetici, http://www.mercatoelettrico.org/En/Default.aspx

The authors acknowledge the help of João Soares to obtain the historical data.

References

1. Hull JC (2002) Fundamentals of futures and options markets, 4th edn. Prentice-Hall, Upper Saddle River, NJ
2. Shahidehpour M, Alomoush M (2001) Restructured electrical power systems – operation, trading, and volatility. Marcel Dekker, Inc, New York
3. Azevedo F, Vale ZA, Oliveira PBM, Khodr HM (2010) A long-term risk management tool for electricity markets using swarm intelligence. Elsevier Electr Power Syst Res 80:424–433
4. Azevedo F, Vale ZA, Oliveira PBM (2007) A decision-support system based on particle swarm optimization for multi-period hedging in electricity markets. IEEE Trans Power Syst 22(3):995–1003
5. Skantze PL, Ilic MD (2001) Valuation, hedging and speculation in competitive electricity markets: a fundamental approach. Kluwer Academic Publishers, Boston, MA
6. Gower JC (1966) Some distance properties of latent root and vector methods used in multivariate analysis. Biometrika 53:325–338
7. Eckart C, Young G (1936) Approximation of one matrix by another of lower rank. Psychometrika 1:211–218
8. Torgerson WS (1952) Multidimensional scaling: theory and method. Psychometrika 17:401–419
9. Torgerson WS (1958) Theory and methods of scaling. Wiley, New York
10. Young G, Householder AS (1938) Discussion of a set of point in terms of their mutual distances. Psychometrika 3:19–22
11. Groenen PJ, Franses PH (2000) Visualizing time-varying correlations across stock markets. J Empir Financ 7:155–172
12. Machado JT, Duarte GM, Duarte FB (2011) Identifying economic periods and crisis with the multidimensional scaling. Nonlinear Dynamic (Springer) 63(4):611–622
13. Martínez-Torres MR, García FJ, Marín SL, Vázquez SG (2005) A digital signal processing teaching methodology using concept-mapping techniques. IEEE Trans Educ 48(3):422–429
14. Rosario D, Romano J (2010) Multidimensional image processing for remote sensing anomaly detection. In: Proceedings of the 2nd international conference on Image Processing Theory Tools and Applications (IPTA), Paris, France, 7–10 July 2010, pp 471 – 476
15. Chen S, Mulgrew B, Grant PM (1993) A clustering technique for digital communications channel equalization using radial basis function networks. IEEE Trans Neural Netw 4:570–578
16. Bingulac SP (1994) On the compatibility of adaptive controllers (Published conference proceedings style). In: Proceedings of the 4th annual Allerton conference circuits and systems theory, New York, pp 8–16
17. Machado JT, Costa AC, Lima MFM (2011) Dynamical analysis of compositions. Nonlinear Dynamic (Springer) 65(4):339–412
18. Borg I, Groenen P (2005) Modern multidimensional scaling – theory and applications, 2nd edn. Springer, New York, NY
19. Martinez W, Martinez A (2005) Exploratory data analysis with MATLAB. Chapman & Hall/CRC, Boca Raton, FL
20. Takane Y, Young FW, Leeuw J (1977) Nonmetric Individual differences multidimensional scaling: an alternating least squares method with optimal scaling features. Psychometrika 42:7–67
21. Cox TF, Cox MA (2001) Multimensional scaling, 2nd edn. Chapman & Hall/CRC, Boca Raton, FL

Chapter 33
PCMAT Metadata Authoring Tool

Paulo Couto, Constantino Martins, Luiz Faria, Marta Fernandes, and Eurico Carrapatoso

Abstract This paper introduces the PCMAT platform project and, in particular, one of its components, the PCMAT Metadata Authoring Tool. This is an educational web application that allows the project metadata creators to write the metadata associated to each learning object without any concern for the metadata schema semantics. Furthermore it permits the project managers to add or delete elements to the schema, without having to rewrite or compile any code.

Keywords Learning objects • Metadata • PCMAT • Web application

33.1 Introduction

The purpose of this paper is to introduce the PCMAT Metadata Authoring Tool. This educational web application was developed within the context of the PCMAT project to give the project's metadata creators a tool that would allow them to write an object's metadata file by providing the metadata values without having to worry about the semantics of the metadata schema.

To achieve its purpose, this paper provides a brief description of the extant learning objects metadata standards, followed by a description of the PCMAT platform, a project funded by the Fundação para a Ciência e Tecnologia (FCT), and of the PCMAT Metadata Authoring Tool.

P. Couto (✉) • C. Martins • L. Faria • M. Fernandes
GECAD – Knowledge Engineering and Decision Support Group,
Institute of Engineering – Polytechnic of Porto, Porto, Portugal
e-mail: pjco@isep.ipp.pt; acm@isep.ipp.pt; lef@isep.ipp.pt; mmaf@isep.ipp.pt

E. Carrapatoso
Faculty of Engineering, University of Porto, Porto, Portugal
e-mail: emc@fe.up.pt

A. Madureira et al. (eds.), *Computational Intelligence and Decision Making: Trends and Applications*, Intelligent Systems, Control and Automation: Science and Engineering 61,
DOI 10.1007/978-94-007-4722-7_33, © Springer Science+Business Media Dordrecht 2013

33.2 Learning Objects Metadata

33.2.1 Learning Objects and Metadata

According to Wiley [1] the term "learning objects" has its roots, possibly, in "Wayne Hodgins' 1994 use of the term in the title of the CedMA working group called Learning Architectures, API's, and Learning Objects", although the concept may have been firstly described by Gerard [2] when, in 1967, explained his vision of "Computer-aided learning, CAL" and the need to create and disseminate CAL materials, namely "integrating books (via microform), sound records, movies, video materials, any recorded form of man's collective experience and creations, into computer-mobilized resources."

Learning objects (LOs) go by many names, each accompanied by its respective definition: "knowledge objects", "instructional component", "pedagogical documents", "educational software components", "online learning materials", or just plainly "resources" [1]. Nonetheless, IEEE's definition of learning object as "any entity, digital or non-digital, that may be used for learning, education or training" [3] is, probably, the most well known, and controversial, due to its quite broad scope. Thus, Wiley's proposed definition of learning object as "any digital resource that can be reused to support learning" [1], as it reduces the scope of learning objects to digital objects, seems to us to be more adequate in defining learning objects as reusable resources available and retrievable over the Internet.

According to the National Information Standards Organization (NISO) [4], metadata is "structured information that describes, explains, locates, or otherwise makes it easier to retrieve, use, or manage an information resource." This definition encapsulates not only what metadata is but also how it can be used, or, in other words, the purpose of metadata. And, in defining metadata as "structured information", it lays out the ground for the need of metadata schemes. Metadata schemes, or schema, are sets of metadata elements that identify the characteristics of a certain type of information resource. A metadata scheme possesses a semantic, meaning the definition of the elements themselves, and content, the values assigned to the elements.

As far as LOs are concerned, there are two established metadata schemes currently in use: the IEEE Learning Object Metadata (LOM) and the Dublin Core Metadata Element Set (DCMES).

33.2.2 IEEE Learning Object Metadata

The IEEE LOM is a multi-part standard, currently consisting of a conceptual data schema [3] and its XML schema binding [5]. This standard defines a structured set of 76 elements, covering a wide variety of characteristics found to be relevant to define a learning object, grouped in the following categories:

1. *general*: information that describes the LO as a whole, as, for example, an identifier, the title, a description, a set of keywords;
2. *life cycle*: information pertaining to the development of the LO;
3. *meta-metadata*: information concerning the actual metadata document and not the described LO;
4. *technical*: information regarding technical requirements and technical characteristics of the LO;
5. *educational*: information about the LO's educational and pedagogic aspects;
6. *rights*: information on the LO's intellectual property rights and conditions of use;
7. *relation*: information that defines the relationship of the described LO with other LOs;
8. *annotation*: space for storing comments on the LO's usage; and
9. *classification*: description of the LO in accordance with different classification systems.

Within each category an element or a group of elements may be repeated, if necessary, and every element is optional. According to the standard, a "LOM instance that contains no value for any of the LOM data elements is a conforming instance" [3].

Nonetheless, if use is made of a LOM data element, this shall be made respecting its structure within the schema and its value should be in accordance with the *datatypes* and *vocabularies* defined in the schema. However, the LOM standard allows the insertion of elements, or even attributes to the LOM elements, other than the ones foreseen within the standard, provided that these are identified by a specific namespace. This possibility allows a community of users to specify which elements and vocabularies they will use, building what is called an *application profile*.

Perhaps due to its status as an international standard, or because it was developed with the purpose of characterising a learning object, or even due to the possibility of developing specific application profiles, LOM has been implemented by a large number of repositories. ARIADNE, SMETE, Learning Matrix, iLumina, MERLOT, HEAL, CAREO, LearnAlberta Online Curriculum Repository and Lydia Inc. are some examples of repositories that implemented the LOM standard [6].

33.2.3 Dublin Core Metadata Element Set

The DCMES "is a vocabulary of 15 properties for use in resource description" [7]. Standardized as ISO Standard 15836:2009, the core Element Set is intended to be "broad and generic, usable for describing a wide range of resources" and comprises the following elements: contributor, coverage, creator, date, description, format, identifier, language, publisher, relation, rights, source, subject, title and type. All these elements are optional and may be repeated if required. Dated 1998, this core set is now part of a larger set of metadata vocabularies known as DCMI Metadata Terms (DCMI-TERMS).

Aware of the fact that the metadata needs of particular communities and applications are very diverse, the Dublin Core Metadata Initiative (DCMI) provides a framework for designing a Dublin Core Application Profile (DCAP). As stated in the Guidelines for Dublin Core Application Profiles [8], a DCAP is a generic construct for designing metadata records that does not require the use of DCMI-TERMS, a "DCAP can use *any* terms that are defined on the basis of RDF, combining terms from multiple namespaces as needed".

"A DCAP is a document (or set of documents) that specifies and describes the metadata used in a particular application" and it consists of the following components:

1. *functional requirements (mandatory)*: describes what a community wants to accomplish with its application;
2. *domain model (mandatory)*: characterizes the types of things described by the metadata and their relationships;
3. *description set profile (mandatory)*: enumerates the metadata terms to be used and the rules for their use;
4. *usage guidelines (optional)*: describe how to apply the application profile, how the used properties are intended to be used in the application context; and
5. *syntax guidelines (optional)*: defines the machine syntax that will be used to encode the data.

Still in accordance with the aforementioned guidelines, application profiles should be "developed by a team with specialized knowledge of the resources that need to be described, the metadata to be used in the description of those resources, as well as an understanding of the Semantic Web and the linked data environment." Thus, following this recommendation, the DC-Education Community (DC-Ed) "decided at the DC-2007 Conference in Singapore to form a new Task Group in order to finish the DC-Education Application Profile by mid-2008" [9]. However, in the last meeting held by the DC-Ed on October 20th, 2010, in Pittsburgh, the development work was still being carried out on the domain model of the DC-Education Application Profile. At this meeting it was foreseen that the complete documentation would be available by July-August 2011 [10].

Notwithstanding, DCMES has been adopted in particular by libraries and archives worldwide, mostly because it is an integral part of the Open Archives Initiative Protocol for Metadata Harvesting (OAI-PMH) [11]. The OAI-PMH allows for compliant metadata from different repositories to be harvested automatically in order to build a centralized point of search. The implementation of this protocol has been quite successful due to the availability of open source freeware like DSpace, Fedora or Greenstone [12].

33.3 PCMAT Metadata Authoring Tool

33.3.1 PCMAT Platform

The PCMAT (Mathematics Collaborative Learning Platform) platform is based on a constructivist approach, and aims assessing the user knowledge and presenting contents and activities adapted to the needs and learning style of the student.

The PCMAT platform allows students to autonomously create and consolidate knowledge, with permanent automatic feedback and support, through instructional methodologies and educational activities explored in a constructivist manner.

The adaptation of the application is based on progressive self-assessment exercises solved by the student that evolve in difficulty and domain topics. The curriculum is defined by the teacher but is dynamically individualized according to the student's current level of knowledge, competences, abilities and learning path. The platform provides contextualized access to tutorials if the students fail a progression step [13].

Others goals of PCMAT include:

1. To define new strategies and architecture for the implementation of an Adaptive Hypermedia Educational platform to support and improve Mathematics in basic schools;
2. To define the student model that describes the information, knowledge, preferences, and learning style of the user;
3. To define the process and tools to produce learning objects and conceive a knowledge representation of the student's attributes concerning the emotional characteristics and develop a set of adaptive and dynamic pedagogical strategies to put forward this hybrid model;
4. Develop the system functionalities for the interaction model (interface adaptation) considering the objectives, profile and knowledge of the student and the Domain and Adaptation model;
5. Improve results and knowledge in Mathematics in Basic Schools.

The system architecture (Fig. 33.1) is based on some strategies already used for the implementation of Adaptive Hypermedia System (AHS), like for example the: Dexter Model; Amsterdam Hypermedia Model (AHM); Adaptive Hypermedia Application Model (AHAM) or; Munich Reference Model [14].

Therefore, in our system, the user requests an assessment by clicking on a link in a Web page. Every assessment/page corresponds to a "concept" in the domain model and the user model (which is an overlay model). The system checks the suitability of the requested page for the current user. The adaptation rules used to check if the page is suitable are defined in the adaptation model. Updates to the Student Model are inferred from the interaction between the user and the application. The correct or wrong answers of the user allow the system to estimate the user's knowledge level about the concept related with the requested content [13].

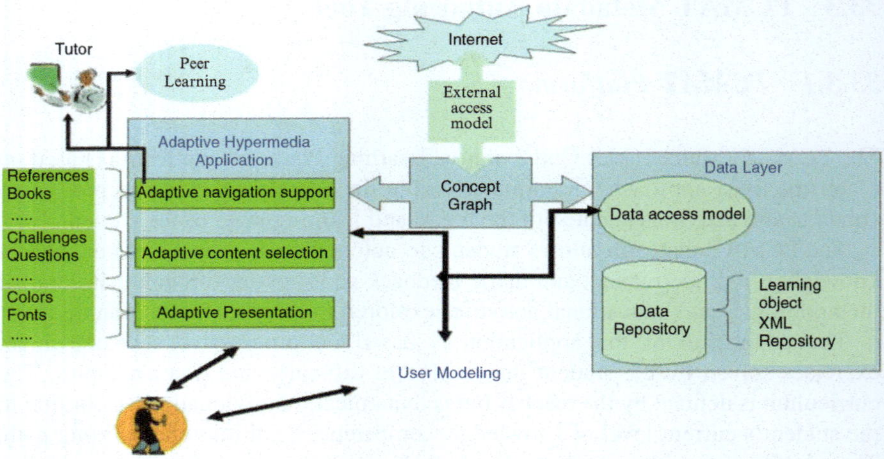

Fig. 33.1 Architecture of the PCMAT platform [13]

33.3.2 PCMAT Metadata Authoring Tool

Inherent to the operation of PCMAT is the existence of a collection of digital learning objects and the need for these to be retrieved manually, to be managed by teachers and content developers, and automatically, to be presented to the students in accordance with the respective learning style and knowledge. To make this possible, a metadata record will be associated to each learning object (LO). This metadata document will contain:

1. *descriptive metadata*: information pertaining to the LO creators, the LO identification, such as title, short description and keywords to help the search and retrieval actions;
2. *administrative metadata*: such as when and how it was created, file type and other technical information, as well as copyright information; and
3. *educational metadata*: as the typical age range (of the intended user), difficulty and interactivity level, as well as the LO underlying concept(s), making it possible for the LO to be retrieved by the system if found suitable to a particular student's knowledge and learning style.

With these requirements in mind we were confronted with the option of developing a PCMAT specific set of metadata or adopting an established standard. For a standard to be adopted, it had: (1) to be extendable, allowing the inclusion of alien elements; and, (2) to permit an XML document as a representation of a metadata instance, thus providing the necessary XML Schema.

In view of the characteristics and development of IEEE LOM and DCMES, it was decided to adopt IEEE LOM as the basis for the development of PCMAT's domain model, mainly because it defines a large set of metadata from which one can

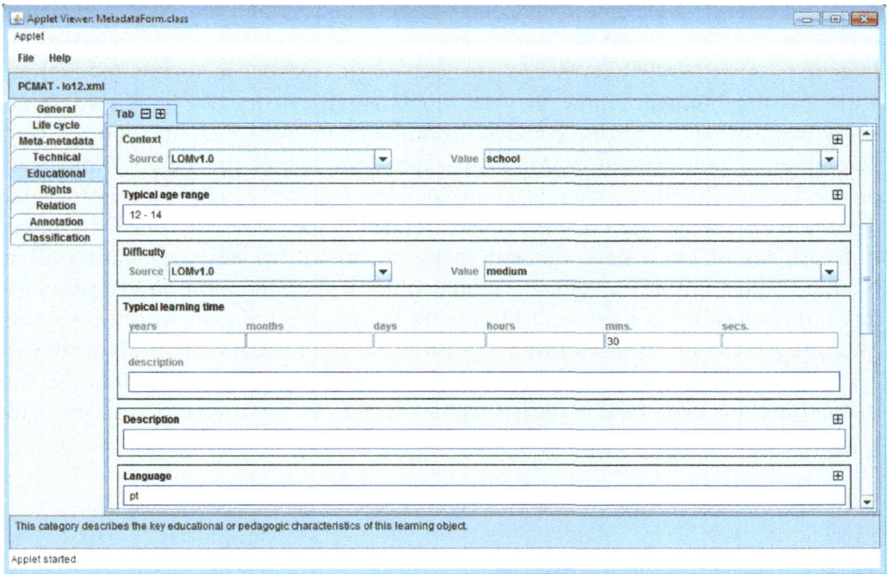

Fig. 33.2 PCMAT Metadata Authoring Tool (screen print)

choose the elements found relevant to characterize a specific set of learning objects and also because it allows the insertion of non-LOM elements or attributes if deemed necessary.

The PCMAT metadata application profile (Fig. 33.2) is based on the Learning Object Metadata standard [3, 5] and on the interpretation provided by the CanCore Guidelines v. 2.0. [15], hence making it possible for the learning objects to be retrieved and reused by other learning platforms.

The development of the PCMAT Metadata Authoring Tool as a web application makes it possible for teachers and content developers to manage on-line, from any computer with an Internet connection, the metadata associated to each learning object, without installing any additional software.

The PCMAT Metadata Authoring Tool was developed in Java SE 6 and presents the metadata creator with nine different forms, each form corresponding to a LOM category. One of the goals present during development was to limit the information to be manually inserted to a minimum while maintaining the created record in compliance with the LOM standard. To achieve this purpose, some elements of the PCMAT metadata application profile will provide information that will seldom be altered and other elements will care for the dynamic creation of the necessary web forms with the least number of fields required to be manually filled. For example, in order to provide the users with the least noise possible during the process of information retrieval, while searching the PCMAT learning objects database, the keywords associated with each LO will be selected from a pre-existing list build from a mathematical thesaurus.

The PCMAT Metadata Authoring Tool is based on an XML document containing an instance of every element that may be present on a metadata file related to a PCMAT LO. Every element of this XML document has certain attributes that influence how the web application will handle it. For example, every element has an attribute called "formField" whose value, "true" or "false", determines if that element is to be considered as part of the PCMAT metadata profile or not. Another example is an attribute called "dataType", whose value determines the drawing of that particular element in the metadata form. This makes it possible that for a new element to be added to the web application all is required is that this new element is added to the XML configuration file, provided these attributes are in place and filled with the required values. This answers an initial requirement for flexibility in order to build a PCMAT metadata profile in accordance with the experience acquired in filling an object's metadata file and in handling this metadata, in retrieving an object to be presented to a student with a certain learning profile and knowledge.

33.3.3 Future Developments

The next steps concerning the PCMAT Metadata Authoring Tool will be putting it to the test of metadata creators. This will help us decide which metadata is relevant to build the PCMAT metadata profile. Tests concerning the accuracy of the introduced metadata, in response to the search and retrieval of the LOs with the difficulty level most adequate to a student's learning style and knowledge, also need to be conducted. Another goal is to develop an application that helps building the XML configuration file.

33.4 Final Remarks

The PCMAT Metadata Authoring Tool is part of the PCMAT platform and its purpose is to help creating the metadata files associated to each PCMAT learning object. This metadata is based on the IEEE LOM standard, to which other elements, deemed necessary to the project, may be added. To this end, the PCMAT Metadata Authoring Tool is not based on a defined metadata profile, but rather on an open XML document to which elements can be added for the authoring tool to handle, this is, to draw the element on a form and to save the respective value in an XML IEEE LOM compliant document.

Acknowledgments This work is supported by FEDER Funds through the "Programa Operacional Factores de Competitividade - COMPETE" program and by National Funds through FCT "Fundação para a Ciência e a Tecnologia" under the project: PCMAT (PTDS/CPECED/108339/2008).

References

1. Wiley D (2002) Connecting learning objects to instructional design theory: a definition, a metaphor, and a taxonomy. In: Wiley D (ed) The instructional use of learning objects. Agency for Instructional Technology, Association for Educational Communications and Technology, Bloomington
2. Gerard R (1967) Shaping the mind: computers in education. In: National academy of sciences: applied science and technological progress: a report to the committee on science and astronautics. U.S. House of Representatives
3. IEEE 1484.12.1-2002, Standard for learning object metadata. The Institute of Electrical and Electronics Engineers, Inc., New York
4. National Information Standards Organization (2004) Understanding metadata. NISO Press, Bethesda, http://www.niso.org/publications/press/UnderstandingMetadata.pdf
5. IEEE 1484.12.3-2005, Standard for learning technology: extensible markup language (XML) schema definition language binding for learning object metadata. The Institute of Electrical and Electronics Engineers, Inc., New York
6. Neven F, Duval E (2002) Reusable learning objects: a survey of LOM-based repositories. In: Proceedings of the 10th ACM international conference on Multimedia (MULTIMEDIA '02). ACM, New York
7. DCMI: Dublin Core Metadata Element Set, version 1.1. DCMI Recommendation, http://www.dublincore.org/documents/dces/
8. DCMI: Guidelines for Dublin Core Application Profiles. DCMI Recommended Resource, http://www.dublincore.org/documents/profile-guidelines/
9. DCMI: Dublin Core Education Application Profile Task Group, http://dublincore.org/educationwiki/DC-Education Application Profile Task Group
10. DCMI: DC-Education Community Meeting: DC-2010, http://dublincore.org/educationwiki/DC-2010 DC-Ed Meeting Notes
11. Open Archives Initiative: the open archives initiative protocol for metadata harvesting, version 2.0, http://www.openarchives.org/OAI/openarchivesprotocol.html
12. Open Archives Initiative: PMH Tools, http://www.openarchives.org/pmh/tools/tools.php
13. Martins C, Couto P, Fernandes M, Bastos C, Lobo C, Faria L, Carrapatoso E (2011) PCMAT: mathematics collaborative learning platform. In: Bajo Pérez J, Corchado JM, Moreno MN, Julián V, Mathieu P, Canada-Bago J, Ortega A, Fernández Caballero A (eds) Highlights in practical applications of agents and multiagent systems. Advances in intelligent and soft computing, vol 89. Springer, Heidelberg
14. Wu H, Houben G, De Bra P (1999) User modeling in adaptive hypermedia applications. In: Proceedings of the interdisciplinary conferentie informatiewetenschap, Amsterdam
15. Friesen N, Fisher S, Roberts A (2004) CanCore guidelines for the implementation of learning object metadata (IEEE 1484.12.1-2002). Athabasca University, Athabasca

Chapter 34
Collaborative Broker for Distributed Energy Resources

João Carlos Ferreira, Alberto Rodrigues da Silva, Vítor Monteiro, and João L. Afonso

Abstract In this work it is proposed the design of a system to handle Distributed Energy Resources (DER), which is a new reality due to Electric Vehicles (EVs), Microgeneration (MG) and the open Electrical Markets (EM). This upcoming reality brings the need of the 'old' central energy control to be installed locally. For that we propose a local energy broker, responsible to handle local energy flow and to exchange energy with 'big' market players, and based on a collaborative approach, to promote user's participation to increase systems knowledge. The energy broker uses an Information Communication Technology (ICT) network to establish a collaborative communication between the involved parts.

Keywords Distributed energy resources • Data mining • Collaborative broker • Electric vehicles • Energy market • Smart grids • Vehicle-to-grid • Microgeneration

34.1 Introduction

New paradigms are emerging, like the Electric Vehicles (EVs), the Microgeneration (MG) and Smart Grids (SG), where there will be a large number of small energy producers. In this context, the Electrical Market (EM) will be the consequence of the deregulation in the use and in the production of electricity,

J.C. Ferreira (✉)
ADDETC – ISEL, 1900-049 Lisbon, Portugal
e-mail: jferreira@deetc.isel.ipl.pt

A.R. da Silva
DEI-IST and INESC-ID, 1000-029 Lisbon, Portugal
e-mail: alberto.silva@acm.org

V. Monteiro • J.L. Afonso
DEI-Centro Algoritmi – Univ. Minho, 4800 Guimarães, Portugal
e-mail: vmonteiro@dei.uminho.pt; jla@dei.uminho.pt

A. Madureira et al. (eds.), *Computational Intelligence and Decision Making: Trends and Applications*, Intelligent Systems, Control and Automation: Science and Engineering 61, DOI 10.1007/978-94-007-4722-7_34,

where power suppliers and consumers are free to negotiate the terms of their contracts. Also the EVs integration on current electrical distribution network, without violating the system's technical restrictions, requires electrical data consumption analysis and smart charging approaches, where the EVs batteries charging or discharging processes needs to be coordinated among the several users, and EV can play an important role in the MG, because they are able to store production excess and deliver energy in the peak of consumption. In this complex scenario, there is a need of an information system to handle electricity change, determine profits or values to pay and help the users on this process. Because there is a common goal (EM participation), the proposed system can benefit from user participation and users community (where users with similar profiles can be identified). These subjects converge under the name of DER (Distributed Energy Resources) as part of an intelligent power system to construct a hybrid, fundamentally different architecture for an Information and Communication Technology (ICT) network enabling the power grid to flexibly accommodate novel devices or clusters of devices. Several initiatives are taking place on this topic: (1) Intelligrid [1], a program in the United States executed by EPRI; (2) Gridwise [2]; CRISP project [3]; (3) European initiative under the program EU_DEEP-project [4] and the activities; and (4) IREDcluster [5]. As described in [6, 7] there are many implications for the grid when making a transition from centralized to decentralized control with merely some central coordination. When compared to hierarchically operated electricity grids with power centrally generated at high voltage levels on a large scale delivering electricity to consumers on lower voltage levels in the network, SG offer a number of challenges for technological research [8]. Our work proposes the creation of a collaborative broker system to handle EV and MG integration on EM, supported by users collaborative process, transaction data integration and central information repository knowledge, where we can store past experience to help on solving problems such as: (1) the excess of energy produced by MG should be distributed along the MG to the EV or to consumers, minimizing the use of the distribution grid, aiming the Energy Storage Systems (ESS), which can store energy and deliver it back to the power grid. These ESSs are controlled by the broker (transactions account); (2) the price of the electricity should be determined taking into account the production capacity, because the MG will be based on renewable sources with different profiles of production; (3) the interaction between the MG and the consumers should be implemented taking into account a maximization strategy to the user. To obtain a significant contribution of electrical energy to the electrical power grid by the producers, the EM should implement attractive benefits to buy and to sell. The control of the flow of energy between the MG, the final consumers and the sellers, is coordinated by the broker (e.g. transaction's account and user's alerts to change consumer behavior, see Sect. 34.5).

The proposed central information repository can store and manage historical data on electricity consumption and production. This information should come in a near future from the smart meter, for this work purposes we simulate consumptions and user behavior [9]. From this central repository it is possible the development of tools to extract knowledge from past electricity exchange log files, EM prices, renewable energy availability and capability, energy consumed or delivered if EV is plugged, and electrical grid constraints. Also, the social networks are integrated as a tool to share and spread useful related information. This central repository will be later, in a SG environment, a fundamental module to store all kinds of SG data and to solve the problems of different data format diversity.

34.2 Microgeneration

Nowadays, the Microgeneration (MG) emerges as a necessity to reduce the greenhouse effect caused by pollutant sources of energy. However, associated with the EM and the increasing in the technological development, the microgeneration has a great potential to the consumers, and to the distribution grid. The main sources of energy in a MG are wind and the sun, and the electrical energy is obtained through micro wind turbines and solar photovoltaic panels. These sources of energy are the most common and the easier to implement in the MG. Presently, this produced energy is provided to the electrical power grid without any concern about the EM or the electrical power grid capability to receive energy. In a smart grid context with ESS, like batteries, EVs, capacitors or flywheels, the energy produced in excess from the MG can be stored in these systems. Posteriorly, the stored energy is used by the electrical grid taking into account: (1) the energy produced by other sources of energy. When is required a great amount of energy, as during the EVs fast charging, the energy provided from electrical grid, can be obtained from the ESSs. It is also important to provide energy to the electrical grid in transitory moments as the consumers' consumptions peaks; (2) the EM, to buy and to sell energy with the best price. This is very important to the EV owners, because they have the potential to discharge the batteries to the electrical grid. This is why the Vehicle-to-Grid (V2G) will play a fundamental role in the smart grids.

In this context, from the point of view of the electrical grid, with the proximity between the MG and the consumers, the flow of power along the transmission and distribution lines will be more reduced, and consequently, the losses are lower. Also with this proximity the possibility to occur failures is more reduced.

34.3 Electric Vehicles

The recent increase of the interest in Electric Vehicles (EVs) represents a gain for independence relatively to the unstable costs of the oil in the international markets, and contributes to reducing the emission of greenhouse gases. However, with this increase it is intended that the methods for charging and discharging batteries are beneficial, not only to their lifespan, but also to preserve the power quality on the electrical grid [10]. During the discharge only a previously agreed (by the electric

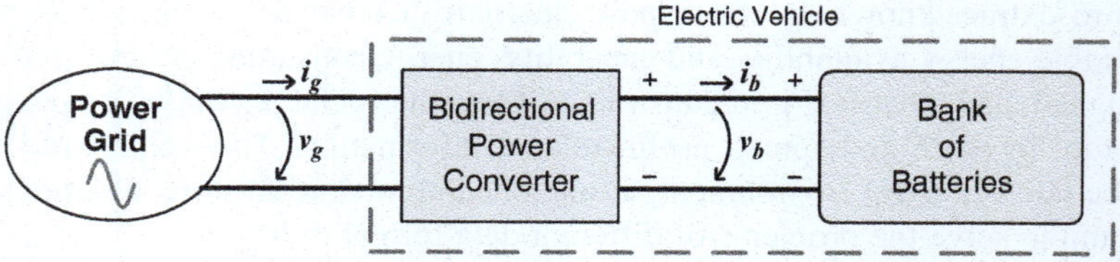

Fig. 34.1 Integration of an EV with a bidirectional power converter in the power grid

vehicle owner) amount of energy will be delivered back to the electrical grid (e.g., if a long trip is planned to occur soon the discharging process is not available).

In typical EVs chargers the energy comes from the electrical grid to the batteries in unidirectional mode, without any coordination. However, in a SG scenario, using bidirectional chargers (V2G concept), the batteries can be charged or discharged from the moment that the vehicle is plugged (Fig. 34.1), according to a compromise between the EV owner and the electrical grid, and controlled by a collaborative broker as described in this paper, aiming a coordination between both.

34.4 Distributed Energy Resources and Energy Markets

Distributed Energy Resources (DER), small-scale power generating technologies close to energy loads, are expected to become an important part of the future power system. MG and EV will play an important role in this process and nearby community will use this power because the network distribution allows it. Since distributed energy resources are installed near the loads, they are likely to be installed on the low-voltage systems, below 25 kV. The distribution systems also account for the higher percentage of system losses compared with the higher voltage transmission systems, causing an improvement of the overall efficiency of the system. DER has the problem of variability (changes in load), uncertainty (supply contingencies) and unpredictability (renewable generation). Main important fact is that EV can store local MG production excess and users' can tune their consumer behavior in part based on MG production, i.e. they can develop a collaborative process based on energy production information they can start/stop washing machines and other equipment that don't have a time constrain obligatory.

Energy market has historically been monopolized and governmentally regulated because of its utmost importance. Like water, energy is essential for life and firm grip on it was a logical choice of policy makers. With general globalization, such monopolized, nontransparent and market detached approach has become economically and politically unacceptable. To leave political influences aside as they surpass the scope of this article, global economic development demanded a change. Despite the efforts to save energy and use it as rational as possible, which are getting more serious every year, the energy consumption is inevitably growing. This is especially significant as the large part of the world is starting its development and great countries, as China and India are demanding their share of life standard. New stakeholders appear in this open market: (1) Broker of electric energy services is an entity or company that acts as a middleman in a marketplace in which those services are priced, purchased, and traded. A broker does not take title on available transactions, and does not generate, purchase, or sell electricity but facilitates transactions between buyers and sellers. If a broker is interested in acquiring a title on electric energy transactions, then it is classified as a generator or a marketeer. A broker may act as an agent between producers, or an aggregation of generating companies, and marketers; (2) Aggregator is an entity or a company that combines customers into a buying group. The group buys large blocks of electric power and other services at cheaper prices. The aggregator may act as an agent (broker) between customers and retailers. When an aggregator purchases power and re-sells it to customers, it acts as a retailer and should initially qualify as a retailer. Our collaborative system has the function of these two entities.

34.5 Collaborative Aggregator Broker for DER

Our investigation proposal is to bring computer science work on software development, Web 2.0, geographic information systems, mobile computation and wireless communication to create a system to support DER energy exchange, define local prices and coordinate energy exchange from local community to big producers. The main modules of the proposed system illustrated on Fig. 34.2, are:

1. Central Repository, stores information about: (1) user energy consumption (amount and time); (2) energy production with available information of power; (3) energy supplier and source (e.g. hydropower, wind power, photovoltaic, etc.); (4) energy prices; and (5) weather information (temperature, wind direction and speed, rain amount, solar radiation, etc.). A proper interface is created for user profile, creation and manipulation. This information data is worked under data mining approach for consumption data analyses. An example of this is the Naïve Bayes (NB) classifier showed. We implemented a weather crawler, based on a web robot to pick weather in web crawler formation from pre-defined sites and store this information on this information repository. Community creation is based on clustering available user profiles, based mainly on geographical position;
2. Information Communication Tools (ICT) are, mainly: communication networks for information exchange; mobile devices for user real time information access (like PDA or IPhone) to receive and send control information for charging the EV batteries and also for system interaction and high-speed digital monitoring, to take care of energy transactions. Since these smart meters aren't deployed yet, we simulate these consumptions, to get working data, details of this see [9];
3. Energy Market function, see Sect. 34.5.2.

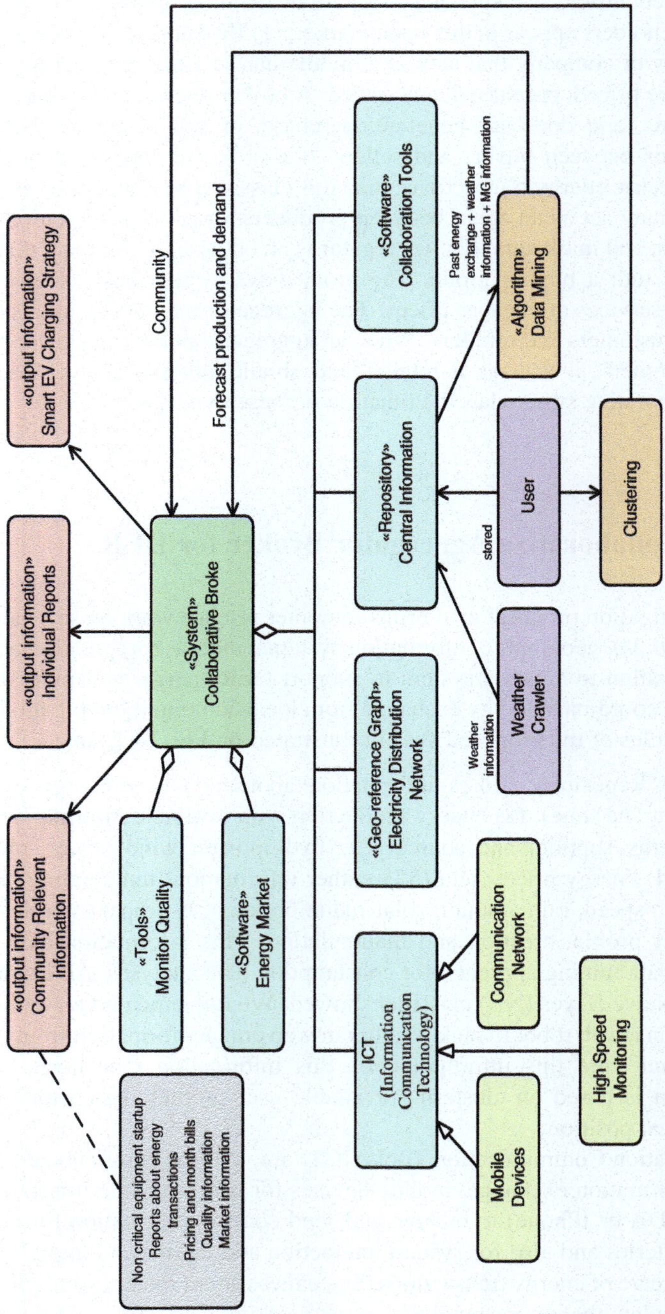

Fig. 34.2 Main system model

4. A collaboration software tool has as goal to help the people involved in a common task. It allows several independent computers working together, through an internet connection. In a SG context it is very important to establish patterns related to the produced and consumed energy from MG and the profiles of the consumers. Collaboration software tool, see Sect. 34.5.3.
5. Geo-reference graph based on electricity distribution network, a description of this is found in [9]. The Geo-Reference Graph for the electrical distribution network allows computational data manipulation, such as distance calculation, identification of power limitation, and identification of user communities.

34.5.1 Repository of Information, Weather Information and Data Mining

A deterministic approach to forecast MG production is complex, because wind power depends on the type of turbine, location (urban versus rural), high, orientation and wind speed. In solar generation the power generation depends on the environmental factors, mainly the irradiation and cell temperature. Each case should be analyzed and this should raise a complex scenario. To avoid this and since we aren't looking for an accuracy prediction, we propose an approach based on data mining. The idea is store the past production with the main weather factors that influences MG production, such as wind speed and direction, temperature and weather condition. A flexible structure needed to store and retrieve different data was created. For this task a database is used to store all transactions (EV charging and discharging and MG production) data, weather information and user profile information. Several approaches using data mining algorithms can be used for knowledge extraction. Past transaction data can be used to try to identify, with clustering approaches, periods of consumption and production, trends with the identification of the main behavior. We implemented a weather crawler, based on a web robot to pick weather information from pre-defined sites, as Portuguese weather site. Details on this can be found at [9] and Fig. 34.3 that shows examples of information for wind and temperature. Manipulating the available data we can perform several reports, like: home energy consumptions, weekly, monthly and annual energy expenses, price variation of electricity, charging periods, among others.

A Naïve Bayes (NB) classifier can be used to relate consumption and MG production to weather information (temperature, wind speed and direction and also humidity with raining information), a small example is shown in Table 34.1. Production capacity is divided in n classes (configurable number, get from clustering analyses of past data). In our implementation n = 10 and classes are defined based on percentage of production capacity; class 0 is zero production, class 1 is performed from 0 to 10% of production and class 10 if we reach maximum production. Wind and temperature were also discretized in a pre-defined class.

Latitude: 40°08′30″N Longitude: 08°52′42″W Profundidate: 9m

Data		Mar totoal (*m*)	Ond. (*m*)	Dir. ond.	Vaga (*m*)	Vento (*nós*)	Dir. vento.	Sets	Temp.Água (°*C*)
2011-07-29 (*6ª feira*)	H00	1.1	0.5	→	1.0	15.6	↓		15.4
	H06	1.0	0.4	→	0.9	13.6	↓		15.4
	H12	0.9	0.3	→	0.8	9.7	↓		15.5
	H18	1.0	0.3	→	0.9	17.5	↓	*	14.5
2011-07-30 (*Sábado*)	H00	1.1	0.3	→	1.0	19.4	↓	*	14.5
	H06	1.1	0.3	→	1.1	19.4	↓	*	14.5
	H12	1.1	0.3	→	1.0	17.5	↓	*	14.5
	H18	1.1	0.3	→	1.1	21.4	↓	*	14.5
2011-07-31 (*domingo*)	H00	1.2	0.3	→	1.1	21.4	↓	*	14.5
	H06	1.1	0.3	→	1.1	19.4	↓	*	14.5
	H12	1.0	0.3	→	1.0	15.6	↓	*	15.5
	H18	0.9	0.2	→	0.8	13.6	↘	*	15.5
2011-08-01 (*2ª feira*)	H00	0.8	0.2	→	0.8	15.6	↓		15.5
	H06	0.8	0.2	→	0.8	15.6	↓		15.5
	H12	0.7	0.2	→	0.7	11.7	↓		15.5
	H18	0.7	0.2	→	0.6	11.7	↓		15.5
2011-08-02 (*3ª feira*)	H00	0.7	0.2	→	0.7	13.6	↓		15.5
	H00	0.7	0.2	→	0.7	13.6	↘		15.5
	H12	0.6	0.2	→	0.6	7.8	↘		16.6

Fig. 34.3 Wind effect and weather information taken from weather sites for Lisbon [11]

Table 34.1 NB (Naïve Bayes) classifier approach for a small example

Day	Weather	Temperature	Wind	Production
1	Sunny	2	3	3
2	Cloudy	1	5	4
3	Rainy	4	1	2
4	Sunny	5	4	9
5	Sunny	3	2	4
6	Rainy	1	2	1
7	Cloudy	3	2	2
8	Cloudy	4	6	5
9	Rainy	3	3	3
10	Sunny	3	2	4
11	Sunny	2	2	To Determinate

$$\text{P (production)} = 0.1(10\ \text{classes})$$

$$\text{P (sun|p3)} = 1/2(\text{appears one in the two examples where production is P3})$$

$$\text{P (sun|p4)} = 2/3(\text{appears twice in the three examples where production is P4})$$

... the same for the others examples

$$\text{P (p1|sun + T2 + W2)} = \text{P(p1)} \times \text{P(sun|p1)} \times \text{P(T2|p1)} \times \text{P(W2|p1)}$$

$$\text{P (p2|sun + T2 + W2)} = \text{P(p2)} \times \text{P(sun|p2)} \times \text{P(T2|p2)} \times \text{P(W2|p2)}$$

...

$$\text{P (p10|sun + T2 + W2)} = \text{P(p10)} \times \text{P(sun|p10)} \times \text{P(T2|p10)} \times \text{P(W2|p10)}$$

Time is also a discrete variable. In our example we simulate one day that have only one class (corresponding to all consumption and production in a day), but in a real case more classes should be added more class. In the literature several authors divide the hour in 20–10 min events, so the number of class goes from $3 \times 24 = 72$ to $6 \times 24 = 144$. Wind speed and directions is correlated to pre-defined classes that characterize local eolic production and temperature is dived in interval classes. In our case we divide wind production capacity in 7 classes (0–6), see Fig. 34.3 and temperature in 5 class (1–5). Renewable production is more dependent in wind speed than solar with temperature. This class is a configuration parameter; more class means more control dependence on these parameters but estimation complexity increase. So Table 34.1 shows a small example how NB algorithm works, showing the probability of an event happening. In this case we want to know for current forecast sun with temperature and wind in class 2. Based on historical data (in case 10 events) NB shows the probability for p1 to p10. For more details see [9]. From this particular example we will have production capacity based on a probability approach. Since we aren't looking for accuracy prediction and most results follow a certain pattern. In this case p3 and p4 have highest probability. If we have installed a capacity of 3 kW this means under these conditions we are able to generate around 1 kW.

34.5.2 Energy Market and Smart Grid Integration

Energy Market participation will require a detail control about electricity flow, and registered users would like to control charging or discharging process, associated with the electricity prices, depending on the supplier and time and local broker picks from a pre-defined server on a standard XML file. In this scenario in Fig. 34.4 is shown a MG with: solar photovoltaic panels; wind turbines; EVs; and ESSs. Beyond the flow of energy is also important the sharing of information, controlled by the broker.

34.5.3 Collaboration Tool, User Profile and Community

User profiles related with their routines are created based in two main characteristics:

1. The EV type, mainly: the batteries technology (as Lead-Acid, Nickel or Lithium); the capacity of batteries (kWh); the batteries characteristics to receive or deliver energy taking into account different current rates of charging or discharging; and the SOC (State of Charge) and SOH (State of Health);
2. Related to the EVs, aiming the user profile and the community, other important characteristic is the data and the time of the travel, and the km planned to perform it. This is particular important to define the energy which the EV owner can sell, or the needed energy to charge the batteries with fast charge.

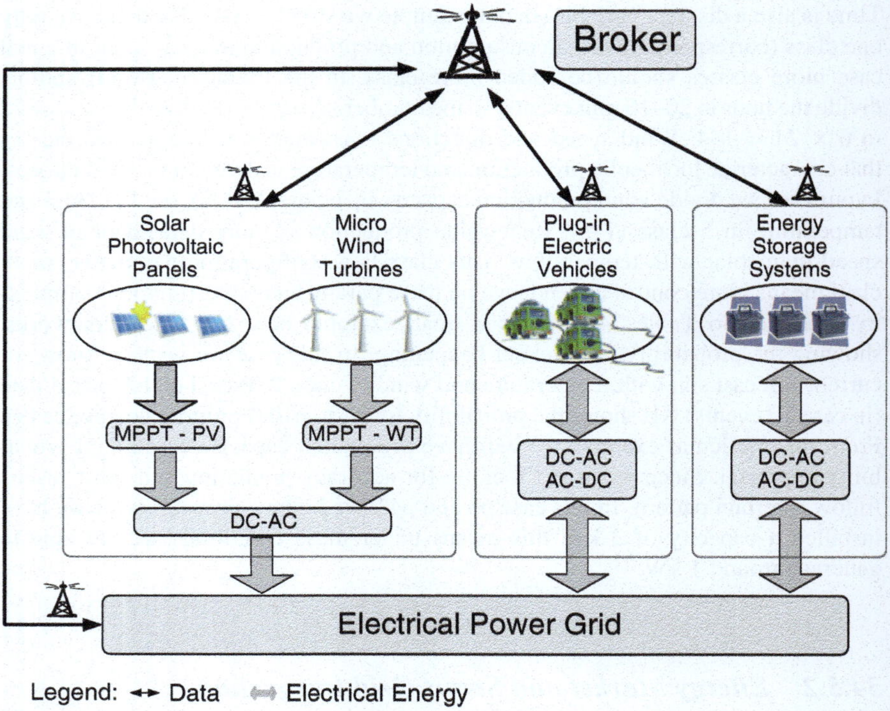

Fig. 34.4 Integration of the MG in electrical grid controlled by a broker

How does a group of interested stakeholders collaborate to create real-world DER (Distributed Energy Resources) pilot programs that benefit multiple stakeholders? The following outlines important questions to address and steps that can be taken towards a common goal. In a SG scenario, there must be a collaborative process between electrical grid, EVs, MG, and EM. In this collaboration, the distribution network uses the shortest path (less impedance) to deliver production excess and EVs, if available, can store this local production excess. If MG doesn't have local EVs, the network can deliver to the nearest EVs neighbor establishing a collaborative process. Also if EV SOC is about drivers' requirements, neighbors can take the EVs energy. Non critical equipment (e.g. washing machine and others) can be started when there is a local production excess. Recording data can show periods of consumption is lower and this machines can be postponed to this period (in SGs it is possible to start this equipment's through real time information). This collaboration is controlled by the central broker, for accounting purposes and the price is established between the local production and local demand. The system will work always above big producers' prices and will always try to minimize energy transitions outside the community, defined by network distribution topology that could be reached and controlled by a collaborative broker and the energy can flow in different ways, as shown in Fig. 34.4.

With the energy flux control, with the SG collaboration, the energy produced by MG can be distributed to EVs or to consumers. Prices should change based on production capacity and energy needs. So the prices of electricity vary in time, but for that we need the measurement capability at the consumers. These meter devices have as inconvenience the price. If the price differentials between hours or time periods are significant, customers can respond to the price structure with significant changes in energy use, reducing their electricity bills if they adjust the timing of their electricity usage to take advantage of lower-priced periods and/or avoid consuming when prices are higher. Customers' load use modifications are entirely voluntary. Market based solutions also give solid background to develop new business opportunities, e.g., to aggregators. The main obstacle to introduce more price granularity at consumer's level is the costs of installing and monitoring the meters.

34.6 Conclusion

This paper describes the work that has been developed to provide a conceptual system to handle DER (Distributed Energy Resources) and acts like an energy market broker. The collaboration approach is a hot topic today with the success of social networks, and applying the same principles to local communities a lot of synergies can be generated towards common goals: reduce the electricity invoice at the end of the month and handle the Microgeneration (MG) as a market player.

This upcoming reality needs new technology that will be available with Smart Grids (SG) full implementation (e.g., energy suppliers must be able to schedule resources, manage aggregation, and communicate both the scarcity and surplus of energy supply over time; energy consumers can remotely control their consumption, starting/stopping equipment).

As a conclusion it can be said that the increasing penetration of MG associated with Electric Vehicles (EVs) in an open market have a great potential. Still a lot needs to be done, such as new legislation, as well as implementation of new metering devices.

Acknowledgments This work is financed by FEDER Funds, through the Operational Programme for Competitiveness Factors – COMPETE, and by National Funds through FCT – Foundation for Science and Technology of Portugal, under the project PTDC/EEA-EEL/104569/2008 and the project MIT-PT/EDAM-SMS/0030/2008.

References

1. Dolle DV (2012, 10 August) Intelligrid architecture – architecture for the intelligent electricity grid of the future. Electricity Innovation Institute. EPRI [Online]. Available at, http://intelligrid.epri.com/docs/Intelligrid_Intro_081804.pdf

2. Gridwise Project (2012, 10 August) Gridwise alliance [Online]. Available at, http://www.gridwise.org/
3. Andrieu C, Fontela M, Enacheanu B, Pham H, Raison B, Besanger Y, Randrup M, Nilsson U, Kamphuis R, Schaeffer G (2010, 29 December) Distributed network architectures D1.7, 2005, 30 August. (2010, 29 December). CRISP [Online]. Available at, http://www.crisp.ecn.nl/deliverables/D1.7.pdf
4. EUDEEP project (2010, 29 December). EUDEEP [Online]. Available at, http://www.eudeep.com/
5. IRED project (2012, 10 August). IRED-Cluster [Online]. Available at, http://www.ired-cluster.org/
6. BenHabib MC, Duarte J, Hommelberg M et al. (2006) Flexible electricity grids – power electronic system and ICT-requirements for novel electricity distribution grids
7. Electricité de France (1986) Le Micro Réseau: 30 ans d'activité. Direction des Etudes et Recherches, Departement FCR, 1986
8. Thanh L, Caire R, Raison B et al. (2008) High level specification of the functionality for novel electricity distribution grid control, INTEGRAL Project
9. Ferreira JC, Silva A, Afonso JL (2011) Agent based approaches for smart charging strategy for electric vehicle. EVTEC 11. In: Proceedings of the 1st international electric vehicle technology conference, Yokohama, Japan, 17–19 May 2011
10. Monteiro V, Ferreira JC, Pinto G, Pedrosa D, Afonso JL (2010) iV2G charging platform. In: Proceedings of the IEEE-ITSC 13th international IEEE conference on intelligent transportation systems, 19–22 Sept 2010, pp 409–414
11. Instituto de Meteorologia de Portugal (2011, 29 July) METEO [Online]. Available at, http://www.meteo.pt/pt/maritima/costeira/

Chapter 35
A Multidimensional Scaling Classification of Robotic Sensors

Miguel F.M. Lima and J.A. Tenreiro Machado

Abstract This chapter analyzes the signals captured during impacts and vibrations of a mechanical manipulator. Eighteen signals are captured and several metrics are calculated between them, such as the correlation, the mutual information and the entropy. A sensor classification scheme based on the multidimensional scaling technique is presented.

35.1 Introduction

The robotic manipulators have several sensors and actuators in order to carry out the desired movements. Due to the multiplicity of sensors, the data obtained can be redundant because the same type of information may be seen by two or more sensors. Due to the price of the sensors, this aspect can be considered in order to reduce the cost of the system. On the other hand, the placement of the sensors is an important issue in order to obtain the suitable signals of the vibration phenomenon. Moreover, the study of these issues can help in the design optimization of the acquisition system. In this line of thought a sensor classification scheme is presented.

Several authors have addressed the subject of the sensor classification scheme. White [14] presents a flexible and comprehensive categorizing scheme that is useful for describing and comparing sensors. The author organizes the sensors according to several aspects: measurands, technological aspects, detection means, conversion

M.F.M. Lima (✉)
CI&DETS and Department of Electrical Engineering, School of Technology, Polytechnic Institute of Viseu, Viseu, Portugal
e-mail: lima@mail.estv.ipv.pt

J.A.T. Machado
Department of Electrical Engineering, Institute of Engineering, Polytechnic Institute of Porto, Porto, Portugal
e-mail: jtm@isep.ipp.pt

A. Madureira et al. (eds.), *Computational Intelligence and Decision Making: Trends and Applications*, Intelligent Systems, Control and Automation: Science and Engineering 61,
DOI 10.1007/978-94-007-4722-7_35,

phenomena, sensor materials and fields of application. Michahelles and Schiele [10] systematize the use of sensor technology. They identified several dimensions of sensing that represent the sensing goals for physical interaction. A conceptual framework is introduced that allows categorizing existing sensors and evaluates their utility in various applications.

Today's technology offers a wide variety of sensors. In order to use all the data from the diversity of sensors a framework of integration is needed. Sensor fusion, fuzzy logic, and neural networks are often mentioned when dealing with problem of combing information from several sensors to get a more general picture of a given situation. The study of data fusion has been receiving considerable attention [4,8]. A survey of the state of the art in sensor fusion for robotics can be found in [5]. Henderson and Shilcrat [6] introduced the concept of logic sensor that defines an abstract specification of the sensors to integrate in a multisensor system.

The recent developments of micro electro mechanical sensors (MEMS), with unwired communication capabilities, allow a sensor network with interesting capacity. This technology was applied in several applications [1], including robotics.

This work presents a development of a sensor classification scheme based on a statistical metrics using the multidimensional scaling technique (MDS).

Bearing these ideas in mind, this chapter is organized as follows. Section 35.2 describes briefly the robotic system enhanced with the instrumentation setup. Section 35.3 presents the main concepts involved in the experiments and Sect. 35.4 shows the experimental results. Finally, Sect. 35.5 draws the main conclusions and points out future work.

35.2 Experimental Platform

The developed experimental platform has two main parts: the hardware and the software components. The hardware architecture is shown in Fig. 35.1 (left). Essentially it is made up of a robot manipulator, a personal computer (PC), and an interface electronic system.

The interface box is inserted between the robot arm and the robot controller, in order to acquire the internal robot signals; nevertheless, the interface captures also external signals, such as those arising from accelerometers and force/torque sensors.

The software package runs in a Pentium 4, 3.0 GHz PC and, from the user's point of view, consists of two applications: the acquisition application and the analysis package. The acquisition application is a real time program for acquiring and recording the robot signals.

After the real time data acquisition, the analysis package processes the data off-line in two phases, namely, pre-processing and processing. The preprocessing phase consists of the signal selection in time, and their synchronization and truncation.

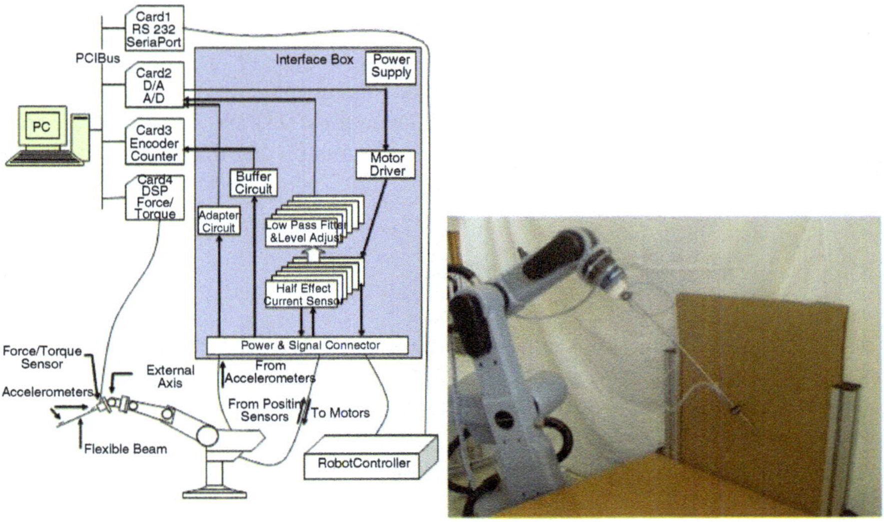

Fig. 35.1 The hardware: block diagram of the architecture (*left*) and the steel rod impact against a rigid surface (*right*)

The processing stage implements several algorithms for signal processing such as the auto and cross correlation, Fourier transform (FT), window Fourier transform, time synchronization, etc.

In the experiments a flexible link is used, consisting of a long and round flexible steel rod clamped to the end-effector of the manipulator. In order to analyze the impact phenomena in different situations two types of beams are adopted: a thin and a gross rod. The robot motion is programmed in a way such that the rods move against a rigid surface. Figure 35.1 (right) depicts the robot with the flexible link and the impact surface.

During the motion of the manipulator the clamped rod is moved by the robot against a rigid surface. An impact occurs and several signals are recorded with a sampling frequency of $f_s = 500$ Hz. The signals come from several sensors, such as accelerometers $\{A_1, A_2\}$, force $\{F_x, F_y, F_z\}$ and torque $\{M_x, M_y, M_z\}$ sensor, position encoders $\{P_1, P_2, P_3, P_4, P_5\}$, and current sensors $\{I_1, I_2, I_3, I_4, I_5\}$.

In order to have a wide set of signals captured during the impact of the rods against the vertical screen thirteen distinct trajectories were defined. For each trajectory the motion of the robot begins in one point, moves against the surface and returns to the initial point. A parabolic profile was used for the trajectories.

Several sets of signals are captured corresponding to the cases: (*i*) without impact, (*ii*) the impact of the rod on a gross screen and (*iii*) the impact of the rod on a thin screen, using either the thin, or the gross rod. Due to space limitations the captured signals are not shown here, but signals with identical behavior can be seen in [7].

35.3 Main Concepts

This section presents a brief description of the fundamental concepts involved in the experiments such as the multidimensional scaling (MDS) and metrics in the time domain, namely the correlation, the mutual information and the entropy.

35.3.1 *Multidimensional Scaling*

MDS is a generic name for a family of algorithms that construct a configuration of points in a low dimensional space from information about inter-point distances measured in high dimensional space. The new geometrical configuration of points, which preserves the proximities of the high dimensional space, allows gaining insight in the underlying structure of the data and often makes it much easier to understand.

The Minkowski distance metric provides a general way to specify the distance $d_{i,j}$, between the objects i and j, for quantitative data in a multidimensional space:

$$d_{ij} = \left(\sum_{k=1}^{m} w_k \left|x_{ik} - x_{jk}\right|^r\right)^{1/r}. \tag{35.1}$$

where m is the number of dimensions, x_{ik} is the value of dimension k for object i and w_k is a weight. For $w_k = 1$, with $r = 2$, the metric equals the Euclidian distance metric, while $r = 1$ leads to the city-block (or Manhattan) metric. In practice, normally the Euclidian distance metric is used but there are several others definitions that can be applied, including for binary data [3].

Typically MDS is used to transform the data into two or three dimensions, and visualizing the result to uncover hidden structure in the data.

35.3.2 *Metrics in the Time Domain*

Several indices can be used to evaluate the relationship between the signal, including statistical, entropy and information theory approaches. These metrics are based on a bidimensional probability density function associated with the two signals $x_1(t)$ and $x_2(t)$ acquired in the same time interval and can be calculated according with the expression:

$$P(x_1, x_2) = \frac{\beta(x_1, x_2)}{\int\int \beta(x_1, x_2) dx_1 dx_2} \tag{35.2}$$

where β is the bidimensional histogram.

The marginal probability distributions of the signals $x_1(t)$ and $x_2(t)$ are denoted as $P(x_1)$ and $P(x_2)$, respectively. The expected values, $E(x_1)$ and $E(x_2)$, and the variances, $V(x_1)$ and $V(x_2)$, are then easily obtained.

The correlation coefficient R [11] is a statistical index that provides a measurement of the similarity between two signals $x_1(t)$ and $x_2(t)$ and is define as

$$R(x_1,x_2) = \frac{E(x_1x_2) - E(x_1)E(x_2)}{\sqrt{V(x_1)V(x_2)}} \tag{35.3}$$

where $E(x_1x_2)$ is the joint expected value.

The mutual information [2,13] is the index that measures the dependence of two variables in the viewpoint of the information theory. The mutual information for the two signals $x_1(t)$ and $x_2(t)$ is:

$$I(x_1,x_2) = \log_2 \frac{P(x_1,x_2)}{P(x_1)P(x_2)} \tag{35.4}$$

The average mutual information between the two signals is given by:

$$I_{av}(x_1,x_2) = \int\int P(x_1,x_2)\log_2 \frac{P(x_1,x_2)}{P(x_1)P(x_2)} dtdt \tag{35.5}$$

The entropy [13] is a statistical measure of randomness. This index applied to the two signals $x_1(t)$ and $x_2(t)$ gives the join entropy [9] between the two signals defined as:

$$H(x_1,x_2) = -\int\int P(x_1,x_2)\log_2 P(x_1,x_2)dtdt \tag{35.6}$$

35.4 Experimental Results

35.4.1 Metrics Applied to the Signals

Figure 35.2a shows the squared correlation coefficient R^2 between the signals captured during the same impact trajectory, for an experiment in the case of (*i*) using the gross rod. The results obtain with R^2 are symmetric relative to the diagonal formed by $R^2(x_i,x_j)$ for $i = j$, where the metric is maximum, as expected. To clearly visualize the results only one side is shown. The correlation between the same families of signals is higher than the correlation between different families. For example, the correlation between the currents and positions is low. The same occurs between the currents and the forces, moments and accelerations. It exists a

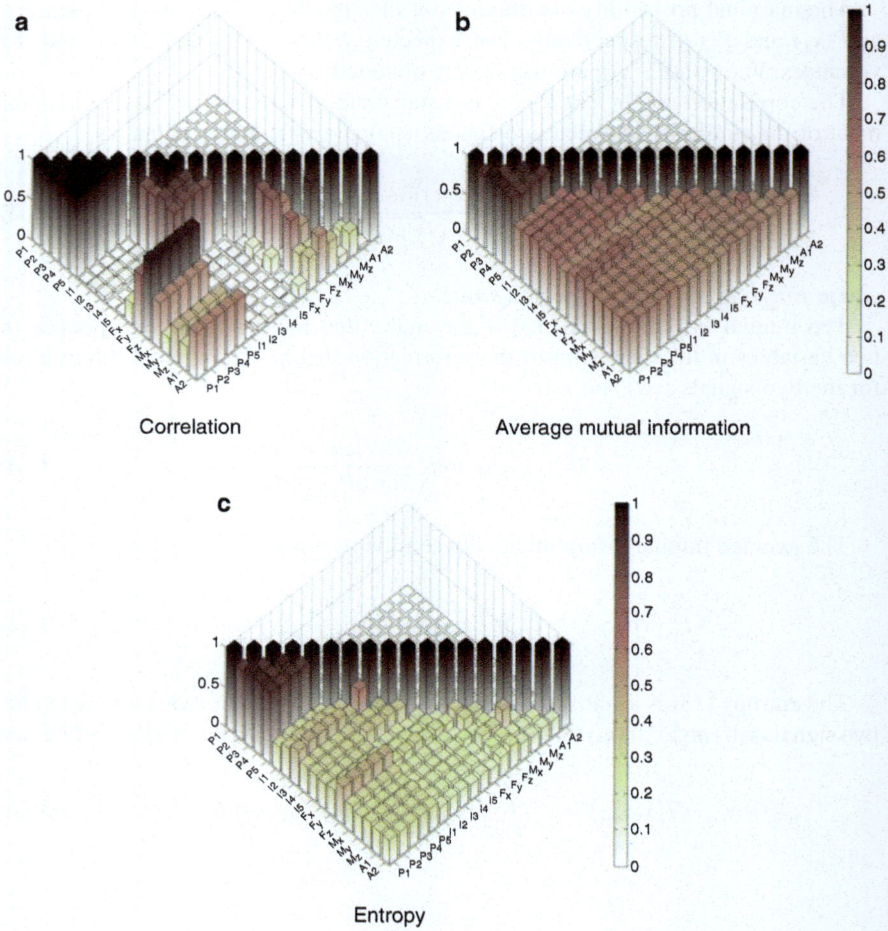

Fig. 35.2 Metrics between the signals for the case (*i*) using the gross rod

strong correlation between the positions and the forces, moments and accelerations that depends, as expected, on the trajectory.

Figure 35.2b shows the average mutual information between the signals for the same experiment used previously for the correlation. Again the results obtain with $I_{av}(x_1, x_2)$ are symmetric relative to the diagonal where the metric is maximum. Due to the same reason referred before only one side is shown. The values presented are normalized. The values of the index $I_{av}(x_1, x_2)$ between the positions are high.

Figure 35.2c shows a chart based on the entropy between the signals for the same experiment used previously for the other metrics. In fact, the values shown are proportional to the inverse of the index $H(x_1, x_2)$ due to the normalization used. Again, the values of this index between the positions are high.

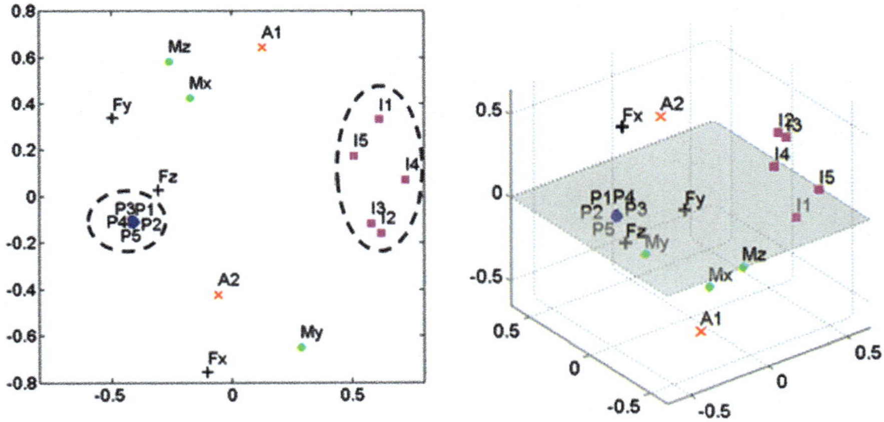

Fig. 35.3 Locus of sensor positioning based on the correlation measure between the signals for the case (*i*) using the gross rod: 2D (*left*); 3D (*right*)

35.4.2 MDS analysis

In order to reveal some hypothetical hidden relationships between the signals the MDS technique is used. Several metric and non-metric scaling MDS criteria were tested. The Sammon [12] metric scaling criterion revealed good results for the correlation. This criterion gives weight to small distances, which helps to detect clusters. On the other hand, for the tests developed, the stress metric scaling criterion, normalized with the sum of squares of the dissimilarities, is the best choice for the mutual information and entropy. Therefore, the referred criteria are adopted in this work. In Fig. 35.3 is shown the 2D (left) and 3D (right) locus of sensor positioning based on the correlation measure between the signals for the case (*i*) using the gross rod. Three groups of signals can be defined. The ellipses depicted in the chart represent two of these groups. The positions $\{P_1, P_2, P_3, P_4, P_5\}$ signals are located close to each other. The electrical currents $\{I_1, I_2, I_3, I_4, I_5\}$ are situated on the left of the chart and near each other. Finally, the remaining signals form a big group composed by the forces $\{F_x, F_y, F_z\}$, moments $\{M_x, M_y, M_z\}$ and the accelerations $\{A_1, A_2\}$ situated at scattered positions away from each other.

Figure 35.4 shows two tests developed to evaluate the consistency of the results obtained by MDS analysis. The value of the stress function *versus* the dimension is shown in Fig. 35.4 (left), which allows the estimation of the adequate p–dimension. An "elbow" occurs at dimension three for a low value of stress, which corresponds to a substantial improvement in fit. Additionally, the Shepard plot (Fig. 35.4 right) shows the fitting of the 3D configuration distances to the dissimilarities.

Figure 35.5 (left) shows the 2D locus of sensor positioning based on the average mutual information measure between the signal for the case (*i*) using the gross rod. The 2D spatial presentation of the MDS is shown due to clarity in the interpretation.

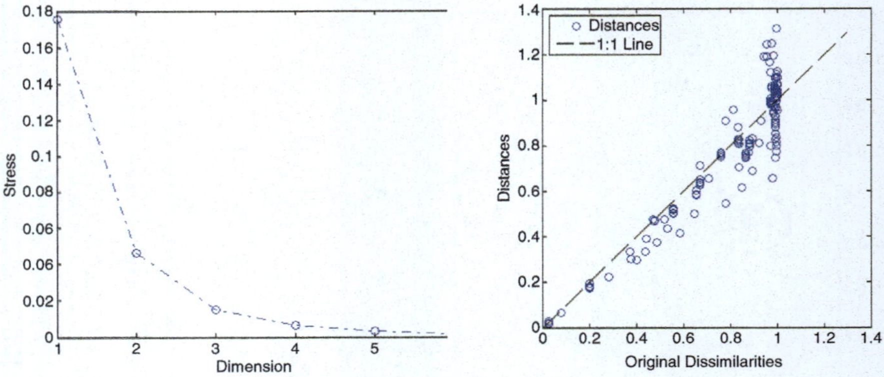

Fig. 35.4 Evaluation of MDS results based on correlation: Stress test (*left*); Shepard plot (*right*)

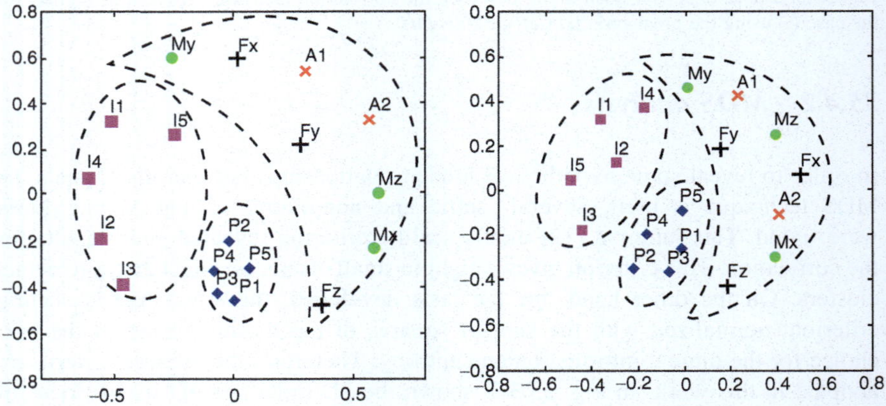

Fig. 35.5 2D Locus of sensor positioning using the signals for the case (*i*) with the gross rod based on: the average mutual information (*left*) and the join entropy (*right*)

Once more three groups of signals can be defined. Again, the positions $\{P_1, P_2, P_3, P_4, P_5\}$ signals are located close to each other and the electrical currents $\{I_1, I_2, I_3, I_4, I_5\}$ are situated on the left of the chart and near each other. Finally, the remaining signals form a big group composed by the forces $\{F_x, F_y, F_z\}$, moments $\{M_x, M_y, M_z\}$ and the accelerations $\{A_1, A_2\}$ situated at scattered positions at top right region. Figure 35.6 shows the stress test (left) and the Shepard plot (right) of the 3D MDS based on the average mutual information. From the stress test we can see that low values of stress are achieved only for high dimensions, which could reveals that the 2D or 3D dimensions are not suitable for representing the locus of the variables. Additionally, the Shepard plot shows the inappropriate fitting of the 3D configuration distances to the dissimilarities. This plot indicates that this metric solution is probably not suitable, because it shows both a nonlinear pattern and a large scatter.

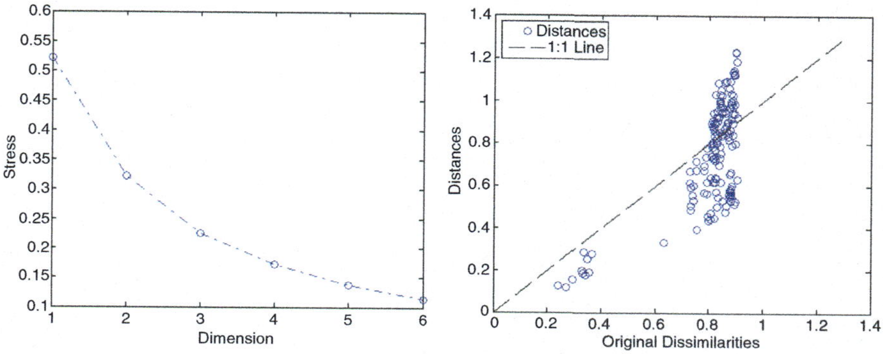

Fig. 35.6 Evaluation of MDS results based on the average mutual information: Stress test (*left*); Shepard plot (*right*)

The former implies that many of the largest dissimilarities would tend to be somewhat exaggerated in the visualization, while moderate and small dissimilarities would tend to be understated. The latter implies that distance in the visualization would normally be a poor evidence of dissimilarity. In particular, a good fraction of the large dissimilarities would be badly understated.

Figure 35.5 (right) shows the 2D locus of sensor positioning based on the join entropy measure between the signals for the same case of the previous examples. Again, the 2D spatial presentation of the MDS is shown due to clarity in the interpretation. The locus of the variables is similar to that obtained for the average mutual information metric. Additionally, the stress test and the Shepard plot of the MDS based on the join entropy presents an identical behavior of that obtained for the average mutual information metric.

In conclusion, from the experiments analyzed for all the three metrics studied, the MDS gives a spatial representation of the signals that can be classified in three groups. However, the tests developed to evaluate the consistency of the results obtained by MDS show that the correlation metric seems to be more appropriate when comparing with the average mutual information and entropy.

35.5 Conclusion

In this work an experimental study was conducted to investigate several robot signals. A new sensor classification strategy based on MDS was proposed. The adopted methodology revealed hidden relationships between the robotic signals and leads to arrange them in three groups.

The classification presented was obtained for an experiment corresponding to one trajectory. In future this approach should be applied for all the 13 trajectories referred before.

References

1. Arampatzis Th, Lygeros J. and Senior Member, Manesis S (2005) A survey of applications of wireless sensors and wireless sensor networks. In: Proceedings of 13th Mediterranean conference on control and automation, Limassol, 2005, pp 719–724
2. Cover T, Thomas J (2006) Elements of information theory. Wiley series in telecommunications and signal processing. Wiley-Interscience, New York, ISBN 0471241954
3. Cox T, Cox M (2001) Multidimensional scaling, 2nd edn. Chapman & Hall/CRC, Boca Raton. ISBN 1584880945
4. Esteban J, Starr A, Willetts R, Hannah P, Bryanston-Cross P (2005) A review of data fusion models and architectures: towards engineering guidelines. Neural Comput Appl (Springer, London) 14(4):273–281
5. Hackett J, Shah M (1990) Multi-sensor fusion: a perspective. In: Proceedings of the IEEE international conference on robotics & automation, Cincinnati, Ohio, pp 1324–1330
6. Henderson T, Shilcrat E (1984) E. Logical sensor systems. J Robot Syst 1:169–193
7. Lima M, Machado J, Crisóstomo M (2007) Experimental signal analysis of robot impacts in a fractional calculus perspective. J Adv Comput Intell Intell Inform, Fuji Technology Press, vol 11(9), ISSN 1343–0130, pp 1079–1085
8. Luo R, Kay M (1990) A tutorial on multisensor integration and fusion. In: Proceedings of the IEEE 16th annual conference of industrial electronics society, Pacific Grove, California, pp 707–722
9. MacKay D (2003) Information theory, inference, and learning algorithms. Cambridge University Press, Cambridge. ISBN 0521642981
10. Michahelles F, Schiele B (2003) Sensing opportunities for physical interaction. Workshop on physical interaction of mobile HCI conference, Udine, Italy.
11. Orfanidis S (1996) Optimum signal processing. An introduction, 2nd edn. Prentice-Hall, Englewood Cliffs
12. Sammon J (1969) A nonlinear mapping for data structure analysis. IEEE Trans Comput C–18 (5):401–409
13. Shannon C (1948) A mathematical theory of communication. The Bell Syst Tech J. July, Oct, vol 27, pp 379–423, pp 623–656
14. White RM (1987) A sensor classification scheme. IEEE Trans Ultrason Ferroelectr Freq Control 34(2):124–126

Chapter 36
Rough Set Theory: Data Mining Technique Applied to the Electrical Power System

C.I. Faustino Agreira, C.M. Machado Ferreira, and F.P. Maciel Barbosa

Abstract This paper presents a study were the Rough Set Theory and Data Mining Technique are applied to the electrical power system. The Data Mining technique classifies the system operation in four possible states: normal, alert, emergency (emergency I and emergency II). The states, that correspond to the normal state can be classified as secure and insecure the remaining ones. In this security studies, the overloads in transmition lines and the violation of the voltage limits are used to classify and rank these contingencies. This technique was applied to the 118IEEE busbar test power network and the results obtained are analyzed. Finally, some conclusions that provide a valuable contribution to the understanding of the power system security analysis are pointed out.

36.1 Introduction

Data mining is the process of analyzing data from different perspectives and summarizing it into useful information. This technique analyzes data from many different dimensions or angles, categorize it and summarize the relationships identified. Technically, data mining is the process of finding correlations or patterns

C.I.F. Agreira (✉)
Departamento de Engenharia Electrotécnica, Instituto Superior de Engenharia de Coimbra, Coimbra, Portugal
e-mail: cif@isec.pt

C.M.M. Ferreira
Departamento de Engenharia Electrotécnica, Instituto Superior de Engenharia de Coimbra, Coimbra, Portugal

INESC, Coimbra, Portugal

F.P.M. Barbosa
INESC Tech, Faculdade de Engenharia da Universidade do Porto, Porto, Portugal
e-mail: fmb@fe.up.pt

A. Madureira et al. (eds.), *Computational Intelligence and Decision Making: Trends and Applications*, Intelligent Systems, Control and Automation: Science and Engineering 61, DOI 10.1007/978-94-007-4722-7_36,

among dozens of fields in large relational databases [1]. Many applications of knowledge discovery and data mining such as rule discovery for semantic query optimization, database integration and decision support, require the knowledge to be consistent with the data [2]. Data mining is an emerging area of computational intelligence that offers new theories, techniques and tools to processes large volumes of data. It has gained considerable attention among practitioners and researchers as evidenced by the number of publications, conferences, and application reports [3]. The growing volume of data that is available in a digital form has accelerated this interest. Data mining has emerged as a contribute tools for data analysis, discovery of new knowledge, and autonomous decision making [3]. The knowledge acquisition process is a complex task, since the experts have difficulty to explain how to solve a specified problem.

Recently, the Rough Sets theory (RST) has been used successfully to handle efficiently problems where large amounts of data are produced [4, 5]. RST constitutes a framework for inducing minimal decision rules. These rules in turn can be used to perform a classification task. Important concepts include the elimination of redundant criteria to give more compact rules. The RST has been conceived as a tool to conceptualize, organize and analyze various types of data, in particular, to deal with inexact, uncertain or vague knowledge. In the Rough Sets analysis the concept of an information system is used to construct the approximation space. It enables representation of data in a useful form of a table. The information system is in fact a finite data table where columns are labeled by attributes and rows are labeled by objects [3]. Attributes are generally classified into conditions and decisions. Usually, a number of condition attributes and a single decision attribute are presented. The study presents a systematic approach to transform examples in a reduced set of rules [3]. These rules can be used successfully to avoid security problems and provides a deeper insight into the influence of parameters on the system performance. The strength of a rule can be quantified using rough membership. The main goal of the rough set analysis is to search large databases for meaningful decision rules and finally acquire new knowledge. This approach is based on four main topics: Indiscernibility, Approximation, Reducts and Decision rules [3]. A reduct is a minimal set of attributes, from the whole attributes set, which preserves the partitioning of the finite set of objects and therefore the original classes. It means that the redundant attributes are eliminated. When the reducts are found, the task of creating definite rules for the value of the decision attribute of the information system is practically performed. Decision rules are generated combining the attributes of the reducts with the values. Decision rules extract knowledge, which can be used when classifying new objects not in the original information system.

This paper is organized as follows. Section 36.1 presents an introduction to the problem. Section 36.2 is devoted to the Rough Set Theory. In Sect. 36.3 is presented the test power network that was analyzed and shows the results obtained using the proposed approach. Finally, in Sect. 36.4, some conclusions that provide a valuable contribution to understand the RST applied to the security analysis of the electric power system.

36.2 The Rough Set Theory

To arrange incomplete and uncertainty data the Rough Set Theory (RST), a new mathematical tool, was used recently [4, 5]. The Rough Set model has several advantages to data analysis. It is just based on the original data and does not need any external information. Moreover, no assumptions about data are necessary and it is suitable to analyze both quantitative and qualitative features. The RST works with lower and upper approximation of a set, as it is represented in Fig. 36.1. The starting point of rough set theory is the indiscernibility relation, which is generated by information about objects of interest. The indiscernibility relation expresses the fact that due to a lack of information it is difficult to discern some objects employing available information or knowledge. The discernibility relation is used for two basic operations in RST i.e. upper $\overline{R}X$ and lower $\underline{R}X$ approximations, which defines crisp and vague manner in the sets. If any concept of the universe can be formed as a union of some elementary sets, it is referred to as crisp (precise). On the contrary, if the concept cannot be presented in such a way, it is referred as vague (imprecise, rough). $\overline{R}X$ is defined as the collection of cases whose equivalence classes are, at least, partially contained in (i.e. overlap with). The set of cases to approximate $\underline{R}X$ is defined as the collection of cases whose equivalent classes are fully contained in the set of cases to approximate [5]. So, there are five regions of interest: $\overline{R}X$ and$\underline{R}X$, and $POS_R(X)$, $BN_R(X)$ and $NEG_R(X)$. These sets are defined as shown below [3]. Let a set $X \subseteq U$, R be an equivalence relation and knowledge. Two subsets base can be associated:

1. R – Lower: $\underline{R}X = U\{Y \in U/R : Y \subseteq X\}$
2. R – Upper: $\overline{R}X = U\{Y \in U/R : Y \cap X \neq \phi\}$
   It means that the elements belong to $\underline{R}X$ set can be with certainly classified as elements of X; while the elements belong to $\overline{R}X$ set can be possibly classified as elements of X. In the same way, $POS_R(X)$, $BN_R(X)$ and $NEG_R(X)$ are defined below [3].
3. $POS_R(X) = \underline{R}X \Rightarrow$ certainly member of X
4. $NEG_R(X) = U - \overline{R}X \Rightarrow$ certainly non member of X
5. $BN_R(X) = \overline{R}X - \underline{R}X \Rightarrow$ possibly member of X

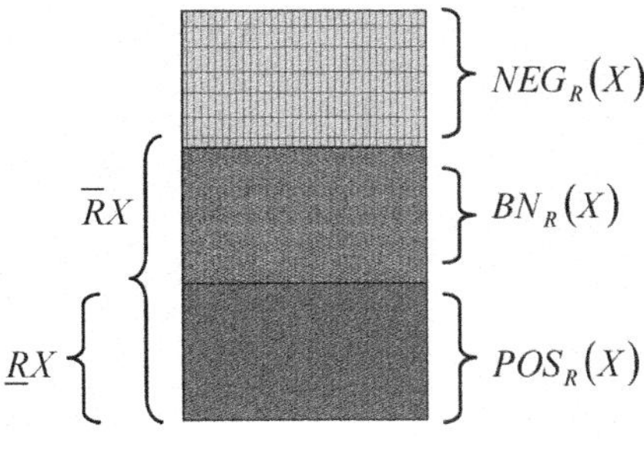

Fig. 36.1 Definition of R-approximation sets and R-regions

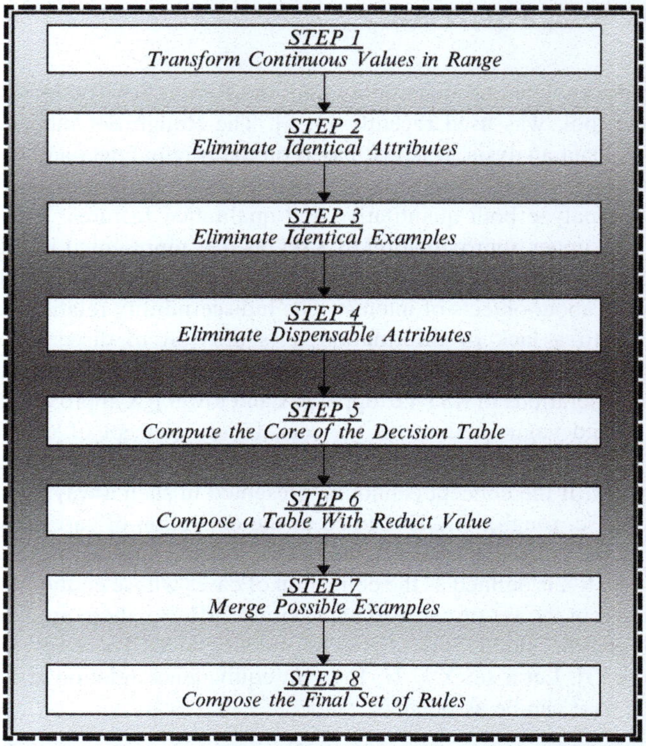

Fig. 36.2 118IEEE Test power network

Due to the granularity of knowledge, rough sets cannot be characterized by using available knowledge. Consequently, with every rough set there are associate two crisp sets, called lower and upper approximation. Logically, the lower approximation of a set consists of all elements that surely belong to the set, whereas the upper approximation of the set is constituted by all elements that possibly belong to the set, and the boundary region of the set consists of all elements that cannot be classified uniquely to the set or its complement, by employing available knowledge [3].

Previous to the presentation of the algorithm it is necessary to define two major concepts in RST, reduct and core. These concepts are important in the knowledge base reduction. The algorithm of the reduction of a decision table is shown in Fig. 36.2 [3]. The process of finding a reduced set of attributes with the similar or approximate classification strength as the original set is called attribute reduction. Some attributes in an information system may be redundant and consequently can be eliminated without losing essential classificatory information. Let R be a family of equivalence relations. The reduct of R, RED(R), is defined as a reduced set of relations that conserves the same inductive classification of set R. The core of R, CORE(R), is the set of relations that appears in all reduct of R, i.e., the set of all indispensable relations to characterize the relation R. As the core is the intersection of all reducts, it is included in every reduct, i.e., each element of the core belongs to some reduct. Therefore, in a sense, the core is the most important subset of attributes, since none of its elements can be removed without affecting the classification strength of attributes. Definitely, the geometry of reducts can be more complex. The core can be empty but there can exist a partition of reducts into a few sets with non-empty intersection. The approximation of classification is a simple extension of the definition of approximation of sets. Namely if $F = \{X_1, X_2, \ldots, X_n\}$ is a family of non empty sets, then $\underline{R}F = \{\underline{R}X_1, \underline{R}X_2, \ldots, \underline{R}X_n\}$ and $\overline{R}F = \{\overline{R}X_1, \overline{R}X_2, \ldots, \overline{R}X_n\}$, are called the $\underline{R}F$ – lower and the $\overline{R}F$ – upper approximation of the family F [3].

Two measures can be defined to describe inexactness of approximate classification. The first one is the extension of the measure defined to describe accuracy of approximation sets. The accuracy of approximation of F by R can defined as [3]:

$$\alpha R(F) = \frac{\sum \mathrm{card}\underline{R}X_i}{\sum \mathrm{card}\overline{R}X_i} \tag{36.1}$$

where card(X) denotes the cardinality of $X = \phi$.

The accuracy of the approximation can be used to measure the quality of approximation of decision classes on the universe U. It is possible to use another measure of accuracy defined by $1 - \alpha R(X)$. Some other measures of approximation accuracy based on entropy or some more specific properties of boundary regions are also used. The choice of a relevant accuracy of approximation depends on the particular data set. The accuracy of approximation of X can be tuned by R [3].

The second measure, called the quality of approximation of F by R, is the following [3]:

$$\gamma R(F) = \frac{\sum \mathrm{card}\underline{R}X_i}{\mathrm{card}\ U} \tag{36.2}$$

The accuracy of classification expresses the percentage of possible correct decision, when classifying objects, employing the knowledge R. The quality of classification expresses the percentage of objects which can be correctly classified to classes F employing knowledge R. By selecting a proper balance between the accuracy of classification and the description size it is expect to find the classifier with the high quality of classification on unseen objects [3] too. One of the most important applications of RST is the generation of decision rules for a given information system for the prediction of classes for new objects which are beyond observation. The rules are presented in an “If condition(s) then decision(s)” format.

36.3 Application Example

As an application example was used the 118 IEEE. test power network (Fig. 36.3), (54 generators, 194 branches and 91 loads) for the security assessment [7]. A first order contingency study was carried out for all the transmission lines, transformers and generators. The input numerical values were obtained using the software

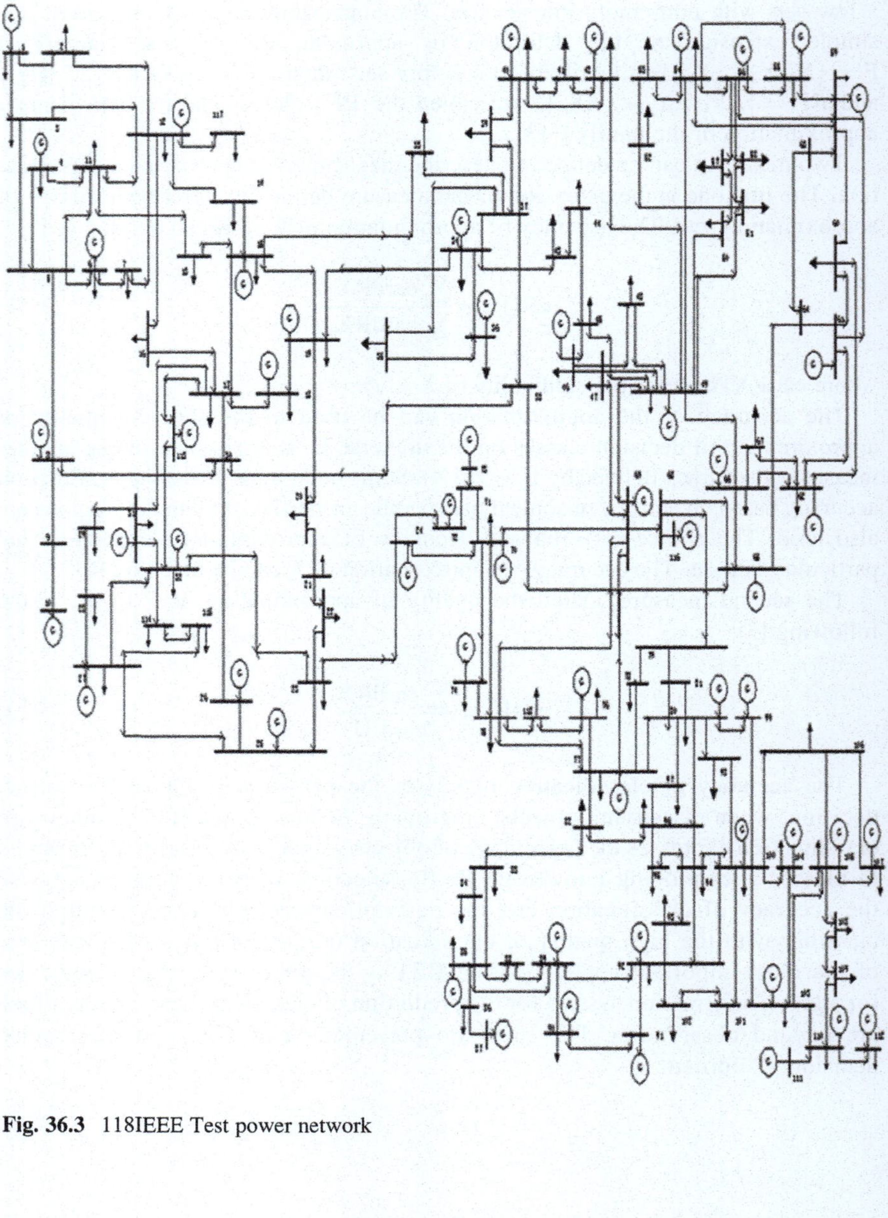

Fig. 36.3 118IEEE Test power network

Table 36.1 Definition of range attributes coding

	Codes			
Attributes	1	2	3	4
A_1	95% <	$95\% < a_1 < 100\%$	$100\% \leq a_1 \leq 110\%$	>110%
A_2	0	$2 \leq a_2$	$3 \leq a_2 \leq 4$	>4
A_3	0,850<	$0{,}850 < a_3 < 1{,}000$	$1{,}000 \leq a_3 \leq 1{,}050$	>1,050
A_4	0<	$2 \leq a_4$	$3 \leq a_4 \leq 5$	>5
S	N	A	E_1	E_2

package SecurMinin2.0 [8]. A first order contingency in transmission lines and transformers was simulated.

In this section are presented the final results produced by the algorithm using the Rough Set theory for the assessment and classification contingencies of an Electric Power System [8–10]. A first order contingency study was carried out and it was obtained a list of 184 contingencies that allows the construction a contingency control database [11].

The specified attributes are as follows:

A_1 – overloads in the transmission lines;

A_2 – number of overload transmission lines;

A_3 – voltage levels;

A_4 – number of busbar with voltage violation;

Dec. – Security

Table 36.1 show the chosen range for the coded qualitative attributes. The condition attributes are coded into four qualitative terms: Low, Medium, High and Full. The decision attribute (Dec.) is divided in four states, Normal, Alert, Emergency I and Emergency II.

Table 36.2 presents a set of information related to a contingency control database.

Table 36.3 show the accuracy of approximation and the quality of classification for all conditions and the attributes in core, for the four different scenarios.

Due to large amount of information only the final rules are presented. A powerful interface between the SecurMining2.0 software package and the ROSE computer program it was used [12, 13]. Using the above computer software packages it was verified that the attributes A_1, A_3 are the Core and the Reduct of the set of contingencies.

The quality of classification for all conditions and the attributes in the core was 0.9286. The precision the approximation of the classification for de contingencies in Emergency I and Emergency II was 1.000 that means 100% of certainty. It is possible to compose the above set of rules using the algorithm described previously. Incorporating the range values the final set of rules and approximate rules that contains the knowledge of a initial database range values were obtained with the SecurMining2.0 software package. The Rules are divided in four sets. The first set contains the rules for Normal State. The second set, contain the rules for Alert State. The third characterize the Emergency I State. Finally the last set of rules characterizes the Emergency State II. In this study we have one approximate rule.

Table 36.2 The attributes represented by the set

	Attributes				
Cont. Nº.	A1	A2	A3	A4	Dec.
1	1	1	3	3	N
2	1	1	4	3	A
3	1	1	4	3	A
4	3	2	4	3	E1
5	1	1	3	3	N
6	4	2	4	1	E2
7	3	2	4	1	E1
8	3	2	4	1	E1
9	4	4	4	4	E2
10	3	2	4	1	E1
11	1	1	3	1	A
12	1	1	3	1	A
13	3	2	4	3	E1
…	…	…	…	…	…
178	1	1	3	3	N
179	1	1	4	2	A
180	1	1	3	2	N
181	1	1	3	1	N
182	1	1	3	1	N
183	1	1	2	1	N
184	1	1	3	1	N

Table 36.3 Definition of range attributes coding

Decision level	Nº of objects	Approximate upper	Approximate lower	Precision the approximate of classification
Normal	106	87	77	0.8450
Alert	46	68	59	0.7201
Emergency I	23	23	23	1.0000
Emergency II	9	9	9	1.0000

Rules for Normal State

1. If (A_1 is L and A_3 is M) then S = N
2. If (A_1 is L and A_3 is H and A_4 is H) then S = N
3. If (A_1 is L and A_4 is L) then S = N

Exact Rules for Alert State

1. If (A_1 is M and A_4 is L) then S = A
2. If (A_1 is L and A_3 is F) then S = A
3. If (A_3 is F and A_4 is H) then S = A

Exact Rules for Emergency I State

1. If (A_1 is H and A_3 is F) then S = E_1.

2. If (A_1 is H) then S = E_1.
3. If (A_2 is L and A_4 is F) then S = E_1.

Exact Rules for Emergency II State

1. If (A1 is F and A4 is F) then S = E2.
2. If (A1 is F and A3 is F and A4 is M) then S = E2.

Approximate Rules

1. If (A_1 is L and A_3 is H and A_4 is L) then S = N or S = A.

36.4 Conclusions

In this paper, an application of the Rough Set Theory on Power systems is proposed. This theory was used to study and analyse the steady–state contingencies and classification. The knowledge acquisition process is a complex task, since the experts have difficulty to explain how to solve a specified problem. The study presents a systematic approach to transform examples in a reduced set of rules. The results that were produced using the Rough Set Theory are a set of rules for the four scenarios proposed, Normal, Alert, Emergency I and Emergency II, showing the importance of the chosen range for the coded qualitative attributes.

The proposed methodology is an efficient filtering tool for contingencies screening and ranking in on-line security studies, since it provides fast, accurate and relevant information about the system security, since it permits the identification of the contingencies considered dangerous. The information supplied by the proposed algorithm, allows a rapid implementation of preventive control measures to the System Operator. This technique of Data Mining is efficient in studies of security analysis in Electrical Power Systems.

Acknowledgments The first author would like to thank "Fundação para Ciência e Tecnologia, FCT", that partially funded this research work through the PhD grant nº: SFRH/BD/38152/2007.

References

1. Data mining: what is data mining. Available on: www.anderson.ucla.edu.
2. Hsu C-N, Knoblock GA (1998) Discovering robust knowledge from databases that change. Data Min Knowl Discov 2(1):69–95
3. Kusiak A (2001) Rough set theory: a data mining tool for semiconductor manufacturing. IEEE Trans Electron Packag Manuf 24(1):44–50

4. Faustino Agreira CI, Pestana R, Machado Ferreira CM, Maciel Barbosa F (2011) Portuguese transmission system contingencies analysis using the rough set theory. 2011 Cigré – assessing and improving power system security, reliability and performance in light of charging energy sources symposium recife, Pernambuco, Brazil, 3–6 April 2011
5. Faustino Agreira CF, Machado Ferreira CM, Maciel Barbosa F (2010) The rough set theory applied to a set of the new severity indices. In: Proceedings of the 2010 11th international conference on probabilistic methods applied to power systems conference, Singapore
6. Pawlak Z (1991) Rough sets–teorical aspects of reasoning about data. Kluwer, Dordrecht
7. Power Systems test Case Archive: 118 Bus Power Flow Test Case, Department of Electrical Engineering, University of Washington, [Online]. Available: http://www.ee.washington.edu/research/pstca/
8. Faustino Agreira CI (2010) Data mining techniques for security study and analysis to the electrical power systems. PhD dissertation, University of Porto
9. Wood J, Wollenberg BF (1996) Power generation operation and control, 2nd edn. Wiley, New York
10. Stevenson WD, Grainger JJ (1994) Power system analysis. McGraw-Hill International Editions, Singapore
11. Faustino Agreira CI, Machado Ferreirav CM, Dias Pinto JA, Maciel Barbosa F (2002) Efficient contingency filtering and ranking algorithm for steady-state security analysis of an electric power system. In: Proceedings of the 38th universities power engineering conference, Staffordshire University, UK
12. Faustino Agreira CI, Machado Ferreira CM, Dias Pinto JA, Maciel Barbosa FP (2004) Electric power systems steady-state security assessment using the rough set theory. In: Proceedings of the 8th international conference on probabilistic methods applied to power systems, Iowa State University, Ames, Iowa, USA, Sept 12–16 2004
13. ROSE2 – Rough sets data explorer. Laboratory of intelligent decision support systems of the Institute of Computing Science, Poznan [Online]. Available: http://www.idss.cs.put.poznan.pl/software/rose

Chapter 37
Tuning a Fractional Order Controller from a Heat Diffusion System Using a PSO Algorithm

Isabel S. Jesus and Ramiro S. Barbosa

Abstract In this paper we use a particle swarm optimization (PSO) algorithm for the tuning of a fractional order controller, applied to a heat diffusion system. PSO is an optimization technique, inspired by social behaviour of bird flocking or fish schooling. It has the better ability of global searching and has been successfully applied in many areas of engineering and science. Simulations are presented assessing the performance of the proposed fractional algorithms.

37.1 Introduction

Particle swarm optimization (PSO) is one of the latest evolutionary techniques developed by Dr. Eberhart and Dr. Kennedy in 1995, inspired by social behaviour of bird flocking or fish schooling. The PSO scheme optimizes searching by virtue of the swarm intelligence produced by the cooperation and competition among the particles of a species. The social system is discussed through the collective behaviours of simple individuals interacting with their environment and each other. Examples of this are the bird flock or fish school. Some applications of PSOs are found in the field of nonlinear dynamical systems, data analysis, electrical engineering, function optimization, artificial neural network training, fuzzy control and many others in real world applications [1, 2].

The concept of differentiation and integration to non-integer order dates to 1695, when Leibniz mentioned it in a letter to L'Hopital. Since then, many scientists developed the area and notable contributions have been made, both in theory and in applications [3, 4]. In fact, fractional calculus (FC) is a generalization of integration and differentiation to a non-integer order $\alpha \in C$, being the fundamental operator ${}_aD_t^\alpha$, where a and t are the limits of the operation [3, 4]. These fractional concepts constitute a useful

I.S. Jesus (✉) • R.S. Barbosa
GECAD – Knowledge Engineering and Decision Support Research Center,
Institute of Engineering – Polytechnic of Porto (ISEP/IPP), Porto, Portugal
e-mail: isj@isep.ipp.pt; rsbisep.ipp.pt

A. Madureira et al. (eds.), *Computational Intelligence and Decision Making: Trends and Applications*, Intelligent Systems, Control and Automation: Science and Engineering 61,
DOI 10.1007/978-94-007-4722-7_37,

tool for describing several physical phenomena in almost all areas of science and engineering, such as heat, magnetism, flow, mechanics or fluid dynamics. Presently, the ability of the FC is being recognized for better modelling and control of many dynamical systems. In fact, during the last years FC has been used increasingly to model the constitutive behaviour of materials and physical systems exhibiting hereditary and memory properties. This is the main advantage of fractional derivatives in comparison with the classical integer-order counterpart, where these effects are neglected.

It is well known that the fractional operator $s^{0.5}$ appears in several types of problems [5]. The transmission lines, the heat flow or the diffusion of neutrons in a nuclear reactor are examples where the half-order operator is the fundamental element. Moreover, diffusion is one of the three fundamental partial differential equations of mathematical physics [6]. Therefore, the control of such systems having in mind FC concepts is an important subject.

In this paper we investigate the control of a heat diffusion system based on fractional algorithms. The adoption of fractional controllers has been justified by its superior performance, particularly when used with fractional dynamical systems, such as the case of the heat system under study.

Bearing these ideas in mind, the paper is organized as follows. Section 37.2 gives the fundamentals of fractional control systems. Section 37.3 gives the approximations to fractional operators. Section 37.4 introduces the heat diffusion system control strategies and discusses the results. Finally, Sect. 37.5 draws the main conclusions and addresses perspectives towards future developments.

37.2 Fundamentals of Fractional Calculus

The mathematical definition of a fractional derivative and integral has been the subject of several different approaches [3, 4]. One commonly used definition for the fractional derivative is given by the Riemann-Liouville definition ($\alpha > 0$):

$$ {}_aD_t^\alpha f(t) = \frac{1}{\Gamma(n-\alpha)} \frac{d^n}{dt^n} \int_a^t \frac{f(\tau)}{(t-\tau)^{\alpha-n+1}} d\tau, \quad n-1<\alpha<n \tag{37.1} $$

where $f(t)$ is the applied function and $\Gamma(x)$ is the Gamma function of x. Another widely used definition is given by the Grünwald-Letnikov approach ($\alpha \in \Re$):

$$ \begin{aligned} {}_aD_t^\alpha f(t) &= \lim_{h\to 0} \frac{1}{h^\alpha} \sum_{k=0}^{\left[\frac{t-a}{h}\right]} (-1)^k \binom{\alpha}{k} f(t-kh), \\ \binom{\alpha}{k} &= \frac{\Gamma(\alpha+1)}{\Gamma(k+1)\Gamma(\alpha-k+1)} \end{aligned} \tag{37.2} $$

where h is the time increment and $[x]$ means the integer part of x.

The "memory" effect of these operators is demonstrated by (37.1) and (37.2), where the convolution integral in (37.1) and the infinite series in (37.2), reveal the unlimited memory of these operators, ideal for modelling hereditary and memory properties in physical systems and materials.

An alternative definition to (37.1) and (37.2), which reveals useful for the analysis of fractional control systems, is given by the Laplace transform method. Considering vanishing initial conditions, the fractional *differintegration* is defined in the Laplace domain, $F(s) = L\{f(t)\}$, as:

$$L\{{}_aD_t^{\alpha}f(t)\} = s^{\alpha}F(s), \quad \alpha \in \Re \tag{37.3}$$

37.3 Approximations of Fractional Operators

In this paper we adopt discrete integer-order approximations to the fundamental element s^{α} ($\alpha \in \Re$) of a fractional-order control (FOC) strategy. The usual approach for obtaining discrete equivalents of continuous operators of type s^{α} adopts the Euler, Tustin and Al-Alaoui generating functions [7].

It is well known that rational-type approximations frequently converge faster than polynomial-type approximations and have a wider domain of convergence in the complex domain. Thus, by using the Euler operator $\omega(z^{-1}) = (1 - z^{-1})/T_c$, and performing a power series expansion of $[\omega(z^{-1})]^{\alpha} = [(1- z^{-1})/T_c]^{\alpha}$ gives the discretization formula corresponding to the Grünwald-Letnikov definition (37.2):

$$D^{\alpha}(z^{-1}) = \left(\frac{1 - z^{-1}}{T_c}\right)^{\alpha} = h^{\alpha}(0) + h^{\alpha}(1)z^{-1} + \ldots = \sum_{k=0}^{\infty} h^{\alpha}(k)z^{-k} \tag{37.4}$$

where T_c is the sampling period and $h^{\alpha}(k)$ is the impulse response sequence given by the expression ($k \geq 0$):

$$h^{\alpha}(k) = \left(\frac{1}{T_c}\right)^{\alpha} \binom{k - \alpha - 1}{k} \tag{37.5}$$

A rational-type approximation can be obtained through a Padé approximation to the impulse response sequence (37.5) $h^{\alpha}(k)$, yielding the discrete transfer function:

$$H(z^{-1}) = \frac{b_0 + b_1 z^{-1} + \ldots + b_m z^{-m}}{1 + a_1 z^{-1} + \ldots + a_n z^{-n}} = \sum_{k=0}^{\infty} h(k)z^{-k} \tag{37.6}$$

where $m \leq n$ and the coefficients a_k and b_k are determined by fitting the first $m + n + 1$ values of $h^{\alpha}(k)$ into the impulse response $h(k)$ of the desired approximation $H(z^{-1})$.

37.4 Heat Diffusion

The heat diffusion is governed by a linear partial differential equation (PDE) of the form:

$$\frac{\partial c}{\partial t} = k_c\left(\frac{\partial^2 c}{\partial x^2} + \frac{\partial^2 c}{\partial y^2} + \frac{\partial^2 c}{\partial z^2}\right) \tag{37.7}$$

where k_c is the diffusivity, t is the time, c is the temperature and (x,y,z) are the space Cartesian coordinates. The system (37.7) involves the solution of a PDE of parabolic type for which the standard theory guarantees the existence of a unique solution [8].

For the case of a planar perfectly isolated surface we can apply a constant temperature C_0 at $x = 0$ and we can analyze the heat diffusion along the horizontal coordinate x. Under these conditions, the heat diffusion phenomenon is described by a non-integer order model, yielding a transfer function of type:

$$C(x,s) = \frac{C_0}{s}G(s)\,, \quad G(s) = e^{-x\sqrt{\frac{s}{k_c}}} \tag{37.8}$$

where x is the space coordinate, C_0 is the boundary condition and $G(s)$ is the system transfer function.

The solution of system (37.8) in the time domain yields:

$$c(x,t) = C_0\, erfc\left(\frac{x}{2\sqrt{k_c t}}\right) = C_0\left[1 - \frac{2}{\sqrt{\pi}}\int_0^{\frac{x}{2\sqrt{k_c t}}} e^{-c^2}dc\right] \tag{37.9}$$

In our study, the simulation of the heat diffusion is accomplished by adopting the explicit numerical integration based on the discrete approximation to differentiation as [9]:

$$c[j+1,i] = r(c[j,i+1] + c[j,i-1]) + (1-2r)\times c[j,i] \tag{37.10}$$

where $r = k_c\Delta t(\Delta x^2)^{-1}$, $\{\Delta x, \Delta t\}$ and $\{i, j\}$ are the increments and the integration indices for space and time, respectively.

Figure 37.1 depicts the polar diagrams for both $G_A(j\omega)$ (*i.e.*, analytical solution according to (37.8)) and $G_N(j\omega)$ (*i.e.*, numerical solution from (37.10)) implementations, when $x = 3.0$ m and $k_c = 0.042\ \mathrm{m^2s^{-1}}$. It is observed a difference between the gains of the theoretical and numerical models, which is caused by the geometric definition of the system.

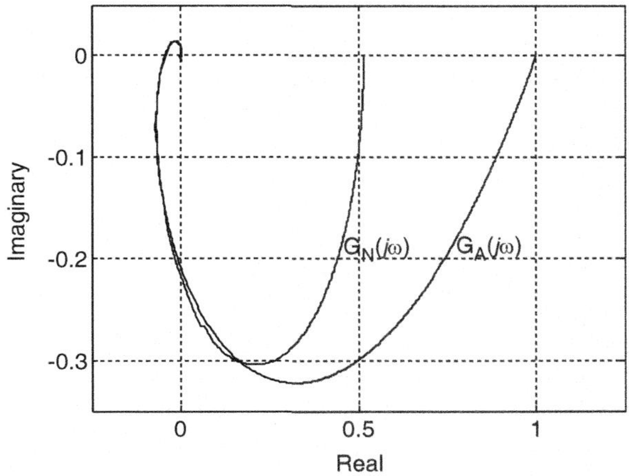

Fig. 37.1 Polar diagrams of $G_A(j\omega)$, and $G_N(j\omega)$, for $x = 3.0$ m and $k_c = 0.042\ m^2\ s^{-1}$

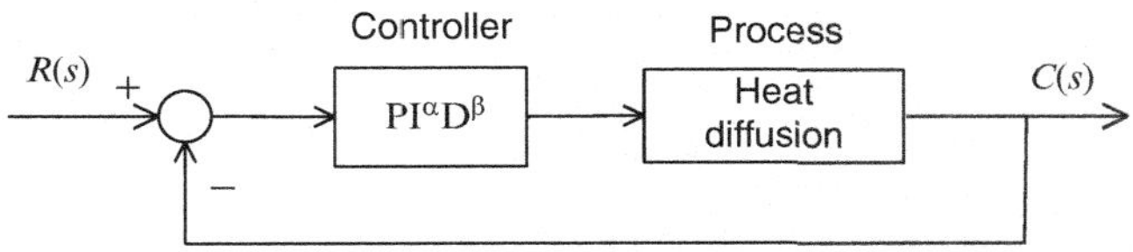

Fig. 37.2 Block diagram of the control system

37.4.1 Control and Optimization Strategies

The generalized PID controller $G_c(s)$ has a transfer function of the form [10]:

$$G_c(s) = K_p + \frac{K_i}{s^\alpha} + K_d s^\beta \tag{37.11}$$

where α and β are the orders of the fractional integrator and differentiator, respectively. The parameters K_p, K_i and K_d are correspondingly the proportional, integral, and derivative gains of the controller. Clearly, taking (α, β) = {(1, 1); (1, 0); (0, 1); (0, 0)} we get the classical {PID, PI, PD, P} controllers, respectively. Other PID controllers are possible, namely: PD$^\beta$ controller, PI$^\alpha$ controller, PID$^\beta$ controller, and so on. The PI$^\alpha$D$^\beta$ controller is more flexible and gives the possibility of adjusting more carefully the closed-loop system characteristics [4].

In previous studies developed by the authors [8], the control of system of Fig. 37.2 was analyzed adopting the classical integer-order PID ((α, β) ≡ (1,1)) tuned by the Ziegler-Nichols open loop (*ZNOL*) heuristics. It was verified that the system under the action of a PID controller does not produce satisfactory results, giving rise to a significant overshoot, and a large settling time.

Moreover, the step response reveals a considerable time delay. The poor results obtained indicate that the method of tuning the system may not be the most adequate for the control of the heat system under consideration. In fact, the system fractional dynamics suggests the adoption of other configurations, namely the use of a fractional PID controller [11, 12]. In this line of thought, in this work we propose a fractional PI^{α} controller tuned through the minimization of the ISE (integral square error) and ITSE (integral time square error) indices, by applying a PSO algorithm in order to achieve a superior control performance of the heat diffusion control system of Fig. 37.2.

It is important to mention that a reliable execution and analysis of a PSO usually requires a large number of simulations to provide that stochastic effects have been properly considered. Therefore, the experiments consist on executing the PSO several times N_{PSO}, in order to generate a combination of controller gains, that leads to a better transient and steady state responses.

In the PSO algorithm, all particles represent a potential solution to a problem, which is performed by adjusting their position taking into account both personal and group experiences [1, 13]. In each iteration, the velocity (v) is actualized by expression (37.12). The new particle position (cp_{k+1}) is found by adding their actual position with the new velocity, as shown in (37.12).

$$\begin{aligned} v_{k+1} &= w{*}v_k + c_1 r_1 (lbp - cp_k) + c_2 r_2 (gbp_k - cp_k), \\ cp_{k+1} &= cp_k + v_{k+1} \end{aligned} \tag{37.12}$$

where w is the inertia weight, c_1 and c_2 are the individual and sociality weights coefficients for modelling attractive forces from the local and global best, respectively; r_1 and r_2 are aleatory numbers between [0, 1], *lbp* is the local best position, *cp* is the current position, and *gbp* is the global best position [13]. The inertia term, forces the particle to move in the same direction as before by adjusting the old velocity. The cognitive term (personal best), forces the particle to go back to the previous best position. On the other hand, the social learning term (global best), forces the particle to move to the best previous position of its neighbours.

The PSO optimizes an objective function by iteratively improving a swarm of solutions, called particles, based on special management of memory. Each particle is modified by referring to the memory of individual swarm's best information. Due to the collective intelligence of these particles, the swarm is able to repeatedly improve its best observed solution and converging to an optimum.

The optimization fitness function corresponds to the minimization of one of these two indices that measure the response error, namely the ISE and the ITSE criteria, defined as:

$$J_{ISE} = \int_0^{\infty} [r(t) - c(t)]^2 dt, \qquad J_{ITSE} = \int_0^{\infty} t[r(t) - c(t)]^2 dt \tag{37.13}$$

We can use other integral performance criteria such as the integral absolute error (IAE) or the integral time absolute error (ITAE). In the present case the ISE and the ITSE criteria have produced the best results and are adopted in

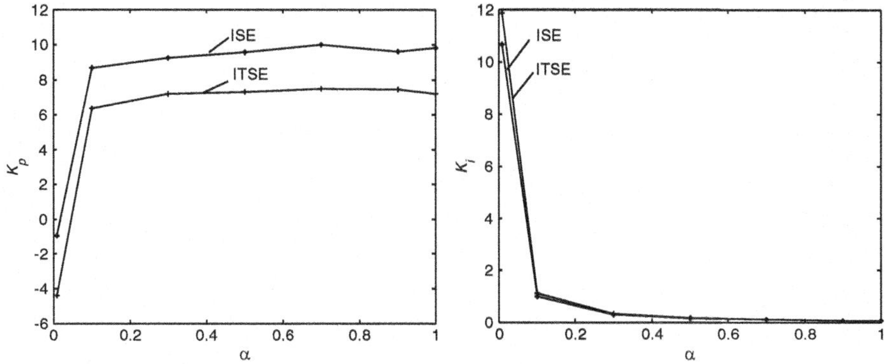

Fig. 37.3 The PI$^\alpha$ parameters (K_p;K_i) *versus* α, for the ISE and ITSE criteria, $x = 3.0$ m, $k_c = 0.042\ \mathrm{m^2s^{-1}}$

the sequel [14]. Furthermore, the ITSE criterion enables us to study the influence of time in the error generated by the system. We establish the following values for the PSO parameters: population number $PN = 40$, $c_1 = c_2 = 2$, $w = 1$ and a maximum number of iterations $I_{Max} = 200$.

37.4.2 Fractional PI Control

In this subsection we analyze the closed-loop system with a fractional PI$^\alpha$D$^\beta$ controller given by the transfer function (37.11) with $\beta = 0$. The fractional-order integrative term K_i/s^α in (37.11) is implemented by using a 4th-order Padé discrete rational transfer function of type (37.6).

In order to assure a zero steady-state error, the fractional integrative term K_i/s^α is implemented by means of an integer integrator [15], yielding:

$$K_i \frac{s^{1-\alpha}}{s}, \quad 0 < \alpha < 1 \tag{37.14}$$

It is used a sampling period of $T_c = 0.1$ s. The PI$^\alpha$ controller is tuned by minimization of the ISE and ITSE (37.13) criteria.

A step reference input $R(s) = 1/s$ is applied at $x = 0.0$ m and the output $c(t)$ is analyzed for $x = 3.0$ m. The heat system is simulated for 500 s. Figure 37.3 illustrates the variation of the fractional PI parameters (K_p; K_i) as function of the integrative order α, when minimizing the ISE and ITSE criteria and for the best case within a sample of 10 experiments. The values of K_p and K_i for both indices remain approximately the same for all α, being the ITSE slightly smaller than the corresponding ISE.

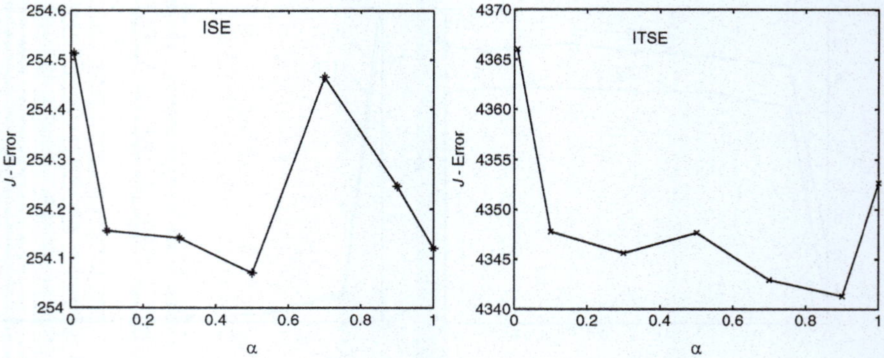

Fig. 37.4 Error J for the ISE and ITSE indices *versus* α, $x = 3.0$ m, $k_c = 0.042$ m^{2}s^{-1}

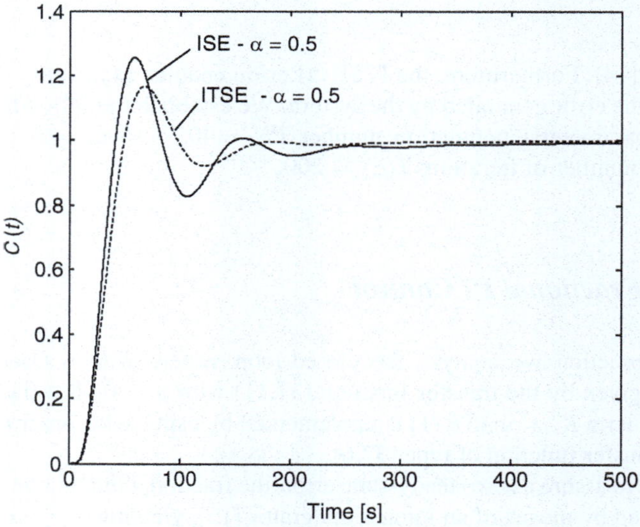

Fig. 37.5 Step responses of the closed-loop system for the ISE and ITSE indices, with a PI$^\alpha$ controller, $x = 3.0$ m, $k_c = 0.042$ m^{2}s^{-1}

Figure 37.4 shows the ISE and ITSE errors as function of α, for the best case within $N_{PSO} = 10$. The graph shows that we have a lower error for $\alpha_{\mathtt{ISE}} = 0.5$ and $\alpha_{\mathtt{ITSE}} = 0.9$. Moreover, the curves indicate a large influence of a weak integrative order on the system dynamics.

Figure 37.5 illustrates the step responses of the closed-loop system, for the PI$^\alpha$, tuned in the ISE and ITSE perspectives for the best cases of $\{\alpha_{\mathrm{ISE}}; \alpha_{\mathrm{ITSE}}\} = \{0.5; 0.9\}$. The controller parameters, corresponding to the minimization of those indices, lead to the values of ISE: $\{K_p, K_i, \alpha\} \equiv \{9.56, 0.15, 0.5\}$ and ITSE: $\{K_p, K_i, \alpha\} \equiv \{7.45, 0.07, 0.9\}$. The step responses with the PI$^\alpha$ (Fig. 37.5) present a

Table 37.1 Time-domain specifications of t_s, t_r, t_p, $ov(\%)$ and error J *versus* α for the step responses of the closed-loop system for the ISE criterion with a PI^{α} controller, $x = 3.0$ m, $k_c = 0.042\ m^2s^{-1}$

α	t_s	t_r	t_p	$ov(\%)$	J
0.01	233.9	21.9	58.9	26.8	254.5
0.1	234.9	22.0	59.1	26.0	254.2
0.3	235.6	22.3	59.4	25.4	254.1
0.5	233.6	21.9	59.0	26.4	254.2
0.7	228.7	21.2	58.1	29.0	254.5
0.9	238.9	22.2	59.1	24.9	254.2
1.0	230.8	21.7	58.8	27.5	254.1

Table 37.2 Time-domain specifications of t_s, t_r, t_p, $ov(\%)$ and error J *versus* α for the step responses of the closed-loop system for the ITSE criterion with a PI^{α} controller, $x = 3.0$ m, $k_c = 0.042\ m^2s^{-1}$

α	t_s	t_r	t_p	$ov(\%)$	J
0.01	158.4	27.1	67.9	17.4	4366.0
0.1	158.2	27.2	67.9	17.2	4347.8
0.3	158.5	27.2	67.9	17.0	4345.6
0.5	159.6	27.3	68.0	16.5	4347.6
0.7	157.5	26.9	67.5	17.6	4342.9
0.9	159.0	27.3	68.0	16.7	4341.3
1.0	160.1	28.1	69.6	16.0	4352.6

reasonable transient behaviour and a zero steady-state error showing the effectiveness of the proposed fractional algorithm when used for the control of the heat system.

Tables 37.1 and 37.2 show the variation of the transient time-domain specifications, namely the settling time t_s, the rise time t_r, the peak time t_p, the percent overshoot $ov(\%)$ and the error J *versus* α, for the closed-loop response tuned through the minimization of the ISE and ITSE indices, respectively. The transient response parameters do not change significantly with the fractional integrative order α.

In conclusion, we get the best controller tuning, superior to the performance revealed by the classical integer-order scheme studied in previous works [14]. These results demonstrate the effectiveness of the fractional-order algorithms when used for the control of fractional-order systems. However, the time delay observed in the step response, indicates that this kind of structure may not be the most adequate for the control of the heat diffusion system. In fact, the system suggests the adoption of a Smith predictor scheme, that is very effective in improving the control of processes having time delays.

37.5 Conclusions

This paper presented the fundamental aspects of application of the FC theory in the control of diffusion systems. In this line of thought, it was studied a heat diffusion system described through the fractional-order operator $s^{0.5}$. The dynamics of the system was analyzed in the perspective of FC and using a PSO algorithm.

We presented a PI^{α} control strategy, that gives better results than those obtained with the conventional PID controller. This result points out the use of a fractional PID algorithm inserted in a Smith Predictor structure.

Moreover, at the PSO algorithm the particles update themselves with the internal velocity. They also have memory, thus retaining part of their previous state, which is important to the algorithm [1], [13]. Furthermore, this memory behaviour of the particles have similarities with that occurs in the fractional order calculus. In fact, the PSO memory enables the algorithm to adjust its searching strategy dynamically in accordance with the current situation.

References

1. Kennedy J, Eberhart RC (1995) Particle swarm optimization. In: Proceedings of the IEEE international conference on neural networks, pp 1942–1948
2. Trelea IC (2003) The particle swarm optimization algorithm: convergence analysis and parameter selection. Inform Process Lett 85:317–325
3. Oldham KB, Spanier J (1974) The fractional calculus: theory and application of differentiation and integration to arbitrary order. Academic Press, New York
4. Podlubny I (1999) Fractional differential equations. Academic Press, San Diego
5. Battaglia JL, Cois O, Puigsegur L, Oustaloup A (2002) Solving an inverse heat conduction problem using a non integer identified model. Int J Heat Mass Transf 44:2671–2680
6. Courant R, Hilbert D (1962) Methods of mathematical physics, partial differential equations. Wiley Interscience II, New York
7. Barbosa RS, Machado JAT, Silva MF (2006) Time domain design of fractional differintegrators using least-squares. Signal Process 86(10):2567–2581
8. Jesus IS, Machado JAT (2008) Fractional control of heat diffusion systems. J Nonlinear Dynamic 54(3):263–282
9. Gerald CF, Wheatley PO (1999) Applied numerical analysis. Addison-Wesley
10. Podlubny I (1999) Fractional-order systems and $PI^{\alpha}D^{\beta}$ controllers. IEEE Trans. on Automatic Control, 44(1):208–213
11. Jesus IS, Barbosa RS (2011) Using a PSO algorithm for tuning a PID^{β} controller applied to a heat system. In: Proceedings of the IFAC – 18th word congress, Milano, Italy, 28 Aug to 2 Sept 2011
12. Jesus IS, Barbosa RS (2011) Fractional PI control of a heat system through a PSO algorithm, In: ISCIES´11 – 2nd international symposium on computational intelligence for engineering systems, Coimbra, Portugal, pp 1–6
13. Eberhart RC, Shi Y (2000) Comparing inertia weights and constriction factors in particle swarm optimization. In: Proceedings of the congress on evolutionary computation, San Diego, pp 84–88
14. Jesus IS, Machado JAT (2007) Application of fractional calculus in the control of heat systems. J Adv Comput Intelligence and Intelligent Inform 11(9):1086–1091
15. Axtell MBM (1990) Fractional calculus applications in control systems. IEEE national aerospace and electronics conference, New York, pp 563–566

Chapter 38
A Tool for Biomedical – Documents Classification Using Support Vector Machines

João Oliveira, Daniel Correia, Carlos Pereira, Paula Veríssimo, and António Dourado

Abstract This paper presents a web tool for text mining of biomedical literature using clustering and support vector machines. The study is specific to the domain of Peptidases, based on curated literature. It has been evaluated the use of ontologies in the text mining and feature selection process, and the preliminary results show that the classifier performance may be improved along with a reduction of the number of features.

Keywords Text mining • Web platform • Support vector machines • Ontologies

38.1 Introduction

In the last years there was a significant increase of research in the biomedical field and an almost exponential growth of the biomedical literature datasets. The largest and most known database, Pubmed, includes more than 20 million citations for

J. Oliveira (✉) • D. Correia
ISEC – Coimbra Institute of Engineering, Coimbra, Portugal
e-mail: a21150146@alunos.isec.pt; a21160499@alunos.isec.pt

C. Pereira
ISEC – Coimbra Institute of Engineering, Coimbra, Portugal

CISUC – Center for Informatics and Systems of University of Coimbra, Coimbra, Portugal
e-mail: cpereira@isec.pt

P. Veríssimo
Department of Biochemistry and Center of Neuroscience and Cell Biology, University of Coimbra, Coimbra, Portugal
e-mail: paulav@ibili.uc.ptdei.uc.pt

A. Dourado
CISUC – Center for Informatics and Systems of University of Coimbra, Coimbra, Portugal
e-mail: dourado@dei.uc.pt

A. Madureira et al. (eds.), *Computational Intelligence and Decision Making: Trends and Applications*, Intelligent Systems, Control and Automation: Science and Engineering 61, DOI 10.1007/978-94-007-4722-7_38, © Springer Science+Business Media Dordrecht 2013

biomedical literature from MEDLINE, online books and other journals from related fields [1]. Due to the large size of this database, it becomes difficult to search and obtain relevant information for particular fields. This is the main reason for the significant investment that has been done concerning to the development of tools for specific domains, such as the presented case study, the Peptidases domain.

This work describes a web platform for intelligent classification of documents, using traditional text mining techniques, k-means clustering and support vector machines. The functionalities of the web platform are briefly described along with the experimental studies for different algorithms. Another platform has been implemented for this domain [2] and here it is intended the analysis of the results impact with the inclusion of n-grams and the gene ontology [3] in the feature selection process. A detailed overview of the available web tools for biomedical literature search is given by Lu [4].

The remaining paper is organized into the following sections: Section 38.2 presents the methodology and techniques in the preprocessing of documents. Section 38.3 includes the Platform description, Sect. 38.4 presents the results and finally, the last section presents conclusions and future developments of the platform.

38.2 Methodology

This section presents the text mining algorithms applied for feature selection, the gene ontology and the algorithms used for unsupervised and supervised document classification.

38.2.1 Text Mining

One of the most important stages of text mining process is documents preprocessing. In this work, the features are extracted from documents abstracts and are represented in a structured way in order to obtain the most significant information. This whole process is illustrated in Fig. 38.1.

The preprocessing starts with the division of the continuous stream of characters, using a Tokenizer, which is simply the decomposition of the document in terms. Usually the delimiters used in this process are the space character, line breaks, tabs and special characters. Then it is used the Stop Word Filter to remove the terms that don't represent a document. These are usually tokens appearing very often in a natural language, including prepositions, articles, adverbs and pronouns. Then a Stemmer is usually applied, in order to reduce the inflected or derived words into their root, allowing to group together the different inflected or derived forms of a word so they could be analyzed as the same word. In order to complete the preprocessing, a term n-gram generator can also be used to create new features that are the joining of two or more other existing features with a space character between them.

One way of improving the process of feature extraction is by integration of domain knowledge in this process. This requires that the domain concepts are moved from

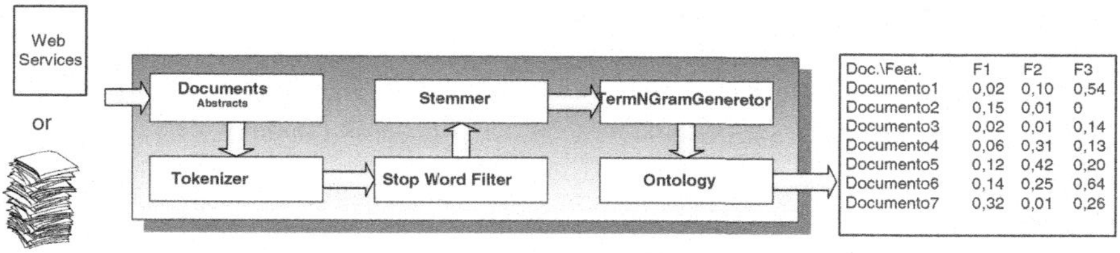

Fig. 38.1 Features extraction process

natural language to a clear structured and unambiguous form, to allow the use of automated processes. These considerations led us to integrate the Gene Ontology at this stage. In the Sect. 38.2.2 we describe how this ontology has been applied.

As already mentioned, it is crucial to have a representation of the documents in a structured way in order to make possible the later use of classification or clustering methods. The followed approach is based on the vector space representation, also known as Bag-Of-Words (BOW), where each line corresponds to a document and each column represents a term. The value contained in a particular cell corresponds to the measure of the importance of a term in the document set.

There are several methods that can be used to measure the relevance of a term, such as, Binary Occurrence, Term Occurrence, Term Frequency (tf), and Term frequency - Inverse Document Frequency (*tfidf*). In our work, we use the method most commonly applied the *tfidf*. In this method the computed value increases proportionally to the number of times a term appears in the document but is decreased by the frequency of the term in the corpus of documents, given by:

$$tfidf = tf_{i,j} * idf$$

$$tf_{i,j} = \frac{f_{i,j}}{fd_j}$$

$$idf = \log\left(\frac{|D|}{ft_i}\right)$$

where $f_{i,j}$ is the number of occurrences of the term in the document *j*; fd_j is the sum of the number of occurrences of all terms in the document *j*; ft_i is the number of all documents that the term *i* appear; |D| is number of documents.

38.2.2 Gene Ontology

The Gene Ontology (GO) project [3] aims to establish a standard for the representation of genes and genes products across species and databases, providing a controlled vocabulary of terms for describing gene product characteristics and gene product annotation data from GO consortium members. GO also defines the relationships between the terms, making it a structured vocabulary. The most general terms represent cellular component, biological process and molecular function. The GO project also offers a set of tools to access and process the data.

The Gene Ontology has been used in several works, such as GoPubMed [5] and TXTGate [6]. GoPubMed is a web server that categorizes the abstracts returned by PubMed according to the Gene Ontology. According to the authors, it offers many advantages like allowing users to quickly navigate through the abstracts by category and automatically shows general ontology terms related to the original query, which often do not even appear directly in the abstract. The authors have also developed a term extraction algorithm that is based on [7] using local sequence alignment of words of the abstract and the words of GO terms.

In the TXTGate, the authors implemented a framework called combining literature indices of selected public biological resources in a flexible text-mining system designed towards the analysis of groups of genes. Sub-clustering and links to external resources allow for in-depth analysis of the resulting term profiles.

This platform also uses the GO ontology with the aim to extract from documents only the terms present in this ontology.

38.2.3 Clustering of Documents

This platform includes unsupervised learning, allowing the classification of non-labeled documents. This type of learning includes clustering, that tries to discover relationships and patterns among the data, grouping the most similar items. Some of the most popular clustering algorithms are K-means, Subtractive Clustering, Expectation Maximization and Fuzzy C-Means. The goal is to include those algorithms in the platform, nevertheless, in this work only the K-means has been used and evaluated.

The K-means, also known as Lloyd's algorithm [8] is an iterative algorithm that aims to find the best possible distribution of points for the clusters (smallest sum of distances of the centers of clusters). The algorithm starts by initializing the centers of the clusters, and from here starts the iterative phase where each point should belong to the cluster to which the distance to its center is lower. After the allocation of points to clusters, centers are recalculated and the process continues until there are not any changes of the centers.

38.2.4 Supervised Classification of Documents

In this work, Support Vector Machines (SVM) [9, 10] have been used for the supervised learning and classification of documents (Fig. 38.2). The SVMs were chosen since they are one of the most popular methods for pattern classification and

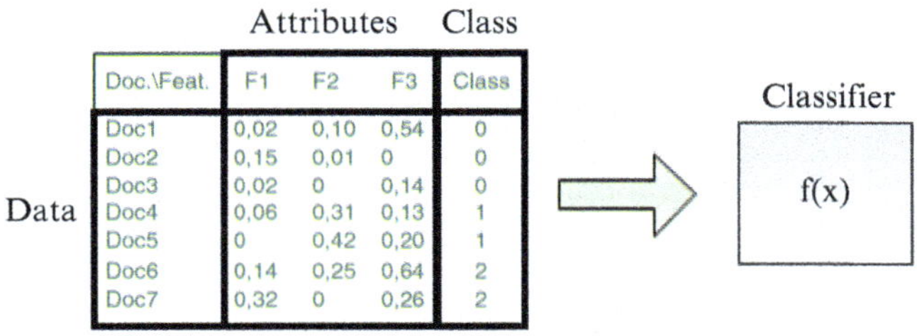

Fig. 38.2 Supervised classification of documents

for categorization of documents [11]. In these cases, the results obtained by SVMs are usually higher than those demonstrated by other techniques such as neural networks.

Two different SVMs were applied. A linear SVM using the "one-against all" strategy is applied to a selected set of documents related to peptidases and later compared to a multiclass "one-against-one" Gaussian kernel classifier.

38.3 Web Platform

DoCluster is a web platform built in ASP.NET. It also uses a wide range of other technologies described below in order to implement its features, including pre-processing of documents, implementation of the learning algorithms and the use of the Gene Ontology. The documents for testing and training purposes can be submitted by the users of the platform. Documents for the machine training can also be retrieved from the NCBI web service [12] that provides access to the documents outside of the regular web query interface of PubMed.

The information extracted from the submitted texts is done using a J# library based in WVTool [13]. In addition to WVtool some other classes have been implemented, which allow grouping multiple terms into one term, filtering terms present in documents and classification.

The main functionalities of our platform include clustering and classification of datasets submitted by users or documents from the NCBI web service. Registered users can also train their own SVM models and use them later to classify new documents. In the clustering and training of SVM models, it is also possible to configure some parameters for feature extraction and algorithms behavior.

After the process of classification or clustering, the results are presented to the users and beyond the simple listing, it is also possible to see the extracted features from the set of documents in order to understand its importance for each document. When the Gene Ontology is used in feature extraction process, the platform includes a link to the AmiGO [14]. This tool consists of a web interface for Gene Ontology database and allows users to explore all the information of each term. The Fig. 38.3 shows the results of clustering documents retrieved from PubMed for the search query "aminopeptidase n". In Fig. 38.4 is possible to see the classification of documents results using a previously trained SVM model.

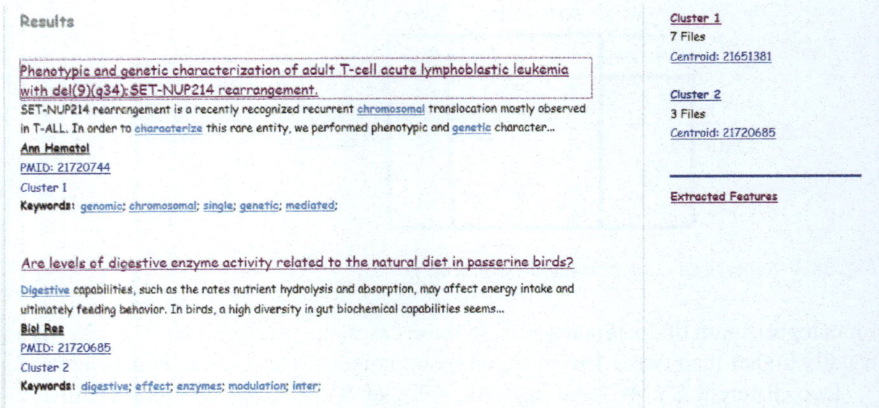

Fig. 38.3 Web platform – results for "aminopeptodase n" query using clustering

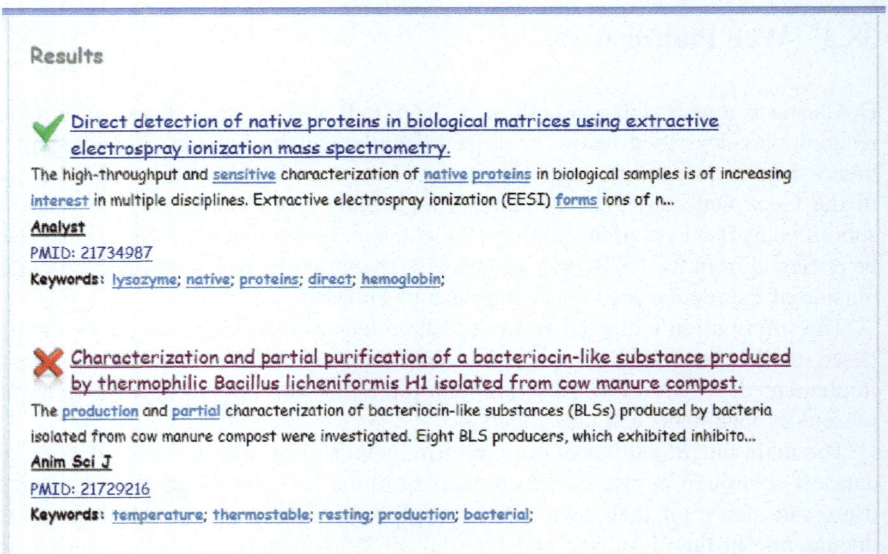

Fig. 38.4 Web platform – Example of classification using SVMs

38.4 Results

Considering the focus of this work in the field of peptidases, the literature datasets used were extracted from Merops database [15]. For the assessment of the classifiers performance, Table 38.1 shows the dataset that includes six classes grouped according to the Merops taxonomy.

Table 38.1 Applied dataset

Classes	Merops ID	# documents
Aminopeptidase N – (1)	M01.001	118
Calpain 2 – (2)	C02.002	163
Chymotrypsin A – (3)	S01.001	96
Angiotensin-converting enzyme – (4)	A02.001	175
Papain – (5)	C01.001	113
Preselinin – (6)	A22.001	139
Total		804

Table 38.2 K-Means without ontology

		Predicted									
Cluster	Assignment	(1)	(2)	(3)	(4)	(5)	(6)	TP	FP	# documents	Precision
1	A02.001	0	0	0	162	0	0	162	0	162	100%
2	M01.001	101	1	0	0	0	1	101	2	103	98.06%
3	A22.001	0	0	0	0	0	103	103	0	103	100%
4	C01.001	17	25	96	13	113	12	113	163	276	40.94%
5	C02.002	0	137	0	0	0	0	137	0	137	100%
6	A22.001	0	0	0	0	0	23	23	0	23	100%
Number of features		8,820									
Quality		79.48%									

The features extraction process used the parameters *SimpleTokenizer*, *Dummy-Stemmer*, *TFIDF*, without Ngrams pruning or minimum size of terms.

The results obtained using K-means without ontology are presented in Table 38.2. The table includes the identified cluster, number of True Positives (TP), False Positives, Precision and Quality. Each cell *(i,j)* corresponds to the number of documents from peptidase *j* allocated to cluster *i*. The "Designation" column gives the ID of the peptidase's documents most present in the cluster.

Precision and Quality are defined by:

$$Precision\,(Cluster\;i) = \frac{TP\,i}{Total\;number\;of\;documents\;i} * 100$$

$$Quality = \frac{TP}{Total\;number\;of\;documents} * 100$$

The results obtained using K-means with ontology are exhibited in Table 38.3. Table 38.4 shows the results obtained with linear SVM and multiclass SVM using the LibSVM library version 3.1 [16]. The following metrics have been used:

$$ACC\,(accuracy) = \frac{TP + TN}{Total\;de\;documentos} * 100\%$$

Table 38.3 K-Means with ontology

		Predicted									
Cluster	Assignment	(1)	(2)	(3)	(4)	(5)	(6)	TP	FP	# documents	Precision
1	S01.001	0	1	89	0	0	0	89	1	90	98.88%
2	M01.001	117	21	7	5	4	60	117	97	214	54.67%
3	C01.001	0	0	0	0	109	0	109	0	109	100%
4	A02.001	0	2	0	170	0	1	170	3	173	98.27%
5	C02.002	0	139	0	0	0	0	139	0	139	100%
6	A22.001	1	0	0	0	0	78	78	1	79	98.73%
Number of features		1,029									
Quality		87.31%									

Table 38.4 Global results

	Nº Features	Accuracy	Recall	Specificity	F-Measure
	SVM with linear kernel, 70% document for training and C = 1				
Without ontology	8,820	91.84%	90.26%	93.42%	91.75%
With ontology	1,029	93.05%	92.84%	93.25%	92.995%
	SVM with Gaussian kernel using cross-validation (five folds) C = 1,000 and gamma = 0.00237 (best parameters without ontology) C = 10 and gamma = 0.03162 (best parameters with ontology)				
Without ontology	8,820	91.668%	90.077%	98.287%	90.573%
With ontology	1,029	98.259%	98.195%	98.654%	98.148%

$$SE\ (sensitivity) = \frac{TP}{TP + FN} * 100\%$$

$$SP\ (specificity) = \frac{TN}{TN + FP} * 100\%$$

$$Precision = \frac{TP}{TP + FP} * 100\%$$

$$Recall = sensitivity$$

$$FM\ (F - Measure) = 2 * \frac{Precision * Recall}{Precision + Recall}$$

The Tables 38.5 and 38.6 show the detailed results of linear SVMs using the "one-against-all" method.

Table 38.5 Linear SVM without GO ontology

Peptidases	TP	TN	FP	FN	SE (%)	SP (%)	ACC (%)
M01.001	30	32	5	3	91.43	85.71	88.57
C02.002	43	41	5	7	85.42	89.58	87.5
S01.001	28	26	0	2	92.86	100	96.43
A02.001	50	50	2	2	96.15	96.15	96.15
C01.001	31	29	2	4	87.88	93.94	90.91
A22.001	39	36	2	5	87.8	95.12	91.46
Number of features		8,820					
Average accuracy		91.84%					

Table 38.6 Linear SVM with GO ontology

Peptidases	TP	TN	FP	FN	SE (%)	SP (%)	ACC (%)
M01.001	32	32	3	3	91.43	91.43	91.43
C02.002	45	44	3	4	91.67	93.75	92.71
S01.001	27	26	1	2	92.86	96.43	94.64
A02.001	50	51	2	1	98.08	96.15	97.12
C01.001	31	29	2	4	87.88	93.94	90.91
A22.001	36	39	5	2	95.12	87.8	91.46
Number of features		1,029					
Average accuracy		93.05%					

38.5 Conclusions

In this work, a web platform for document search and classification has been implemented using ontologies, clustering and support vector machines. The use of the gene ontology provided an important improvement to the algorithms performance, both in clustering or supervised classification.

The platform consists of an ongoing work and additional features like the inclusion of other classifiers and ontologies are being done. In the case of the clustering algorithms, the k-means forces the definition of the number of clusters, which is difficult to set, so other clustering techniques like subtractive clustering will be implemented in the near future.

In terms of domain knowledge the inclusion of other ontologies will be considered, since with the current feature extraction process, some non-relevant terms are still used. In addition, new preprocessing filters will be tested.

Acknowledgments This work was supported by FCT (Foundation for Science and Technology) and FEDER through Program COMPETE (QREN) executed under the project FCOMP-01-0124-FEDER-010160 (PTDC/EIA/71770/2006), designated BIOINK - Incremental Kernel Learning for Biological Data Analysis. The platform used services from third parties to whom we thank and reference.

References

1. NCBI Pubmed Database (2011) http://www.ncbi.nlm.nih.gov/pubmed/
2. Correia D, Campos D, Pereira C, Verissimo P, Dourado A (2010) A platform for intelligent search and classification of biomedical literature. Paper presented at the workshop on applications of computational intelligence, WACI'10 - Workshop Application of Computacional Intelligence, Coimbra
3. Ashburner M, Ball CA, Blake JA, Botstein D, Butler H, Cherry JM, Davis AP, Dolinski K, Dwight SS, Eppig JT, Harris MA, Hill DP, Issel-Tarver L, Kasarskis A, Lewis S, Matese JC, Richardson JE, Ringwald M, Rubin GM, Sherlock G (2000) Gene ontology: tool for the unification of biology. The gene ontology consortium. Nat Genet 25(1):25–29, citeulike-article-id:212874
4. Lu Z (2011) PubMed and beyond: a survey of web tools for searching biomedical literature. Database 2011. doi:10.1093/database/baq036
5. Doms A, Schroeder M (2005) GoPubMed: exploring PubMed with the gene ontology. Nucleic Acids Res 33(2):W783–W786. doi:10.1093/nar/gki470
6. Glenisson P, Coessens B, Van Vooren S, Mathys J, Moreau Y, De Moor B (2004) TXTGate: profiling gene groups with text-based information. Genome Biol 5(6):1–12. doi:10.1186/gb-2004-5-6-r43
7. Delfs R, Doms A, Kozlenkov A, Schroeder M (2004) GoPubMed: ontology-based literature search applied to Gene Ontology and PubMed. In: German Conference on Bioinformatics, pp 169–178
8. Lloyd SP (1982) Least squares quantization in PCM. IEEE Trans Inf Theory 28(2):129–137. doi:10.1109/TIT.1982.1056489
9. Cortes C, Vapnik V (1995) Support-vector networks. Mach Learn 20(3):273–297. doi:10.1007/bf00994018
10. Vapnik VN (1995) The nature of statistical learning theory. Springer, New York
11. Joachims T (2002) Learning to classify text using support vector machines: methods, theory and algorithms. Kluwer Academic Publishers, Norwell
12. NCBI Entrez Programming Utilities (2011) http://www.ncbi.nlm.nih.gov/entrez/query/static/eutils_help.html
13. Wurst M (2007) The word vector tool-user guide/operator reference/developer tutorial. Can be found in here: http://ftp.heanet.ie/disk1/sourceforge/r/project/ra/rapidminer/2.%20Text%20Plugin/4.0beta/rapidminer-wvtool-4.0beta-tutorial.pdf
14. Carbon S, Ireland A, Mungall CJ, Shu S, Marshall B, Lewis S, Hub tA, Group tWPW (2009) AmiGO: online access to ontology and annotation data. Bioinformatics 25(2):288–289. doi:10.1093/bioinformatics/btn615
15. Rawlings ND, Barrett AJ, Bateman A (2010) MEROPS: the peptidase database. Nucleic Acids Res 38(Database issue):D227–233. doi:10.1093/nar/gkp971
16. Chang C-C, Lin C-J (2011) LIBSVM: a library for support vector machines. ACM Trans Intell Syst Technol 2(3):1–27. doi:10.1145/1961189.1961199

Chapter 39
Conflicts Management in Retail Systems with Self-Regulation

Bruno Magalhães and Ana Madureira

Abstract Self-Regulation and retail problem solving using Multi-Agent Systems are considered two promising areas but relatively little explored. The regulation in these environments must be able to handle with the implicit dynamism and variation in a complex area such as retail systems with a minimum human interference. Agents must be able to change their behaviour based on rules previously formatted, using an autonomic process that does the maintenance of a knowledge base which is well defined and consistent, to be possible to meet all the objectives and to provide support on the need of change.

39.1 Introduction

With the increasing competition as result of globalization of markets, comes the need for continuing innovation in products, services and retail processes. Retail systems are central to this process, aimed to reducing costs, optimizing processes, control of the business, strengthening the policy of quality customer service, reducing delivery times, the ability to answer to external requests, among others. Retail systems must be able to handle dynamic nature of demand and market performance. Increased competitiveness is one of the responsible factors for increased research and development of new approaches to support these systems, since the evolution required in terms of features is debated constantly, to the needs implicit in business practices.

B. Magalhães (✉)
School of Engineering – Polytechnic of Porto (ISEP/IPP), Porto, Portugal
e-mail: 1040272@isep.ipp.pt

A. Madureira
GECAD – Knowledge Engineering and Decision Support Research Center,
School of Engineering – Polytechnic of Porto (ISEP/IPP), Porto, Portugal
e-mail: amd@isep.ipp.pt

A. Madureira et al. (eds.), *Computational Intelligence and Decision Making: Trends and Applications*, Intelligent Systems, Control and Automation: Science and Engineering 61,
DOI 10.1007/978-94-007-4722-7_39, © Springer Science+Business Media Dordrecht 2013

The main objective of this work is the analysis and implementation of a mechanism-based Self-Regulation in a Multi-agent autonomic system, providing support in decision, making on complex issues in a retail environment, based on knowledge previously collected and represented – Self-Regulation Retail System (SelfRetail). Thus, the main contributions of this work are:

- Applicability and suitability to the paradigms of Autonomic Computing and Multi-Agent Systems to the context of the characteristics needed to solve the problem;
- Proposal for modelling and specification of a system with Self-Regulation in conflict management in retail;
- The formulation of a model to represent domain knowledge;
- Standardize the mechanisms to support implementation of components characterized by paradigms addressed through modules that can be adaptable to different activities or knowledge;
- Study of performance and system support required by the restrictions in the retail business area;
- Test and simulate the system, making some comparisons with aspects of the functionalities offered in today's retail system.

The work reported in this paper is concerned with the resolution of Retail challenges taking advantages from Multi-agent Systems and Autonomic Computing paradigm.

39.2 Self-Regulation and Artificial Intelligence Paradigms

Self-Regulation can be defined as the process by which systems tend to reach a particular objective with no external interference. All the mechanisms dictating its behaviour is internal to the system e.g. are autonomous. This field of research has received much attention through Autonomic Computing paradigm [1–5].

The operation of systems is one of the key aspects in the success of a business area such as Retail, which has systematically increased competition. The formulation of a permanent operating system in terms of functionality and operations support are no longer seen as a tool of storage/data management, making it a set of practices or procedures, to be achieved objectives and benefits from the experience of the system. One of the singular aspects between the different retailers in terms of success is the ability to adjust and analyse real data captured by its systems, which represents the state of the business as well as the performance of services and strategies in place. Due to these facts, there is an obligation on the existence of quality assurance of data collection, it is necessary to avoid problems that happens in the system caused by different factors like the level of complexity implicit in a retail environment, that can compromise the operability in business due the loss of efficiency and ownership in decision making.

Autonomic Computing (AC) is an IBM Grand Challenge proposed in 2001 by Paul Horn, Senior Vice-President of IBM Research. Horn argues that the IT industry's focus on constant expansion will soon reach its breaking point: massive data centre's are built in organic, ad hoc ways, resulting in a heterogeneous composition whose maintenance costs in terms of qualified staff, time and capital will soon exceed corporate capabilities. AC proposes a broad new field of research related to the automation of IT management processes, drawing inspiration from the human autonomous nervous system. From its inception, the concept revolves around four self-* properties, in which research efforts may be categorized: Self-Configuring, Self-Healing, Self-Optimizing and Self-Protecting. Although the names of these properties are fairly self-explanatory, there is one inherent and implicit concept of significant importance: pro-activeness. This is what separates this area of research from some of the functionalities which are already being integrated with existing software systems. Software systems managing IT resources without human supervision, called Autonomic Managers, are expected to continuously and autonomously respond to changes, and continuously seek ways to improve efficiency or counter negative environment changes. Many studies have already been made around this area. These range from Software Engineering concerns to address this new development paradigm [1, 2] all the way down to industry integration [6, 7]. Important techniques have been tapped into from areas such as Service-Oriented Architectures [7–107], Multi-Agent Systems [10], Grid Computing [2, 11] and Control Theory [9, 129].

39.3 SelfRetail System Prototype

During the last two decades efforts have been directed on the developing of intelligent systems which applicability would help to solve these problems inherently distributed. Self-regulation is within the context of the specification of architectures or autonomic systems capable of addressing the change, recover from conflicts and integrate the solution in the new socio-economic and operational context effectively and efficiently. It is intended to review and analyse the Self-Regulation in Conflict Management in Retail and subsequently the definition and implementation of mechanisms that provides to the retailer loyalty in terms of data for their decisions and strategy building. The system SelfRetail (Self-Regulation Retail System) is a Multi-Agent system providing autonomous capacity of Self-Regulation. The purpose of the system is the implementation of mechanisms to control the retail environment with support for the decision. This support is provided based on various agents with different characteristics, which in its autonomous operation can generate a predefined amount of business rules that can be interpreted and enforced. The rules are generated based on an ontological scheme of knowledge representation, something that allows for easy maintenance and modification in order to have greater adaptability and robustness on the need of change. This research comes from the need to provide retail systems with the existing capacity to react and adapt to changes or amendments, with minimal human intervention, allowing the retailer to focus on higher value added activities.

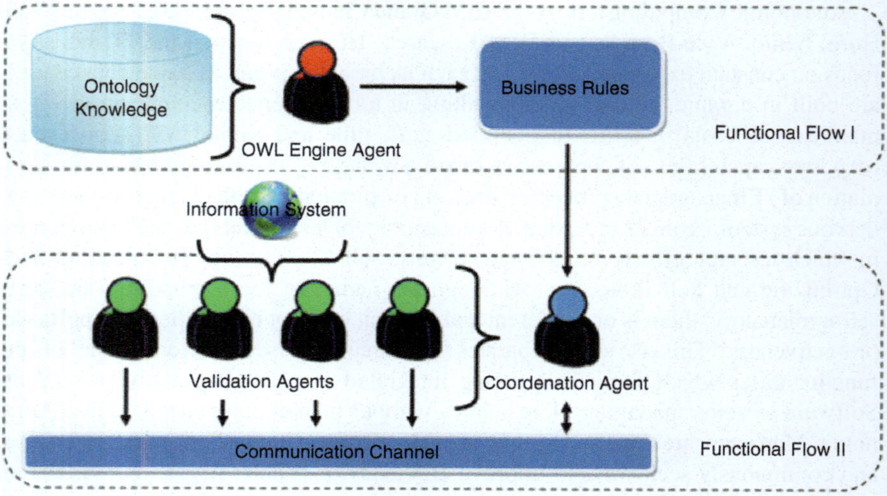

Fig. 39.1 Agents organization model

39.3.1 SelfRetail Model

The implemented system assumes architecture based on multi-agent systems with autonomic an mechanism that allows adaptability to any unforeseen changes in the process of regulating the business of the retailer based on a core of knowledge represented by schemas. The SelfRetail system is based on three different types of agents.

As shown in Fig. 39.1, it is possible to identify the constituent agents of the architecture. The OWL Engine Agent is an agent that operates isolated from the other agents (Model with a single agent), appearing in Flow I (Maintenance and declaration of a business model). The input in terms of information to the processing of the agent after detection of an event is primarily obtained from the ontology knowledge. The information resulting from this process is a set of business rules, plus a set of categorizations of the same rules, so there is a control practices in performance by the retailer. The Coordination Agent and Validation Agent are the elements of Federation Model, with the dependence of the second according to the first. These agents are part of Flow II (Using the model to support the process of decision making). The Coordination Agent controls the entire process of validation. Validation Agents are the resources in the system for processing a particular package of events and rules. The connection between the Federation and the single model are the generated business rules that are set to a repository. The Coordination Agent, who controls all components of the Federation model, has the function to capture these rules and prepare them for being executed in a particular group of events.

39.3.2 SelfRetail System Architecture

The proposed system assumes an architecture based on Multi-Agent Systems with integrated mechanisms of autonomous self-regulation that will provide the adaptability and robustness of regulating the retail business, with a core of knowledge represented by schemes or structures (Fig. 39.2).

The computational agents that should be part of the system are dependent on the features associated with the defined requirements. There is an agent named OWL Engine Agent which will be responsible for all operations on the ontology that holds the knowledge. Another agent known as Coordination Agent will have the responsibility for coordinating and managing requests for validation of business rules published in a dynamic repository. This agent will be the connection point with the Validation Agent. The Validation Agent is replicated itself into groups of N resources and has the function to receive and process fragmented requests on a previously generated rule.

The way of interaction of the different agents and components in the architecture will differ on their functions and characteristics. Typically, the autonomous elements of the architecture should be part of a community of agents where each has its particular behaviour. In a global way, all the agents should operate in a synchronized manner to the achievement of the basic purposes of Self-Regulation.

In terms of functional flows, the system can be separated into two with different specific aims. Note that although this division, these flows are part of the mechanism of self-regulation because of their interconnection and mutual dependence. The two flows are:

- Flow 1 – Maintenance and declaration of a business model.
- Flow 2 – Using the model to support decision making.

The first flow concerns with the creation and maintenance of the knowledge base to provide the functionality throughout the system. In the knowledge base is stored all the information that will be required to support the validation of the constraints on the business. The process should consist of multiple components to share the most necessary activities around the maintenance mechanism of knowledge. To access the core knowledge, there should be a layer of business services that provide a predefined set of methods and functions to manipulate the information. The services layer will provide consistency and control for the concurrency in terms of read and write.

In terms of the structure for knowledge representation, this is one of the most important aspects due to its relation with the entire context of the system and its components. The layout of the structure is based in an ontology which will store the logical knowledge used to describe each business rule used. As result, the decision for this aspect will provide flexibility in terms of knowledge handling as an ontology is a well-defined structure for representation and it provides mechanisms for necessary changes. The ontology will be responsible for maintain a structure of the information scope intended to be approached and it is expected to have an

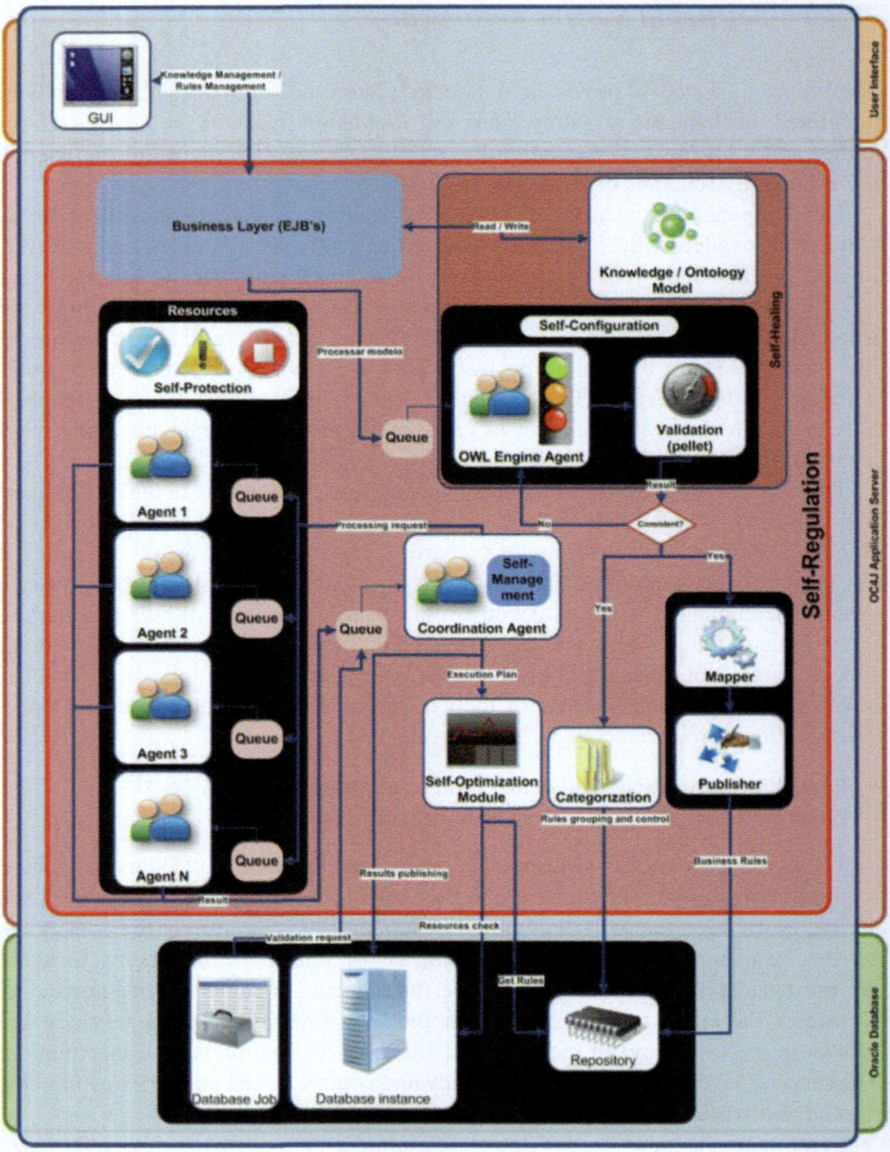

Fig. 39.2 SelfRetail architecture

intermediary agent (computational agent) for the validation and application of knowledge in that scope. It is projected to store information about the knowledge captured so that the business rules could be generated. These business rules should be the result of the mapping between specific codes with the ontology knowledge composed of properties, classes and restrictions.

The second point as a flow of the system operation is related with the application of business rules. The connection between the first point (maintenance and declaration of a business model) and the second (use the model to support in the decision making) is held indirectly through business rules created by the intermediary (OWL Engine Agent). These rules are published to a database repository each time the ontology is changed (the change of an ontology is the modelling of knowledge). The repository will be used as a place to share rules with constant parameterization and that can be accessed by different entities.

In this context, the second flow is triggered whenever exists a request for validation of the environment according to the rules published on the repository. The request is routed to a computational agent that manages all the requests for processing (Validation Agent). The main function of this agent is to ensure the planning and management about the life cycle of the request, i.e., it must requests resources (Validation Agents) for processing and forward them the request fragments by whether they are available. The rules are established to be distributed on each package of events (fragments) so that each Validation Agent could perform the validation. Communication is done by a message containing the rule that must be validated. These agents should also be able to answer questions made by the coordinator so that it can be aware about the status of the processing required at the time.

At the end of processing, each resource sends the result to the Coordinator Agent so he can join all the information in order to inform the system (via the database) that the request was fully processed. As result of the interconnection of all the features and behaviours presented in this overview, the reference to the need of a Self-Regulation component for the retail system is now more understandable. This component allows the definition of groups of processes that must be executed for a given environment under control. The component of Self-Regulation does not refer to a specific feature or mechanism, but rather a set of flows that work together in order to guide business in the direction planned by the retailer. By making adjustments/adaptation of the generic business rules, there will be an adjustment of the plan on the change, which will have an impact on the system behaviour since it acts based on the knowledge it has, something that provides the retailer the mandatory alignment of the information systems with the business.

39.4 Representation and Interpretation of the Ontology Knowledge

The knowledge stored in the ontology is the basis for the entire process of regulating by the SelfRetail system. The purpose of the representation of information through a knowledge structure allows the system to participate in supporting the decisions of the retailer. This factor provides abstraction to the retailer in some important restrictions that must be validated; otherwise it could compromise all the business decisions. It is not supposed to have a complete and evolved ontology with the representation of all the existing rules. The evolution in the representation of knowledge will emerge over time, with the experience and the on-going study of how to represent it.

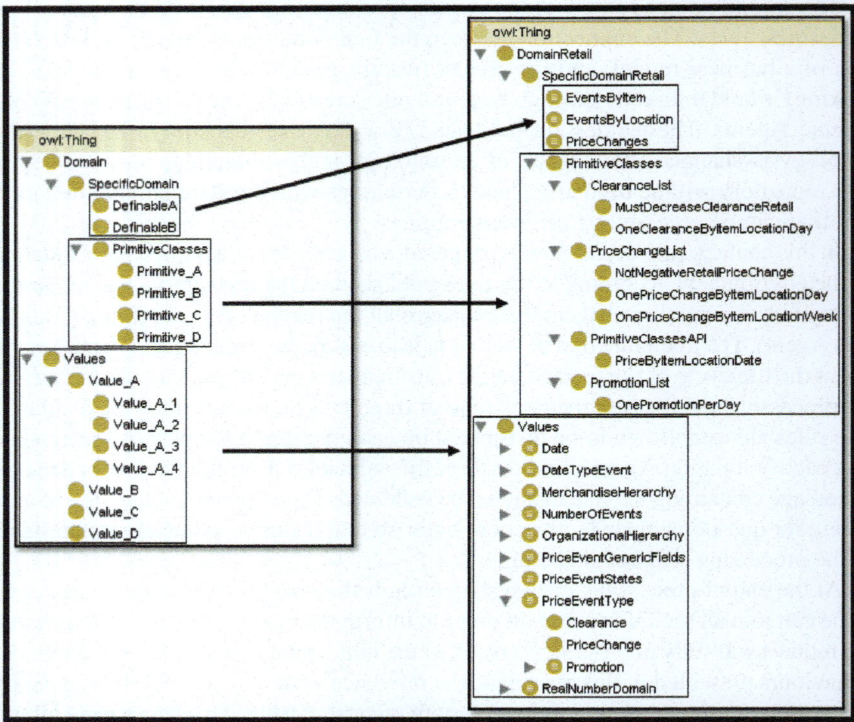

Fig. 39.3 Ontological structure

39.4.1 Base Structure of the Ontology

The developed system is implemented to support a specific ontological structure. This structure must be conventional to a predefined scheme presented in Fig. 39.3

The fact that the system is prepared for the scheme presented does not invalidate the possibility of relying on other schemes. In Fig. 39.3 is represented the knowledge representation scheme used for the business area addressed in this project and system SelfRetail.

39.4.2 Interpretation with Classification Logic

The classification logic used in the implementation of the ontology is based initially on the description of a definable number of classes or primitive through the establishment of relations between properties and values. The final result is a set of objects with differences, which identify distinct entities and that can be validated with the use of inference engines. It is required a set of properties for the characterization of entities covered by each value. These properties are part of the created ontology to support the knowledge representation in the system SelfRetail.

The defined properties are part of the ontology context and do not have to be used for characterize all values for all entities. By default, the relationship is established at the higher level, thus allowing a situation of no use for specific characterization. The relation is pointing to the general values area, so it is not took into account on the process of mapping.

39.5 Conclusions

The proposed solution is based on multiple computational paradigms which provide an important contribution to the evolved designed architecture. The system has capabilities to solve the different problems identified on the analysis phase for the Oracle Retail. With some more time in terms of operation, it will guarantee the complete validation of the most important behaviours on the retailer information system in terms of conflict management resulting from business rules.

As a whole, the system provides the Self-Regulation of the environment where different computational agents have different tasks to perform according to multiple events triggered by various activities. The final result of each process has implicitly the idea of cooperation between all the entities to control the retailer information system using knowledge. All this aspects are part of the group that support the Retailer Decision Making.

Acknowledgments This work is supported by FEDER Funds through the "Programa Operacional Factores de Competitividade – COMPETE" program and by National Funds through FCT "Fundação para a Ciência e a Tecnologia" under the project: FCOMP-01-0124-FEDER-PEst-OE/EEI/UI0760/2011.

References

1. EMA (2006) Practical autonomic computing: roadmap to self managing technology. A white paper prepared for IBM. Enterprise Management Associates (EMA)
2. Ganek A (2006) Overview of autonomic computing: origins, evolution, direction. In: Hariri S, Parashar M (eds) Autonomic computing – concepts, infrastructure, and applications. CRC Press, Boca Raton, pp 3–18
3. Horn P (2001) Autonomic computing: IBM's perspective on the state of information technology. IBM Research. Retrieved on October 2011 from http://www.research.ibm.com/autonomic/manifesto/autonomic_computing.pdf
4. IBM (2006) Autonomic computing white paper. An architectural blueprint for autonomic computing. IBM Research Retrieved on October 2011 from http://people.cs.kuleuven.be/~danny.weyns/csds/IBM06.pdf

5. IBM (2005) Autonomic computing manageable resource interface specification and touchpoint implementation guide, version 0.51
6. Parashar M (2006) Autonomic grid computing: concepts, requirements, and infrastructure. In: Hariri S, Parashar M (eds) Autonomic computing – concepts, infrastructure, and applications. CRC Press, Boca Raton, pp 49–70
7. Schwan K, Cooper B, Eisenhauer G, Gavrilovska A, Wolf M, Abbasi H, Agarwala S, Cai Z, Kumar V, Lofstead J, Mansour M, Seshasayee B, Widener P (2006) AutoFlow: autonomic information flows for critical information systems. In: Salim H, Manish P (eds) Autonomic computing – concepts, infrastructure, and applications. CRC Press, Boca Raton, pp 275–303
8. Rafaeli S, Adams R, Milojicic D, Iyer S, Brett P, Talwar V (2006) Scalable management – Technologies for management of large-scale, distributed systems. In: Salim H, Manish P (eds) Autonomic computing – concepts, infrastructure, and applications. CRC Press, Boca Raton, pp 305–328
9. Bhat V, Parashar M, Kandasamy N (2006) Autonomic data streaming for high-performance scientific applications. In: Hariri S, Parashar M (eds) Autonomic computing – concepts, infrastructure, and applications. CRC Press, Boca Raton, pp 413–433
10. Chess D, Hanson J, Kephart J, Whalley I, White S (2006) Dynamic collaboration in autonomic computing. In: Salim H, Manish P (eds) Autonomic computing – concepts, infrastructure, and applications. CRC Press, Boca Raton, pp 253–274
11. Kephart J, Chess D (2003) The vision of autonomic computing. Computer magazine. IEEE Comput Soc 36(1):41–50
12. Abdelwahed S, Kandasamy N (2006) A control-based approach to autonomic performance management in computing systems. In: Hariri S, Parashar M (eds) Autonomic computing – concepts, infrastructure, and applications. CRC Press, Boca Raton, pp 149–167
13. Bezek A, Gams M, Bratko I (2006) – Multi-agent strategic modeling in a robotic soccer domain. In: Proceedings of the fifth international joint conference on autonomous agents and multi-agent systems, Japan
14. Bordini R, Braubach L, Dastani M, Seghrouchni A, Gomez-Sanz J, Leite J, O'Hare G, Pokahr A, Ricci A (2006) A survey of programming languages and platforms for multi-agent systems. Informatica 30(1):33–34
15. Briot JP, Meurisse T, Peschanski F (2006) Architectural design of component-based agents: a behavior-based approach. PROMAS 2006:71–90
16. Burgard W, Moors M, Stachniss C, Schneider F (2005) Coordinated multi-robot exploration. IEEE Trans Robot 21:376–386
17. Breitman K, Leite J (2004) Ontology as a requirements engineering product. J Auton Agents Multiagent Syst 8(3):203–236
18. Cervenka R, Trencansky I, Calisti M, Greenwood D (2004) AML: agent modeling language. Toward industry-grade agent-based modeling, Published In: (Lecture notes in computer science), vol 3382/2005: Agent-Oriented Software Engineering V: 5th international workshop, AOSE 2004, New York, NY, USA, 19 July
19. Dalheimer M, Pfreundt FJ, Merz P (2005) Agent-based grid scheduling with Calana. In: Proceedings of parallel proceedings and applied mathematics (PPAM 2005), Poznan. Poland, 11–14 Sept. Published in Lecture Notes in Computer Science, vol 3911. pp 741–750
20. Dorf D (2010) The new rules of retail. Oracle. An Oracle white paper. Retrieved on October 2011 from http://www.oracle.com/us/industries/retail/oracle-retail-new-rules-wp-080053.pdf
21. Drummond N, Horridge M (2005) A practical introduction to ontologies & OWL. The University of Manchester
22. Franklin S, Graesser A (1996) Is it an agent or just a program?: a taxonomy for autonomous agents. In: Intelligent agents III. Agent theories, architectures and languages (ATAL'96). Berlin, Germany
23. Gualtieri A, Ruffolo M (2005) An ontology-based framework for representing organizational knowledge. In: Proceedings of I-KNOW'05 Graz, Austria, 29 June–1 July

24. Guarino N, Poli R (1995) Formal ontology, conceptual analysis and knowledge representation. Int J Human Comput Stud – Special issue: the role of formal ontology in the information technology archive. Guarino N, Poli R. (eds.): 43(5–6):625–640. Academic Press, Inc
25. Gruber TR (1993) A translation approach to portable ontology specifications. Knowledge Systems Laboratory. Computer Science Department Stanford University Stanford
26. Guarino N (1997) Semantic matching: formal ontological distinctions for information organization, extraction, and integration. In: proceedings of International Summer School on Information Extraction (SCIE '97): a multidisciplinary approach to an emerging information technology. Lecture Notes in Artificial Intelligence. Springer, Berlin, pp 139–170
27. Guarino N (1999) The role of identity conditions in ontology design. Spatial information theory cognitive and computational foundations of geographic information science. In: Proceedings of international conference COSIT'99, Stade, Germany, 25–29 Aug 1999. Lecture Notes in Computer Science, vol 1661. Springer, Berlin, pp 221–234
28. Horling B, Lesser V (2004) A survey of multi-agent organizational paradigms. Knowl Eng Rev 19(4):281–316
29. Liu H, Parashar M (2006) A programming system for autonomic self-managing applications. In: Salim H, Manish P (eds) Autonomic computing – concepts, infrastructure, and applications. CRC Press, Boca Raton, pp 211–215
30. Luck M, McBurney P, Shehory O, Willmott S (2005). Agent technology: computing as interaction (A Roadmap for Agent Based Computing). AgentLink
31. Madureira A (2006). Sistemas de Gestão de Processos de Negócio, Lesson presented on academic evaluation proofs for Coordinator Professor. ISEP (In Portuguese)
32. Oracle (2009a) Oracle retail merchandising – implementation guide, Release 12.0.3. Retrieved on October 2011 from http://docs.oracle.com/cd/E05442_01/merchandising/pdf/1203/merch-1203-impg.pdf
33. Oracle (2009b) Oracle retail price management – Operations guide, Release 13.0 Retrieved on October 2011 from http://docs.oracle.com/cd/E12440_01/rpm/pdf/130/rpm-130-og.pdf
34. Oracle (2007a) Optimize planning and merchandising decisions. Oracle. Retrieved on October 2011 from http://www.oracle.com/us/industries/retail/046280.pdf
35. Oracle (2007b) Managing at the speed of reality. EMEA, Oracle Retail. Retrieved on October 2011 from http://www.supplychainstandard.com/liChannelID/5/Articles/1902/Managing+at+the+speed+of+reality.html
36. Panait L, Luke S (2005) Cooperative multi-agent learning: the state of the art. J Auton Agents Multi-Agent Syst 11(3):387–434
37. Pinheiro PJ (2006) Pricing strategies. Technical Report Enabler/Wipro retail
38. Ramos C (2002) Agentes inteligentes e sistemas cooperativos. ISEP. Porto (Portugal) (In Portuguese)
39. Sweitzer J, Draper C (2006) Architecture overview for autonomic computing. In: Hariri S, Parashar M (eds) Autonomic computing – concepts, infrastructure, and applications. CRC Press, Boca Raton, pp 71–98

Chapter 40
Adaptive e-Learning Systems Foundational Issues of the ADAPT Project

Eduardo Pratas and Viriato M. Marques

Abstract This paper presents some foundational issues for the design and implementation of adaptive e-learning systems and in particular their application to the ADAPT project. As a matter of fact, the current Learning Management Systems (LMS's) lack pedagogy and interactivity, giving rise to e-learning models that strongly reside on student's own motivation. Some Artificial Intelligence (AI) models and techniques can help to overcome this problem and turn future LMS's into (almost) human teachers.

40.1 Introduction

Human teachers maintain a direct contact with the students. This allows the evaluation of their performance and learning profile, answer to their doubts and questions, follow their performance, fit examples and give general orientations to each one. In this context e-learning systems pose some problems such as allowing less more than creating e-learning course contents accessible by an Internet site.

AI techniques – in particular Fuzzy Systems (FS), Case-based Reasoning (CBR), Page Rank (PR) and Link-Mining (LM) – can help to overcome these problems. By combining the output of learning-style questionnaires, learning preferences and previous course contents classification, FS can be used to infer and vaguely describe the way that subjects should be presented to each student; CBR can mimic a teacher's experience by storing past stories of success and failure to use or avoid in the future; PR can be used to find examples and complementary documents that proved to be relevant in previous courses; LM may help to understand or detect changes in each student behavior. Together, these techniques can be used as part of

E. Pratas (✉) • V.M. Marques (✉)
ISEC – Instituto Superior de Engenharia de Coimbra, IPC – Polytechnic Institute of Coimbra, Coimbra, Portugal
e-mail: eduardo.pratas@gmail.com; viriato@isec.pt

A. Madureira et al. (eds.), *Computational Intelligence and Decision Making: Trends and Applications*, Intelligent Systems, Control and Automation: Science and Engineering 61,
DOI 10.1007/978-94-007-4722-7_40,

an Intelligent Tutoring System (ITS), capable of replacing a human teacher, to some extent, in a LMS environment.

Next sections are organized as follows: Sect. 40.2 describes the general aspects of an ITS; Sect. 40.3 introduces the CBR paradigm and its use for managing and adapting course story boards; Sect. 40.4 introduces learning styles and describes the learning preferences concept; Sect. 40.5 focuses on contents classification; Sect. 40.6 describes fuzzy systems, the Mamdani inference and its application for mapping learning styles and learning preferences into course contents presentation; Sect. 40.7 briefly describes PR and LM as a complementary tool for document retrieval and student performance analysis. Section 40.8 concludes the paper and presents some future work.

40.2 Intelligent Tutoring Systems

ITS's are computer applications based on AI techniques that try to capture the profile and knowledge of the student and identify the causes of his errors. ITS's try to select adequate contents, examples and problems whose presentation can include multimedia and simulation techniques. They try to replace the human teacher and create an action sequence that simulates an educational attitude. For example, a study from Carnegie Mellon University revealed that, in a test of computer programming, the students that learned using ITS obtained higher classifications than the other ones, also solving complex problems in a shorter time. A comparative study and some prototypes can be found in [1]. The architecture of an ITS consists of four modules:

- The Domain Module stores the course contents and data, as well as problem solving strategies represented, for instance, by rules, scripts or templates that contain correct sequence actions. This module stores the domain knowledge;
- The Student Module evaluates the student acquired knowledge, his difficulties and reasoning capacities. It compares his actions and options with those described by the Domain Module. An ITS is capable of identifying simple problems such as the carry from the most significant bit in an addition operation, or complex ones such as a procedure error in an action sequence for the setup of cruiser guns [2];
- The Tutor Module decides what approach should be followed in order to present a given subject. For example, for a beginner it may choose to show some solved problems before presenting an unsolved one; next questions could be of increasing complexity; or it may choose to let the student freely explore a given subject. This module stores the pedagogy;
- The User Interface can range from simple dialog boxes to voice recognition. Simulation and virtual reality techniques can provide an approximate real environment.

In what follows, CBR and PR are directly linked to the Tutor Module, i.e. pedagogy; LM is more directly related to the Student Module; Taxonomies are used in the Domain Module. In a LMS context, this module, together with the User Interface Module, should include interactive learning objects to motivate the student activity.

40.3 Case-based Reasoning and Story Boards

40.3.1 Case-Based Reasoning

Case-based Reasoning is a paradigm that is based on cases that represent past occurrences. Each case consists of a description and its associated solution. The resolution of a problem consists of finding a past case similar to the current case and presents its solution directly or by adapting the past case to the new context. The operation of a CBR system is well depicted by the four phase cycle of Aamodt and Plaza [3]:

1. *Retrieve*: find the case or cases that are relevant to solve the present case;
2. *Reuse*: use the information that the case has to solve the current problem. The solution may be the same or have to be adapted;
3. *Review*: test if the proposed solution is valid. If invalid, it performs its correction or gives the system the correct solution (that it may not know yet);
4. *Retain*: save the knowledge contained in part or the totality of the present case, if it contains relevant information.

In the ADAPT platform, a case represents the path – sequence of pages – accessed by each student along a course. The platform makes use of the CBR solutions to determine the presentation sequence and the content type that is going to be viewed by each (new) student.

40.3.2 Story Boards

The path followed by each student (*story board*), containing the accessed pages with all the possible navigation links, is represented by a graph structure. The nodes of these graph structures point to the pages that contain the subjects of a given course. However, each subject may have various presentation forms, each one more *adapted* to the learning profile of each student. The graph edges represent links to other contents that are accessible from the present node.

Each student begins his course by filling a form whose main goal is the determination of his learning-style. If there is a previously known story board (case) that matches the student profile, this story board is used. For system startup, some cases are created by the teacher, containing all the study material. Over time, these graph structures can change according to the performance achieved by the students that use the platform. This means that an individual graph structure is created for each student.

40.4 Learning Styles and Learning Preferences

Over the years various models of learning styles have been proposed to determine the learning preferences of a student. Each model classifies the various preferences of an individual, representing them on a scale. These preferences may be related, for example, to the way students prefer to receive information or how a student prefers to process information. There are different learning style models such as [4]:

- Models based on personality (Myers-Briggs Type Indicator) [5];
- Models based on information processing (Schmeck's [6] and Kolb´s [7]);
- Models based on Social Interaction (Reichmann & Grasha [8]);
- Models based on multidimensional factors (Keefe [9]).

All the above mentioned learning style models define the learning-style of an individual by analyzing a group of his learning preferences. On the contrary, the VARK – Visual, Aural/Auditory, Read/Write, Kinesthetic – questionnaire provides users with a profile of their learning preferences on how to receive information. For this reason it's not considered a learning style model but a questionnaire that measures a specific learning preference. The models considered for the implementation of the ADAPT platform are described next.

40.4.1 The Myers-Briggs Model

One of personality assessment instruments used for the classification of individuals is the Myers-Briggs Type Indicator. In this model each individual has a predisposition to be *sensory (S)* or *intuitive (N), thinking (T)* or *feeling (F), extraversion (E)* or *introversion (I)* and *judgment (J)* or *perceptual (P)*. These features are grouped into four dimensions: *Sensory/Intuitive, Thinking/Feeling, Extraversion/Introversion* and *Judging/Perceptual* [10]. The combinations of preferences into these four dimensions originate 16 different personality profiles, with different characteristics.

40.4.2 The Kolb Model

The Kolb's model considers learning as a process in which knowledge is created through the transformation of experience [11]. The essence of the model consists of describing the learning cycle by making a description of how an individual generates knowledge from his experience, the concepts that determine his behavior in novel situations, as well as how that individual modifies these concepts in order to improve his learning. To achieve this goal, the model defines a repetitive cycle of learning, consisting of four consecutive stages: Concrete Experience (CE), Reflective Observation (RO), Abstract Conceptualization (AC) and Active Experimentation (AE).

40.4.3 The Felder-Silverman Model

The Felder and Silverman model [12] classifies the learning style of a student into four dimensions of learning preferences: Sensory/Intuitive, Visual/Verbal, Active/Reflective, Sequential/Global.

Felder and Silverman developed the model of learning style, for two reasons: the former was to find out differences in learning styles among the most important engineering students; the second was to provide a basis for engineering teachers to develop an approach that would meet the learning needs of all students [13]. The four dimensions of learning preferences in the Felder-Silverman Model are:

- *Active/Reflective* – active students that learn by experimentation or group work *versus* reflective learners that learn by thinking about things and like working alone;
- *Sensory/Intuitive* – the sensory student that is concrete, practical, oriented toward facts and procedures *versus* the intuitive student, conceptual, innovative, oriented toward theories and meanings;
- *Visual/Verbal* – the visual learner that prefers visual representations of presented material such as pictures, diagrams and flowcharts, *versus* the verbal student that prefers written and spoken explanations;
- *Sequential/Global* – the sequential student that is "linear", orderly and learns through small incremental steps, *versus* the global student, holistic, system thinker, that learns in large jumps.

40.4.4 VARK and Learning Preferences

The VARK questionnaire created by Neil Fleming and Colleen Mills [11] allows to determinate the learning preference of an individual. The term *learning preference* refers to the way that a person has a greater willingness to receive the information. The authors of the questionnaire believe that people receive information through the senses and that the brain selects some of this information, ignoring the rest. People choose the information they want to watch according to their interests, but the way the information is received has a great influence too. The VARK questionnaire considers four different types of learning: (V) visual, (A) aural, (R) read/write and (K) kinesthetic.

The VARK model suggests study strategies for each type of learner. For example, study strategies for V learners might include pictures, videos, slides graphs, etc. Good strategies for A learners would be watching tutorials, remembering examples, hearing stories, etc. A study strategy for R learners would be the use of lists headings, definitions, etc. Finally, K learners like to use all their senses, sight, touch, hearing, etc. A study strategy for them would be the use of examples of principles, applications, exhibits, samples, photographs, etc.

40.5 Contents Classification and Management

40.5.1 Bloom Taxonomies

The Bloom taxonomy is a classification of different levels of objectives and competencies defined by teachers, so that students might reach a certain cognitive level [22]. The Bloom's taxonomy was revised later by Anderson and a group of researchers from cognitive psychology [14]. The two versions of the taxonomy are described as Bloom's Taxonomy and Bloom's Revised Taxonomy. The results of this revision were changes made in terminology and structure that replaced the names of Bloom's original model by verbs, because this is the way that learning objectives are usually described. Processes characterized by the taxonomy correspond to learning outcomes in each category that represents what an individual as learned.

Table 40.1 lists the categories and the cognitive processes associated with each one. The categories are ordered by increasing complexity of one to six, being one the most simple and six the most complex. The purpose of the use of taxonomies in the ADAPT platform is to classify the contents of a given discipline according to its level of difficulty and complexity in order to introduce these contents to each student according to their cognitive level.

Table 40.1 Cognitive process of Bloom revised taxonomie

Categories	Cognitive process
1. Remembering	Recognize
2. Understanding	Interpret, exemplify, summarize
3. Applying	Executing, implementing
4. Analyzing	Analyze, organize, assign
5. Evaluating	Check, criticize
6. Creating	Generate, plan, produce

40.5.2 SCORM

E-learning platform files are located in compressed files called e-learning content packages. These packages include HTML documents, media files, style sheets and all scripting components. LMS's use these packages to provide learning contents for their users. This creates the need for reusing e-learning content packages. In order to take care of this issue standards like SCORM (Sharable Content Object Reference Model) are used.

SCORM is a collection of standards and specifications for web-based e-learning platforms. It defines communications between the content on the client side and a host system called the run-time environment that is usually supported by a learning management system. A SCORM package is a compressed archive (e.g. zip), built in a predefined manner. One of the more important parts of the content is an XML-file called manifest-file that contains all the information about what is in the package, what it is for and how it should be used. In order to allow reusability, the content of a complete package is divided into smaller parts called Sharable Content Objects (SCO). A SCO is the smallest reusable unit in the context of SCORM. It has all the necessary files to use its content and, additionally, software to communicate with a SCORM compatible LMS.

The usage of LOM (Learning Object Metadata) is recommended by the SCORM specification. LOM is a data model, usually encoded in XML, used to describe learning objects and other digital resources used to support learning. LOM groups the data elements into nine categories [15]: General, Lifecycle, Meta-Metadata, Technical, Educational, Rights, Relation, Annotation and Classification. LOM was the first of the learning standards to become an official standard. It was widely accepted from the beginning and it was integrated into almost all e-learning standards.

In the ADAPT platform learning objects use XML files to describe their taxonomical level and their type of content.

40.6 Fuzzy Systems and Contents Mapping

40.6.1 Fuzzy Systems

Fuzzy logic theory is a generalization of classical logic and allows solving problems that arise from the crisp classification of true or false. In Fuzzy Logic, a proposition may vary in its degree of truth between 0 and 1, what allows it to be partly true and partly false, simultaneously. Fuzzy logic arose from the fuzzy set theory that introduces the concept of degree of membership of an element to a set. This concept allows the definition of linguistic terms such as, for example, *slow* or *fast*, *small*, *medium* or *large*, by means of sets of values, discrete or continuous, giving rise to the so called *computing with words*.

40.6.2 The Mamdani Inference

Fuzzy Sets and Fuzzy Logic were created by Zadeh [16]. Mamdani [17] applied it to control problems, creating the well-known Mamdani inference. In this method the decision process is based on fuzzy inference rules such as "If A then N", in which both the antecedent and consequent are linguistic terms represented by fuzzy sets [18]. The Mamdani inference is composed of four steps: *fuzzification*, *inference*, *aggregation* and *defuzzification*.

- *Fuzzification*: This step consists of computing the membership value of a crisp value according to the membership function that defines a given linguistic term.
- *Inference*: After performing fuzzification, the membership functions of the inference rules consequents are evaluated by cutting them by the membership value of their antecedent (or the minimum/maximum of their antecedent if it contains an AND/OR, respectively);
- *Aggregation:* The aggregation phase takes place only if there is more than one fired rule with the same linguistic term in its conclusion. In this case, the maximum of the membership values of these consequent linguistic terms, is taken;
- *Defuzzification:* In this step the resulting membership values of the conclusion terms are transformed back into crisp values. There are many methods for performing this operation, being COA – Center of Area – one of the more common.

40.6.3 Contents Mapping

One of the main goals of the ADAPT platform is to adapt content presentation to each student, according to his learning style. The Felder-Silverman model is used to determinate the learning style of the student. As mentioned before, the Felder-Silverman model has four dimensions ranging from -11 to $+11$ and, after answering the model questionnaire, each student has a score associated with each dimension. The contents mapping is done by using the poles of the first three dimensions of the model as input variables of a Fuzzy Mamdani Inference, to determine the type of contents to be shown. The two poles of the fourth dimension are used to determine the sequence of the pages that are going to be presented. The variables in the consequents represent the degree of membership of each student to each learning preference, according to the VARK model.

40.7 Page Rank and Link Mining

"In this paper, we present Google, a prototype of a large-scale search engine which makes heavy use of the structure present in hypertext." – so begins the paper by Sergey Brin and Lawrence Page that presented Google to the world [19]. Google makes use of the link Web Structure in order to rank its search results. This algorithm is called PageRank (PR).

Link Mining (LM) is a more general research area that includes link analysis [20], hypertext, relational learning and inductive logic. One of the topics in LM is link-based object ranking.

In a LMS context, a content is ranked as important if many students used it, and obtained good outcomes. By tracking each student activity, classification techniques can be used to predict if a student behavior has changed along the time. If it is the case, new learning-style questionnaires and/or cognitive evaluation tests are presented. If the change is confirmed, the story board is adapted to fit the student preferences and capabilities. This is the heart and final goal of ADAPT.

40.8 Conclusions

The approaches presented in this paper are used in ADAPT – Adaptive e-leaning System – which is under development.

The project clearly fits some of the goals of the confidence agreement between Ministério da Ciência Tecnologia e do Ensino Superior (MCTES/Portuguese Government) and the Presidents of the Portuguese Polytechnic Institutes, namely in what concerns to the distance learning commitment in the Comunidade de Países de Língua Portuguesa (CPLP).

The primary goal of ADAPT is the development of a full functional adaptive intelligent LMS. Susceptible of spin-offs creation, ADAPT may have a considerable impact in CPLP once Brazil and Angola seem really committed in e-learning implementation.

Learning is an important issue as recognized by the Lisbon Strategy: *"The Lisbon Strategy is the EU's response to facing the challenges of globalization, demographic change and the knowledge society. [...]. There are three overall objectives: improving the quality and effectiveness of education and training systems; facilitating access to education and training systems; and opening up EU education and training systems to the wider world"* [21].

E-learning clearly fits the 2nd and the 3rd goals; ADAPT clearly fits the former.

Acknowledgments The authors wish to thank FCT – Fundação para a Ciência e Tecnologia - by funding the ADAPT Project – PTDC/CPE-CED/115175/2009 and FEDER – Eixo I of Programa Operacional Factores de Competitividade (POFC)/QREN (COMPETE: FCOMP-01-0124-FEDER-014418).

References

1. Benjamin B et al (1956) The taxonomy of educational objectives. The classification of educational goals, Handbook I: cognitive domain. David mckay Company, Inc., New York, pp 8–10

2. Ritu D, Sugata M (1999) Learning styles and perceptions of self. Int Educ J 1(1):61–71
3. Aamodt A, Plaza E (1994) Case-based reasoning: foundational issues, methodological variations, and system approaches. Artif Int Commun 7:39–52
4. Myers, Isabel Briggs with Peter B. Myers (Original edition 1980; Reprint edition 1995) Gifts differing: understanding personality type. Davies-Black Publishing, Palo Alto, 248 p. ISBN 0-89106-074-X
5. Zadeh LA (1996) Fuzzy sets, fuzzy logic and fuzzy systems – selected papers by Zadeh LA 1965–1996, Advances in fuzzy systems – applications and theory, vol 6. World Scientific, Singapore, 19–34
6. Myers IB, McCaulley MH (1985) Manual: a guide to the development and use of the Myers-Briggs type indicator. Consulting Psychologists Press, Palo Alto
7. Schmeck RR (1983) Learning styles of college students. In: Dillon R, Schmeck R (eds) Individual differences in cognition. Academic, New York
8. Kolb D (1984) Experiential learning: experience as the source of learning and development. Prentice-Hall, Englewood
9. Reichmann SW, Grasha AF (1974) A rational approach to developing and accessing the construct validity of a student learning style scale instrument. J Psychol 87:213–223
10. Mamdani EH, e Assilian S (1975) An experiment in linguistic synthesis with a fuzzy logic controller. Int J Man Mach Stud 7:1–13
11. SHAI – Stottler Henke Artificial Intelligence (2003) White papers, www.shai.com
12. Felder RM, Silverman LK (1988) Learning and teaching styles in engineering education [Electronic Version]. Eng Educ 78(7):674–681
13. European Comission – Education and Training (2009) http://ec.europa.eu/education/lifelong-learning-policy/doc28_en.htm. Last accessed on 10 Sept 2012
14. Anderson et al (2001) A taxonomy for learning, teaching, and assessing: a revision of Bloom's taxonomy of educational objectives. Longman, New York, pp 67–69
15. Felder RM, Spurlin J (2005) Applications, reliability, and validity of the index of learning styles [Electronic Version]. Int J Eng Educ 21(1):103–112
16. VARK – A guide to learning styles (2010) http://www.vark-learn.com/english/index.asp. Last accessed on 10 Sept 2012
17. Jensen D, Goldberg H (1998) Papers of the AAAI fall symposium on AI and link analysis, AAAI Press, Menlo Park
18. Mamdani EH (1973) Aplications of fuzzy algorithms for control of simple dynamic plant. Proc IEEE 12:1585–1588
19. BolzanW, Giraffa L (2002) Estudo Comparativo sobre Sistemas Tutores Inteligentes Multiagentes Web, Technical report Nº24, PUCRS Faculdade de Informática, BrasilJ. Clerk Maxwell, A treatise on electricity and magnetism, 3rd ed, vol 2. Clarendon, Oxford, 1892, pp 68–73
20. Friesen N, McGreal R (2005) CanCore: best practices for learning object metadata in ubiquitous computing environments. IEEE Computer Society, Washington, DC
21. Brin S, Page L (1998) The anatomy of a large-scale hypertextual web search engine, Seventh international World-Wide Web conference (WWW 1998), Brisbane, Australia, April 14–18
22. Keefe JW (1989) Learning style profile handbook: accommodating perceptual, study and instructional preferences, vol II. National Association of Secondary School Principals, Reston

Chapter 41
Recognizing Music Styles – An Approach Based on the Zipf-Mandelbrot Law

Viriato M. Marques and Cecília Reis

Abstract What makes a musical theme a good one? As an art, music evaluation is essentially a matter of aesthetical criterions and cultural inheritance. In fact, classifying melodies or complete musical pieces as good or bad is a particularly challenging task, although central for the production of artificial music by means of Evolutionary Computation, as the fitness function exactly consists of evaluating the quality of the generated melodies or complete scores. This paper presents a possible approach for the recognition of musical pieces style and, ultimately, their author. Certainly not enough from the point of view of fitness function implementation, this work is a step towards this goal, susceptible of further development and complementation by other techniques. The approach is based on the Zipf-Mandelbrot law, a probability distribution from the family of power laws that also describes word frequency in a given language or cities population all over the world.

41.1 Introduction

Evolutionary Computation (EC) is a subfield of Artificial Intelligence (AI) that deals with optimization problems. EC uses iterative progress, such as growth or development in a population. This population is then selected in a guided random search using parallel processing to achieve the desired goal. Such processes are

V.M. Marques (✉)
Engineering Institute, Coimbra Polytechnic Institute (ISEC/IPC), Coimbra, Portugal
e-mail: viriato@isec.pt

C. Reis
Engineering Institute, Porto Polytechnic Institute (ISEP/IPP), Porto, Portugal
e-mail: cmr@isep.ipp.pt

A. Madureira et al. (eds.), *Computational Intelligence and Decision Making: Trends and Applications*, Intelligent Systems, Control and Automation: Science and Engineering 61,
DOI 10.1007/978-94-007-4722-7_41,

often inspired by biological mechanisms of evolution [1, 2]. Therefore, EC is a general term for several computational techniques which are based, to some degree, on the evolution of biological life in the natural world. Among other paradigms, EC includes Genetic Algorithms (GA's).

The basic concept of GA is the simulation of processes in the natural system, necessary for evolution, specifically those that follow the principles first laid down by Charles Darwin regarding the survival of the fittest [3, 4]. A GA moves individuals – represented by chromosomes – of one population to a new one, by using a selection of the fittest, together with the genetics – inspired operators of crossover, mutation and, less usually, inversion. As the fittest individuals tend to be more frequently selected, the overall result is that the algorithm tends to converge towards optimal or almost optimal solutions, represented by the fittest individuals. The evaluation of their fitness is carried out by a function called fitness function that measures how good each individual is in terms of the sought solution.

This approach gives raise to some kind of creativity, as new – i.e. previously inexistent – individuals result from these operations. And that's why AG's have been used for the automatic generation of musical pieces. However, their application in this field gives raise to the following major problem: how can the fitness function evaluate if a musical theme is good? Which features should it take into account?

To overcome this problem, Interactive Evolutionary Computation (IEC) has already been used, with some interesting results [5–8]. In IEC the fitness function is replaced by a human being, that interactively evaluates the results obtained by the GA over its iterations. But this means that each genotype must be converted to its corresponding fenotype – i.e. the audible individual – in order to be possible to evaluate it. Besides not being a completely automated process, IEC requires a great effort from its users. To overcome this problem, some solutions such as keeping populations small or pre filtering the worst individuals, are used.

For this pre filtering process, as well as for the implementation of the fitness function in the original GA's, models such as neural networks and composition rules, have been used. However, there's still a long way to go in this aspect as, as stated before, music evaluation is essentially a matter of aesthetical criterions and cultural inheritance: classifying melodies or complete musical pieces as good or bad is a particularly challenging task. The problem resides on the determination and selection of features: what determines if a musical theme or complete score is good or bad?

This paper presents the work developed by the authors in this field: some musical features are devised and extracted from musical scores of distinct types (and authors); those that more approximately fit the Zipf-Mandelbrot law – that, according to [9, 10] applies to the distribution of some features obtainable from music scores – are selected; then, a classifier is trained using as inputs the values of these features; finally some new examples of the same styles are applied and the performance is evaluated. The results show that, for globally identifying the kind of music under consideration. . .it works!

41.2 Music

41.2.1 Basic Principles

The fundamental characteristics of a musical sound are pitch, duration and timbre. In the Western (European) Music style, pitches are organized into diatonic scales composed by seven notes: C, D, E, F, G, A, B (C...) that repeat along each musical instrument range. The term interval designates the pitch distance between any two notes. A scale is an ordered sequence of musical intervals [9, 10]. Between adjacent notes of a diatonic scale, the intervals can be of 1 tone or 1 semitone – i.e. 1/2 tone – as shown in Fig. 41.1. The first note of a scale may be anyone as long as the intervals between adjacent notes remain the same: this ensures that it always sounds the same way, although using higher or lower pitches. This system gives rise to the tonality concept, i.e. the same musical piece can be played using distinct musical notes: since the intervals between them remain the same, the piece will sound the same way. Twelve tonalities are defined, corresponding to the seven diatonic notes plus the five sharps ♯ (or flats).

The diatonic scale of Fig. 41.1 is the C major scale. Its relative minor harmonic scale begins with an A and progresses with the following notes, adding a sharp to its 7th degree, G#. Melodies based on major scales generally sound joyful, while those based on minor scales tend to be sad (in fact nobody knows why). Tonality and major/minor modes have been the pillars of Western Music for centuries.

The duration and absence of sounds is defined by notes and rests. In its basic form, the duration of each note is just 1/2 of its precedent. In a musical score, the duration of the notes define the rhythm.

The timbre allows the distinction between different musical instruments and is the basis of the instrumentation phase in music composition.

Basically, the composition of a musical piece comprises two or three phases: (1) melody generation: create a sequence of notes called a theme, beautiful and "making sense" (this involves pitches and rhythm); (2) harmonize it, i.e. "fill" the score by writing vertically, so that various sounds can be heard simultaneously (in a simple way, can be thought as arranging an accompaniment for a singer); (3) if it is intended for various instruments, give each one a part (instrumentation).

The study was carried out on musical scores harmonized for piano, i.e., the complete music including melody and accompaniment, right and left hand.

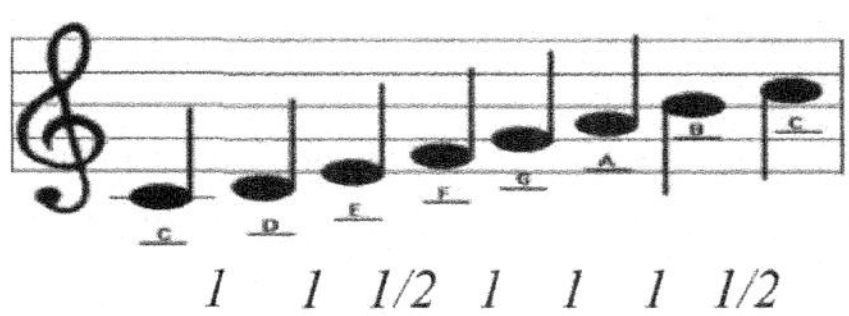

Fig. 41.1 The diatonic scale of C major

41.2.2 Physics and Music

The pitch of each note is defined by its (fundamental) frequency. For instance, A4 (the 4 standing for the 5th A occurring in a piano keyboard) corresponds to 440 Hz. An ascending octave – for example, from A4 to A5 – is defined by frequency duplication. Consequently, $f_2 = 2f_1$ where f_2 = frequency of the x_n note and f_1 = frequency of x_{n-1} note.

A chromatic scale consists of a sequence of 12 notes all separated by a semitone. Briefly, it contains all the ♯ and ♭. As the first and the last notes are separated by an octave, the frequency relationship for any half-tone in an ascending chromatic scale is defined by $f_2 = \sqrt{[12]2}f_1$. This tuning system, consecrated by J.S.Bach in his masterpiece "Das Wohltemperierte Klavier", is called Twelve-tone Equal Temperament. These are the basic relations that allow the synthesis of pitches, from musical scores to the audio domain, as it happens with MIDI – Musical Instruments Digital Interface – sound reproduction and other file formats.

41.2.3 Music Styles

The history of Western Music is typically divided into the following periods: Medieval (500–1400); Renaissance (1400–1600); Baroque (1600–1750); Classical (1750–1800); Romantic (1800–1900); Modern and Contemporary (1900. . .) [10–13].

The study was carried out on musical pieces of J.S. Bach (1685–1750), Scott Joplin (1867–1917) and John P. De Sousa (1854–1932). Bach belongs, of course, to the pure Baroque; Joplin and de Sousa both belong to the Romantic/Modern era: although absolutely distinct – Joplin with his famous ragtime's such as Maple Leaf and de Sousa with his military marches such as Stars and Stripes – it is expectable that it will be harder to distinguish between them both, then between them and Bach.

41.3 The Zipf-Mandelbrot Law

Zipf's law is an empirical statistical law of the family of Discrete Power Laws. It states that given a support of a certain natural language, the frequency of any word is inversely proportional to its rank in the frequency table. The most famous example of Zipf's law is, perhaps, the word frequency table for the English Language. But besides this one there are numerous examples of its applicability, such as, for example, world cities population or the number of visits to Internet sites [14, 15].

G.Zipf was an American linguist and philologist, Chairman of the German Department and University Lecturer at Harvard University. His law was subsequently

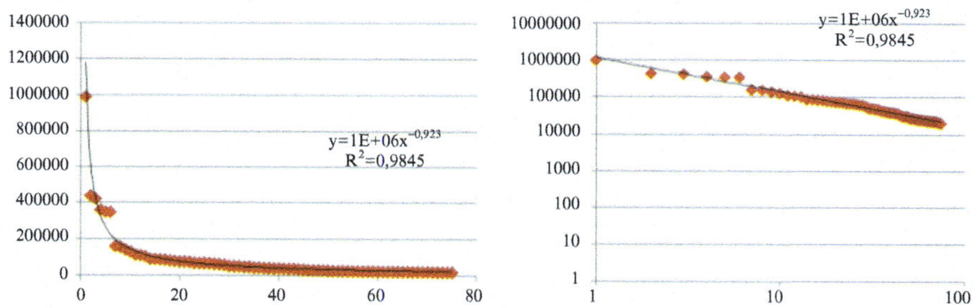

Fig. 41.2 Zipf's Law: direct plot and logarithmic scales

generalized by the mathematician B. Mandelbrot. Formally Zipf's law is expressed by

$$f(k,N,s) = \frac{1/k^s}{\sum\limits_{n=1}^{N}(1/n^s)} \tag{41.1}$$

where N = number of items, k = rank of each item, s = exponent characterizing the distribution, f = relative frequency of each item ranked by k. Figure 41.2 shows two graphical aspects of Zipf's law for a given English word count: the first one is obtained by directly plotting word count versus its rank; the second is obtained by taking the logarithmic values of both axis, so obtaining a straight line.

In the context of this paper, n refers to the possible ranks of a given feature in a musical score and k to a specific rank of that feature. But the exponent s is unknown for each feature: the goal is exactly to compute it, expecting that it will be distinct for the same feature for distinct music styles. If this is the case, then the exponents that characterize a set of features can be used as classifier inputs. Notice that the value of s is such that it maximizes r-squared value (i.e., the square of Pearson correlation coefficient) between the predicted and the real values of n.

41.4 Music File Formats and Data Pre-Processing

In this approach one must deal with musical scores, not with sound files. There are many file formats for storing music scores. The better known is certainly MIDI. However, this format is quite complex as it can store data about musical instruments, distinct tracks, expression, etc. It is very powerful as it is intended for the reproduction, as humanly as possible, of simple or very complex music pieces. Software for MIDI processing – and not only – includes Finale, Sibelius and AnvilStudio, the last one being freeware. Besides this, and under some circumstances, the MIDI technology changes some musical properties of the original

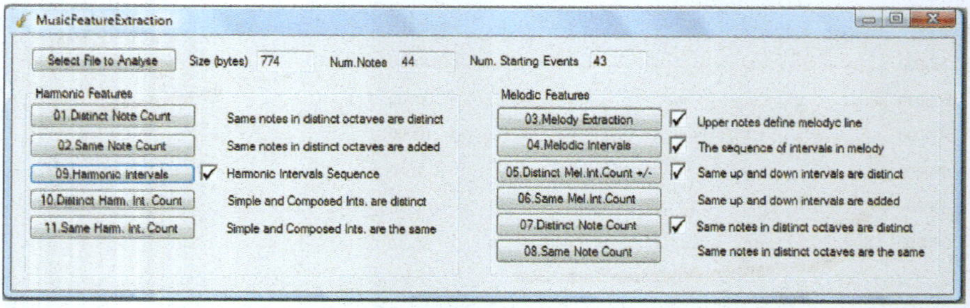

Fig. 41.3 The application for score transposing and feature extraction

score. For our purposes this is not good: on the contrary, we need a file structure as simple as possible that stores the notes, their duration and nothing else, if possible.

One solution could be the XML notation that is being used for some years. However, the site KernScores (http://kern.ccarh.org/) provides thousands of music scores in various formats (including MIDI, XML and pdf). One of these, Melisma, is simple and powerful enough for our goal. Basically each file begins by declaring the tonality and mode; each of the following lines comprises three numbers: the first one is the note code; the second one stands for its starting instant and the third one for its finishing instant.

Melisma files of some Bach chorales, Joplin ragtime's and de Sousa military marches were downloaded from KernScores site. For our purpose the only data pre-processing operation needed consisted of transposing the scores to the same tonality: C major for major mode pieces and A minor for minor mode pieces.[1] An application was developed (Fig. 41.3) that, besides this task, also accomplishes all the necessary feature extractions as explained in the following section.

41.5 Music Features

41.5.1 Feature Extraction

Once the pre processing was done, a set of features was extracted for each file using the above referred application. Only harmonic and melodic features were used, i.e., the rhythmical aspects were not taken into account. These features were:

- Harmonic:

  1. Notes-1: same notes in distinct octaves are distinct
  2. Notes-2: same notes in distinct octaves are the same

[1] Another approach could be followed, consisting of replacing each note by its scale degree.

Table 41.1 Note count for a Bach Choral

NoteCode	Note	NoteCount	NoteCode	Note	NoteCount	NoteCode	Note	NoteCount
55	G4	42	62	D5	18	67	G5	9
60	C5	34	64	E5	18	65	F5	8
52	E4	21	43	G3	14	45	A3	7
36	C3	20	40	E3	10	47	B3	5
59	B4	20	48	C4	10	50	D4	5
57	A4	19	38	D3	9	35	B2	3
53	F4	18	41	F3	9	58	Bb4	1

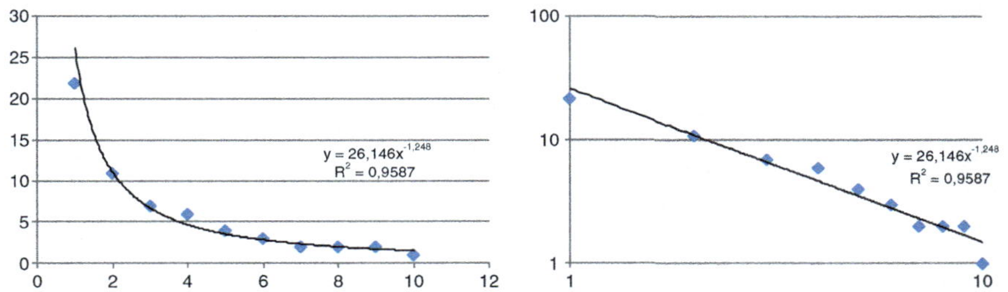

Fig. 41.4 Zipf law for the feature Melodic Intervals-1 (Bach Choral)

3. Intervals-1: simple and composed intervals of the same type are counted as distinct
4. Intervals-2: simple and composed intervals of the same type are counted as the same

- Melodic:

1. Notes-1: same notes in distinct octaves are distinct
2. Notes-2: same notes in distinct octaves are the same
3. Intervals-1: simple and composed intervals of the same type are counted as distinct
4. Intervals-2: simple and composed intervals of the same type are counted as the same

Table 41.1 shows an example of the harmonic feature Notes-1 for the Bach Choral "Aus meines Herzens Grunde". Notice that note count is ranked from its highest to its lowest value.

Figure 41.4 shows an example of the Melodic Intervals-1 feature and the ideal Zipf distribution, for the computed value of exponents, using linear and log scales, for the Bach Choral of Table 41.1.

Notice that in order to process the melodic features, in first place the melody must be extracted from the whole score. This is far from being a trivial task as it is not granted that in each instant the melody is always the highest pitch (note). In fact, there are lots of possible occurrences to consider. Some examples follow: (1) the

Table 41.2 s and r^2 for the features of some musical pieces

	Harmony							
Composer	F1s	F1R	F2s	F2R	F10s	F10R	F11s	F11R
JPSousa	1.32	0.77	1.99	0.66	1.35	0.82	1.25	0.95
JPsousa	1.13	0.77	1.28	0.69	1.39	0.83	1.46	0.58
SJoplin	1.19	0.76	1.36	0.82	1.32	0.71	1.06	0.72
SJoplin	1.20	0.75	1.13	0.76	1.11	0.72	0.96	0.59
Bach	0.87	0.59	1.23	0.41	1.23	0.69	1.18	0.62
Bach	1.04	0.72	1.10	0.59	1.31	0.75	1.05	0.66
	Melody							
Composer	F5s	F5R	F6s	F6R	F7s	F7R	F8s	F8R
JPSousa	1.29	0.92	1.50	0.93	1.15	0.79	1.27	0.48
JPsousa	1.18	0.94	1.35	0.94	1.19	0.86	1.13	0.71
SJoplin	1.29	0.92	1.41	0.93	1.17	0.82	1.41	0.81
SJoplin	1.27	0.91	1.42	0.88	1.22	0.80	1.52	0.76
Bach	1.25	0.96	1.35	0.92	0.48	0.91	0.47	0.91
Bach	1.47	0.95	1.79	0.98	1.10	0.84	1.12	0.76

musical piece may begin with an introduction that can't be considered to belong to the melody yet; (2) the melody is interrupted, i.e., a rest occurs; so, which is the note where it starts again if many others, belonging to the "accompaniment", occur meanwhile? These and some other problems have been solved by analyzing, for each case, the specific occurrence and trying to overcome it in a so general as possible way.

41.5.2 Feature Selection

Data such as the depicted in Table 41.1, resulting from the feature extractions of the various musical scores, was then processed in order to compute the value of the exponent s that would give a maximum r-squared when fitted with its most similar Zipf's distribution. Some results are presented in Table 41.2.

Among the eight processed features, the top five – those with the highest values of r^2 – were selected as inputs for a classifier. The label is, of course, the composer/music style.

41.6 Mining the Data

Data-mining (DM) was performed over the data using Rapidminer software (http://rapid-i.com). In order to clearly visualize the results, the decision tree classifier was selected based on the ID3 algorithm with previous discretization of the inputs, i.e., the five selected features. The resulting decision tree is shown in Fig. 41.5.

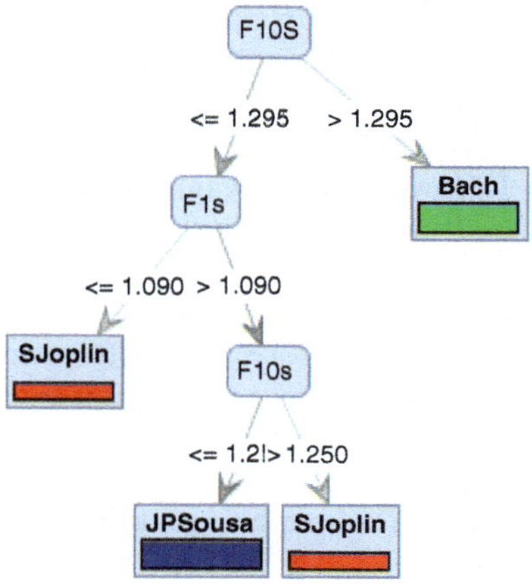

Fig. 41.5 The decision tree for composer/music style recognition

Meta Data View   Data View   Plot View

ExampleSet (6 examples, 6 special attributes, 5 regular attributes)

Row No.	Name	Composer	confidence(...	confidence(...	confidence(...	prediction(Com...
1	VenusTransit	JPSousa	1	0	0	JPSousa
2	WashingtonPost	JPSousa	1	0	0	JPSousa
3	Chor002	Bach	0	1	0	Bach
4	Chor224	Bach	0	1	0	Bach
5	Solace	SJoplin	0	0	1	SJoplin
6	Winers	SJoplin	0	0	1	SJoplin

Fig. 41.6 Some results obtained by applying new examples to the decision tree of Fig. 41.5

As you can see, the resulting tree is (perhaps) surprisingly simple and evidences some of the expected characteristics: (1) as Bach is quite distinct from the other two composers, a single feature is enough to recognize his chorals: the harmonic intervals count; (2) to discriminate between Joplin and de Sousa, some more effort is needed, as expected, due to some more expectable similarities between them.

The classifier was then tested with new examples of Bach chorales, ragtime and military marches. Fig. 41.6 shows a situation where a precision of 100% was achieved. K-nearest neighbors and neural networks were also tested, with similar results.

41.7 Conclusions and Future Work

Although the examples used in this work were quite distinguishable from the human hearing point of view, the case is that the obtained model fulfilled the expectations.

Future work includes the application of this approach to more styles, particularly to more similar styles.

So far, rhythmic features have not been explored. This is another subject to take into account.

However, we began this paper by describing the difficulties experienced in the implementation of fitness functions for music evaluation in the context of Genetic Algorithms. In this context, and although this work can show some progress into this direction, there is, naturally, much to be done! First, a generated melody must be evaluated, alone, including its rhythmic aspects. In this phase, some composition rules and sequence pattern detection may help. Then, the harmony must be created and evaluated too.

The overall result, a complete musical piece, beautiful, fitting the tonal Western style, is far from being reached. Meanwhile, we'll keep trying.

References

1. Goldberg D (1989) Genetic algorithms in search optimization and machine learning. Addison-Wesley, Boston
2. Bäck T, Fogel DB, Michalewicz Z (1997) Handbook of evolutionary computation. Institute of Physics Publishing/Oxford University Press, Bristol
3. Davis L (1991) Handbook of genetic algorithms. Van Nostrand Reinhold, New York
4. Mitchell M (1997) An introduction to genetic algorithms. MIT Press, Cambridge, MA
5. Marques V, Reis C, Machado J (2010) Interactive evolutionary computation in music – a straightforward approach for naïve melody generation. In: Proceedings of IEEE-SMC'10 – Systems, Men and Cybernetics, Istanbul, Turkey
6. Takagi H (2001) Interactive evolutionary computation: Fusion of the capabilities of EC optimization and human evaluation. In: Proceedings of the IEEE, vol 89(9), Sept, pp 1275–1296. http://ieeexplore.ieee.org/xpl/login.jsp?tp=&arnumber=949485&url=http%3A%2F%2Fieeexplore.ieee.org%2Fxpls%2Fabs_all.jsp%3Farnumber%3D949485#AdditionalDetails
7. Schutte S, Eklund J, Axelsson J, Nagamachi M (2004) Concepts, methods and tools in Kansei Engineering. Theor Issues Ergon Sci 5(3):214–232
8. Semet Y (2002) Interactive evolutionary computation: a survey of existing theory. University of Illinois, Illinois
9. Zamacois J (1986) Teoria de la música, Livros 1 e 2. Editora Labor, Barcelona
10. Stephan R (1978) Enciclopédia Meridiano Fischer vol 7 – Música. Editora Meridiano, Lisboa, Portugal
11. Stehman J (1979) História da Música Europeia, 2ª ed. Editora Bertrand, Lisboa, Portugal
12. Bennet R (1997) Uma Breve História da Música, Cadernos de Música da Universidade de Cambridge, Rio de Janeiro, Jorge Zahar Editor
13. Deutsche Grammophon (1997) História da Música. Editora Edilivro, Lisboa, Portugal
14. Wentian L (2003) Zipf's law everywhere. Glottometrics 5:14–21
15. Adamic L, Huberman B (2002) Zipf's law and the internet. Glottometrics 3:143–150

Chapter 42
A Platform for Peptidase Detection Based on Text Mining Techniques and Support Vector Machines

Daniel Correia, Carlos Pereira, Paula Veríssimo, and António Dourado

Abstract This paper presents a web platform for the detection of peptidases and motifs search from Merops database. The methodology for peptidases detection uses text mining techniques combined with Support Vector Machines (SVM). Preliminary results using two types of SVMs, the C-Support Vector Classification (C-SVC) and One-class SVM, show the feasibility of the methodology. Despite of the best results obtained with C-SVC the One-class SVM can be an alternative solution if only positive examples are available for training.

Keywords Protein classification • Text mining • Support vector machines • Web platform

---

D. Correia (✉)
Department of Informatics Engineering and Systems,
Coimbra Institute of Engineering, Portugal
e-mail: a21160499@alunos.isec.pt

C. Pereira
Department of Informatics Engineering and Systems,
Coimbra Institute of Engineering, Portugal

Department of Informatics Engineering,
University of Coimbra, Portugal
e-mail: cpereira@isec.pt

P. Veríssimo
Department of Biochemistry and Center of Neuroscience and Cell Biology,
University of Coimbra, Coimbra, Portugal
e-mail: paulav@ibili.uc.pt

A. Dourado
Department of Informatics Engineering, University of Coimbra, Portugal
e-mail: dourado@dei.uc.pt

A. Madureira et al. (eds.), *Computational Intelligence and Decision Making: Trends and Applications*, Intelligent Systems, Control and Automation: Science and Engineering 61,
DOI 10.1007/978-94-007-4722-7_42,

42.1 Introduction

The peptidases are a class of enzymes that catalyze chemical reactions, allowing the decomposition of protein substances into smaller molecules. They are involved in several processes crucial for the correct functioning of organisms. Its detection characterization is central to a better understanding of their role in biological systems.

One of the main objectives of this work is to offer to the scientific community a web platform that allows the detection of peptidases. The methodology uses text mining techniques and two types of SVM algorithms, C-SVC and One-class SVM.

The use of text-mining techniques for protein classification has been suggested in [1], however this work combines the methodology of n-gram counts with state-of-the art Support Vector Machines (SVMs). In order to investigate the detection ability of peptidases, a dataset formed by sequences of different classes from Merops has been used.

The remainder of this paper is organized as follows. The next Section presents the methodologies used for the platform development, processes for extraction and the representation of the characteristics of the sequences. Section 42.3 briefly describes the two types of Support Vector Machines used. Section 42.4 presents the experiments and the results. The Sect. 42.5 presents the functionalities offered by the platform. The last Section presents the conclusions and future improvements to the platform.

42.2 Methodology

In this section we give an overview of the methodology used for feature extraction and some brief details of the implementation are presented (see Sect. 42.2.1). The sequence representation is discussed in Sect. 42.2.2.

42.2.1 Features Extraction

Regardless of the specific algorithms that can be applied, the Text Mining process always includes a pre-processing step, aiming to obtain a structured representation of the information contained in the documents. In the case study of this work, the extraction process has been adapted to extract features from protein sequences, as shown in Fig. 42.1. In order to implement the extraction of the characteristics of each sequences the WVTool library [2] has been used. This Java library provides a simple and flexible implementation to create a vector representation of the sequences. In order to be able to use this library in our platform, developed in ASP.NET, it was necessary to build a .NET assembly. For that, the IKVM [3] has been used, since it offers support for running Java applications in the Common Language Runtime (CLR) that is the runtime environment of the .NET framework.

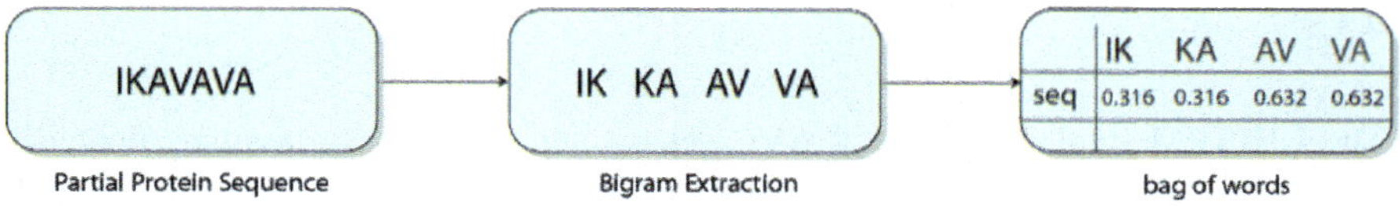

Fig.42.1 Feature extraction process

The extraction process involves setting the size (n) of the substrings $(n - grams)$. The substrings are constructed by separating the continuous flow of the constituent amino acids. It is considered that the presence of consecutive combinations of n sequences of amino acids in proteins is potentially relevant for classification.

The frequency of each n-gram in a particular sequence corresponds to the extracted feature. For example, considering the sequence IKAVAVALILA, the $n - grams$ of 2 (bigram) would be IK, KA, AV, VA, AV, VA, AL, LI, IL, and LA.

42.2.2 *Sequence Representation*

The representation of the sequences in a structured manner is necessary for its successful use by the classification algorithm. The used approach consists in the vector representation [4] also known as bag-of-words (BOW), which is formed by a two-dimensional vector, where each line corresponds to a given sequence and each column represents one of the n-gram of that sequence. The value of each cell corresponds to the relevance of the feature.

There are several methods for calculating the relevance of terms, such as binary occurrence, number of term occurrences, term frequency $(tf)$and term frequency – inverse document frequency $(tf - idf)$In this work, the term frequency (tf)has been chosen and is defined according to:

$$tf_{ij} = \frac{f_{i,j}}{fd_j}$$

Where $f_{i,j}$is the number of occurrences of feature $i$in the sequence j, and $fd_j$is the sum of the number of occurrences of all the features of sequence j.

42.3 Support Vector Machines

Support Vector Machines [5] are supervised learning type algorithms used for classification and regression, based on statistical learning theory [6]. There are various implementations and types of SVMs, being the LIBSVM [7] one of the most used library. Two types of SVMs have been tested in this work: the C-Support Vector Classification and the One-class SVM. Next, both techniques are briefly presented.

42.3.1 C-SVC

The LIBSVM implements the C-SVC binary and multiclass classifier. For each training instance $(x_i, y_i), i = 1, \ldots.., l$ where $x_i \in R^n$ and $y \in \{-1, 1\}$ The C-SVC classifier solves the following optimization problem [7]:

$$\min_{w,b,\varepsilon} \frac{1}{2} w^t w + C \sum_{i=1}^{l} \varepsilon_i$$

with the following restrictions:

$$y_i(w^t \emptyset(x_i) + b) \geq 1 - \varepsilon_i$$

$$\varepsilon_i \geq 0, i = 1, \ldots, l$$

The decision function is represented by:

$$sgn(w^t \emptyset(x) + b) = sgn(\sum_{i=1}^{l} y_i \alpha_i K(x_i, x) + b))$$

42.3.2 One-Class

The One-class is characterized by using only positive examples to train the machine. Schölkopf et al. [8] proposed this approach where the positive examples are separated from the origin by the decision hyperplane.

For each training instance $(x_i, y_i), i = 1, \ldots.., l$ where $x_i \in R^n$ (without any reference to the class), this algorithm solves the following optimization problem [7].

$$\min_{w,b,\varepsilon} \frac{1}{2} w^T w - \rho + \frac{1}{vl} \sum_{i=1}^{l} \varepsilon_i$$

subject to

$$w^T \emptyset(x_i) \geq \rho - \varepsilon_i$$

$$\varepsilon_i \geq 0, i = 1, \ldots, l$$

The boundary decision surface is given by:

$$sgn(\sum_{i=1}^{l} \alpha_i K(x_i, x) - \rho)$$

42.4 Experiments and Results

In order to evaluate the performance of the classifiers, a dataset consisting of 20778 sequences, having approximately 10% of the sequences present in the Merops 9.4 [9] and 2710 sequences from SCOP [10], both being randomly collected. This benchmark dataset was previously used and described by Pereira et al. [11]. In order to evaluate the results using the One-class algorithm the dataset presented was divided in two sets, train and test as showed in Table 42.1. The results with the C-SVC were obtained using cross-validation.

The features were extracted from the dataset using the text mining techniques previously described. Seven types of extractions have been defined for all the possible combination of the three n-grams (*i*) unigram, (*ii*) bigram and (*iii*) trigram (see Table 42.2).

For evaluating the performance of the classifiers, we use four metrics: accuracy, sensitivity or recall, specificity and precision defined according to:

$$accuracy = \frac{tp + tn}{tp + fp + tn + fn}$$

$$sensitivity = \frac{tp}{tp + fn}$$

$$specificity = \frac{tn}{tn + fp}$$

$$precision = \frac{tp}{tp + fp}$$

where t_p is the number of true positives t_n is the number of true negatives, f_p is the number of false positives and f_n is the number of false negatives.

Table 42.1 Dataset division for One-class algorithm

	#Train sequences	#Test sequences
Aspartic (A) peptidases	721	127
Cysteine (C) peptidases	2474	437
Glutamic (G) peptidases	7	2
Metallo (M) peptidases	5427	958
Asparagine (N) peptidases	35	6
Serine (S) peptidases	5896	1040
Threonine (T) peptidases	502	88
Peptidases of unknown (U) catalytic type	296	52
Non peptidase		2710

Table 42.2 Extractions details (number of features and n-grams combinations)

Extraction Type	# Features
n-gram 1 (unigram)	24
n-gram 2 (bigram)	473
n-gram 3 (trigram)	3,378
n-gram 1 and 2	497
n-gram 1 and 3	3,402
n-gram 2 and 3	3,851
n-gram 1, 2 and 3	3,875

Table 42.3 Best results and parameters obtained for the two algorithms studied ((a) c-svc and (b) One-class)

(a)	c	γ	Accuracy (%)	Sensitivity (%)	Specificity (%)	Precision (%)	F-Measure (%)
n-gram (1)	2^3	2^2	96.434	98.793	80.701	97.153	97.966
n-gram (2)	2^1	2^1	98.277	99.181	92.251	98.842	99.011
n-gram (3)	2^0	2^0	97.892	99.303	88.487	98.291	98.794
n-gram (1–2)	2^7	2^{-3}	98.360	99.290	92.140	98.830	99.059
n-gram (1–3)	2^0	2^0	98.200	99.491	89.594	98.456	98.971
n-gram (2–3)	2^0	2^{-1}	98.056	99.181	90.554	98.592	98.886
n-gram (1–2–3)	2^0	2^0	98.248	99.319	91.107	98.675	98.996

(b)	nu	γ	Accuracy (%)	Sensitivity (%)	Specificity (%)	Precision (%)	F-Measure (%)
n-gram (1)	0.2	2^2	81.94	79.89	83.99	83.30	81.56
n-gram (2)	0.1	2^0	91.59	89.85	93.32	93.08	91.44
n-gram (3)	0.1	2^{-11}	92.62	93.73	91.51	91.70	92.70
n-gram (1–2)	0.1	2^1	91.18	89.15	93.21	92.92	91.00
n-gram (1–3)	0.1	2^1	89.23	86.35	92.10	91.62	88.91
n-gram (2–3)	0.1	2^{-13}	91.88	92.44	91.33	91.42	91.93
n-gram (1-2-3)	0.5	2^0	91.00	88.56	93.43	93.10	90.77

The F-measure score is based on the precision and recall, defined below

$$(x_i, y_i), i = 1, \ldots\ldots, lF - measure = 2 \times \frac{Precision \times Recall}{Precision + Recall}$$

The results with the C-SVC were obtained using the all dataset shown in Table 42.3, with Gaussian kernel and five folds cross-validation. Preliminary results using this type of SVM were obtained through a grid search for the parameters C (from 10^{-2}, 10^{-1} to 10^4) and γ (from 0.2 to 0.8 with increments of 0.2). For the One-class algorithm, the results were obtained using the dataset division presented in Table 42.3 with a Gaussian kernel.

The preliminary results for the various types of n-grams using One-class SVM, where obtained again through a grid search, where γ parameter ranged from 0.01 to 0.91 and n*u* ranged from 0.1 to 0.9 with increments of 0.1 (see Fig. 42.2).

In order to extend the parameter selection criteria and to prove the superior performance of C-SVC, a larger common grid search for both algorithms has been

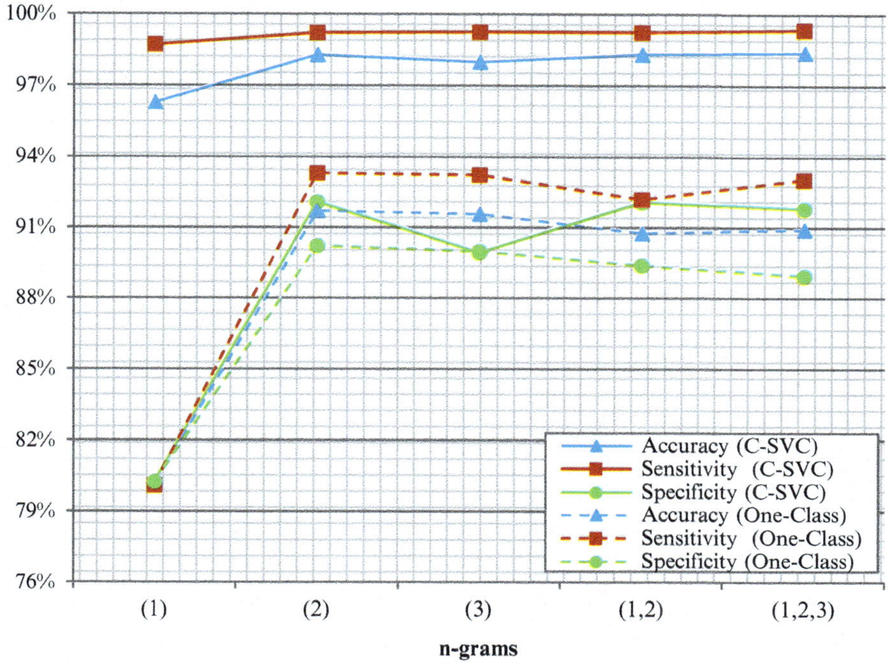

Fig.42.2 Preliminary results obtained through a grid search for the two algorithms ((**a**) c-svc and (**b**) One-class)

applied using the following range of search: for parameter C (from $2^{-4}, 2^{-3}$ to 2^{11}), γ (from $2^{-13}, 2^{-12}$ to 2^{2}) and nu (from 0.1, 0.2 to 0.9). In Fig. 42.3, it can be seen the best results obtained for both algorithms.

The trend of the results in both algorithms was the same. In C-SVC, we have improved all the results with this new grid search. Again, the One-class exhibited an inferior performance for all n-grams compared to C-SVC. Nevertheless, the One-class proves to be also a feasible methodology when the available dataset doesn't indicate the positive and negative labels.

The followed experiments for both algorithms studied in this paper are reported in more detail in Table 42.3.

42.5 Platform

The BioinkSearch web platform was developed using the ASP.NET framework. In order to build a friendly graphical user interface, other technologies like Cascading Style Sheets (CSS) and jQuery JavaScript library were used. The platform main functionalities are the motif search and peptidase detection. The Fig. 42.4 shows the main window of the platform.

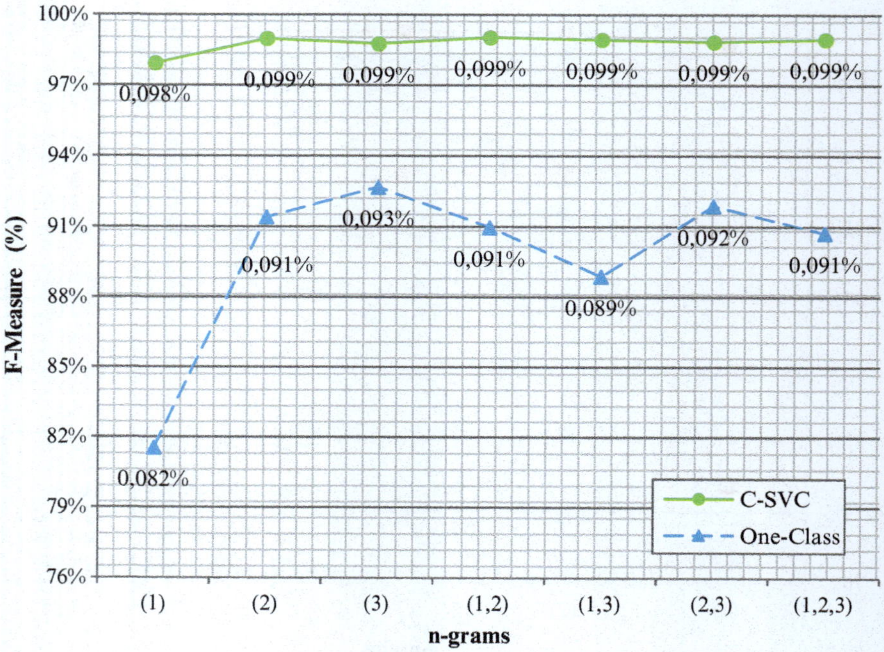

Fig.42.3 Best f-measure results with c-svc and One-class for the various combinations of n-grams

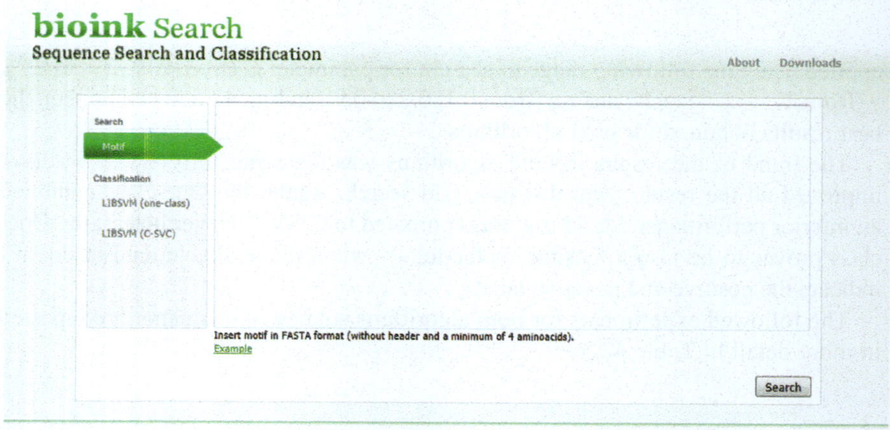

Fig.42.4 General aspect of BioinkSearch interface

42.5.1 Motif Search

Motifs are recurrent patterns of nucleotides or amino acids usually having some biological significance. This platform implements a Motif search – represented in Fig. 42.5. The search is performed on sequences described in the Merops database.

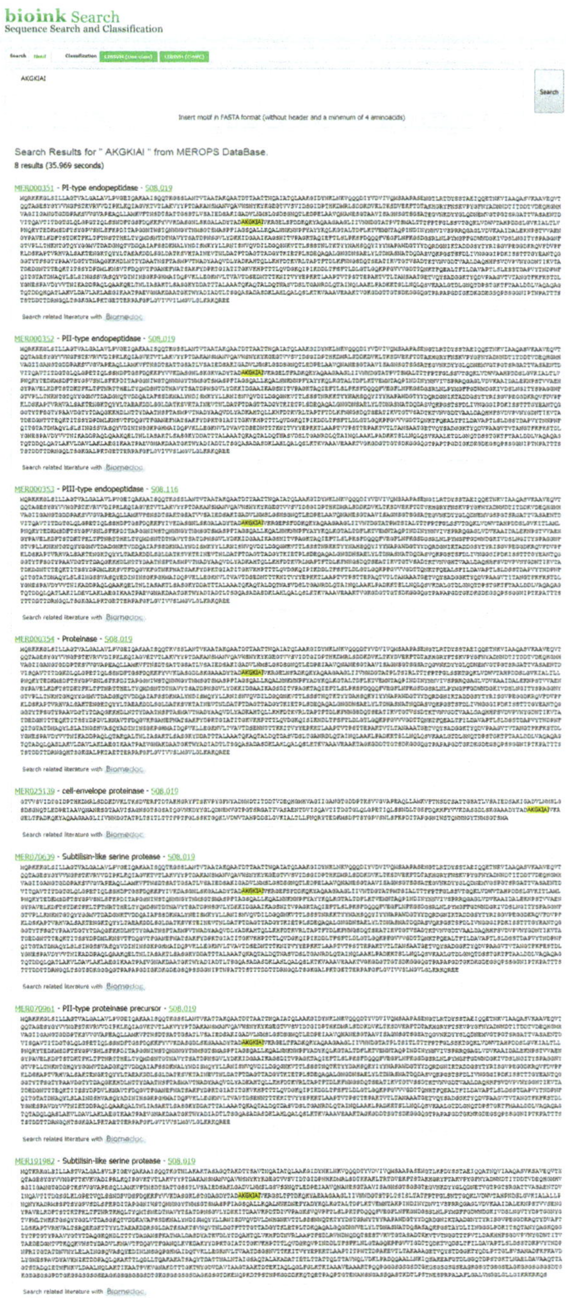

Fig.42.5 Motif Search Interface

42.5.2 Peptidase Detection

Users may submit a sequence or a set of sequences in FASTA format to assess whether or not they are peptidases. The two types of the described SVMs, previously trained, can be applied (Fig. 42.6). Together with the prediction results, users may search literature using another platform for literature analysis.

42.6 Conclusions

We have presented a web platform aiming to support the research process of peptidases detection, which is a fundamental issue in the field of biological sciences. A state-of the-art technique has been used, combining text-mining techniques with support vector machines. This approach proved to be able to successfully detect peptidases. Our results with C-SVC are superior to those obtained with the One-class approach. Nevertheless, the One-class SVM can be an alternative solution if only positive examples are available for training.

Future studies will consider the application of other kernel-based classifiers and larger datasets. We will also need to extend our methods in order to deal with larger n-grams. This web platform is still being improved and new functionalities are being included.

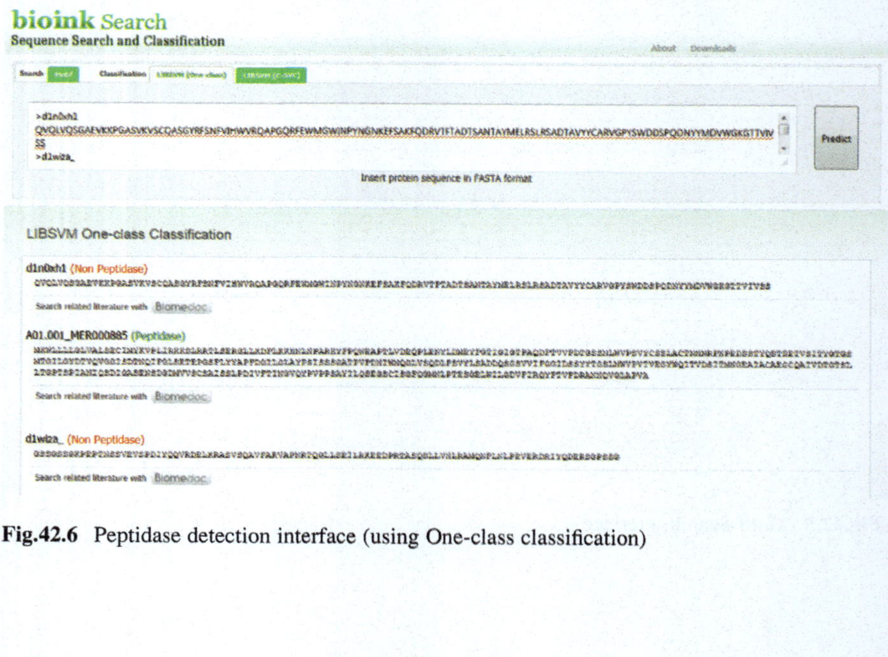

Fig.42.6 Peptidase detection interface (using One-class classification)

Acknowledgments This work was supported by FCT (Foundation for Science and Technology) and FEDER through Program COMPETE (QREN) under the project FCOMP-01-0124-FEDER-010160 (PTDC/EIA/71770/2006), designated BIOINK – Incremental Kernel Learning for Biological Data Analysis

References

1. Cheng BYM, Carbonell JG, Klein-Seetharaman J (2005) Protein classification based on text document classification techniques. Protein Struct Funct Bioinform 58(4):955–970. doi:10.1002/prot.20373
2. Wurst M. Word Vector Tool. http://sourceforge.net/projects/wvtool
3. Frijters J IKVM.NET: A JVM for the Microsoft .NET Framework. http://www.ikvm.net.
4. Salton G, Wong A, Yang CS (1975) A vector space model for automatic indexing. Commun ACM 18(11):613–620. doi:10.1145/361219.361220
5. Cortes C, Vapnik V (1995) Support-vector networks. Mach Learn 20(3):273–297. doi:10.1007/bf00994018
6. Vapnik VN (1995) The nature of statistical learning theory. Springer, New York
7. Chang C-C, Lin C-J (2011) LIBSVM: a library for support vector machines. ACM Trans Intell Syst Technol 2(3):1–27. doi:10.1145/1961189.1961199
8. Schölkopf B, Platt JC, Shawe-Taylor J, Smola AJ, Williamson RC (2001) Estimating the support of a high-dimensional distribution. Neural Comput 13(7):1443–1471. doi:10.1162/089976601750264965
9. Rawlings ND, Barrett AJ, Bateman A (2010) MEROPS: the peptidase database. Nucleic Acids Res 38(Database issue):D227–233. doi:10.1093/nar/gkp971
10. Murzin AG, Brenner SE, Hubbard T, Chothia C (1995) SCOP: a structural classification of proteins database for the investigation of sequences and structures. J Mol Biol 247(4):536–540. doi:10.1006/jmbi.1995.0159
11. Pereira C, Morgado L, Correia D, Verissimo P, Dourado A (2011) Kernel machines for proteomics data analysis: algorithms and tools. Paper presented at the European Network for Business and Industrial Statistics. ENBIS 2011, Coimbra, Portugal

Chapter 43
Optimal Configuration of Uniplanar-Unilateral External Fixators in Tibia Fractures

Luis Roseiro and Augusta Neto

Abstract The use of external fixation devices in orthopedic surgery is very common in open tibial fractures. There are several hypotheses of the structure configuration: uniplanar-unilateral; uniplanar-bilateral; biplanar and multiplanar. The stability on the focus of the fracture is also related to the type of configuration of the external fixators. The unilateral-uniplanar configuration was considered in this study. A simplified finite element model for the tibia and external fixator was developed in order to determine the stiffness in the focus of the fracture. A genetic algorithm was defined to minimize the displacement of the fracture focus (objective function), varying the positions of the mechanical components of the external fixator. Different types of loading were studied and the obtained results are presented and discussed.

43.1 Introduction

The bone fracture can be defined as a situation in which there is loss of bone continuity in addition to a bone separation into several fragments. Its seriousness can vary. Some spontaneous factures can be treated without even ever being diagnosed, whereas others require urgent medical treatment. Due to the implications for citizens, fracture treatments have and will continue to have enormous socioeconomic consequences. In fact, although the curing process of a fracture is a phenomenon with biological features, it is directly related to medical treatment carried out,

L. Roseiro (✉)
Mechanical Engineering Department, Coimbra Institute of Engineering, Rua Pedro Nunes – Quinta da Nora, 3030-199 Coimbra, Portugal
e-mail: lroseiro@isec.pt

A. Neto
Mechanical Engineering Department – Polo II, Faculty of Science and Technology, University of Coimbra, Rua LR Santos – Pinhal de Marrocos, 3030-788 Coimbra, Portugal
e-mail: augusta.neto@dem.uc.pt

A. Madureira et al. (eds.), *Computational Intelligence and Decision Making: Trends and Applications*, Intelligent Systems, Control and Automation: Science and Engineering 61, DOI 10.1007/978-94-007-4722-7_43,

namely how rapid the bone can heal, and then the so-called bone callus, and return to normality. It is the bone callus that re-establishes integrity, continuity and stiffness of the bone member, enabling a return to normality. Usually, fixation of the bone or bone fragments is the most suitable technique to ensure necessary stability for the treatment of fractures, which can be done internally or through external fixation systems. The choice of stability methodology of fractures depends on the type of fracture and the available means for treatment.

The external fixtures are systems that combine advantages of low cost and easy application, with the possibility to access soft tissues. These consist of an assembly of several mechanical components so as to maintain stability and stiffness of the bone structure. The fixators are attached to the bone with linking pins. The structure and function of each external fixator depends essentially on the shape of its components. There are various types of external fixators and each one can be used in a certain type of fracture. The most complex kind of external fixator allows for a multiplanar fixation with any configuration, which allows its application in almost all clinical situations. However, these have high costs and limited access to soft tissues. The mechanical stability of an external fixator depends on several factors. The simplest and versatile external fixator is the tubular system type AO. It is also the most common used and this can be found in Ramos et al. [1]. This system is composed of linking pins to the bone, longitudinal beams and connecting elements between pins and the beams. According to surgeons' criterion, the installation of the pins can vary in distance and inclination thus enabling the assembly of different configurations. There are various hypotheses about the configuration of these types of fixators: uniplanar and unilateral, uniplanar and bilateral, biplanar or multiplanar. The degree of stability of the focus of fracture is directly related to the type of configuration used.

The healing of the bone is sensitive to mechanical stability of the fixators. In a study carried on 45 patients with diaphysis fracture treated with unilateral external fixators, Kershaw et al. [2] related the weight discharge on the fractured tibia to the existence of micro-movements on the focus of the fracture. It was observed that the existence of micro-movements on the focus of fracture reduced the time of consolidation of the tibia. Later, in a study done by Emami et al. [3] on 68 patients with treated fractures through unilateral external fixators, the failure of the system of fixators was connected to the likeliness that the discharge of its own weight was higher than the mechanical stability of the fixator. The need of 25 surgical revisions to patients showed the interest in studying the resistance and stiffness of these systems of fixation. Epari et al. [4] observed the existence of a relation between the stiffness of the external fixators and the resistance and stiffness of the bone callus formed after 9 weeks on the focus of the fracture. If on one hand the external fixator has a resistance/stiffness correlation allowing complete immobility of the focus of the fracture, on the other hand the existence of micro-movements suggests faster healing and recuperation. Therefore, pre-surgical planning with a position program of the fixator components can play an important role in clinical goals surgeons may have. However, this is only possible if surgeons have at their disposal tools allowing them to foresee the behavior of the fixator system – the bone. This study aims to contribute to the development of these types of tools. By using a model of external fixators type AO in the uniplanar-unilateral configuration with a beam, we attempted to obtain an ideal position of the mechanical components of the fixator in case of cross-sectional fractures of the tibial diaphysis. To do this, a 3D geometry of the obtained tibia from the computerized axial tomography (CAT Scan) was used so as to develop a simplified model of the set tibia – external fixator. A finite element model together with the use of genetic algorithms made possible the determination of an ideal position of various situations of the considered loads. It was observed that the stiffness of the fixators depends on the position they are placed on the pins with relation to the focus of the fracture and the distance of the beam to the bone.

43.2 Simplified Model of Tibia – External Fixator

The tibia used in this study was obtained from a CAT scan made for a healthy, 22 year old male. From the CAT scan, the geometrical model 3D of the tibia was obtained, which is illustrated in Fig. 43.1a. The cortical and trabecular components of the tibia have been separated.

Due to the complexity of the tibial bone and its differences from human to human, a methodology with a simplified model with equivalent behavior was chosen. This type of methodology can facilitate the implementation of the technique from patient to patient. The updating of the simplified model was made from the modal analysis, using the finite element method. Modal analysis of the tibia was carried out in NASTRAN/PATRAN with 10 node tetrahedral elements, for a total of 57,674 elements and 90,500 nodes. The finite element mesh is illustrated in Fig. 43.1b. The cortical bone has been considered with elasticity modulus E = 17 GPa and Poisson coefficient $\upsilon = 0.3$. The trabecular bone has been considered with elasticity modulus E = 7 GPa and Poisson coefficient $\upsilon = 0.2$. The specific mass of bone was considered with 1,800 kg/m^3.

A 1D finite element model was then created with the section defined from the cross section to the average level of the tibia diaphysis. The model was defined with 55 quadratic finite elements following the Timoshenko beam theory. Figure 43.2 shows the section considered for the beam.

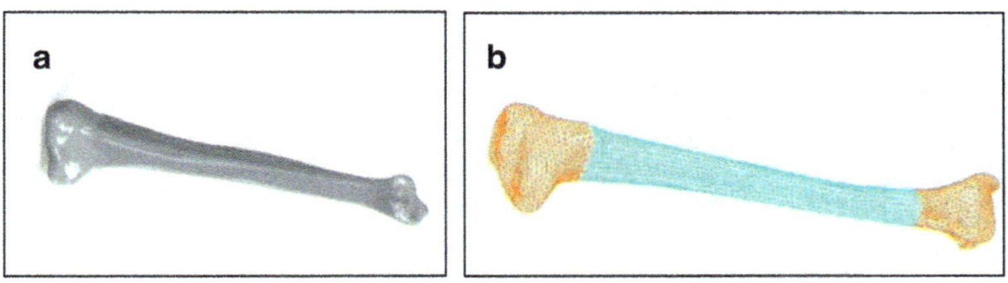

Fig. 43.1 3D model of the tibia (**a**) and tibia finite element mesh (**b**)

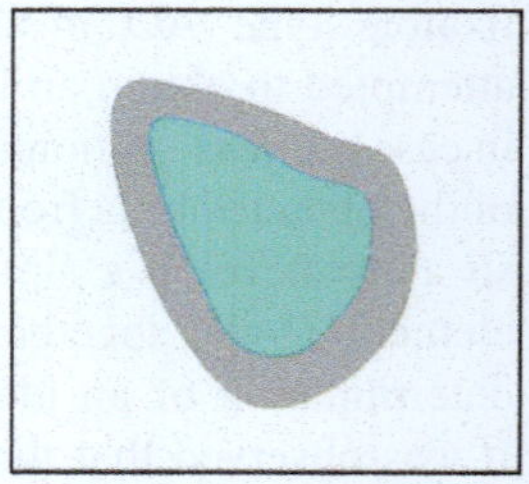

Fig. 43.2 Beam section for 1D finite element model

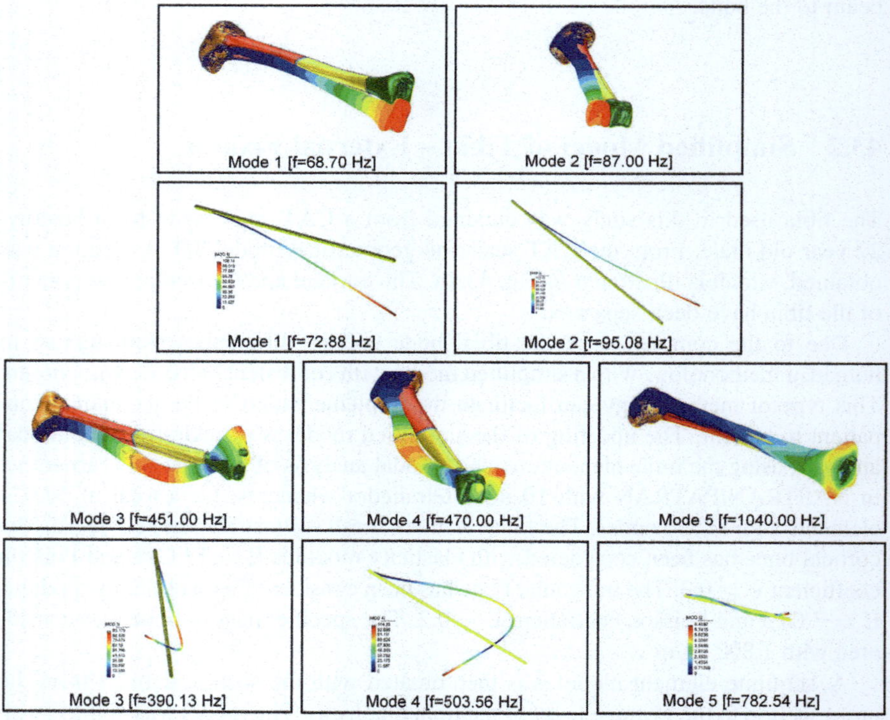

Fig. 43.3 Modes and frequency of vibration between 1D and 3D finite element models

Figure 43.3 compares the values of the frequencies of vibration between the 3D model and the simplified model 1D. The results obtained show a good a proximity between the two models.

For the purpose of this study a transverse fracture (90°) in the central tibia diaphysis is considered. According to Wong et al. [5], the model of the tibia can be considered with rigid fixation in the contact area with the knee, and free in the foot connection, where the loads are applied in a region of 20% equivalent to the bottom surface of the tibia.

The beam model of the tibia was considered with the focus of fracture simulated with a 4 mm opening on the central area and a symmetrical distribution of components

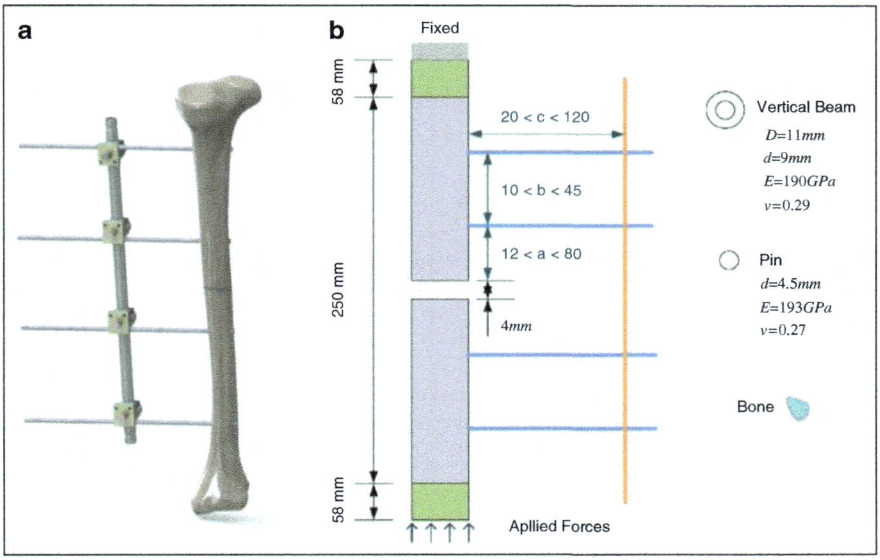

Fig. 43.4 Tibia – external fixator models: (**a**) 3D model (**b**) simplified model

in relation to the focus. The tibia analyzed had a length of 366 mm and the constant section shown in Fig. 43.2. The fixator components were considered to have a symmetrical distribution from the focus of the fracture. Figure 43.4 illustrates the simplified model utilized in the optimization process. The mechanical and geometrical properties of the external fixator components are shown in the same figure.

The finite element model was defined from the simplified model using quadratic Timoshenko beam elements with 6 degrees of freedom (Neto and Leal [6]). This element was integrated in the library of the finite elements program FEAP (Taylor [7]). The finite element methodology developed has enabled us to obtain the properties associated with the numerical modeling of beams with material, geometry and arbitrary section. The construction of the stiffness matrix used in the program of finite elements was based on the variational asymptotic method presented by Yu et al. [8].

43.3 Optimization Methodologies

The optimization process tries to obtain the optimal configuration for five situations (load cases). As an initial draft the following positions of the pins and vertical beam were considered: a = 15 mm, b = 45 mm and c = 25 mm. The value of the forces involved in each load case is presented in Table 43.1. The first three cases consider the axial force together with the load on the fixator plane. The fourth case is the same as the third case but with an additional force on the perpendicular plane to the fixator. The fifth case only considers torsion.

Table 43.1 Load cases and types

	Load case				
Load type	A	B	C	D	E
Axial [N]	10	–	10	10	–
Force in the fixator plane [N]	–	10	10	10	–
Force perpendicular to the fixator plane [N]	–	–	–	0.5	–
Torsion [N.mm]	–	–	–	–	100

Table 43.2 Genetic parameters and operators

Type	Value
Initial population	100
Iterations	20
Elitism	10%
Crossover probability	75%
Mutation probability	2.5%
Doping roullet	20%

The optimal position of the mechanical components of the external fixator is obtained from a real-coded genetic algorithm programmed in Fortran. The objective function of the genetic algorithm is the minimization of the norm of the resulting displacement in the fracture focus according to

$$F_{obj} = \min \sqrt{u_x^2 + u_y^2 + u_z^2}$$

where u_x, u_y and u_z represents the displacement in the x, y and z directions, respectively. Three variables are considered for the design of the algorithm, namely the distances a, b and c illustrated in Fig. 43.4.

The optimization process was done using genetic algorithms (GAs) methodology. GAs are search and optimization techniques inspired by Darwin's theory of natural evolution (Goldberg, [9]). GAs start with a population of chromosomes and each one of them represents a solution in the search space. Each solution is evaluated, using a fitness function which demonstrates the merit of the respective individual (in this case calculated by the finite element model). Based on the fitness, a random selection is made in which the best chromosomes have a higher chance of being selected. The successive application of genetic operators (crossover and mutation) to the selected individuals generates tendentiously more fitting populations. When the algorithm attains a pre-established criterion, it stops. The considered settings and genetic operators for the developed genetic algorithm are presented in Table 43.2.

43.4 Results and Discussion

Table 43.3 summaries the numerical results obtained with the optimization process for the five load cases studied. Figure 43.5 illustrates the distribution of the norm of the resulting displacement for load case A in the initial and final project positions.

Table 43.3 Results obtained in the optimization process

		Final project [mm]				
Variable	Initial project [mm]	A	B	C	D	E
a	15	12	14	14	13	13
b	45	28	40	45	40	20
c	25	20	20	20	20	20
Percentage decrease of displacement on the facture focus [%]		*80.5*	*1.5*	*5.0*	*3.4*	*69.0*

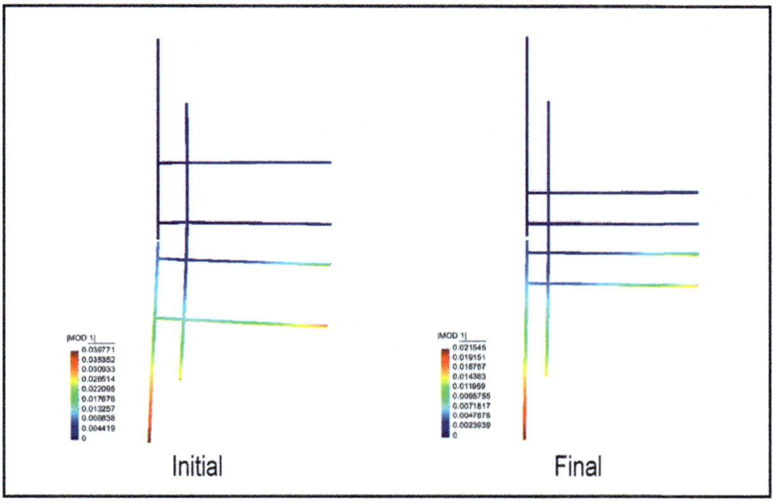

Fig. 43.5 Distribution of the norm of the resulting displacement for load case A in the initial and final project

The results obtained show that the optimal position of the parallel beam is as close as possible to the bone for all situations. With respect to the first pin, the indication is to place it as close as possible to the fracture. Concerning the second pin, it should be placed away from the first pin in case of flexion loads, and in axial loads and torsion loads, the second pin should be placed near the first pin.

43.5 Conclusions

The positions of fixators are usually defined by the surgeon and based on his/her own experience. Although from a clinical point of view, there are several factors that can determine the fixators' positions, a model of optimization can contribute to the best clinical decision. In this kind of work we used a simplified model of an external fixator to optimize the position of its components. The results obtained allow us to conclude that it is possible to move in the direction of the development of tools that make it enable us to determine the optimum position of fixators, thus serving as a useful aid to the surgeon.

References

1. Ramos L, Rotbande I, Shehata I, Knackfuss I (1999) Contribuição ao estudo mecânico do fixador externo tubular AO. Rev Bras Ortop 34(2):134–138
2. Kershaw C, Cunningham J, Kenwright J (1993) Tibial external fixation, weight bearing and fracture movement. Clin Orthop Rela Res 293(9):28–36
3. Emami A, Mojoberg B, Karlstrom G, Larssom S (1995) Treatment of closed tibial shaft fractures with unilateral external fixation. Injury 26:299–303
4. Epari D, Kassi J, Schell H, Duda G (2007) Timely fracture-healing requires optimization of axial fixation stability. J Bone Jt Surg 89:1575–1585
5. Wong C, Mikkelsen P, Hansen L, Darvann T, Gebuhr P (2010) Finite element analysis of tibial fractures. Dan Med Bull 57(5):A4148
6. Neto M, Yu W, Leal R (2008) Generalized Timoshenko modeling of composite beam structures: sensitivity analysis and optimal design. Eng Optim 40(10):891–906
7. Taylor R (2000) FEAP – A finite element analysis program, Version 7.2 Example Manual, Department of Civil and Environmental Engineering, University of California at Berkeley, Berkeley, CA, USA
8. Yu W, Hodges D, Volovoi V, Cesnik C (2002) On Timoshenko-like modeling of initially curved and twisted composite beams. Int J Solid Struct 39:5101–5121
9. Goldberg D (1989) Genetic algorithms in search, optimization and machine learning. Addison-Wesley, Reading

Chapter 44
Identification of the Forces in the Suspension System of a Race Car Using Artificial Neural Networks

Luis Roseiro, Carlos Alcobia, Pedro Ferreira, Abderrahmane Baïri, Najib Laraqi, and Nacim Alilat

Abstract The suspension system together with the chassis of the car play an important role in the performance of a race car. The knowledge of the forces that are transmitted through the car's suspension system allows a better adjustment and optimization of various structural components. In this study, an identification method, with a neural network based methodology, was tested using experimental data in order to identify and quantify the horizontal and vertical forces in the wheel hub, which are transmitted to the triangle of the suspension and to the chassis of the car. The obtained results are promising and show that it is possible to use this methodology for real-time monitoring of horizontal and vertical forces acting on the triangle suspension of the formula student car.

44.1 Introduction

The car chassis optimization is a matter that has been widely treated properly from several different points of view, with many papers and works analyzing the optimization of parts or systems of the vehicle. These have included instance joints [1], suspension elements [2, 3], wheels [4], seats [5] crashworthiness and frame [6], and so on. In race car engineering, out of F1 environment, most of the teams do not have the time or resources in order to develop intense theoretic optimization tasks in car designing and setting up. In general we can assess that the main design objective of a race car frame is increasing structural stiffness in order to improve the general

L. Roseiro (✉) • C. Alcobia • P. Ferreira
Mechanical Engineering Department, Coimbra Institute of Engineering, Rua Pedro Nunes – Quinta da Nora, 3030-199 Coimbra, Portugal
e-mail: lroseiro@isec.pt; calcobia @isec.pt; pmsf@isec.pt

A. Baïri • N. Laraqi • N. Alilat
LTIE, EA 4415, GTE, Université Paris Ouest, 50 Rue de Sèvres, F-92410 Ville d'Avray, France
e-mail: abairi@u-paris10.fr; nlaraqi@u-paris10.fr; nlaraqi@u-paris10.fr

A. Madureira et al. (eds.), *Computational Intelligence and Decision Making: Trends and Applications*, Intelligent Systems, Control and Automation: Science and Engineering 61, DOI 10.1007/978-94-007-4722-7_44,

Fig. 44.1 Phisec formula student car

dynamic response, knowing that a highly deformable frame will induce uncontrolled and non-adequate movements.

It is well known that the ultimate part that defines a race car handling, acceleration and braking abilities are the tires. The suspension system controls the way tires are in contact with the track to provide maximum grip [7]. Nevertheless, a suspension system cannot work properly if it is not connected to a good frame, which is the element that connects the rest of the vehicle. An appropriate design means that the chassis interacts properly with all other systems in the car, but also achieves certain requirements regarding stiffness, strength, total weight and distribution, dimensions, construction feasibility, cost and regulations.

Beyond controlling the structure global stiffness, an adequate balance is needed between the front and the rear chassis areas to avoid dynamic weight transfer that would increase under-steer or over-steer behavior. Total weight must remain low in order to be as near as possible to the compulsory minimum weight. Body weight distribution must also assure that the center of gravity is low enough to minimize corner inertias and that its position is longitudinal and transversally centered. Additionally, the frame design has to be compatible with the other chassis systems and elements (suspension, steering, engine, transmission system, brakes, seat, etc.).

This work is about a methodology used in a Formula Student racing car (Phisec – Formula Student car of Coimbra Institute of Engineering, shown in Fig. 44.1) to identify and quantify the horizontal and vertical forces in the wheel hub, which are transmitted to the triangle of suspension and then to the chassis of the car. The Formula Student competition consists of a competition of world-class engineers that challenges university students from around the world to design and build a single-seat racing car. This competition provides an exercise in real context, with the creation of a fictitious company in which students have to develop a prototype race car in terms of design and manufacturing, together with the procedures involved in the setting of an actual manufacturing company car involving various technical, financial management and marketing. All components of the car play an important role, and the knowledge and interpretation of the forces involved in the movement of the car in various conditions allows for adjustment and optimization of the design of the car and the improvement of their performance. Like other

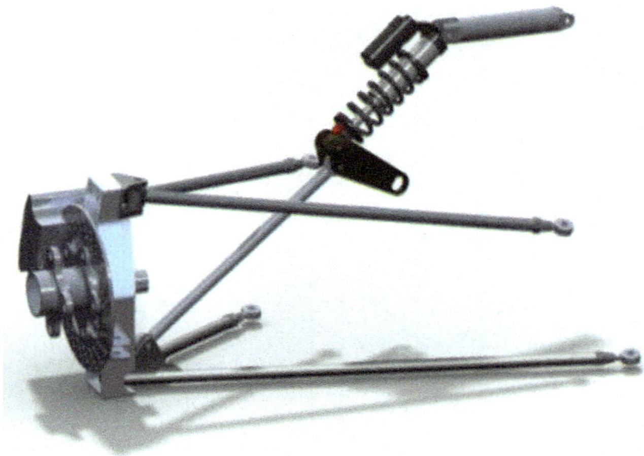

Fig. 44.2 3D Model of Phisec suspension triangle

formula student cars, the Phisec suspension systems are independent and consist of a double wishbone with two sets of spring/dampers. Each wishbone has two tubular beams with "rod-end" terminals that establish the connection to the chassis car on one side and to the connector block to the wheel (called "pushrod") on the other.

To complement the two triangles, a tubular beam, commonly known as "upright", links the triangle to the spring/damper system. The suspension triangle is still made up of a steering arm which is responsible for varying the angle of the wheel in the vertical plane. Figure 44.2 shows the suspension triangle in study.

In static analysis the forces developed in the suspension triangle can be considered fully absorbed by the spring/damper system through the pushrod. On the other hand, in a dynamic situation the forces are also absorbed by the wishbones. It should be noted that all the forces are ultimately transmitted to the chassis.

Figure 44.3 illustrates the forces and the reactions developed in braking. The forces depend on the friction, weight and speed of the car and has the opposite direction to the movement. The reaction forces are absorbed by the wishbones; one of them is subjected to tensile forces and the other one to compression forces.

In this study, a two-type identification method with a neural network based methodology was tested using experimental data collected from strain gages and load cells. The main objective is to estimate the horizontal and vertical forces based on the information of the deformation in the structural components of the suspension triangle. The other objective is to identify the deformation in the structural components according to the horizontal and vertical forces that can be applied. The development of these types of methodology (these types of methodologies ou this types of methodology) will be useful, not only to take a real-time monitoring of the behavior of the car, but also in order to optimize the triangle suspension and chassis of the car to the different conditions of use on the track.

Fig. 44.3 Forces in suspension triangle

44.2 Experimental Methodology

In order to obtain the necessary data for the implementation of the methodology, a mechanical system was developed and built to fit one suspension unit from the formula student car, including the damping system. The horizontal and vertical loads on the suspension triangle are applied through two mechanical screw type systems. The forces were applied in the hub and the data monitored and acquired with two compression load cells from "AEP Transducers", type TSTM, with a maximum rate of 5 kN.

The monitoring of the behavior of the five structural elements of suspension triangle was made with strain gage measurements. The triangle suspension was instrumented with five unidirectional strain gauges, one on each structural member. The gages used are from Vishay (Micro Measurements Group) reference 125UR-06-CEA-350, grids made with constantan alloy, with auto-compensation on temperature for steel and a resistance of 350 Ω. The nominal gain factor of calibration certificate is $2.1 \pm 0.5\%$. The signal transportation from the strain gages was done with shielded cable. The position of the gages in the structural components was determined based on the information from a developed finite element model.

Figure 44.4 shows the experimental setup with the horizontal load cell and one of the strain gages used.

The data were acquired using a National Instruments acquisition board, reference NI USB-9162, with a LabView programming. The graphic interface of the developed acquisition program is shown in Fig. 44.5. The gages were connected to the board in quarter-bridge configuration and the load cells are connected in full-bridge.

A total of 1,310 data sets were acquired, with the range of variation of values that is illustrated in Table 44.1, where C1 represents the vertical force, C2 the horizontal force, and E1, E2, E3, E4, E5 the strain measured for each couple of forces.

Fig. 44.4 Suspension triangle with load cells and strain gage used

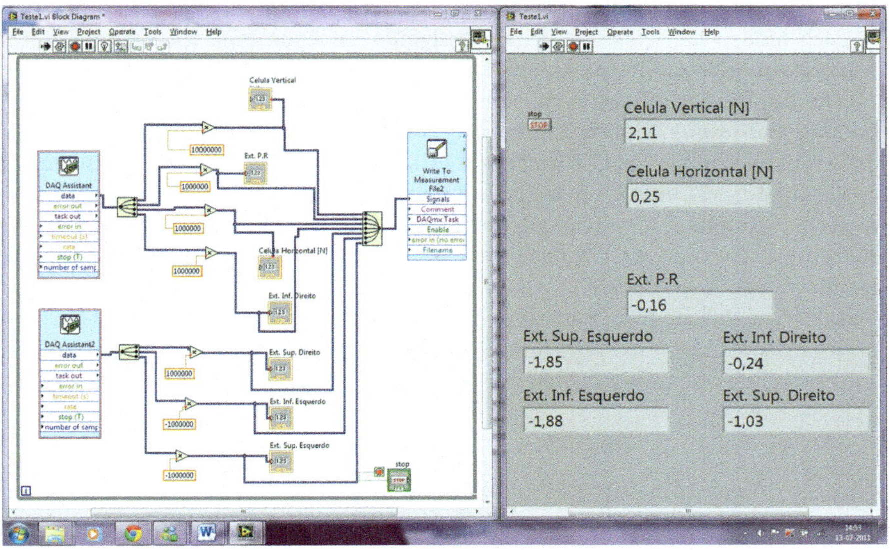

Fig. 44.5 LabView acquisition program

Table 44.1 Acquired data

Designation		Minimum value	Maximum value
Vertical force [N]	C1	1.8	728.1
Horizontal force [N]	C2	−1.7	900.5
Deformation με	E1	−72.8	3.4
	E2	−41.0	10.0
	E3	−29.5	3.1
	E4	−1.9	165.4
	E5	−3.2	34.6

44.3 Neural Networks Identification Methodology

44.3.1 Why Neural Networks?

It has been recognized since early that neural networks offer a number of potential benefits for application in the field of engineering, particularly for pattern recognition problems. Some appealing features of neural networks are their ability for learning through examples, they do not require any a priori knowledge and can approximate arbitrary well any non-linear continuous function [8]. Among the several architectures used in practice, feedforward type neural networks, shown in Fig. 44.6, have been considered suitable for the identification problem of the signature analysis.

A feed forward neural network consists of several layers; each one with some processing elements, called neurons, linked to each other by weights. The weights determine the nature and the strength of the connection between the neurons. The number of nodes considered in the input and output layers depend on the specifications of the problem. The number of hidden layers, the number of neurons in each hidden layer, as well as the activation function type for each neuron is selected according to the experience and some convergence criterions.

The application of artificial neural network consists of two stages, namely training and testing. During the training stage an input-to-output mapping, using the available sample data, is presented to the network. The network evaluates its own output based on the presented input and compares this value with the target (presented) output. The actual output error (the sum of squares error function in this study) is used to adjust the node weights so that the error can be reduced. The learning stage stops once a cross validation pre-set error threshold is reached and the node weights are frozen at this point. During the testing stage, data that have not been presented to the network in the learning stage are provided as input and the corresponding output is calculated using the fixed node weights.

44.3.2 Neural Networks Considered

In this work the Software Satistica have been used with SANN – Automatic Network Search (ANS) to implement the neural networks methodology [9]. The training of the neural network have been performed with a second order type

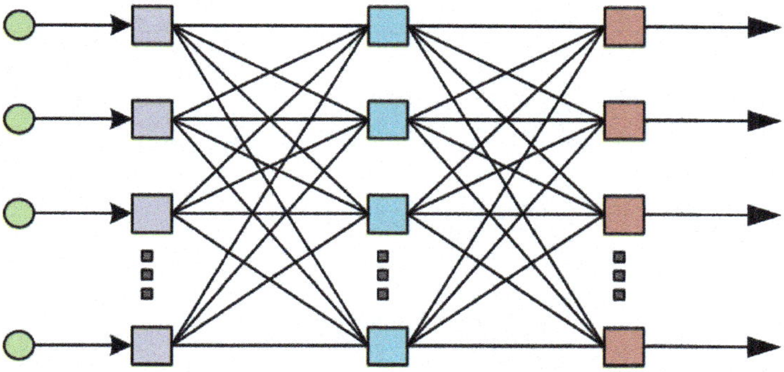

Fig. 44.6 Feedforward neural networks

algorithm, the BFGS Broyden, Fletcher, Goldfarb, and Shanno [10]. This algorithm has very fast convergence but needs high memory due to the storing of the Hessian matrix.

From the acquired data 70% were used to train the networks, 15% for the validation process and 15% to test the networks. The selection of the data for training, validation and test were random according to the Statistica program. Only three layers have been considered for the design of the net.

Two types of identification process were tested with the neural network methodology:

- **DNN – Direct Neural Network** – in this case the objective was to identify the local deformation in each one of the structural members (in the areas where the strain gages have been placed) due to the horizontal and vertical forces applied. The neural network have two neurons in the first layer (the inputs are the forces C1 and C2) and five neurons in the output layer (the outputs are the deformations E1 to E5);
- **INN – Inverse Neural Network** – this case represents the final objective; estimate the horizontal and vertical forces in the suspension triangle according to data collected from the strain gages. Then, the first layer has five neurons (deformations E1 to E5) and the output layers two neurons (C1 and C2).

44.3.3 Best Neural Networks Obtained

For the two types of identification process a total of 200 different neural networks have been tested, in which the number of hidden neuron and the type of activation function vary. The best neural networks obtained and their performances are presented in Table 44.2. For the Direct Neural Network, the best ANN uses the hyperbolic tangent function in the hidden layer and the logistic function in the output layer. This network has a mean test performance of 98.67%. For the Invers Neural

Table 44.2 Acquired data

		Type of identification	
		DNN	INN
Neural network		MLP 2-4-5	MLP 5-8-2
Neurons in hidden layer		4	8
Training epochs		334	230
Activation functions	Hidden layer	*Tanh*	*Exp*
	Output layer	*Logistic*	*Tanh*
Performance [%]	Training	98.68	99.84
	Cross validation	98.88	99.84
	TEST	98.67	99.86

Network, the best net uses the exponential function in the hidden layer and the hyperbolic tangent function in the output layer.

44.4 Conclusions

In this study, a two-type identification method, with a neural network based methodology, was tested using experimental data collected from strain gages and load cells. A suspension triangle from one formula student competition car (Phisec), made with five structural components was instrumented and put to work in laboratory conditions. Collected data was used to train and test the neural networks. Two types of neural network identifications were tested:

- The identification of the deformation in the structural components according to the horizontal and vertical forces;
- The quantification of the horizontal and vertical forces based on the information of the deformation in the five structural components. In both cases high performance neural networks were obtained.

The obtained results show that it is possible to develop a methodology based on neural networks for real-time monitoring of horizontal and vertical forces acting on the triangle suspension of the formula student. These data will be extremely useful in the optimizing of the triangle suspension and chassis of the car to the different conditions of use on the track.

References

1. Nikolaidis E, Zhu M (1996) Design of automotive joints: using neural networks and optimization to translate performance requirements to physical design parameters. Comput Struct 60(6):989–1001
2. Jazar GN, Alkhatib R, Golnaraghi M (2006) Root mean square optimization criterion for vibration behaviour of linear quarter car using analytical methods. Veh Syst Dyn 44(6):477–512

3. Sun L, Cai X, Yang J (2007) Genetic algorithm-based optimum vehicle suspension design using minimum dynamic pavement load as a design criterion. J Sound Vib 301:18–27
4. Jahed H, Farshi B, Eshraghi A, Nasr A (2008) A numerical optimization technique for design of wheel profiles. Wear J 264:1–10
5. Siefert A, Pankoke S, Wölfel HP (2008) Virtual optimisation of car passenger seats: simulation of static and dynamic effects on drivers' seating comfort. Int J Ind Ergon 48:410–424
6. Zhang Y, Zhu P, Chen G (2007) Lightweight design of automotive front side rail based on robust optimisation. Thin-Walled Struct 45:670–676
7. Mántaras DA, Luque P, Vera C (2004) Development and validation of a 3-dimensional kinematic model for the McPherson steering and suspension mechanisms. J Mech Mach Theory 39(6):603–619
8. Hornik K, Stinchcombe H, White H (1989) Multilayer feedforward networks are universal approximators. Neural Network 2:183–192
9. StatSoft Inc. STATISTICA – Data Analysis Software System. www.statsoft.com
10. Nocedal J, Wright J (2006) Numerical optimization, 2nd edn. Springer, Berlin/New York

Chapter 45
Combinational Logic Circuits Design Tool for a Learning Management System

Cecília Reis and Viriato M. Marques

Abstract This paper presents a genetic algorithm based tool for combinational logic circuits synthesis specifically developed to be integrated in a learning management system. With this application the students can learn digital systems design by testing pre-defined circuits and also by projecting new circuits.

45.1 Introduction

The basic description of a Learning Management System (LMS) is: a software application that automates the administration, tracking, and reporting of training events [1]. However, it is not that simple. A robust LMS should be able to do the following:

- centralize and automate administration;
- use self-service and self-guided services;
- assemble and deliver learning content rapidly;
- consolidate training initiatives on a scalable web-based platform;
- support portability and standards;
- personalize content and enable knowledge reuse.

The current LMS lack pedagogy and interactivity, giving rise to e-learning models that strongly reside on student's own motivation.

Technical teaching is traditionally composed of lectures, knowledge application exercises and laboratory classes. The teacher maintains a direct contact with the

C. Reis (✉)
Engineering Institute, Porto Polytechnic Institute (ISEP/IPP), Porto, Portugal
e-mail: cmr@isep.ipp.pt

V.M. Marques
Engineering Institute, Coimbra Polytechnic Institute (ISEC/IPC), Coimbra, Portugal
e-mail: viriato@isec.pt

A. Madureira et al. (eds.), *Computational Intelligence and Decision Making: Trends and Applications*, Intelligent Systems, Control and Automation: Science and Engineering 61,
DOI 10.1007/978-94-007-4722-7_45,

students thus allowing the evaluation of their performance and psychological profile, answer doubts and questions, fit examples and give general orientations to each one. In this context, e-learning systems pose some problems: LMS's allow little more than creating e-learning course contents accessible by an Internet site, offering some additional services such as chat and on-line sessions, all in a single system of comfortable usage by means of Internet access. Contents development makes use of learning-objects; their reutilization is the goal of specifications such as SCORM, AICC and IMS [2, 3].

However, pedagogy is supposed to be included in the contents, not embedded in the LMS. Contents and example adaptation to student's questions is difficult or absent; the role of such LMS's is essentially passive: the initiative belongs to the student. Nevertheless, learning styles, pedagogy, interactivity, autonomy and creativity must integrate such systems. In this context it is of major importance to provide the LMS's with the following features:

1. Catch the performance and learning style of students as well as success and failure stories;
2. Adapt the (shown) contents to each student cognitive level and preferential learning profile;
3. Provide high interactivity, including a query answering system;
4. Be able to learn "how to teach each student" based on previous cases of success and failure described by (previously used) course scripts.

In this line of though the current paper proposes a Digital Circuits Synthesis Tool (DCSTool) which includes online synthesis capabilities for solving What-if problems posed by students in an interactive way. The module also incorporates a creative capability based on Evolutionary Computation techniques. Therefore, the results evaluation (fitness) can be carried out automatically or in an interactive way, this way resulting in an Interactive Evolutionary Computation technique.

45.2 Evolutionary Computation (EC)

Evolutionary Computation (EC) is a subfield of artificial intelligence, more particularly computational intelligence that involves combinatorial optimization problems. EC uses iterative progress, such as growth or development in a population. This population is then selected in a guided random search using parallel processing to achieve the desired end. Such processes are often inspired by biological mechanisms of evolution [4, 5]. Therefore, EC is the general term for several computational techniques which are based, to some degree, on the evolution of biological life in the natural world. Evolutionary techniques mostly involve metaheuristic optimization algorithms such as:

- Evolutionary Algorithms: genetic algorithms, evolutionary programming, evolution strategy, genetic programming and learning classifier systems.
- Swarm Intelligence: ant colony optimization and particle swarm optimization.

Fig. 45.1 Evolutionary computation algorithm

1. Initialize the population
2. Calculate the fitness of each individual in the population
3. Reproduce selected individuals to form a new population
4. Perform evolutionary operations such as crossover and mutation on the population
5. Loop to step 2 until some condition is met

Evolutionary Algorithms (EAs) are search methods that take their inspiration from natural selection and survival of the fittest in the biological world. EAs differ from more traditional optimization techniques in that they involve a search from a "population" of solutions, not from a single point. Each iteration involves a competitive selection that weeds out poor solutions. The solutions with high "fitness" are "recombined" with other solutions by swapping parts of a solution with another. Solutions are also "mutated" by making a small change to a single element of the solution. Recombination and mutation are used to generate new solutions that are biased towards regions of the space for which good solutions have already been seen. Figure 45.1 presents a pseudo-code for an evolutionary algorithm.

Swarm Intelligence (SI) is artificial intelligence based on the collective behavior of decentralized, self-organized systems. The expression was introduced by Gerardo Beni and Jing Wang in 1989, in the context of cellular robotic systems [6]. SI systems are typically made up of a population of simple agents interacting locally with one another and with their environment. The agents follow very simple rules, and although there is no centralized control structure dictating how individual agents should behave, local interactions between such agents lead to the emergence of complex global behavior. Natural examples of SI include ant colonies, bird flocking, animal herding, bacterial growth, and fish schooling.

45.2.1 Genetic Algorithms

Genetic Algorithms (GAs) are adaptive heuristic search algorithm based on the evolutionary ideas of natural selection and genetic. The basic concept of GAs is designed to simulate processes in natural system necessary for evolution, specifically those that follow the principles first laid down by Charles Darwin of survival of the fittest [7, 8].

In 1859 the British naturalist, Charles Darwin, published a book that would change the way humans view the world. In this book entitled 'The Origin of Species', Darwin proposed that humans, and in fact all creatures, were not put on this planet by God and made unchanging, but rather that they evolved from other creatures. Over time, creatures change to adapt to their environment to survive and thrive [9].

GA were invented by John Holland in the 1960s and were developed by Holland, his students and colleagues at the University of Michigan. Holland's 1975 book Adaptation in Natural and Artificial Systems presented the genetic algorithm as an abstraction of biological evolution and gave a theoretical framework for adaptation under the GA [10]. Holland's GA is a method for moving from one population of chromosomes to a new population by using a kind of natural selection together with the genetics – inspired operators of crossover, mutation, and inversion. Essentially, they are a method of "breeding" computer programs and solutions to optimization or search problems by means of simulated evolution. Processes loosely based on natural selection, crossover, and mutation are repeatedly applied to a population of binary strings which represent potential solutions. Over time, the number of above-average individuals increases, and highly-fit building blocks are combined from several fit individuals to find good solutions to the problem at hand.

Each chromosome consists of genes [11]. The selection operator chooses those chromosomes in the population that will be allowed to reproduce. Crossover exchanges subparts of two chromosomes, roughly mimicking biological recombination between two organisms. Mutation randomly changes the values of some locations in the chromosome. Inversion reverses the order of a contiguous section of the chromosome.

45.3 Digital Circuits Synthesis Tool

The DCSTool is based on a Genetic Algorithm.

45.3.1 The Implemented Genetic Algorithm

A GA strategy is adopted to design combinational logic circuits. The circuits are specified by a truth table and the goal is to implement a functional circuit with the least possible complexity. Several sets of logic gates have been defined, as shown in Table 45.1, being Gset 1a and 1b the simplest ones and Gset 6 the most complex.

For each gate set the GA searches the solution space, based on a simulated evolution aiming the survival of the fittest strategy. In general, the best individuals of any population tend to reproduce and survive, thus improving successive generations. However, inferior individuals can, by chance, survive and also reproduce [4]. In our case, the individuals are digital circuits, which can evolve until the solution is reached (in terms of functionality and complexity).

45.3.1.1 Circuit Encoding

In the GA scheme the circuits are encoded as a rectangular matrix A (row $\times$ column $= r \times c$) of logic cells as represented in Fig. 45.2.

Table 45.1 Gate sets

Gate set	Logic gates
Gset 1a	{NAND,WIRE}
Gset 1b	{XOR,WIRE}
Gset 2	{AND,XOR,WIRE}
Gset 3	{AND,OR,XOR,WIRE}
Gset 4	{AND,OR,XOR,NOT,WIRE}
Gset 5	{AND,OR,XOR,NOT,NAND,WIRE}
Gset 6	{AND,OR,XOR,NOT,NAND,NOR,WIRE}

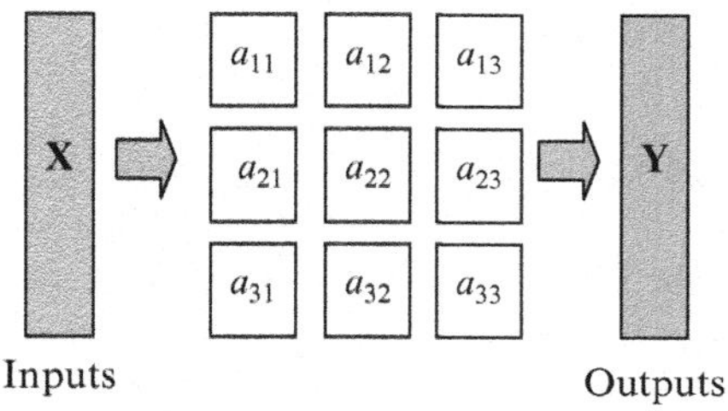

Fig. 45.2 Example of a matrix 3 × 3 to represent a circuit

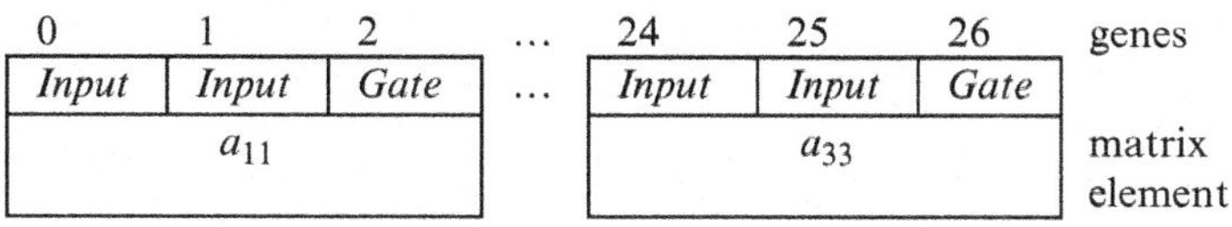

Fig. 45.3 Chromosome for the 3 × 3 matrix of Fig. 45.2

Each cell is represented by three genes: *<input1><input2><gate type>*, where *input1* and *input2* are one of the circuit inputs, if they are in the first column, or one of the previous outputs, if they are in other columns. The gate type is one of the elements adopted in the gate set. The chromosome is formed by as many triplets of this kind as the matrix size demands. For example, the chromosome that represents a 3 × 3 matrix is depicted in Fig. 45.3.

A 3 × 3 matrix A representing a circuit with input X and output Y.

45.3.1.2 The Genetic Operators

The initial population of circuits (strings) is generated at random. The search is then carried out among this population. The three different operators used are reproduction, crossover and mutation, as described in the sequel.

In what concern the reproduction operator, the successive generations of new strings are reproduced on the basis of their fitness function. In this case, it is used a tournament selection [4] to select the strings from the old population, up to the new population.

For the crossover operator, the strings in the new population are grouped together into pairs at random. Single point crossover is then performed among pairs. The crossover point is only allowed between cells to maintain the chromosome integrity.

The mutation operator changes the characteristics of a given cell in the matrix. Therefore, it modifies the gate type and the two inputs, meaning that a completely new cell can appear in the chromosome. Moreover, it is applied an elitist algorithm and, consequently, the best solutions are always kept for the next generation.

To run the GA we have to define the number of individuals to create the initial population P. This population is always the same size across the generations, until the solution is reached.

The crossover rate CR represents the percentage of the population P that reproduces in each generation. Likewise MR is the percentage of the population P that mutates in each generation.

45.3.1.3 The Fitness Function

The calculation of the fitness function F is divided in two parts, f_1 and f_2, which measure the functionality and the simplicity, respectively. In a first phase, we compare the output Y produced by the GA-generated circuit with the required values Y_R, according with the truth table, on a bit-per-bit basis. By other words, f_1 is incremented by one for each correct bit of the output until f_1 reaches the maximum value f_{10}, that is, when we have a functional circuit. Once the circuit is functional, in a second phase, the GA tries to generate circuits with the least number of gates. This means that the resulting circuit must have as much genes <gate type> ≡ <wire> as possible. Therefore, the index f_2, that measures the simplicity (the number of null operations), is increased by one (zero) for each wire (gate) of the generated circuit, yielding:

$$f_{10} = 2^{ni} \times no \tag{45.1a}$$

$$f_1 = f_1 + 1 \text{ if bit } i \text{ of } \mathbf{Y} = \text{bit } i \text{ of } \mathbf{Y_R}, i = 1, \ldots, f_{10} \tag{45.1b}$$

$$f_2 = f_2 + 1 \text{ if } \textit{gate type} = \textit{wire} \tag{45.1c}$$

$$F = \begin{cases} f_1, & F < f_{10} \\ f_1 + f_2, & F \geq f_{10} \end{cases} \tag{45.1d}$$

where ni and no represent the number of inputs and outputs of the circuit.

45.3.2 The Interface

The tool for automatic synthesis of combinatorial circuits presents an interactive graphical interface that allows the user/student to enter the information needed to design the circuits. (Fig. 45.4).

The student can choose a circuit that is already registered (circuits that were previously projected through the system), as shown in Fig. 45.5.

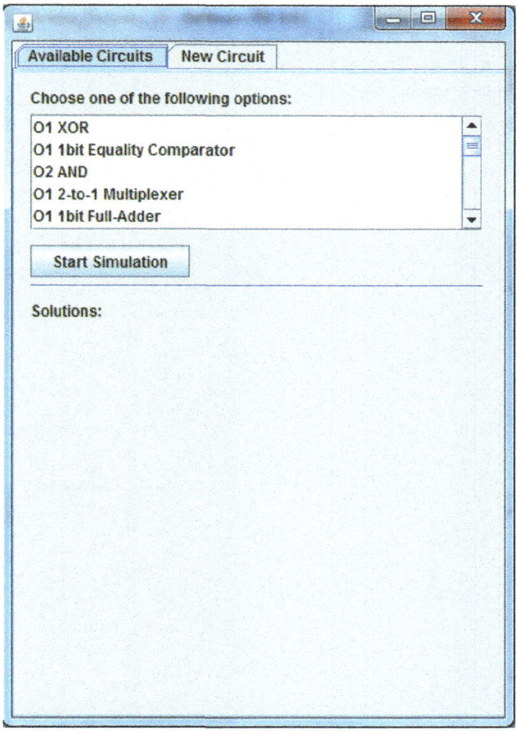

Fig. 45.4 Graphical interface – available circuits

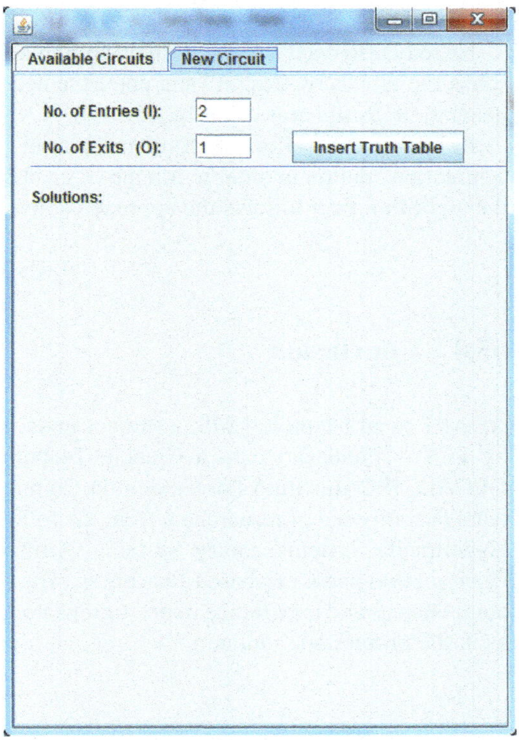

Fig. 45.5 Graphical interface – new circuit

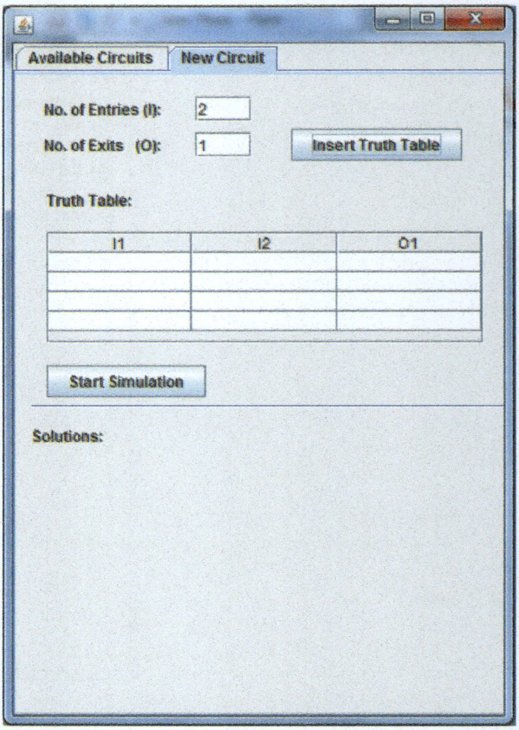

Fig. 45.6 Graphical interface – new circuit truth table

He may also want to design a new combinatorial circuit, in which case he should select the tab "New Circuit" and enter the necessary information for the system to generate it, as illustrated in Fig. 45.6.

In both cases, after selecting the circuit, the student must press the "Start Simulation" button in order to run the GA and obtain the solutions. For new circuits the algorithm then follows the approach above described.

45.4 Conclusion

ADAPT is an adaptive LMS, a project under development, approved for funding by FCT – Fundação para a Ciência Tecnologia. The involved Institutions are GECAD, IPC (Instituto Politécnico de Coimbra), ISEP and UTAD. The research team is composed of researchers from the following domains: Education, Adaptive Hypermedia Systems, Fuzzy Systems, Artificial Neural Networks, Evolutionary Computation and Case-based Reasoning. The team includes recognised leadership competences and experience in prestigious Portuguese teaching institutions, as well as in the professional domain.

The DCSTool is a component of the ADAPT project, an interactive evolutionary tool for digital circuits synthesis based on previous research, now to integrate an LMS platform.

As concept proof and first application domain of ADAPT, a course on Digital Systems has been selected, for the following reasons: (a) Consolidated knowledge domain, well delimited; (b) Well defined synthesis techniques. The tests will be carried out at ISEP and/or ISEC.

Acknowledgements The authors wish to thank FCT-Fundação para a Ciência e Tecnologia; Eixo I do Programa Operacional Factores de Competitividade (POFC) do QREN- COMPETE: FCOMP-01-0124-FEDER-014418; GECAD-Grupo de Engenharia do Conhecimento e Suporte à Decisão.

References

1. Ellis RK (ed) (2009) Learning circuits field guide to learning management systems learning circuits' "field guide to learning management systems". American Society for Training & Development (ASTD), Alexandria
2. Felder RM (1988) Learning and teaching styles in engineering education. J Eng Educ 78(7):674–681
3. Kolb A (2005) Learning styles and learning spaces: enhancing experiential learning in higher education. Acad Manag Learn Educ 4(2):193–212
4. Goldberg D (1989) Genetic algorithms in search, optimization and machine learning. Addison-Wesley, Reading
5. Bäck T, Fogel DB, Michalewicz Z (1997) Handbook of evolutionary computation. Institute of Physics Publishing/Oxford University Press, Bristol
6. Beni G, Wang J (1989) Swarm intelligence in cellular robotic systems. In: Proceedings of the NATO advanced workshop on robots and biological systems, Tuscany, Italy, 26–30 June 1989
7. Davis L (1991) Handbook of genetic algorithms. Van Nostrand Reinhold, New York
8. Mitchell M (1997) An introduction to genetic algorithms. MIT Press, Cambridge
9. Darwin C (1859) On the origin of species by means of natural selection, or, the preservation of favoured races in the struggle for life. J. Murray, London
10. Holland JH (1975) Adaptation in natural and artificial systems. The University of Michigan Press, Ann Arbor
11. Michalewicz Z (1992) Genetic algorithms + data structures=evolution programs, 3rd Revised and Extended edn. Springer, Berlin/Heidelberg/New York

Chapter 46
Labeling Methods for the General Case of the Multi-objective Shortest Path Problem – A Computational Study

J.M. Paixão and J.L. Santos

Abstract This paper is devoted to the study of the classical labeling technique for solving the multi-objective shortest path problem (MSPP), which is an extension of the shortest path problem (SPP) resulting from considering more than one cost function (criteria) for the arcs. Different versions are computationally tested making use more than 9,000 test problems involving random, complete and grid networks. The computational experience shows that the network structure plays a key aspect conditioning the algorithm performance. We also conclude that the labeling correcting version seems to have the best performance and is able to solve large size instances.

Keywords Multiple objective programming • Combinatorial optimization • Shortest path problem • Labeling algorithm • Non-dominated path.

46.1 Introduction

The shortest path problem (SPP) is a well known combinatorial optimization problem intensively studied [3, 4, 12] since the pioneer works published in the middles of the last century [2, 7, 11, 16]. Given a network $G = (N, A, c)$, where N is the set of nodes, A is the set of oriented arcs and c is a cost function defined over A, the SPP consists of finding a path linking, at least cost, two specific nodes.

J.M. Paixão
Operations Research Centre, Department of Statistics and Operational Research, Faculty of Sciences, University of Lisbon, Lisbon, Portugal
e-mail: jpaixao@fc.ul.pt

J.L. Santos (✉)
CMUC, Department of Mathematics, University of Coimbra, Coimbra, Portugal
e-mail: zeluis@mat.uc.pt

A. Madureira et al. (eds.), *Computational Intelligence and Decision Making: Trends and Applications*, Intelligent Systems, Control and Automation: Science and Engineering 61,
DOI 10.1007/978-94-007-4722-7_46, © Springer Science+Business Media Dordrecht 2013

When c is a k-vectorial function ($k \geq 2$), one has the multi-objective shortest path problem (MSPP) firstly addressed in the literature in a paper published in 1974, [21]. Until now, most of the work done relatively to the MSPP is focused on the bi-objective case ($k = 2$) [9, 10], where the solutions verify some properties yielding to the development of specific techniques for solving the problem.

Most of the algorithms presented in the literature for the MSPP correspond to an extension of the labeling technique known as very efficient for the single objective case. The generalization of this technique to deal with several criteria must have into account that, in general, there is not a path exhibiting the best value for all the criteria. Hence, for the MSPP, we are looking for non-dominated (ND) paths in the network (also called of Pareto optimal or efficient solutions), that is, paths for which it is not possible to find a better value for one criterion without getting worse for some of the other criteria.

In the present paper, we report an extensive computational study carried out in order to identify the best out 27 versions of the labeling algorithm for the MSPP. Those versions correspond to combinations of the two labeling techniques (correcting/setting), with five different data structures used for storing the set of candidate labels and four different operators considered for ordering the labels. We investigate the performance of the different solution methods on three different kinds of networks – random, complete and grid.

Details about this computational experience are given in Sect. 46.5. The labeling algorithm is discussed in Sect. 46.3 while in Sect. 46.4, some theoretical results are presented and the correctness of the algorithm is proved. The Sect. 46.2 contains the mathematical background and, the final section summarizes the conclusions.

46.2 Definitions and Basic Results

A network is denoted by $G = (N, A, c)$, where $N = \{1, \ldots, n\}$ is the set of nodes and $A \subseteq N \times N$ is the set of oriented arcs. Let k be the number of criteria and c: $A \rightarrow \mathrm{R}^k$ be a k-dimensional function vector. Each arc $(i, j) \in A$ is associated with a vector cost $c(i, j) = \mathbf{c}_{i,j} = (c^1{}_{i,j}, \ldots, c^k{}_{i,j})$ such that $c^l{}_{i,j} \geq 0, \forall\, l \in \{1, \ldots, k\}$.

A path $p = \langle\, v_0, a_1, v_1, \ldots, a_r, v_r \,\rangle$, from the node v_0 to v_r, is an alternating sequence of nodes and arcs. By convention, $\langle\, v_0 \,\rangle$ is considered as a null path ($r = 0$).

The set of all paths from i to j is denoted by $P_{i,j}$ and the set of all paths in the network is denoted by $P_G = \cup_{\mathrm{i,j} \in N} P_{i,j}$. A cycle is a path with non repeated nodes except the initial and terminal ones which are coincident; that is, $v_0 = v_r$.

With no loss of generality, we consider that N has an initial node s and a terminal node t such that $P_{s,t} \neq \emptyset$. P will be used instead of $P_{s,t}$.

Multiple arcs (arcs between the same pair of nodes) are not allowed. As a consequence, p can be denoted only by the sequence of its nodes, $\langle\, v_0, v_1, \ldots, v_r \,\rangle$.

A vectorial objective function f is defined over P_G, as follows:

$$f : P_G \rightarrow \mathrm{R}^k$$
$$p \rightarrow f(p) = (f_1(p), \ldots, f_k(p)),$$

where

$$f_l(p) = \sum_{(i,j)\in p} c^l{}_{i,j}, \forall l \in \{1, \ldots, k\}.$$

The concatenation operator, $\diamond$, joins two paths $p = \langle v_0, v_1, \ldots, v_{r\_p} \rangle$ and $q = \langle u_0, u_1, \ldots, u_{r\_q} \rangle$ such that $v_{r\_p} = u_0$. Then, $p \diamond q = \langle v_0, v_1, \ldots, v_{r\_p} = u_0, u_1, \ldots, u_{r\_q} \rangle$.

Now, let us recall that, in order to solve the MSPP, one looks for the set of non-dominated (ND) paths from s to t, mathematically described as follows:

Definition 2.1. Let p and q be two paths with the same initial and terminal node. One says that p dominates q or q is dominated by p ($p <_D q$) if and only if

$$f(p) \neq f(q) \text{ and } f(p) \leq_{\mathrm{R}^k} f(q) (\text{that is}, f_l(p) \leq f_l(q), l = 1, 2, \ldots, k).$$

Definition 2.2. Let p be a path in $P_{i,j}$, $i, j \in N$. If there is no path $q \in P_{i,j}$ such that $q <_D p$, then p is called non-dominated, efficient or Pareto optimal path. The set of ND paths from i to j is denoted by $\bar{D}_{i,j}$ and $\bar{D}$ will be used for $\bar{D}_{s,t}$.

For the single objective case ($k = 1$), the SPP can be easily solved when the *optimality principle* holds, meaning that every shortest path is formed by shortest sub-paths. A necessary and sufficient condition for the SPP to verify the optimality principle is that the network has no cycles with negative cost, [1].

For the multi-objective case, the optimality principle corresponds to saying that every ND path is formed by ND sub-paths. However, for the optimality principle to be verified in this case, only the following sufficient condition is known: for each criterion, the network has no cycle with negative cost (see [15] for a proof and [19] for a counterexample for the reciprocal condition).

46.3 The Labeling Algorithm

46.3.1 *Mono-objective Shortest Path*

A sketch of the algorithm is given in the Algorithm 1. In few words, it scans the nodes of a subset $X \subseteq N$ in order to update the labels assigned to their successors. For $i \in X$, with label π_i, the procedure computes $\pi_i + c_{i,j}$ for each j such that $(i, j) \in A$. If $\pi_i + c_{i,j} < \pi_j$, then π_j is updated and the vertex j joins the set X.

```
{p_i*: best path from s to i found at this moment}
{ π_i: label of node i, that is, f (p_i*)}
{X: set of "unscanned" nodes}
X ← {s}; π_s ← 0; π_i ← ∞, ∀ i ∈ N-{s}
while X ≠ Ø do
   π_i ← the label of some node i ∈ X; X ← X \ {i}
   for all (i, j) ∈ A do
      if π_i + c_ij < π_j
      then π_j ← π_i + c_ij; p_j* ← p_i* ◇ ⟨ i, j ⟩; X ← X ∪ {j}
   end_for all
end_while
```

Algorithm 1: Mono-objective Labeling Algorithm

One strategy for selecting the node of X to be scanned is to pick the one with the least label value. This is named in the literature as label setting (LS) technique since if the optimality principle holds then each node is scanned at most one time. The proof for this, lays on the monotonicity property exhibited by the objective function, that is, $f(p \diamond \langle i, j \rangle) \geq f(p)$; for a complete proof, see [1].

Alternatively, one may select the node to scan just following, for instance, a FIFO policy (first in, first out). In this case, the same node may be selected more than once and, therefore, the procedure stops only when the tree constituted by the shortest paths from s to $i \in N$, is obtained. Using this technique, known as the label correcting (LC) algorithm, one needs to compute the shortest tree rooted at s.

46.3.2 Data Structures for the Set of Candidates

The label setting approach requires the ordering of X in order to pick up the node with the smallest label value. In this work, we consider the following ones: *List* [1], *Dial* buckets [5, 6] and *Heap* [1]. On Dial data structure, we usually consider a circular vector of dimension $M = 1 + \max\{c_{i,j}: (i, j) \in A\}$ to hold the buckets, because at most M buckets are not null at each iteration. Table 46.1 presents the worst case complexity for the basic set operations on X.

For the label correcting version, we use a FIFO (First In, First Out) policy for selecting the nodes from X which is represented by a queue where new nodes are added to the tail and the selected node is removed from the head. We also adopted a slightly modification of this structure (DEQUEUE), proposed by Pape [17, 18], where a new node is inserted at the head, rather than at the tail, if the corresponding label is less than the label for the node currently positioned at the head.

Table 46.1 Worst case complexity order for the basic set operations

X	Find minimum	Remove minimum	Insert element	Remove element	Memory space
not ordered	$O(\|X\|)$	$O(1)$	$O(1)$	$O(\|X\|)$	$\|X\|$
List (simple)	$O(1)$	$O(1)$	$O(\|X\|)$	$O(\|X\|)$	$\|X\|$
List (double)	$O(1)$	$O(1)$	$O(\|X\|)$	$O(1)$	$2\ \|X\|$
Heap (simple)	$O(1)$	$O(\log_2(\|X\|))$	$O(\log_2(\|X\|))$	$O(\|X\|)$	$\|X\|$
Heap (double)	$O(1)$	$O(\log_2(\|X\|))$	$O(\log_2(\|X\|))$	$O(\log_2(\|X\|))$	$2\ \|X\|$
Dial (simple)	$O(M)$	$O(1)$	$O(1)$	$O(\|X\|/M)$	$\|X\| + M$
Dial (double)	$O(M)$	$O(1)$	$O(1)$	$O(1)$	$2(\|X\| + \mathrm{M})$

46.3.3 Generalization for the Multi-objective Case

In the generalization the labeling algorithm for the MSPP, we must notice that the concept of *best* path from s to t is not appropriate since we may have several "*best*" paths from s to any node. So, for each $i \in N$, we define $\prod_i$ as the set of temporarily non-dominated paths from s to i. Whenever a new s-i path p is found, if there is no path in $\prod_i$ dominating p, the set is updated by adding the new path and eventually removing paths that became dominated by p.

The Algorithm 2 shows a sketch of the multi-objective labeling algorithm, where the logical function $DT(q, \prod_j)$, performs a *dominance test* on $q \in P_{s,j}$ relatively to the temporary ND paths in $P_{s,j}$. Hence, $DT(q, \prod_j) =$ TRUE if and only if there is a path $w \in P_{s,j}$ for which $(w <_D q)$.

> $\{\Pi_i$: set of temporarily ND paths from s to $i\}$
> $\{X$: set of paths corresponding to "*unscanned*" labels$\}$
> $X \leftarrow \{\langle s \rangle\}$; $\Pi_s \leftarrow \{\langle s \rangle\}$; $\Pi_i \leftarrow \emptyset, \forall\, i \in N\text{-}\{s\}$
> **while** $X \neq \emptyset$ **do**
> $p \leftarrow$ a path of X; $X \leftarrow X \setminus \{p\}$
> $i \leftarrow$ the terminal node of p
>   **for all** $(i, j) \in A$ **do**
>   **if** $DT(p \diamond \langle i, j \rangle, \Pi_j) =$ FALSE
>   **then** $q \leftarrow p \diamond \langle i, j \rangle$; $X \leftarrow X \cup \{q\}$
> $\Pi_j \leftarrow (\Pi_j \cup \{q\}) - \{u \in \Pi_j : q <_D u\}$
>   **end_for all**
> **end_while**

Algorithm 2: Multi-objective Labeling Algorithm

Finally, let us remind that for the label setting version of the algorithm, the finding of the minimum label value in X is needed. Note that this cannot be accomplished by using the operator $\leq_{R^k}$ which is not a total order relation in R^k, [8] and requires the definition of other operators as described in the next section.

46.4 Determining the Minimum Label in X

The label setting version of the algorithm is based upon the condition that the label selected, at each iteration, for the scanning procedure will become definitively non-dominated (ND). In this order, we introduce an operator R in $P_{i,j}$ $(i, j \in N)$ and an auxiliary function h_R: $P_G \to$ R satisfying the following binary relations:

- $\leq_R \subset P_{i,j} \text{ x } P_{i,j}$, such that $p \leq_R q \Leftrightarrow h_R(p) \leq h_R(q)$;
- $<_R \subset P_{i,j} \text{ x } P_{i,j}$, such that $p <_R q \Leftrightarrow h_R(p) < h_R(q)$.

The Theorem 4.1, stated next, proves the correctness of the label setting algorithm when $<_R$ and $\leq_R$ hold the following properties:

Property 1: [dominance] $p <_D q \Rightarrow p <_R q, \forall\, p, q \in P_{i,j}$.

Property 2: [monoticity] $p \leq_R p \diamond \langle j, l \rangle, \forall\, p \in P_{i,j}$ and $\forall\, (j, l) \in A$.

Theorem 4.1. Let $<_R$ and $\leq_R$ be the relations above mentioned, holding the dominance and the monoticity properties. If $\in X$ is a path selected for scanning by the label setting such that $p \leq_R q, \forall\, q \in X$, then p is a (definitive) ND path.

Proof: Suppose that $p \in P_{s,i}$ is a dominated path, then there is a path $w \in P_{s,i}$ such that $w <_D p$. On the other hand, since $p \in X$, the path w only can be generated after the selection of p, for scanning. This means that w is obtained by an extension of one of the paths in X. That is, $w = w_1 \diamond w_2$, with $w_1 \in X$. Therefore, $w_1 \leq_R w <_R p$ contradicting the hypothesis that $p \leq_R q, \forall\, q \in X$. ◆

There are several operators that can be defined over P for which the properties 1 and 2 are verified. One of them is the lexicographic order used in [13–15]:

$$p \leq_{lex} q \Leftrightarrow \begin{cases} f(p) = f(q) \\ or \\ \exists_x \in \{1, \ldots, K,\}, & \begin{cases} f_x(p) < f_x(q) \\ f_y(p) = f_y(q), \forall y < x \end{cases} \end{cases}$$

$$p <_{lex} q \Leftrightarrow f(p) \neq f(q) \text{ and } p \leq_{lex} q$$

The auxiliary function associated with this operator is given in the next table where the order operators tested in this work are also presented. For h_{lex}, B is defined as $\max\{f_l(q)\text{: } 1 \leq l \leq k, q \in P_G, \text{ and } q \text{ has no cycles}\}$.

R	*lex*	*sum*	*norm*	*max*
h_R	$\sum_{l=1}^{k} f_l(p) \times B^{k-l}$	$\sum_{l=1}^{k} f_l(p)$	$\sum_{l=1}^{k} (f_l(p))^2$	$\max_{1 \leq l \leq k} \{f_l(p)\}$

If a Dial structure is used for the set X, the parameter M depends on the operator. The next table summarizes the upper bound values associated to each operator where $M_c = \max\{c^l_{i,j}\text{: } (i, j) \in A \text{ and } 1 \leq l \leq k\}$.

R	*lex*	*sum*	*norm*	*max*
M_R	$(B^k - 1) \times M_c + 1$	$k \times M_c + 1$	$2M_c \sum_{l=1}^{k} f_l(p) + k \times M_c^2$	$M_c + 1$

The upper bounds obtained for "$R = lex$" and "$R = norm$" are too large to be computationally practicable. Hence, when using a Dial structure for the label setting algorithm, we opted for taking $M_{lex} = M_c + 1$ and consider the elements of X in v_l lexicographically sorted while we did not consider the operator "R_{norm}".

The next Lemmas show that the non-negativity assumption for the costs is sufficient for the operators *sum*, *norm* and *lex* holding the properties 1 and 2. However, property 1 is not guarantee for "$R = max$". In fact, considering p and q two paths with $f\ (p) = (3,5)$ and $f\ (q) = (4,5)$, we have $p <_D q$ but $h_{max}(p) = h_{max}(q) = 5$.

Lemma 4.2: If $c^l_{i,j} \geq 0, \forall\ (i, j) \in A$ and $l \in \{1, \ldots, k\}$, then $<_{sum}$, $<_{norm}$ and $<_{lex}$ hold property 1.

Lemma 4.3: If $c^l_{i,j} \geq 0, \forall\ (i, j) \in A$ and $l \in \{1, \ldots, k\}$, then property 2 is verified for $<_{sum}$, $<_{max}$, $<_{norm}$ and $<_{lex}$.

The proof for the above results relies on the fact that if $c^l_{i,j} \geq 0, \forall\ (i, j) \in A$ and $l \in \{1, \ldots, k\}$, then $h_R(p \diamond \langle j, l \rangle) \geq h_R(p), \forall\ p \in P_{i,j}$ and $\forall\ (j, l) \in A$.

As shown in Theorem 4.5, the correctness of the label setting algorithm can be proved for the case when the operators hold the following slightly different versions of the dominance and the monotonicity properties:

Property 3: [weak dominance] $p <_D q \Leftarrow p \leq_R q, \forall\ p, q \in P_{i,j}$.

Property 4: [strict monoticity] $p <_R p \diamond \langle j, l \rangle, \forall\ p \in P_{i,j}$ and $\forall\ (j, l) \in A$.

Lemma 4.4 states the sufficiency of the strict positivity assumption for the costs for guarantying that properties 3 and 4 hold for all the operators R.

Lemma 4.4: If $c^l_{i,j} > 0, \forall\ (i, j) \in A$ and $l \in \{1, \ldots, k\}$, then $\leq_R$ and $<_R$ ($R \in \{lex, max, sum, norm\}$) satisfy properties 3 and 4, respectively.

Theorem 4.5: Let $\leq_R$ and $<_R$ be the relations above mentioned, holding properties 3 and 4. If $p \in X$ is a path selected for scanning by the label setting such that $p \leq_R q, \forall\ q \in X$, then p is a (definitive) ND path.

Proof: Suppose that $p \in P_{s,i}$ is a dominated path, then there is a path $w \in P_{s,i}$ such that $w <_D p$. On the other hand, since $p \in X$, the path w only can be generated after the selection of p, for scanning. This means that w is obtained by an extension of one of the paths in X. That is, $w = w_1 \diamond w_2$, with $w_1 \in X$. Therefore, $p \leq_R w_1 <_R w$ contradicting the hypothesis that $w <_D p$ because in this case $w \leq_R p$. ◆

Finally, notice that R operators can also be used in the DEQUEUE structure.

46.5 Computational Results

In this section, we report the computational experiments carried out in order to evaluate the performance of the labeling algorithm. We tested the different combinations summarized in the next table, making up a total number of 27 code implementations, on three different kinds of networks: random, complete and grid.

		R operator			
Versions	Data structure	*lex*	*max*	*norm*	*sum*
SETTING	List	X	X	X	X
	Dial	X	X	–	X
	Heap	X	X	X	X
CORRECTING	FIFO	–	–	–	–
	DEQUEUE	X	X	X	X

where *List*, *Dial* and *Heap* was implemented as simple or double data structure.

The computational experiments were carried out on a 3.00 GHz Intel(R) Pentium(R) 4 processor with 1,024 KB cache size and 384 MB of RAM, running over Linux operating system (Suse 9.3 version), at the Laboratory for Computational Mathematics of Centre for Mathematics of the University of Coimbra. All the codes were written in the C language.

The computational experience was conducted in two stages: the first one use small size instances for eliminating the low performance codes while in the second one we aimed at identifying the most efficient codes for the large size instances.

The test instances were generated for three different types of network (Random, Complete, and Grid) with the costs for the arcs randomly generated in [1, 1,000], using a uniform distribution. For each type of network, we considered variations on the number of nodes (n) and on the number of criteria (k), producing 12 classes of instances each one of them involving several groups accordingly to the values of n and k. Altogether, we considered a total number of 180 groups of instances, as it is shortly described in the next table (d is the average density node).

Class	Network type	Network parameters range	Groups
RandN-small	Random	$n \in \{i{*}250: i \in \mathrm{N}, 1 \leq i \leq 20\}; d = 3; k = 3$	20
RandK-small	Random	$n = 1{,}000; d = 3; k \in \{i: i \in \mathrm{N}, 2 \leq i \leq 20\}$	19
CompN-small	Complete	$n \in \{5i: i \in \mathrm{N}, 1 \leq i \leq 20\}; d = n\text{-}1; k = 3$	20
CompK-small	Complete	$n = 25; d = n\text{-}1; k \in \{\mathrm{i}: i \in \mathrm{N}, 2 \leq i \leq 20\}$	19
GridN-small	Square Grid	$n \in \{i^2: i \in \mathrm{N}, 5 \leq i \leq 20\}; d \cong 4; k = 3$	16
GridK-small	Square Grid	$n = 49; d \cong 4; k \in \{i: i \in \mathrm{N}, 2 \leq i \leq 20\}$	19
RandN-large	Random	$n \in \{1{,}000i: i \in \mathrm{N}, 1 \leq i \leq 20\}; d = 6; k = 6$	20
RandK-large	Random	$n = 5{,}000; d = 6; k \in \{i: i \in \mathrm{N}, 2 \leq i \leq 9\}$	8
CompN-large	Complete	$n \in \{10i: i \in \mathrm{N}, 1 \leq i \leq 14\}; d = n\text{-}1; k = 6$	14
CompK-large	Complete	$n = 100; d = n\text{-}1; k \in \{i: i \in \mathrm{N}, 2 \leq i \leq 9\}$	8
GridN-large	Square Grid	$n \in \{i^2: i \in \mathrm{N}, 5 \leq i \leq 13\}; d \cong 4; k = 6$	9
GridK-large	Square Grid	$n = 100; d \cong 4; \in \{i: i \in \mathrm{N}, 2 \leq i \leq 10\}$	9

From an empirical statistical analysis done in [20], it was shown that a minimum of 50 instances should be taken as a representative set of computational test cases for assessing the performance of the MSPP algorithms on a specific type of network with the same size and number of criteria. Hence, we generated 50 different instances for each sub-group, making up a total number of 9,050 test problems which are available in: *http://www.mat.uc.pt/~zeluis/INVESTIG/MSPP/mspp.htm*

Table 46.2 reports the average CPU time (in seconds) obtained in the first stage for FIFO code. One may verify that CPU time increases faster with k than with n on random and complete networks; the opposite situation occurs on grid networks.

Table 46.3 reports the computational performance of the different versions for the labeling algorithm using the label correcting version with a FIFO rule (LC-FIFO) as benchmarking algorithm. Hence, the results for all the other versions are presented relatively to the corresponding values in the last column of the table showing the computational time required by LC-FIFO for solving the largest instances considered for each class of test problems.

From Table 46.3, FIFO attests to be the most efficient version since the large majority of the values are greater than 1 (they are slower than the benchmarking algorithm). In terms of computational efficiency, DEQUEUE comes next, being even faster than FIFO for the CompN-small instances.

In terms of CPU time, the LS algorithm is in general less efficient than the LC one. The worst performance is by far achieved when using a *List* data structure. Therefore, *List* versions are not considered in the second stage of the computational experience. Concerning to the R operators, R_{lex} produces the worst results, either for the LC or the LS versions, and will not be more tested.

Now, relatively to the *Heap* and the *Dial* versions, one cannot find a significant difference between them being reasonable to test them in the second phase.

Finally, excluding R_{lex}, in general the single link data structure proved slightly better than double link. This is due to the relatively low percentage of temporarily ND labels that turn out to be dominated at a further iteration. Hence, we consider only the single link data structure, reducing the number of versions from 27 to 9.

Table 46.4 reports the average CPU time (in seconds) obtained in the second stage for FIFO code having a similar pattern like in the Table 46.2.

Table 46.5 presents the results obtained with the reduced set of versions for the large size test problems and is organized likewise the Table 46.3. From those values, one can take the same kind of conclusion; particularly the FIFO version is consistently quite competitive and the LC as well also the DEQUEUE version with h_{max}.

Concerning to the operators, let us refer that "*max*" performs better that "*sum*" and "*norm*", for the random and complete networks; the contrary occurs for the grid networks. Relatively to the data structures for the LS algorithm, it is clear that the Dial structure proves slightly better than the Heap structure.

Table 46.2 CPU time (FIFO version) and number of ND solutions for the small size instances

	n					k				
n and k	1,000	2,000	3,000	4,000	5,000	4	8	12	16	20
CPU time	0.003	0.007	0.016	0.026	0.043	0.057	0.047	0.194	0.506	1.062
ND s-t paths	5.70	6.12	6.42	7.28	7.34	8.42	23.52	41.02	59.52	75.30
	RandN-small					RandK-small				
n and k	20	40	60	80	100	4	8	12	16	20
CPU time	0.001	0.008	0.032	0.087	0.182	0.005	0.062	0.301	1.039	3.102
ND s-t paths	14.46	27.74	37.14	48.60	58.12	37.64	170.80	392.08	690.26	1067.68
	CompN-small					CompK-small				
n and k	81	169	256	324	400	4	8	12	16	20
CPU time	0.002	0.034	0.204	0.559	1.328	0.001	0.022	0.068	0.144	0.309
ND s-t paths	85.28	289.04	547.26	869.64	1095.40	99.68	608.48	1024.44	1372.12	1906.40
	GridN-small					GridK-small				

Table 46.3 Small size instances – average CPU time ratios relatively to the FIFO version

		Ratio							
Class		List		Heap		Dial		DEQUEUE	FIFO[a]
	h	(simple)	(double)	(simple)	(double)	(simple)	(double)		(CPU)
	lex	47.998	44.257	2.195	2.409	1.284	1.368	**1.143**	0.043
RandN-	*norm*	32.687	34.548	1.678	1.934	–	–	**1.122**	
-small	*sum*	34.776	36.917	1.661	1.937	1.179	1.260	**1.085**	
	max	23.589	24.506	1.636	1.890	1.125	1.237	**1.075**	
	lex	24.854	24.054	1.645	1.746	1.255	1.280	**1.094**	1.062
RandK-	*norm*	14.012	14.515	1.373	1.496	–	–	**1.089**	
-small	*sum*	14.395	14.943	1.359	1.492	1.154	1.180	**1.066**	
	max	11.121	11.862	1.369	1.499	1.134	1.172	**1.089**	
	lex	1.820	1.559	1.370	1.349	1.194	**1.173**	1.246	0.182
CompN-	*norm*	1.239	1.236	1.061	1.158	–	–	**0.903**	
-small	*sum*	1.308	1.313	0.950	1.173	1.002	**0.919**	1.165	
	max	1.184	1.157	1.115	1.123	0.953	**0.886**	0.947	
	lex	2.036	2.003	1.230	1.251	1.172	1.167	**1.073**	3.102
CompK-	*norm*	1.697	1.700	1.131	1.206	–	–	**1.019**	
-small	*sum*	1.754	1.735	1.128	1.209	1.069	1.070	**1.012**	
	max	1.587	1.647	1.130	1.182	1.035	1.043	**1.015**	
	lex	4.183	3.879	1.840	2.172	2.001	1.754	**1.603**	1.328
GridN-	*norm*	2.471	2.371	1.084	1.332	–	–	**1.059**	
-small	*sum*	2.443	2.568	1.203	1.307	1.196	1.269	**1.095**	
	max	2.167	2.162	1.143	1.208	1.113	1.187	**1.062**	
	lex	1.901	1.887	1.362	1.449	**1.213**	1.274	1.271	0.309
GridK-	*norm*	1.603	1.637	1.212	1.194	–	–	**1.050**	
-small	*sum*	1.623	1.574	1.187	1.222	1.186	**1.180**	1.124	
	max	1.526	1.549	1.229	1.301	**1.124**	1.202	1.171	

[a]- FIFO CPU time (in seconds) for the largest problem within the class

Table 46.4 CPU time (FIFO version) and number of ND solutions for the large size instances

	n				k			
n and k	5,000	10,000	15,000	20,000	3	5	7	9
CPU time	9.428	29.151	51.800	75.919	0.451	4.516	18.697	49.224
ND s-t paths	71.42	82.50	84.80	100.08	16.60	50.72	95.74	154.84
	RandN-large				RandK-large			
n and k	50	80	110	140	3	5	7	9
CPU time	0.613	7.752	51.355	300.740	0.212	5.916	95.638	594.135
ND s-t paths	265.24	509.86	811.92	1170.34	58.12	369.12	1173.00	2665.04
	CompN-large				CompK-large			
n and k	49	81	121	169	3	5	7	9
CPU time	0.007	0.218	5.048	131.848	0.004	0.229	3.543	27.850
ND s-t paths	329.20	1545.72	6183.24	17622.75	130.02	1363.92	6139.22	15037.04
	GridN-large				GridK-large			

Table 46.5 Large size instances – average CPU time ratios relatively to the FIFO version

		Average ratio			Largest ratio[a]			CPU[b]
Class	h	Heap	Dial	DEQUEUE	Heap	Dial	DEQUEUE	FIFO
RandN-	h_{norm}	1.184	–	**1.010**	1.163	–	**1.009**	
-large	h_{sum}	1.190	1.087	**1.010**	1.160	1.072	**1.008**	75.919
	h_{max}	1.126	1.035	**1.002**	1.093	1.022	**0.996**	
RandK-	h_{norm}	1.197	–	**0.987**	1.125	–	**1.007**	
-large	h_{sum}	1.218	1.061	**0.989**	1.127	1.067	**1.008**	49.224
	h_{max}	1.145	1.006	**0.984**	1.078	1.027	**0.999**	
CompN-	h_{norm}	1.113	–	**0.988**	1.186	–	**0.980**	
-large	h_{sum}	1.154	1.133	**0.986**	1.239	1.232	**0.990**	300.740
	h_{max}	1.109	1.141	**1.066**	1.166	1.173	**0.977**	
CompK-	h_{norm}	1.064	–	**0.980**	1.204	–	**1.006**	
-large	h_{sum}	1.093	1.071	**0.980**	1.231	1.230	**1.005**	594.135
	h_{max}	1.045	1.027	**0.974**	1.168	1.169	**0.996**	
GridN-	h_{norm}	1.493	–	**1.109**	1.920	–	**1.393**	
-large	h_{sum}	1.578	1.454	**1.218**	1.928	1.862	**1.403**	131.848
	h_{max}	1.716	1.640	**1.488**	2.163	2.191	**1.853**	
GrigK-	h_{norm}	1.561	–	**1.110**	1.852	–	**1.126**	
-large	h_{sum}	1.603	1.502	**1.104**	1.866	1.820	**1.133**	27.850
	h_{max}	1.797	1.726	**1.504**	2.169	2.113	**1.769**	

[a] – FIFO ratio CPU time for the largest problem within the class
[b] – FIFO CPU time (in seconds) for the largest problem within the class

46.6 Conclusion

In this paper, we described and tested 27 variants of the labeling algorithm for the multi-objective shortest path problem (MSPP). Those different implementations result from considering the two labeling techniques (setting and correcting), with several combinations for data structure used for the set of unscanned labels and for the ordering operators on the set of paths.

The computational experience was carried out making use of a large and representative set of test problems randomly generated, consisting of around 9,000 instances for MSPP, available at a public site. As far as we know, this is the first comprehensive computational study on algorithms for the MSPP, since it involves three types of network (random, complete and grid) and considers a reasonable range for the number of criteria on large size networks.

The computational results show that the labeling algorithm is able to solve large size MSPP instances, within a reasonable computing time. As far as the performance of the versions is concerned, our experience shows that depends on the type of network. For grid networks, the LC-FIFO (label correcting version with a FIFO policy for selecting the labels) proved to be consistently the most efficient one. For the random and complete networks, the alternative label correcting version (LC-DEQUEUE) performs better for some instances.

Acknowledgments This work was supported in part by FCT through POCTI – Research Units Pluriannual Funding to CMUC (*Centro de Matemática da Universidade de Coimbra*) and CIO (*Operations Research Center of the University of Lisbon*), and grant POCTI/MAT/139/2001 cofounded by the EU program FEDER.

References

1. Ahuja R, Magnanti TL, Orlin J (1993) Network flows – theory, algorithms, and applications. Prentice-Hall, Inc., Upper Saddle River
2. Bellman R (1958) On a routing problem. Q Appl Math 16:87–90
3. Cherkassky BV, Goldberg AV, Radzik T (1993) Shortest paths algorithms: theory and experimental evaluation. Math Program 73:129–174
4. Deo N, Pang C (1984) Shortest path algorithms: taxonomy and annotation. Networks 14(2): 275–323
5. Dial R (1969) Algorithm 360: shortest path forest with topological ordering. Commun ACM 12:632–633
6. Dial R et al (1979) A computational analysis of alternative algorithms and labeling techniques for finding shortest path trees. Networks 9:215–348
7. Dijkstra EW (1959) A note on two problems in connexion with graphs. Numer Math 1:269–271
8. Ehrgott M (1997) Multiple criteria optimization – classification and methodology. Shaker Verlag, Aachen
9. Ehrgott M, Gandibleux X (2000) A survey and annotated bibliography of multiobjective combinatorial optimization. OR-Spektrum 22:425–460
10. Ehrgott M, Gandibleux X (2002) Multiple criteria optimization: state of the art annotated bibliographic survey. Kluwer Academic Publishers, Boston
11. Ford J, Fulkerson D (1962) Flows in networks. Princeton University Press, Princeton
12. Gallo G, Pallotino S (1988) Shortest path methods. Ann Oper Res 13:3–79
13. Guerriero F, Musmanno R (2001) Label correcting methods to solve multicriteria shortest path problems. J Optim Theory Appl 111:589–613
14. Hansen P (1980) Bicriterion path problems, In: Fandel G, Gal T (eds) Multiple criteria decision making: theory and application. Lectures notes in economics and mathematical systems, vol 177. Springer, Heidelberg, pp 109–127
15. Martins E (1984) On a multicriteria shortest path problem. Eur J Oper Res 16:236–245

16. Moore E (1959) The shortest path through a maze. In: Proceedings of the international symposium on the theory of switching. Harvard University Press, pp 285–292
17. Pape U (1974) Implementation and efficiency of Moore-algorithms for the shortest route problem. Math Program 7:212–222
18. Pape U (1980) Algorithm 562: shortest paths lengths. ACM Trans Math Softw 6:450–455
19. Santos J (2003) Optimização vectorial em redes. Ph.D. thesis, Departamento de Matemática, Universidade de Coimbra
20. Santos J, Paixão J, Rosa M (2005) A statistical analysis on the number of non-dominated paths in the multiobjective shortest path problem. In: Proceedings international network optimization conference, Lisbon, Portugal, pp 21–23
21. Vincke P (1974) Problèmes multicritères. Cahiers du Centre d'Études de Recherche Opérationelle 16:425–439

The manufacturer's authorised representative in the EU is Springer Nature Customer Service Centre GmbH, Europaplatz 3, 69115 Heidelberg, Germany. If you have any concerns regarding our products, please contact ProductSafety@springernature.com

Printed and bound by CPI Group (UK) Ltd, Croydon, CR0 4YY

22/09/2026

02220980-0007